AP®
Biology
Prep Plus
2018–2019

PUBLISHING

New York

Special thanks to the following for their contributions to this text: Laura Aitcheson, Becky Berthiaume, Matthew Callan, Andrea Carver, Brandon Deason, Sumir Desai, Christopher Durland, M. Dominic Eggert, Tim Eich, Tyler Fara, Elizabeth Flagge, Dan Frey, Bella Furth, Adam Grey, Allison Harm, Katy Haynicz-Smith, Duncan Honeycutt, Andre Jessee, Rebecca Knauer, Jeffrey Koetje, Liz Laub, Terry McMullen, Mark Metz, Jenn Moore, Kristin Murner, Nicholas Nguyen, Aishwarya Pillai, Elijah Schwartz, Jill Sherman, Linda Brooke Stabler, Oscar Velazquez, Bonnie Wang, Shayna Webb-Dray, Lee Weiss, Lauren White, Allison Wilkes St. Clair, Dan Wittich, Michael Wolff, Jessica Yee, and Nina Zhang.

Published by Kaplan Publishing, a division of Kaplan, Inc.
750 Third Avenue
New York, NY 10017

Printed in the United States of America

10 9 8 7 6 5 4 3 2 1

ISBN-13: 978-1-5062-0333-1

TABLE OF CONTENTS

PART 1: INTRODUCTION

PART 2: FOUNDATIONS OF LIFE

PART 3: STRUCTURES OF LIFE

PART 4: PROCESSES OF LIFE

PART 5: TRANSFORMATIONS OF LIFE

PART 1

Introduction

CHAPTER 1

Inside the AP Biology Exam

INTRODUCTION TO THE AP BIOLOGY EXAM

There's a good way and a bad way to skip the Introduction to Biology class in college. Many students take the bad way, which consists of going to sleep ridiculously late every night with the Xbox controller still wedged in their sweaty hands, setting the alarm for 1:30 pm, then waking up and asking a roommate, "What did I miss?" This is not exactly the sort of behavior that will land you on the dean's list.

Then there's the good way: skip the whole Introduction to Biology experience entirely—hundreds of students crammed into an auditorium, the tiny dot that is the professor just visible down in front of an ocean of seats—by getting a good score on the Advanced Placement (AP) Biology exam.

✓ AP Expert Note

Depending on the college, a score of 3, 4, or 5 on the AP Biology exam will allow you to leap over the freshman intro course and jump right into more advanced classes.

These advanced classes are usually smaller in size, better focused, more intellectually stimulating, and—simply put—just more interesting than a basic course. If you are just concerned about fulfilling your science requirement so that you can get on with your study of pre-Columbian art or Elizabethan music or some such nonbiological area, the AP exam can help you there, too. Pass the AP Biology exam and, depending on the requirements of the college you choose, you may never have to take a science class again.

Test Prep ≠ Studying

If you're holding this book, chances are you are already gearing up for the AP Biology exam and probably nearing completion of the AP Biology course. Your teacher has spent the year cramming your head full of the biology know-how you will need to have at your disposal. There is more to the AP Biology exam than biology know-how, however. You have to be able to work around the challenges and pitfalls of the test—and there are many—if you want your score to reflect your abilities. You see, studying biology and preparing for the AP Biology exam are not the same thing. Rereading your textbook may be helpful, but it's not enough.

That's where this book comes in. We'll show you how to marshal your knowledge of biology and put it to brilliant use on Test Day. We'll explain the ins and outs of the test structure and question formats, so you won't experience any nasty surprises. We'll even give you answering strategies designed specifically for the AP Biology exam.

OVERVIEW OF THE TEST STRUCTURE

Advanced Placement exams have been around for half a century. While the format and content have changed over the years, the basic goal of the AP program remains the same: to give high school students a chance to earn college credit or advanced placement. To do this, a student needs to do two things:

- Find a college that accepts AP scores
- Score well enough on the exam

The first part is easy, because most colleges accept AP scores in some form or another. The second part requires a little more effort. If you have worked diligently all year in your course work, you've laid the groundwork. The next step is familiarizing yourself with the test.

What's on the Test

Two main goals of the College Board (makers of the AP Biology exam) are (1) to help students develop a conceptual framework for modern biology, and (2) to help students gain an appreciation of science as a process. To this end, the AP Biology course is designed to expose the student to four main ideas.

Big Idea 1: The process of evolution drives the diversity and unity of life.

Big Idea 2: Biological systems utilize free energy and molecular building blocks to grow, to reproduce, and to maintain dynamic homeostasis.

Big Idea 3: Living systems store, retrieve, transmit, and respond to information essential to life processes.

Big Idea 4: Biological systems interact, and these systems and their interactions possess complex properties.

These four big ideas are referred to as evolution, cellular processes, genetics and information transfer, and interactions, respectively. The four big ideas encompass the core principles, theories, and scientific processes that guide the study of life. Each of these big ideas is broken down into enduring understandings and learning objectives that will help you to organize your knowledge.

This approach to scientific discovery is about thinking, not just memorization. It's about learning concepts and how they relate, not just facts. Because of this, *the College Board has increased the emphasis on themes and concepts and placed less weight on specific facts in both the AP Biology course and exam.* The chapters in the review section of this book are intended to take advantage of this design by focusing on concepts and synthesizing information from different concepts to help you better understand, and learn, the AP Biology course and exam content.

Now that you know what's on the test, let's talk about the test itself. The AP Biology exam consists of two sections, or, more precisely, two sections and one intermission. Section I has a Part A and a Part B. In Part A, there are 63 multiple-choice questions with four answer choices each. In Part B, there are six grid-in items that integrate scientific thinking and mathematical skills. For each grid-in item, you will need to calculate the correct answer and then enter it into a grid section on the answer sheet. You will have 90 minutes to complete all 69 questions. This section is worth 50 percent of your total score.

After this section is completed, there will be a 10-minute break. During the break you will not be able to consult teachers, other students, or textbooks. Furthermore, you may not access any electronic or communication device, which means cell phones, computers, and calculators are off limits.

After the break, there's a 10-minute, recommended "reading period." This doesn't mean you get to pull your favorite novel out of your backpack and finish that chapter you started earlier. Instead, you're given 10 minutes to pore over Section II of the exam, which consists of two long free-response questions and six short free-response questions that are worth 50 percent of your total score. You then have 80 minutes to answer all of these questions. The phrase "long free-response" means roughly the same thing as "large, multistep, and involved." Although the two long free-response questions are worth a significant amount each and are broken into multiple parts, they usually don't cover an obscure topic. Instead, they take a fairly basic biology concept and ask you several questions about it. Sometimes diagrams are required or experiments must be set up properly. It's a lot of biology work, but it is fundamental biology work. The short free-response items are typically illustrative examples or concepts that you are expected to explain or analyze, providing appropriate scientific evidence and reasoning. A typical response for the short free-response items will be about a paragraph or two in length. You have approximately 20 minutes for each of the long free-response questions and six minutes for each of the short-response questions.

HOW THE EXAM IS SCORED

AP scores are based on the number of questions answered correctly. *No points are deducted for wrong answers.* No points are awarded for unanswered questions. Therefore, you should *answer every question, even if you have to guess.*

When the 180 minutes of testing are up, your exam is sent away for grading. The multiple-choice part is handled by a machine, while qualified graders—a group that includes biology teachers and professors, both current and former—grade your responses to Section II. After an interminable wait, your composite score will arrive by mail. (For information on rush score reports and other grading options, visit collegeboard.org or ask your AP Coordinator.) Your results will be placed into one of the following categories, reported on a five-point scale:

5 = Extremely well qualified

4 = Well qualified

3 = Qualified

2 = Possibly qualified

1 = No recommendation

Some colleges will give you credit for a score of 3 or higher, but it's much safer to get a 4 or a 5. If you have an idea of where you will be applying to college, check out the schools' websites or call the admissions offices to find out their particular rules regarding AP scores.

REGISTRATION AND FEES

You can register for the exam by contacting your guidance counselor or AP Coordinator. If your school doesn't administer the exam, contact the Advanced Placement Program for a list of schools in your area that do. There is a fee for taking AP exams, the current value of which can be found at the official exam website listed below. For students with acute financial need, the College Board offers a fee reduction equal to about one-third of the cost of the exam. In addition, most states offer exam subsidies to cover all or part of the remaining cost for eligible students. To learn about other sources of financial aid, contact your AP Coordinator.

For more information on all things AP, contact the Advanced Placement Program:

Phone: (888) 225-5427 or (212) 632-1780

Email: apstudents@info.collegeboard.org

Website: https://apstudent.collegeboard.org/home

CHAPTER 2

Strategies for Success

HOW TO USE THIS BOOK

Kaplan's *AP Biology Prep Plus* contains precisely the information you will need to ace the test. There's nothing extra in here to waste your time—no pointless review of material that's not on the test, no rah-rah speeches. We simply offer the most potent test-preparation tools available.

Book Features

Specific AP Biology Strategies

This chapter features an extended discussion of general test-taking strategies, as well as strategies tailored specifically to the AP Biology exam and the types of questions it contains. In addition, chapter 23 of this book is devoted exclusively to free-response question strategy and sample questions.

Customizable Study Plans

We recognize that every student is a unique individual and that there is not a single recipe for success that works for everyone. To give you the best chance to succeed, we have developed three customizable study plans, each offering guidance on how to make the most of your study time, based on your characteristics as a student. With the instructions offered in the following section of this chapter, you'll be able to select and customize the study plan that is right for you, maximizing your chances of earning the score you need.

Comprehensive Review

The 20 content chapters of this book (chapters 3 through 22) are designed to cover every concept tested on the AP Biology exam. However, unlike the textbook used in your class, this book focuses exclusively on the material you are required to know and examples you can use in free-response questions. Each content chapter includes a series of Learning Objectives that identify key takeaways on each topic and help to organize the chapter. Throughout the text, important terms are highlighted in bold; these terms are compiled in the Glossary, which you can find online. The most commonly tested topics, ones that appear on virtually every AP Biology exam, are demarcated with High Yield icons to help you recognize when information is absolutely essential to know.

Pre- and Post-Quizzes

Every content chapter features a pre-quiz ("Test What You Already Know") and a post-quiz ("Test What You Learned"). Every Learning Objective in a chapter is tested in both quizzes, with test-like Stand-Alone multiple-choice questions that give you a good sense of how these topics will be tested on the exam. You can use the results of these quizzes to guide your studying and chart your progress in mastering the Learning Objectives. Additional quizzes can also be found in your online resources, which come free with the purchase of this book.

AP Biology Lab Investigations

Several of the content chapters feature sections that begin with "AP Biology Lab" followed by a number and a title. These sections review information that is relevant to the 13 official laboratory investigations that the College Board recommends for students in all AP Biology classes. Whether or not you completed these labs as part of your course, you'll find the information in these sections helpful; they're full of high-yield content applicable to multiple-choice, grid-in, and free-response questions.

Full-Length Practice Tests

In addition to all of the practice questions included in each content chapter (as well as the free-response practice questions featured in chapter 23), this book includes two full-length practice tests. Taking a practice AP exam gives you an idea of what it's like to answer biology questions for three hours. Granted, that's not a fun experience, but it is a helpful one. Practice exams give you the opportunity to test and refine your skills, as well as a chance to find out what topics you should spend some additional time studying. And the best part is that it doesn't count! Mistakes you make on our practice exams are mistakes you won't make on the real test. We generally advise that you take one practice exam near the beginning or middle of your preparation and save the other one for a few days before your exam. To score your tests and see what number out of 5 you would achieve for a similar performance on Test Day, check out the answers and explanations and score calculators in your online resources.

Book Layout

This book is divided into seven parts. The first part (which includes this chapter and the previous one) features introductory information about the exam, exam preparation, and test-taking strategies. The last part focuses on practice and includes the chapter on free response questions as well as the two full-length practice exams. The parts in the middle review all of the content.

CHOOSING THE BEST STUDY PLAN FOR YOU

There's a lot of material to review before Test Day, so it's essential to have a good game plan that optimally uses the study time you have available. Toward that end, we have developed three distinctive approaches—Comprehensive Review, Targeted Review, and Time Crunch Review—that you can select from and then further customize to suit your studying needs. How do you know which plan to choose? Read through the following descriptions and check off all of the boxes that apply to you.

Comprehensive Review	Targeted Review	Time Crunch Review
☐ I have two months or more before the test ☐ I want to review all of the content that will be on the test ☐ I didn't take an official AP class this year or my AP class didn't cover a lot of the material effectively ☐ I have a lot of of free time that I can devote to studying	☐ I have less than two months before the test ☐ I want to focus on specific topics that are areas of opportunity for me ☐ I generally know what my strengths and weaknesses are ☐ My AP class covered most of the material effectively, except for a few topics ☐ I have some free time that I can devote to studying	☐ I have less than one month before the test ☐ I want to review the most important topics that will be tested ☐ I'm not sure about my strengths and weaknesses ☐ I have very little free time and/or I am preparing for other AP tests

Some boxes may be more important in helping you select which study play is right for you. For instance, if you only have two weeks before Test Day, Comprehensive Review is probably not a viable option. Nevertheless, most students find they benefit from following the plan with the most checked boxes. After you've made your selection, you can tear out the perforated study plan page from the front of the book, separate the bookmark that contains your choice of plan, and use it to keep track both of your place in the book and your progress in the plan.

If you select the Targeted Review plan, you'll have some additional choices to make about which content you want to focus on in each chapter. Use the blank lines on the Targeted Review bookmark to fill in the names of topics and/or the numbers of Learning Objectives that you wish to target. If you're unsure about what your optimal areas of opportunity are, then you should use the results of a practice test or the "Test What You Already Know" chapter quizzes to identify them. You don't need to choose all your targeted topics now; you can decide on these as you progress. In addition, you can further customize any of the study plans by skipping over chapters or sections that you've already mastered or by adjusting the recommended time scale to better suit your schedule.

SPECIFIC QUESTION STRATEGIES

The AP Biology Exam can be challenging, but with the right strategic mindset, you can get yourself on track for earning the 3, 4, or 5 that you need to qualify for college credit or advanced placement. Before diving into strategies specific to the test, let's review some general strategies that will aid you on any standardized test.

✔ **AP Expert Note**

General Test-Taking Strategies

1. **Pacing.** Because many tests are timed, proper pacing allows someone to attempt every question in the time allotted. Poor pacing causes students to spend too much time on some questions to the point where they run out of time before completing all the questions.
2. **Process of Elimination.** On every multiple-choice test you ever take, the answer is given to you. If you can eliminate answer choices you know are incorrect and only one choice remains, then that must be the correct answer.
3. **Knowing When to Guess.** The AP Biology exam does not deduct points for wrong answers, while questions left unanswered receive zero points. That means you should always guess on a question you can't answer any other way.
4. **Recognizing Patterns and Trends.** The AP Biology exam doesn't change greatly from year to year. Sure, each question won't be the same, and different topics will be covered from one administration to the next, but there will also be a lot of overlap from one year to the next. Because of this, certain patterns can be uncovered. Learning about these trends and patterns can help students taking the test for the first time.
5. **Taking the Right Approach.** Having the right mindset plays a large part in how well people do on a test. Those students who are nervous about the exam and hesitant to make guesses often fare much worse than students with an aggressive, confident attitude.

These points are generally valid for standardized tests, but they are quite broad in scope. The rest of this section will discuss how these general ideas can be modified to apply specifically to the AP Biology exam. These test-specific strategies and the factual information reviewed in this book's content chapters are the one-two punch that will help you succeed on the exam.

Multiple-Choice Questions

The multiple-choice questions are numbered, but that does not mean you must answer the questions in the given order. In fact, it's highly unlikely that the questions will be presented to you in a confidence-inspiring, point-building, time-saving order. There is good news, though: you are free to navigate the section in a manner that highlights your strengths and downplays your weaknesses. All of the multiple-choice questions carry the same weight, so you don't get extra credit for correctly answering a hard question. In the end, colleges will never know which questions you answered correctly. All they will know is that you were smart enough to spend your time where it was more likely to turn into points.

Of the 63 multiple-choice questions, there are two distinct question types: Stand-Alone and Data.

Stand-Alones

These questions typically make up a little over half of the AP Biology exam. Each Stand-Alone question covers a specific topic, and then the next Stand-Alone hits a different topic. The question stem may be as short as a single sentence, but it is not uncommon to see multiple paragraphs for a single question! In addition to words, many question stems will be accompanied by an equation, table, graph, or figure. The next question is a typical Stand-Alone.

22. The human immune system involves two types of specific immune responses: cell-mediated and humoral. Which of the following provides a correct distinction between these two types of immune responses?
    (A) In cell-mediated immunity, T cells only respond to antigens presented on the surface of cells, while in humoral immunity, B cells may respond to free antigens as well.
    (B) In humoral immunity, T cells produce antibodies that bind to antigens, while in cell-mediated immunity, B cells perform this function.
    (C) Humoral immunity only provides a mechanism by which infected cells are recognized, while cell-mediated immunity recognizes and destroys infected cells.
    (D) Humoral immunity may be triggered by nonspecific antigens, while cell-mediated immunity always involves a specific response of antibody to antigen.

You get some information to start with, and then you're expected to answer the question. The number of the question (22) makes no difference because there's no order of difficulty on the AP Biology exam. Tough questions are scattered between easy and medium questions.

Data Questions

Just as the name suggests, a group of two to five questions is preceded by data in one form or another. The data may be a simple sentence or two, but usually it is something more complex, such as:

- A description of an experiment (50–200 words), often with an accompanying illustration
- A graph or series of graphs
- A large table
- A diagram

The next question is a sample Data question.

Questions 61–63

The following graphs present the frequency of size classes for tail length of *Felis domesticus* as measured at four different sites around the world.

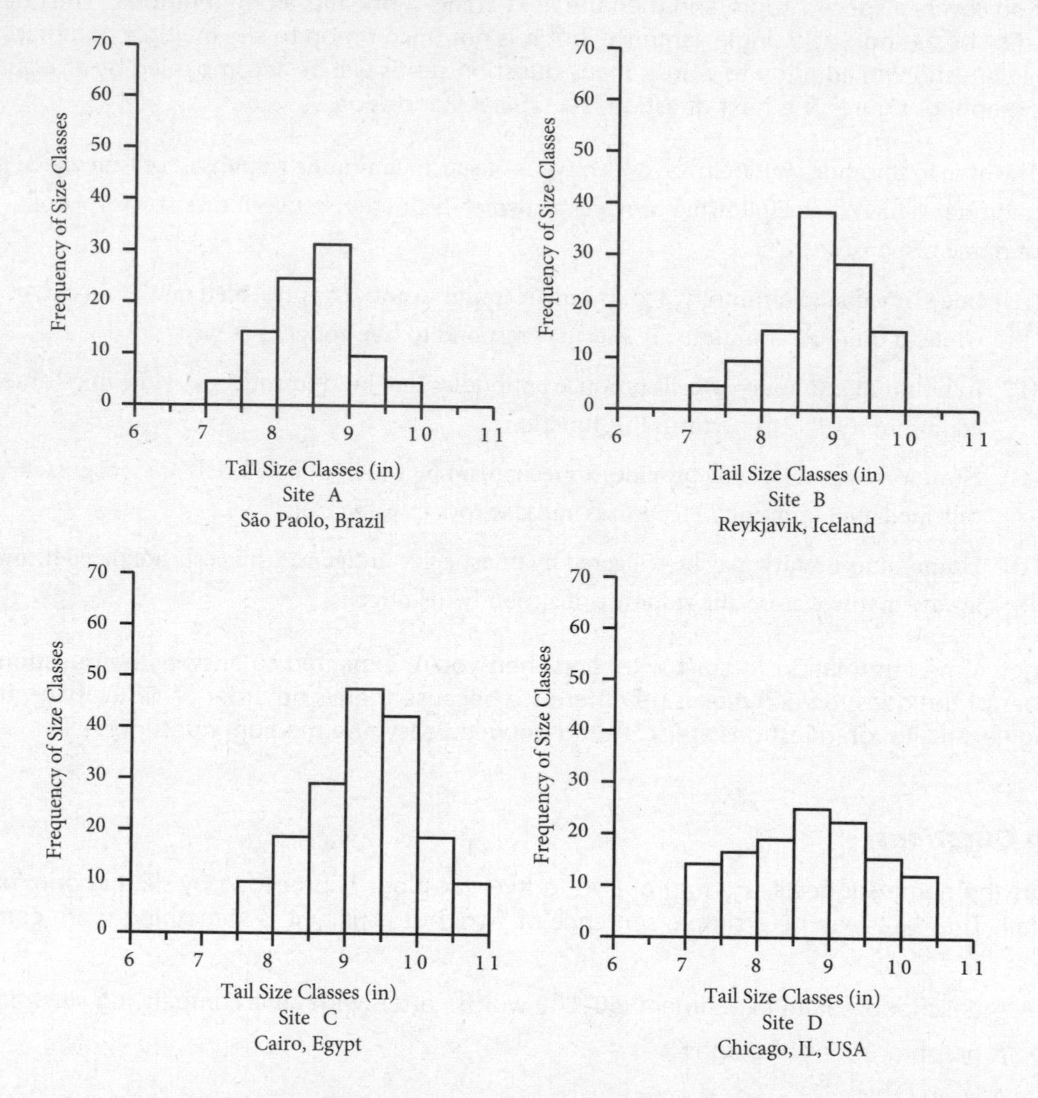

61. According to the data, at Site C, the most common size class of tails among *Felis domesticus* is

(A) 7–8 inches

(B) 8–8.5 inches

(C) 9–9.5 inches

(D) 9.5–10 inches

Some Stand-Alone questions will require you to analyze data; the difference between Stand-Alone and Data questions is the number of questions associated with the data. In a Stand-Alone question, you will analyze the data and then answer a single question. On Data questions you'll get multiple questions referencing the same data.

Those are the two types of questions you'll see in Part A. Combine this knowledge with the fact that you must answer 63 multiple-choice *and* 6 grid-in questions in 90 minutes, and you'll see why it is imperative to take control of the test. Section management may not come to mind when you think about Biology, but it will be tested on the AP Biology exam!

✔ **AP Expert Note**

You don't need to get every multiple-choice question right on the AP Biology exam. To get a 4 or 5, you need to get a large portion, but not all, of the questions right. If you don't have enough time to get to every question, make sure that the questions you skip are the longest, most involved ones. That's a great use of your limited resource: time.

There's more to it than just tackling questions in the right order, however. The more you know about the question types, the better equipped you will be to handle them.

✔ **AP Expert Note**

Educated Guessing

Many times you can eliminate at least one answer choice from a problem. It may seem insignificant, but it gets you closer to the correct answer and it can significantly increase your chances of guessing correctly. You won't get every guess right, but over the course of the test, this form of educated guessing will improve your score.

Stand-Alone Questions

It's easier to talk about what isn't in the Stand-Alone questions than what is there.

- There's no order of difficulty; that is, questions don't start out easy and gradually become tougher.
- There are no two questions connected to each other in any way.
- There's no pattern as to what biology concepts appear when.

The Stand-Alones look like a bunch of disconnected biology questions, one following another, and that's just what they are. A genetics question may follow an ecology question, which may follow a question about metabolism.

There's no overall pattern, so don't bother looking for one. Nevertheless, just because the section is randomly ordered doesn't mean you have to approach it on the same random terms. Instead, you can create two lists as you go about your studying, one that includes all the concepts you've mastered and another that includes all the concepts you find more of a challenge.

When you get ready to tackle the Stand-Alones, keep these two lists in mind. On your first pass through the section, answer all the questions that deal with concepts you like and know a lot about. Quickly glance over any tables, figures, or images and look for terminology in the question stem to clue you into the concept being tested. It should not take very long for you to figure out whether you have the factual chops needed to answer it. If you do, answer that problem and move on. If the question is on a subject that's not one of your strong points, skip it and come back later.

The overarching goal is to correctly answer the greatest possible number of questions in the time available. To do this, focus on your strengths during the first pass through the section. Some questions might be very difficult, even in a subject you're familiar with. Take a minute or so on a tough question, and if you can't come up with an answer, make a mark by the question number in your test booklet and move on. The first pass is about picking up easy points.

Once you've swept through and snagged all the easy questions, take a second pass and try the tougher ones. These tougher questions may cover subjects you're not strong in, or they may just be very difficult questions on subjects you are familiar with. Odds are high that you won't know the answer to some of these questions, but don't leave them blank. You should always take a stab at eliminating some answer choices, and then make an educated guess.

To select the correct answer on an AP Biology exam question, you will need to know the relevant science but, even if certain facts elude you, you can still increase your odds of choosing correctly by keeping the following two key ideas in mind.

Comprehensive, Not Sneaky

Some tests are sneakier than others. They have convoluted writing, questions designed to trip you up mentally, and a host of other little tricks. Students taking a sneaky test often have the proper facts, but get the question wrong because of a trap in the question itself.

✔ AP Expert Note

The AP Biology test is NOT a sneaky test. The test works hard to be as comprehensive as it can be, so that students who only know one or two biology topics will soon find themselves struggling.

Understanding these facts about how the test is designed can help you answer questions on it. The AP Biology exam is comprehensive, not sneaky. You've probably taken an AP Biology course, so trust your instincts when guessing. If you think you know the right answer, chances are you dimly remember the topic being discussed in your AP course. The test is about science, not traps, so trusting your instincts will help more often than not.

You don't have much time to ponder every tough question, so trusting your instincts can help keep you from getting bogged down and wasting time on a problem. You might not get every educated guess correct but, again, the point isn't about getting a perfect score. It's about getting a good score, and surviving hard questions by going with your gut feelings is a good way to achieve this.

On other problems, though, you might have no inkling of what the correct answer should be. In that case, turn to the second key idea.

Think "Good Science"!

The AP Biology test rewards good biologists. The test wants to foster future biologists by covering fundamental topics and sound laboratory procedure. What the test doesn't want is bad science. It doesn't want answers that are factually incorrect, too extreme to be true, or irrelevant to the topic at hand.

Yet these "bad science" answers invariably appear, because it's a multiple-choice test and you must have three incorrect answer choices around the one right answer. So, if you don't know how to answer a question, look at the answer choices and think "good science." This may lead you to find some poor answer choices that can be eliminated.

You would be surprised how many times the correct answer on a multiple-choice question is a simple, blandly worded fact like, "Cells come in a variety of sizes and shapes." No breaking news there, but it is good science: a carefully worded statement that is factually accurate.

> ✔ **AP Expert Note**
>
> **Thinking about good science in terms of the AP Biology exam can help you in two ways:**
>
> 1. It helps you cross out extreme answer choices, choices that are factually inaccurate, and choices that are out of place.
> 2. It can occasionally point you toward the correct answer, because the correct answer will be a factual piece of information sensibly worded.

Neither the "good science" nor the "comprehensive, not sneaky" strategy is 100 percent effective every time, but they do help more often than not. On a tough Stand-Alone question, these techniques can make the difference between an unanswered question and a good guess.

Data Questions

Data questions require a slightly greater initial time investment, but don't let that intimidate you! Once the data is understood, you may find that you can answer the questions rather quickly. Because most of the new information is in the shared introduction, you'll probably notice that the question stems are actually a little shorter than those of the average Stand-Alone. As you navigate through the multiple-choice questions, treat the Data questions in much the same way you would the Stand-Alone questions, with awareness of your strengths and weaknesses and your overall goal of getting more questions correct. If you see a Punnett square and you love heredity, then dive right in. However, if the topic is one that is more likely to induce anxiety than correct answers, then skip it and return after your first pass through the section.

The key to getting through Data questions on the exam is to be able to quickly analyze and draw conclusions from the data presented.

> **✔ AP Expert Note**
>
> **At least one—and most likely several—of the Data questions you see on Test Day will deal with experiments. Make sure you understand all the basic points of an experiment—testing a hypothesis, setting up an experiment properly to isolate a particular variable, and so on—so that you will be able to breeze through these questions when you come to them.**

Questions with Graphs

Most graph questions require a bit of biology knowledge to determine what the right answer is, but some graph questions only test whether or not you can read a graph properly. If you can make sense of the vertical and horizontal axes, then you can determine what the correct answer is. Granted, very, very few graph questions are this easy but, even so, it's nice to have a slam-dunk question or two. Therefore, if you see a graph, look at the problem and see if you can answer the question just by knowing how to read a graph.

Grid-In Questions

Grid-in items appear in Part B of Section I after the multiple-choice questions. Six questions are presented that require you to apply your scientific thinking and mathematical skills to calculate a response. Then you "grid in" your responses in the grid provided on your answer sheet, which will resemble the one shown here.

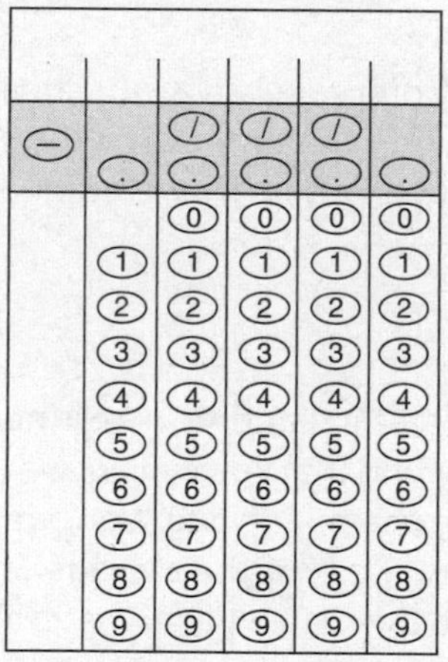

The next question is a sample grid-in item.

121. In a certain species of fruit fly, the allele for red eyes *R* is dominant to the white allele *r*. A scientist performed a cross between a red-eyed fruit fly and a white-eyed fly. When crossed, 62 offspring result. Of these 62 offspring, 43 have red eyes and 19 have white eyes. Calculate the chi-square value for the null hypothesis that the red-eyed parent was heterozygous for the eye-color gene. Give your answer to the nearest tenth.

Just like Data questions, grid-in items require you to analyze the presented information carefully. You must identify the given information, determine the correct formula for solving the problem, make your calculations, and then correctly grid in the answer. Obviously, there is more room for error on the grid-in questions and this should be factored in to your decision to "answer now" or "save for later." Keep in mind, all things being equal, if you need to guess, you'll have better odds guessing on a multiple-choice question, where the answer is actually given to you. As a case in point, would you rather pick the answer to the chi-square question above from four answer choices or grid in your own answer?

To help with the more quantitative problems on the AP Biology exam, you will be allowed to use basic four-function calculators (with square root). In addition, because these questions focus on applying math and not just recalling information, a formula list will be given to you on Test Day. A sample formula list is provided in Appendix A of this book.

Free-Response Questions

Of course, the multiple-choice and grid-in questions only account for 50 percent of your total score. To get the other 50 percent, you have to tackle the free-response questions. Because free-response questions are so distinctive, a separate chapter in the book is devoted to free-response strategy and sample questions.

✔ AP Expert Note

See chapter 23 for information on making the most of the 10-minute reading period, planning your free responses effectively, and scoring as many points as possible. Chapter 23 also includes several sample free-response questions for you to practice with, along with grading rubrics that allow you to determine how many points you would earn on Test Day.

Be sure to use all the strategies discussed in this chapter (and chapter 23) when taking the chapter quizzes and practice exams. Trying out the strategies during practice will get you comfortable with them and make it easier for you to put them to good use on the real exam.

COUNTDOWN TO TEST DAY

Three Days Before the Test

Take a full-length practice test under timed conditions. Use the techniques and strategies you've learned in this book. Approach the test strategically, actively, and confidently.

> ✔ **AP Expert Note**
>
> **DO NOT take a full-length practice test if you have fewer than 48 hours left before the test. Doing so will probably exhaust you and hurt your score on the actual test.**

Two Days Before the Test

Go over the results of your practice test. Don't worry too much about your score or about whether you got a specific question right or wrong. The practice test doesn't count, but examine your performance on specific questions with an eye to how you might get through each one faster and better on the test to come.

The Night Before the Test

DO NOT STUDY. Get together an "AP Biology Exam Kit" containing the following items:

- A watch (as long as it doesn't have internet access, have an alarm, or make noise)
- A few No. 2 pencils (pencils with slightly dull points fill the ovals better; mechanical pencils are NOT permitted)
- A pen with black or dark blue ink (for the free-response questions)
- Erasers
- Your 6-digit school code (home-schooled students will be provided with their state's or country's home-school code at the time of the exam)
- Photo ID card
- Your AP Student Pack
- If applicable, your Student Accommodation Letter, which verifies that you have been approved for a testing accommodation such as braille or large-type exams

Know exactly where you're going, how you're getting there, and how long it takes to get there. It's probably a good idea to visit your test center sometime before the day of the test so that you know what to expect—what the rooms are like, how the desks are set up, and so on.

Relax the night before the test. Read a good book, take a long, hot shower, watch something you'll enjoy. Get a good night's sleep. Go to bed early and leave yourself extra time in the morning.

The Morning of the Test

First, wake up on time. After that:

- Eat breakfast. Make it something substantial, but not anything too heavy or greasy.
- Don't drink a lot of coffee if you're not used to it. Bathroom breaks cut into your time, and too much caffeine is a bad idea.
- Dress in layers so that you can adjust to the temperature of the testing room.
- Read something. Warm up your brain with a newspaper or a magazine. You shouldn't let the exam be the first thing you read that day.
- Be sure to get there early. Allow yourself extra time for traffic, mass transit delays, and/or detours.

During the Test

Don't be shaken. If you find your confidence slipping, remind yourself how well you've prepared. You know the structure of the test; you know the instructions; you've had practice with—and have learned strategies for—every question type.

If something goes really wrong, don't panic. If you accidentally misgrid your answer page or put the answers in the wrong section, raise your hand and tell the proctor. He or she may be able to arrange for you to regrid your test after it's over, when it won't cost you any time.

After the Test

You might walk out of the AP Biology exam thinking that you blew it. This is a normal reaction. Lots of people—even the highest scorers—feel that way. You tend to remember the questions that stumped you, not the ones that you knew. We're positive that you will have performed well and scored your best on the exam because you followed the Kaplan strategies outlined in this chapter. Be confident in your preparation, and celebrate the fact that the AP Biology exam is soon to be a distant memory.

PART 2

Foundations of Life

CHAPTER 3

Biological Molecules

LEARNING OBJECTIVES

In this chapter, you will review how to:

- **3.1** Recall the elements and molecules important to life
- **3.2** Explain why life requires free energy
- **3.3** Recognize the role of water in biological processes
- **3.4** Differentiate between common monomers and polymers
- **3.5** Determine how structure affects properties of polymers

Foundations

TEST WHAT YOU ALREADY KNOW

1. The diagram below shows part of the carbon cycle as it occurs on land. Which of the following processes is occurring at the box with the "A" ?

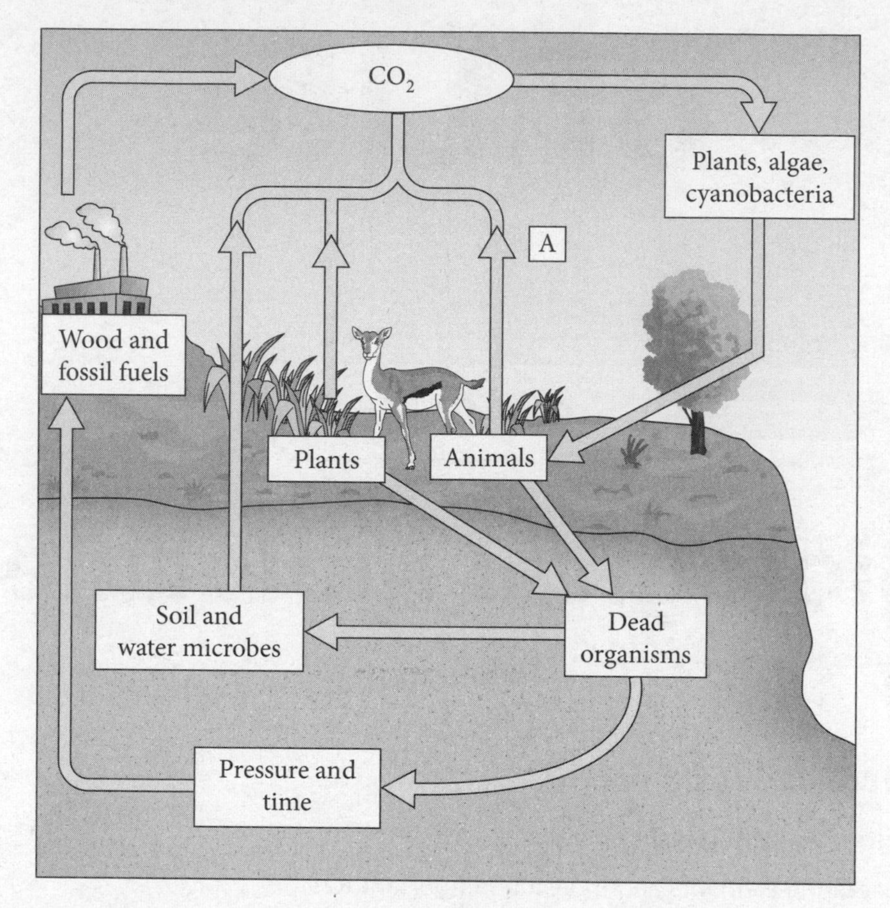

(A) Organic matter is broken down into simpler molecules by microorganisms.

(B) Carbon dioxide reacts with water to produce glucose as the primary food source for these organisms.

(C) Organic matter from the remains of dead animals is converted by extreme pressure into coal and oil.

(D) Nutrients inside a cell, such as glucose, are converted into usable forms of energy, such as ATP.

2. When a person exercises, muscles in the body convert energy stored in complex molecules into kinetic energy. This action releases a large amount of heat, which is transferred to the surroundings. Which law of thermodynamics best explains why this dissipation of heat occurs?

   (A) The first law of thermodynamics, because this is an example of energy being created

   (B) The second law of thermodynamics, because this is an example of energy being created

   (C) The first law of thermodynamics, because this is an example of increasing entropy

   (D) The second law of thermodynamics, because this is an example of increasing entropy

3. Researchers are comparing capillary action in two different species of plants by monitoring water flow with water-based dyes. The first species is a flowering plant in which the walls of the xylem are lined with negatively charged molecules. The second species is a wilted *Myrothamnus flabellifolia*, a plant that becomes brown and shriveled under drought conditions and is revived to its original green color and appearance once water becomes available. The researchers discovered that the walls of the water-conducting vessels in desiccated *Myrothamnus* are lined with hydrophobic molecules. Which of the following capillary activities will most likely be observed by the researchers?

   (A) The dye will reach the same height in both plants because the chemical nature of the vessel walls does not affect capillary action.

   (B) The dye will reach a higher point in *Myrothamnus* because the hydrophobic walls increase the surface tension of water.

   (C) The dye will reach a higher point in the flowering plant because the vessel walls in *Myrothamnus* do not interact with water and have lower adhesion forces.

   (D) No movement of water will be observed in the wilted *Myrothamnus* because the hydrophobic lining interferes with the cohesion of water molecules.

4. A scientist finds a novel nucleic acid in a seawater sample and hopes to determine its origin. The nucleic acid consists of a single strand and does not have characteristics of a viral nucleic acid. Which of the following is most likely the composition of the molecule?

   (A) The molecule contains a pentose sugar, a phosphate group, and the nitrogenous bases adenine, thymine, cytosine, and guanine.

   (B) The molecule contains a pentose sugar, a phosphate group, and the nitrogenous bases adenine, uracil, cytosine, and guanine.

   (C) The molecule contains a tetrose sugar, a phosphate group, and the nitrogenous bases adenine, uracil, cytosine, and guanine.

   (D) The molecule contains a pentose sugar, a fatty acid, and the nitrogenous bases adenine, uracil, cytosine, and guanine.

5. A researcher examines the structure of a protein that has been extracted from the membrane of a eukaryotic cell. She finds that the surface of one domain of the protein consists entirely of hydrophobic amino acids, whereas the surface of another domain consists entirely of hydrophilic amino acids. What can she most reasonably conclude about this protein?

   (A) This protein is probably found deep within the cell and must have been mixed with the membrane material in error during sample preparation.

   (B) The protein is probably found embedded completely within the phospholipid bilayer.

   (C) The protein is probably bound to the outer surface of the cell membrane, facing the extracellular matrix.

   (D) The protein is probably embedded in the cell membrane with portions extending either into the extracellular matrix or the cytoplasm.

Answers to this quiz can be found at the end of this chapter.

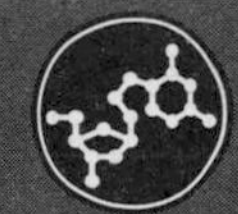

MATTER AND ENERGY

3.1 Recall the elements and molecules important to life

3.2 Explain why life requires free energy

The world around us follows a hierarchy of organization. All life on Earth is connected, from the smallest individual units of matter (atoms and molecules) to complex organisms. In this chapter, we will discuss the basic building blocks that compose the living world. Although some organisms are more complex than others, different levels of organization do not correlate with levels of complexity. An individual **cell** with its multitude of chemical reactions is just as dynamic and complex as an entire community of species. This chapter will review the most crucial concepts of the building blocks of life.

Elements Essential to Life

Ninety-nine percent of all living matter is made up of only four elements. These elements are nitrogen (N), carbon (C), hydrogen (H), and **oxygen** (O). Phosphorus (P) and sulfur (S) account for almost all of the remaining 1 percent of living matter. All six elements are important in biochemistry, especially carbon (C).

> ✔ **AP Expert Note**
>
> **The mnemonic device N'CHOPS (for nice chops!) can be used to remember the main elements of life.**

Most molecules that contain carbon (C) are known as **organic molecules** or organic compounds. (There are some exceptions: for instance, carbon dioxide and hydrogen cyanide have the chemical formulas CO_2 and HCN, respectively, but are generally not considered to be organic.) Life wouldn't exist without organic compounds. Carbon earned its place as a staple in biology because it can bond to other atoms or to itself in four different equally spaced directions, allowing complex molecules of almost unlimited size and shape to be formed. The structural properties of the carbon atom have allowed for the formation of molecular compounds—such as DNA and enzymes—that have unique chemical identities and functions. The message to take home is that there are only a few elements important to biology (N'CHOPS) and that carbon is the main element of life because of the variety of organic compounds it can form.

> ✔ **AP Expert Note**
>
> **Macromolecules such as carbohydrates, lipids, proteins, and nucleic acids are examples of organic compounds that serve as building blocks in all living organisms. These compounds are discussed in detail later in this chapter.**

Free Energy in Living Systems

All living things require the capture or harvest of free energy from the environment to grow, reproduce, and maintain dynamic homeostasis. However, living systems must also follow the laws of thermodynamics, which means that entropy is always increasing. To balance the movement toward entropy, organisms must take in more energy than is used by the organism.

Though you do not need to memorize the specific steps of the reactions that help living things to capture and use free energy, it is important to remember that endergonic (energy-consuming) and exergonic (energy-releasing) reactions are typically connected. For example, the catabolic breakdown of ATP to ADP is an energetically favorable exergonic reaction. This release of free energy is coupled to endergonic reactions such as the synthesis of glutamine, an essential amino acid.

THE IMPORTANCE OF WATER

3.3 Recognize the role of water in biological processes

A water molecule is formed between two H atoms bonded covalently to a single O atom. The oxygen molecule tends to control all of the electrons by keeping them away from the hydrogen atoms, giving the oxygen a slightly negative charge, which is balanced by slightly positive charges on the hydrogen atoms. This is an example of a polar covalent bond. The water molecule looks like Mickey Mouse because it has a big head (the oxygen atom) and two big ears (the hydrogen atoms).

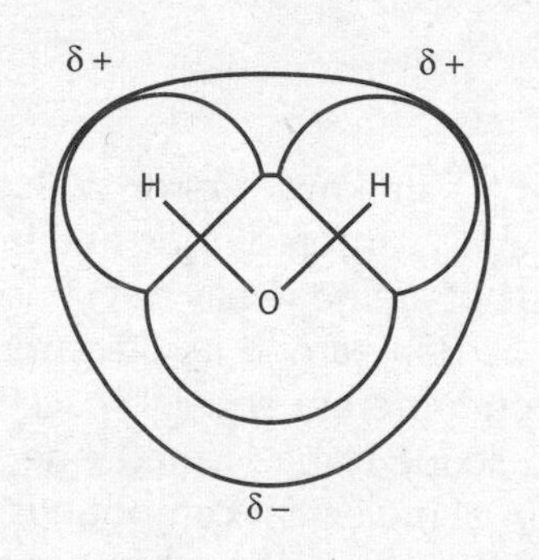

The head has a slightly negative charge and the ears have a slightly positive charge, which makes water a polar compound. The shape and charge of this molecule give it unique properties.

Water is the only substance on Earth that commonly exists in all three physical states (gas, liquid, and solid). The substance has a high specific heat so it serves as a temperature stabilizer for other compounds and vaporizes at a relatively high temperature.

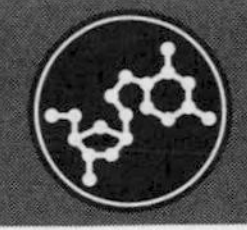

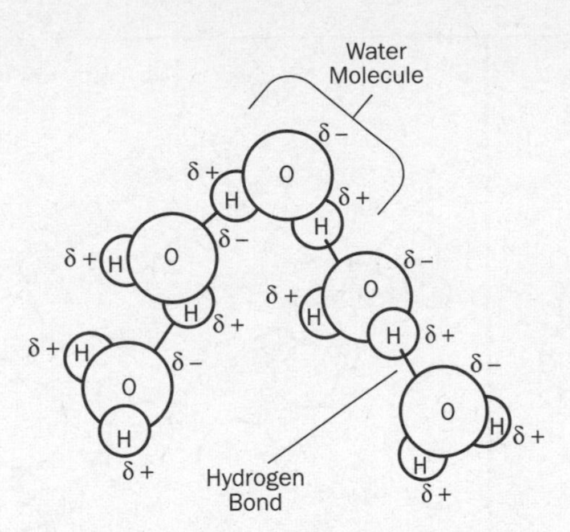

Water plays a key role in hydrolysis, condensation, and other chemical reactions that are essential to life. It is also fundamental to the biological activity of nucleotides, carbohydrates, lipids, and proteins. There are two specific characteristics of water that are particularly important to remember. First, the polar nature of water makes it "sticky." The positive charges on the hydrogen atoms cling to the negative charges on the oxygen atoms between molecules; i.e., water molecules attract one another, causing water to have high surface tension. This is what makes water bead on windshields and form round raindrops. Even matter with greater density can float on top of water if it doesn't break the surface tension.

This surface tension is also the force behind capillary action, by which water (and anything dissolved in it) will climb up a thin tube or move through the spaces of a porous material until it is overcome by gravity. It is as if each water molecule drags along the one behind it, as well as any **nutrients** dissolved in the water. Capillary action plays an important role in moving nutrients and other metabolites through living things. Plant roots, for example, take in water from the soil, full of **minerals** and dissolved nutrients; capillary action draws the water and its load through the plant against gravity.

✓AP Expert Note

Water expands rather than contracts when it freezes. One consequence of this is that solid ice floats on liquid water.

Second, the polar nature of water makes it a good solvent. Capillary action would be a lot less useful if water lacked this property. Fortunately, water molecules are happy to surround positive and negative ions and readily dissolve other polar compounds. In addition, hydrogen bonds can form between the hydrogen and oxygen atoms of the water molecule and surrounding molecules.

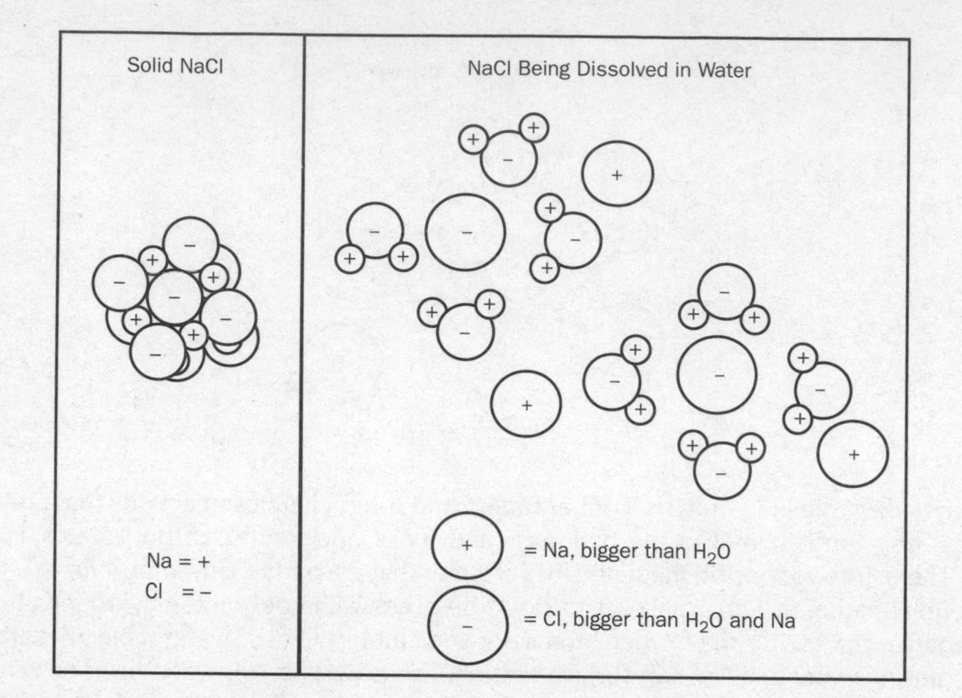

Water's role as a solvent is another reason it is so important for living things. Chemicals that are dissolved by the water in blood can be carried around the body rapidly and easily; this is how sugar gets to muscles, for example. Each living cell is, in large part, a **membrane** surrounding chemicals dissolved in water. It is only by being dissolved in water that these chemicals can participate in many of the biological reactions that keep living things alive.

✔ AP Expert Note

Remember that many organic compounds are nonpolar or have nonpolar regions and won't mix with water. Such nonpolar regions are consequently called hydrophobic (literally meaning "water-fearing"), while polar and ionic parts of molecules are hydrophilic ("water-loving").

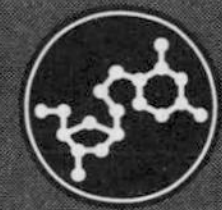

BIOLOGICAL MONOMERS AND POLYMERS

3.4 Differentiate between common monomers and polymers

3.5 Determine how structure affects properties of polymers

While water molecules contribute to the majority of a cell's total mass, the majority of a cell's dry mass consists of four classes of biological macromolecules: carbohydrates, lipids, proteins, and nucleic acids. These macromolecules play a role in cell structure and carry out functions necessary for the survival and growth of living organisms. For instance, the breakdown of macromolecules provides energy for cellular activities.

Carbohydrates, proteins, and nucleic acids are examples of **polymers** that are built from smaller molecular units, or **monomers,** connected by covalent bonds. Lipids are not considered polymers because they are not composed of monomers. While some lipids are made up of smaller molecules like fatty acids, those subunits are not directly bonded to one another as in proper polymers.

Cells assemble polymers via a process called dehydration synthesis (condensation), in which energy is consumed to form a covalent bond and release water. During dehydration synthesis, either two hydrogen atoms from one monomer and an oxygen atom from another monomer are combined, or a hydrogen atom from one monomer and a hydroxyl (OH) group from another monomer are combined. The reverse process is hydrolysis. Hydrolysis reactions use water to break down polymers and release energy—one monomer gains a hydrogen atom and the other monomer gains a hydroxyl group.

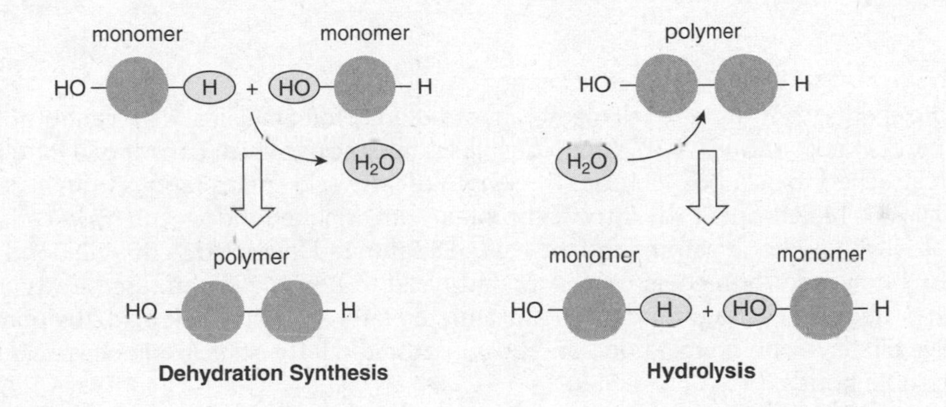

Carbohydrates

Carbohydrates are a source of immediate energy in living systems and serve as structural components. The three primary classes of carbohydrates are monosaccharides, disaccharides, and polysaccharides.

Monosaccharides are simple sugars that have the formula $(CH_2O)_n$, where *n* represents the number of carbons in the backbone and ranges from three to seven. Monosaccharides are classified based on the number of carbon atoms (triose, tetrose, pentose, hexose, or heptose), as well as on the position of their carbonyl group (an oxygen double-bonded to a carbon). A monosaccharide with an aldehyde group (carbonyl group at the end of the carbon chain) is an aldose, and a monosaccharide with a ketone group (carbonyl group in the middle of the carbon chain) is a ketose. Glucose and galactose are examples of aldoses, and fructose is an example of a ketose; all three are hexoses because they each have six carbons.

Disaccharides are formed when two monosaccharides undergo dehydration synthesis to form a glycosidic bond. Common disaccharides include lactose (glucose and galactose), maltose (two glucose molecules), and sucrose (glucose and fructose).

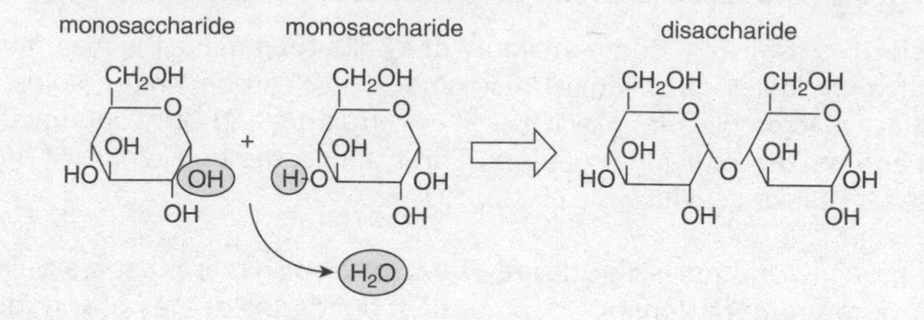

Polysaccharides are composed of many monosaccharides linked by glycosidic bonds. In plants, both starch and cellulose are polysaccharides composed of glucose molecules, but they differ in structure and function. The difference is the linkage of the glucose bonds formed during dehydration synthesis: starch contains alpha linkages and cellulose contains beta linkages. Starch serves as the main energy storage material in plants, while cellulose provides structural support. To use energy stored in starch, organisms must first break it down into glucose. Animals, however, are unable to hydrolyze cellulose because they do not have the enzymes that break down the beta linkages. In animals, glycogen (a glucose polymer similar to starch, but containing branches) is a form of energy storage and chitin (analogous in structure to cellulose) provides support.

Lipids

Lipids store energy, contribute to cell membrane structure, protect against desiccation, and provide building blocks for hormones. A subgroup of lipids is fatty acids, which consist of a long hydrocarbon chain attached to a carboxyl group. The length of fatty acid chains ranges from 4–36 carbons but is usually 12–18 carbons. If the fatty acid contains only single bonds in the hydrocarbon chain, it is saturated with hydrogen atoms and thus called a saturated fatty acid. A double bond between carbons in the hydrocarbon chain causes the fatty acid to become unsaturated. Fatty acids may differ in their degree of unsaturation. Monounsaturated fatty acids like oleic acid (the primary fatty acid in olive oil) have one double bond and polyunsaturated fatty acids like linoleic acid have two or more double bonds.

The configuration of the double bonds affects the behavior of fatty acids (and the fats that contain them). In the cis configuration, hydrogen atoms at a double bond are in the same plane (on the same side), and in the trans configuration, they are in two different planes (on opposite sides). Unlike single bonds, which allow atoms to rotate freely, double bonds restrict rotation and lock the orientation of atoms. A cis double bond causes a bend in the chain, which prevents fatty acids from packing tightly against one another, so they are liquid at room temperature and have a relatively low melting point. Saturated fatty acids on the other hand pack tightly together, so they are solid at room temperature and have a relatively high melting point.

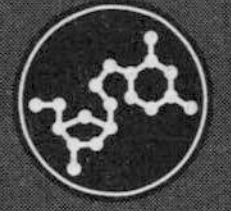

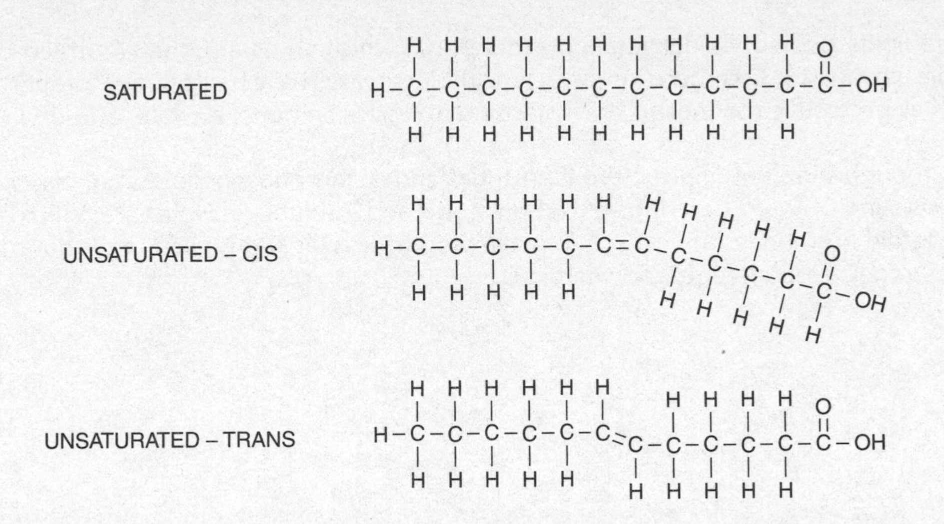

Fats called triglycerides are another subgroup of lipids. Triglycerides are formed via dehydration synthesis. Hydroxyl groups on a glycerol backbone react with the carboxyl groups of three fatty acids to form ester bonds and release three molecules of water. Primarily stored in **adipose** (fat) tissue, triglycerides are a major form of energy storage.

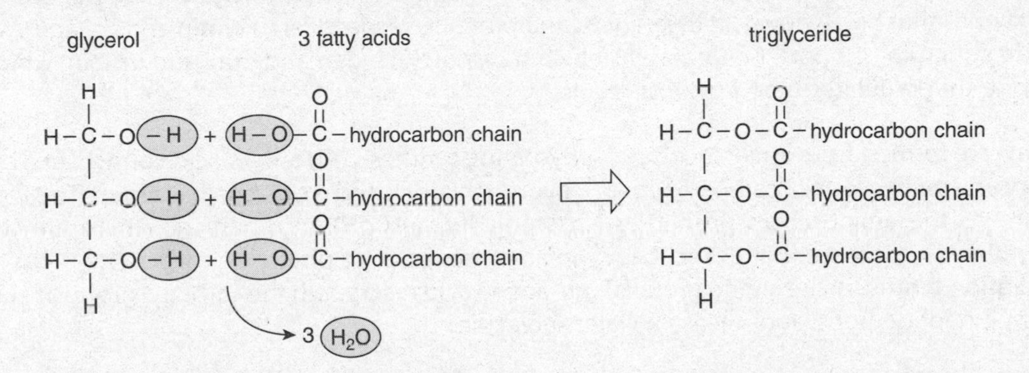

Phospholipids, like fats, are composed of glycerol and fatty acid chains. However, instead of three fatty acids, phospholipids have two fatty acids and a modified phosphate group. The phosphate group or head is negatively charged, polar, and hydrophilic, whereas the fatty acids or tails are unsaturated, uncharged, nonpolar, and hydrophobic. This makes phospholipids amphipathic (containing both hydrophilic and hydrophobic regions).

As the major components of the **plasma membrane**, phospholipids are arranged in a bilayer with the heads facing outward toward the intracellular and extracellular fluids and the tails facing inward to prevent them from contacting water. The phospholipids' chemical and physical characteristics cause the membrane to be fluid, semipermeable, and to require low energy to remain stable. The separation of the intracellular and extracellular fluids is vital for cell communication and metabolism.

polar head
nonpolar tails
phospholipid bilayer

Foundations

Waxes are lipids formed via dehydration synthesis, in which an ester bond is formed between a long-chain alcohol (12–32 carbons) and a fatty acid. Since waxes have hydrophobic properties, plants use them as protective coatings to prevent losing excessive amounts of water and drying out.

Steroids, though different in structure than triglycerides and phospholipids, are also considered lipids. Consisting of four fused rings of carbon atoms and insoluble in water, sterol lipids form the basis of steroid hormones. Cholesterol, for example, absorbs fat and is used to produce hormones such as estrogen, testosterone, and vitamin D.

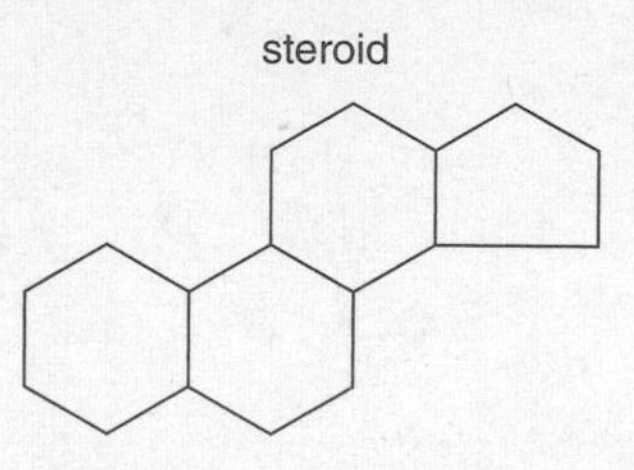

Proteins

Making up half the total dry weight of cells, **proteins** function as enzymes, structural materials, membrane transport, and signaling molecules. Examples include **lactase**, **lipase**, and **pepsin**, which are enzymes that break down lactose, lipids, and proteins, respectively; keratin and collagen, which provide structural support; hemoglobin, which transports oxygen and iron; and insulin, which is a hormone that regulates blood glucose levels.

Proteins are formed from amino acids via dehydration synthesis. An amino acid consists of a central carbon, an amino group, a carboxyl group, a hydrogen atom, and a side chain R group. The R group determines the amino acid's properties (size, polarity, and pH). In organisms, there are twenty α-amino acids; each is either acidic, basic, polar, or nonpolar at physiological pH (7.2–7.4). During dehydration synthesis, the amino group of one amino acid reacts with the carboxyl group of another to form a peptide bond and release a water molecule.

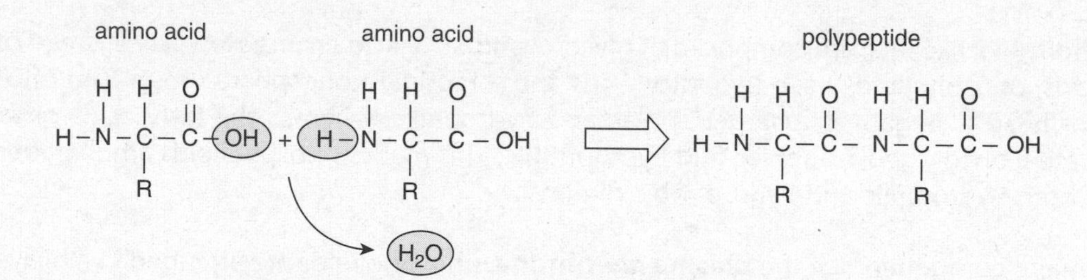

A chain of amino acids is a polypeptide, which typically ranges from 50 to 1,000 amino acids. Since polypeptides have two distinct ends, they have directionality. The amino acid sequence is written and read from the amino end (N-terminus) to the carboxyl end (C-terminus). The properties and order of the amino acids determine the structure and function of the polypeptide. For instance, proteins with more basic amino acids will have an overall positive charge in a neutral solution, while proteins with more acidic amino acids will have an overall negative charge.

The sequence of amino acids in a polypeptide chain is the protein's primary structure and is determined by the DNA of the gene that encodes the protein. The protein's secondary structure refers to local folds along the chain due to interactions between the atoms of the backbone. The most

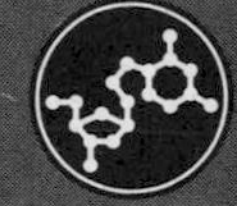

common types of secondary structures are the α helix and the β pleated sheet, which involve hydrogen bonds between the carbonyl O of one amino acid and the amino H of another. A protein's shape and stability are influenced by the interactions between a protein and its immediate environment. For example, a protein in an aqueous environment will fold so that the hydrophilic R groups are at the surface and the hydrophobic R groups are on the inside of the protein. This three-dimensional folding, due to interactions between the R groups of amino acids, is a protein's tertiary structure. R group interactions include hydrogen bonds, ionic bonds, dipole-dipole interactions, London dispersion forces, hydrophobic interactions, and covalent disulfide bonds. The orientation and arrangement of multiple polypeptide chains or subunits constitute a protein's quaternary structure. Hemoglobin, which consists of two α and two β subunits, is an example of a protein with quaternary structure. Protein structure is important to its function: if a protein loses its shape (is denatured) at any structural level, its function will likely be diminished and possibly lost entirely.

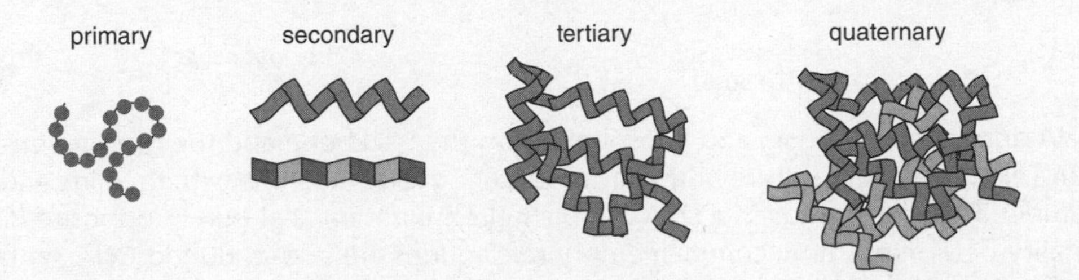

Nucleic Acids

Nucleic acids encode biological information in nucleotide sequences and provide information for **protein synthesis**. Nucleotides are composed of three components: a pentose sugar (**deoxyribose** or **ribose**), a **nitrogenous base** that is a purine (**adenine** and **guanine**) or pyrimidine (**thymine, cytosine**, or **uracil**), and one or more phosphate groups attached to the 5' carbon of the sugar.

Nucleotide monomers join by dehydration synthesis to form nucleic acids: the 3' carbon hydroxyl group of the sugar combines with a hydrogen of the phosphate group of another nucleotide to form a phosphodiester bond. Like proteins, nucleic acids have distinct ends, defined by the 3' and 5' carbons of the sugar in the nucleotide.

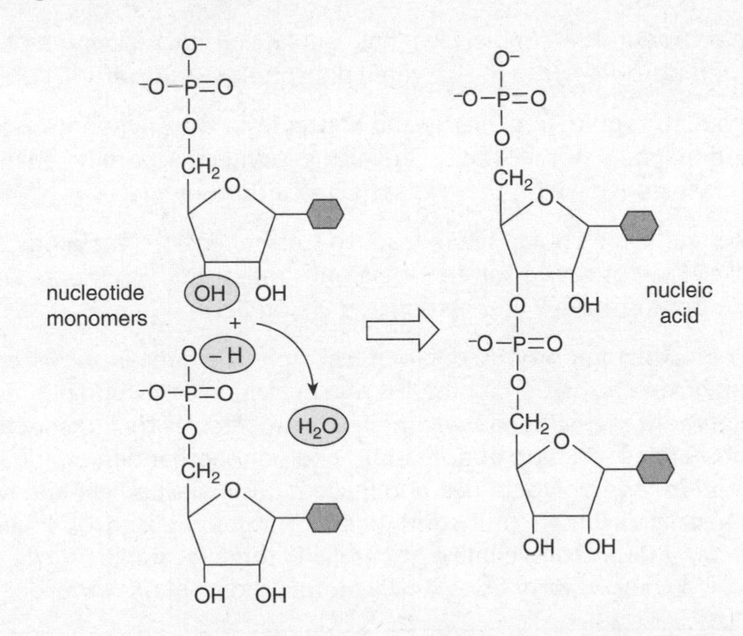

Foundations

There are two types of nucleic acids: deoxyribonucleic acid (**DNA**) and ribonucleic acid (**RNA**). DNA carries the genetic information living organisms need to function, grow, and reproduce. Living organisms then use RNA to carry instructions from DNA for protein synthesis. The structural differences between DNA and RNA account for their differing functions. DNA contains deoxyribose and thymine and is double-stranded, while RNA contains ribose and uracil and is single-stranded. However, remember that viruses, unlike organisms, can have double- or single-stranded DNA or RNA.

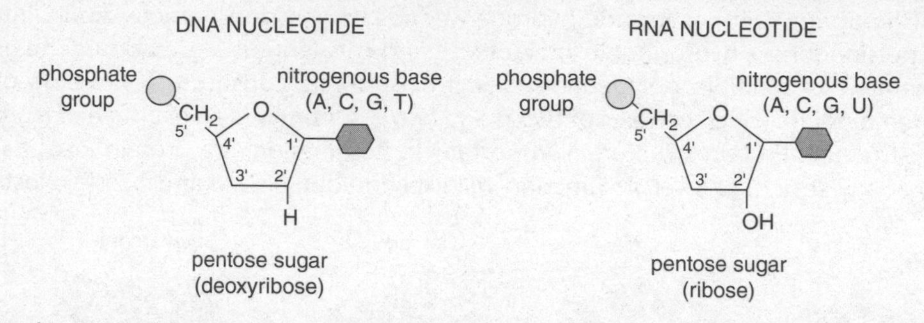

In the DNA double helix, sugars and phosphate form the backbone and the nitrogenous bases of each strand extend inward and are bound by base pairing rules (adenine with thymine and cytosine with guanine). The two strands in a DNA double helix are antiparallel (run in opposite directions). Directionality determines how complementary nucleotides are added during DNA synthesis and the direction in which DNA is transcribed to RNA (from 5' to 3'). RNA occurs in different forms including messenger RNA, ribosomal RNA, and transfer RNA.

✓AP Expert Note

Nucleic acids are discussed in further detail in chapter 16, Molecular Genetics.

RAPID REVIEW

If you take away only 4 things from this chapter:

1. Organic compounds are molecules that contain carbon. All living matter is made up of nitrogen, carbon, hydrogen, oxygen, phosphorus, and sulfur (N'CHOPS).
2. Living systems require free energy and matter from the environment to grow, reproduce, and maintain homeostasis. Organisms survive by coupling chemical reactions that increase entropy with those that decrease entropy.
3. The water molecule's polar nature leads to surface tension (enabling capillary action) and makes it an effective solvent. It expands rather than contracts when it freezes. These properties make water essential to life on Earth.
4. The four most common types of biological molecules are carbohydrates, lipids, proteins, and nucleic acids. Carbs, proteins, and nucleic acids include polymers composed of simpler monomer subunits, which give the molecules their distinctive properties: monosaccharides make up disaccharides and polysaccharides, amino acids make up monopeptides and polypeptides, and nucleotides make up RNA and DNA. Lipids are not technically polymers, but some of them are composed of smaller molecules: triglycerides (fats and oils) contain glycerol and three fatty acids, while phospholipids contain glycerol, two fatty acids, and a modified phosphate group.

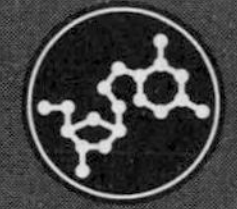

TEST WHAT YOU LEARNED

1. Adenosine triphosphate (ATP) can be synthesized from adenosine diphosphate (ADP) according to the following reaction:

$$ADP + P_i \rightarrow ATP + H_2O$$

The energy change occurring during this reaction is shown in the graph:

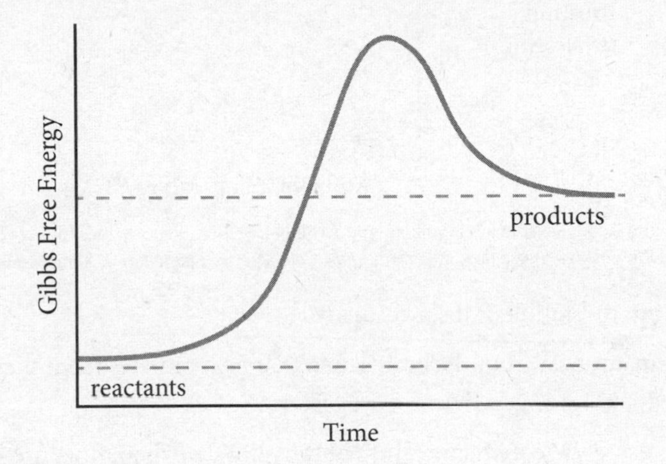

Which statement best describes the formation of ATP?

(A) The formation of ATP is endergonic, and energy is released during the process.

(B) The formation of ATP is exergonic, and energy is absorbed during the process.

(C) The formation of ATP is endergonic, and energy is absorbed during the process.

(D) The formation of ATP is exergonic, and energy is released during the process.

2. Reindeer live in arctic regions and have large bodies covered by thick, insulating fur. A scientist studying how reindeer adapt to warm temperatures observed a change in respiratory patterns as temperature increased. The frequency of reindeer panting as a function of ambient temperature is shown in the graph.

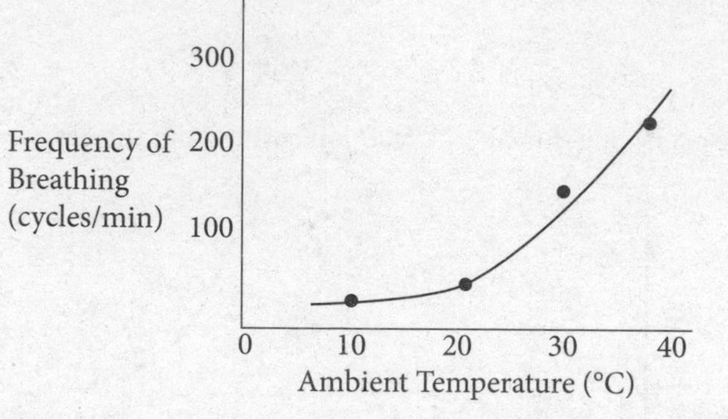

Adapted from Øyvind Aas-Hansen, Lars P. Folkow, and Arnoldus S. Blix, "Panting in reindeer (*Rangifer tarandus*)," *American Journal of Physiology—Regulatory, Integrative and Comparative Physiology* 279, no. 4 (October 2000): R1190.

What is the best interpretation of these results?

(A) Panting dissipates heat from the body when that heat radiates from the open nose and mouth.

(B) Panting decreases the core temperature of the reindeer when moisture from the lining of the nose and mouth evaporate into the air.

(C) Water in saliva has a high heat capacity and absorbs excess heat from the environment.

(D) Panting removes water from the surface of the tongue, allowing reindeer to absorb more heat from the water they drink.

3. Animals store glucose in the liver as a polymer called glycogen. Plants store glucose as a different polymer called starch. Cellulose is another polymer that is an important part of producing strong plant cell walls. Although cellulose is also a polymer of glucose, plant foods with high cellulose content, such as celery, do not provide many calories when eaten by humans. Which statement best explains why humans cannot effectively extract energy from cellulose, despite it being a polymer of glucose?

(A) The glucose monomers in cellulose are connected by different bonds than those in glycogen and starch. Human digestive enzymes cannot break down these bonds.

(B) The monomers in glycogen, starch, and cellulose differ slightly. Humans cannot break down the form of glucose found in cellulose.

(C) Cellulose is structured in a way that makes it impossible for mammals to digest.

(D) Cellulose is a larger polymer than glycogen and starch, with more associated water molecules that protect it from digestion.

4. Glycogen and triacylglycerols are both used by animals to store energy. If scientists were examining a sample of storage molecules, what is the best way to distinguish whether those molecules consisted of glycogen or triacylglycerols?

    (A) Use spectrometry and spectroscopy to determine whether the molecules in the tissue contained oxygen.

    (B) Examine whether the molecules were hydrophilic or hydrophobic and determine how many calories per gram they contain.

    (C) Determine the identity of the material by noting whether it was found in skeletal muscle.

    (D) Determine the mass of the material and compare it to reference values.

5. Photosynthesis is a chemical process by which plants use sunlight to convert carbon dioxide and water to an energy source, as shown in the reaction below:

$$6\ CO_2 + 12\ H_2O + \text{sunlight ¡ } C_6H_{12}O_6 + 6\ O_2 + 6\ H_2O$$

This reaction usually requires the pigment chlorophyll. Which of the following best explains what occurs during this reaction?

    (A) Carbon dioxide undergoes reduction and water undergoes oxidation. In the process, glucose is made as the primary food source for the plant and oxygen gas is released as a byproduct.

    (B) Carbon dioxide undergoes oxidation and water undergoes reduction. In the process, glucose is made as the primary food source for the plant, and oxygen gas is released as a byproduct.

    (C) Carbon dioxide undergoes reduction and water undergoes oxidation. In the process, oxygen gas is made as the primary food source for the plant, and glucose is released as a byproduct.

    (D) Carbon dioxide undergoes oxidation and water undergoes reduction. In the process, oxygen gas is made as the primary food source for the plant, and glucose is released as a byproduct.

Answer Key

Test What You Already Know

1. **D** **Learning Objective:** 3.1
2. **D** **Learning Objective:** 3.2
3. **C** **Learning Objective:** 3.3
4. **B** **Learning Objective:** 3.4
5. **D** **Learning Objective:** 3.5

Test What You Learned

1. **C** **Learning Objective:** 3.2
2. **B** **Learning Objective:** 3.3
3. **A** **Learning Objective:** 3.5
4. **B** **Learning Objective:** 3.4
5. **A** **Learning Objective:** 3.1

Detailed solutions can be found in the Answers and Explanations section at the back of this book.

REFLECTION

Test What You Already Know score: __________

Test What You Learned score: __________

Use this section to evaluate your progress. After working through the pre-quiz, check off the boxes in the "Pre" column to indicate which Learning Objectives you feel confident about. Then, after completing the chapter, including the post-quiz, do the same to the boxes in the "Post" column. Keep working on unchecked Objectives until you're confident about them all!

Pre	Post	
❑	❑	**3.1** Recall the elements and molecules important to life
❑	❑	**3.2** Explain why life requires free energy
❑	❑	**3.3** Recognize the role of water in biological processes
❑	❑	**3.4** Differentiate between common monomers and polymers
❑	❑	**3.5** Determine how structure affects properties of polymers

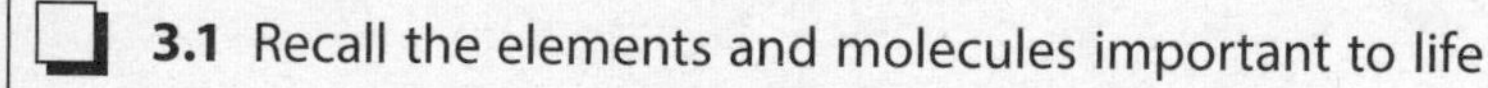

FOR MORE PRACTICE

Complete more practice online at kaptest.com. Haven't registered your book yet? Go to kaptest.com/booksonline to begin.

CHAPTER 4

The Origin of Life and Natural Selection

LEARNING OBJECTIVES

In this chapter, you will review how to:

- **4.1** Evaluate evidence for origin hypotheses
- **4.2** Recall Darwin's theory
- **4.3** Explain how natural selection impacts evolution

Foundations

TEST WHAT YOU ALREADY KNOW

1. One important hypothesis that addresses the origin of life on Earth is the RNA World Hypothesis. Which of the following statements provides the greatest support for the RNA World Hypothesis?
   (A) RNA is a more stable molecule than DNA.
   (B) The first self-replicating structures were RNA-based protobionts.
   (C) RNA was the first chemical polymer to form on Earth.
   (D) The first living cells used reverse transcription to produce a DNA genome from RNA.

2. White sand lizards vary in their color, which may be an adaptation to their environment. Those living in lighter environments may blend in better when their color is lighter, and those with darker coloration may fare better on darker backgrounds. Researchers used a nontoxic paint to color the backs of lizards with similar natural colorations and placed them in environments that matched their color ("matched") or that conflicted with their color ("mismatched"). The figure below shows the proportion of lizards in each treatment group that were recaptured (i.e., still alive) after 16 days.

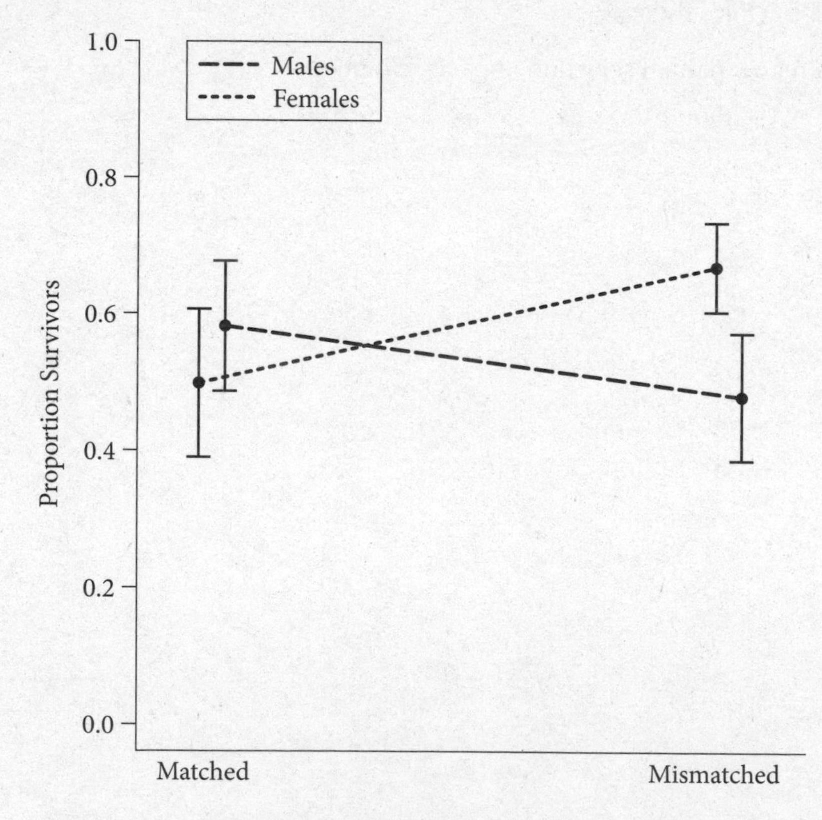

Adapted from Kayla M. Hardwick et al., "When Field Experiments Yield Unexpected Results: Lessons Learned from Measuring Selection in White Sands Lizards," *PLoS ONE* 10, no. 2 (2015): e0118560.

Assume that males and females reproduce at a similar rate. According to Darwin's theory of natural selection, which group of individuals most likely had the highest fitness?

(A) Female lizards painted to resemble their background had the highest fitness.

(B) Male lizards painted to resemble their background had the highest fitness.

(C) Female lizards painted a different color from their background had the highest fitness.

(D) Male lizards painted a different color from their background had the highest fitness.

3. A scientist studying bird migration places bands around the legs of each bird that visits her research site each year in order to examine whether increased water temperatures affect survival. In this way, she is able to document a pattern that larger birds are increasingly more likely to return than smaller birds. What conclusion can most reasonably be drawn from these data?

(A) Larger birds probably have access to more food, allowing them more energy to migrate successfully.

(B) The birds may become larger in future generations due to selective pressure.

(C) Smaller birds may reproduce at younger ages so that they can reproduce successfully despite having shorter lifespans.

(D) The birds may cease migrating so that more of them can survive.

Answers to this quiz can be found at the end of this chapter.

ORIGIN HYPOTHESES

4.1 Evaluate evidence for origin hypotheses

One of the most enduring debates in biology concerns the origin of life. While evolution is universally accepted among biologists, there is still disagreement concerning how the whole process began. Evolution, as we currently understand it, depends not merely on the existence of heritable information in the form of genes, but also on the numerous enzymes and other cellular structures that facilitate the replication of those genes and the transmission of them to offspring. The question then emerges: how did genes, enzymes, and cells come about in the first place?

There are some generally accepted points of agreement. Scientists estimate that the Earth is about 4.5 billion years old and that, in the beginning, it was a very inhospitable place. When Earth first came into existence, there was very little or no atmospheric oxygen, and the surface of the Earth was bombarded by intense ultraviolet radiation. Around 3.9 billion years ago, there were heavy rains and violent storms, which led to the production of basic inorganic chemical building blocks from the soil and the accumulation of energy needed to drive reactions for producing simple organic molecules.

The prevailing theory of the origin of life, sometimes known as the "organic soup" model, is that these organic molecules became more and more complex until amino acids and nucleic acids were formed. Once strings of nucleic acids formed, they could self-replicate within the "soup." These self-replicating structures organized into **protobionts**, which were droplets of segregated chemicals. Chemicals continued to organize until the first identifiable cells, the first unicellular organisms, came into being.

Of course, there are points of dispute concerning the details of this process, such as whether the initial reactions that produced organic molecules actually occurred in solution, as proposed by the organic soup model, or on solid reactive surfaces, such as clay particles. There is also uncertainty over whether the first self-replicating molecules were deoxyribonucleic acid (DNA) or ribonucleic acid (RNA). The RNA World Hypothesis theorizes that RNA molecules were the first self-replicators, serving as both heritable information and as functional enzymes. The current paradigm of life, in which DNA is the primary source of heritable information and proteins the primary biological catalysts, only emerged later, according to this view.

Evidence for origin theories comes from biochemistry, laboratory experiments, and models of early Earth environments. However, the strongest evidence can be found in the fossil record. The first prokaryote fossils have been found in geological deposits thought to be 3.5 billion years old. The oldest eukaryotic cells are from deposits that are about 2 billion years old, while the oldest fossils of multicellular organisms are from deposits that are 1.25 billion years old.

Molecular and genetic evidence demonstrates that all life on Earth shares a common ancestor. For example, all eukaryotic cells share common traits, such as the presence of a cytoskeleton, a nucleus, membrane-bound organelles, linear chromosomes, and endomembrane systems, elements that have likely been conserved from some of the earliest life-forms.

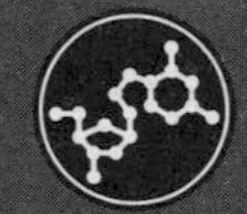

DARWIN AND NATURAL SELECTION

4.2 Recall Darwin's theory

4.3 Explain how natural selection impacts evolution

Early Evolutionary Ideas and Darwinism

Although some early scientists believed that species are immutable and do not evolve or change, others believed that evolution occurs. Theories of evolution developed long before the time of **Charles Robert Darwin** (1809–1882), some even emerging in the ancient world. However, prior to Darwin, even the scientists who accepted the idea that species change over time did not have a solid idea for a mechanism that could explain the changes they observed.

The most well-known hypothesis prior to Darwin was that of **Jean-Baptiste de Lamarck** (1744–1829), who suggested that organisms pass on acquired traits in an attempt to reach a more perfect form. For example, by Lamarck's logic, if a mother works out at the gym and becomes strong and healthy, she will pass her acquired strength and health to her children, and so on. Lamarck's ideas concerning the inheritance of acquired characteristics have generally been discredited, though some of them are being reconsidered in light of research into epigenetics, the study of heritable factors other than genes (such as molecular modifications to DNA that influence gene expression).

Darwin presented his postulates for his **theory of evolution** by **natural selection** in his work ***The Origin of Species*** (1859). There have been slight modifications to the theory based on more up-to-date knowledge about genetics and molecular biology but, for the most part, Darwin's theories are accepted today. Virtually all scientists consider evolution by means of natural selection to be an established fact.

Darwin had several original postulates for his theory of evolution by natural selection:

1. Individuals vary in their characteristics within a population. This means that all giraffes have long necks, but their necks aren't all the exact same length.
2. The variations observed in populations are inherited. When a big dog has puppies, they tend to be big, and a little dog's puppies tend to be little.
3. A considerable number of individuals in a population seem to die as they compete for limited resources in the environment. This is where the term "survival of the fittest" emerged; "fit" organisms are simply the ones that don't die off, those with characteristics that make them more likely to survive to reproductive age within their specific environments. These characteristics could include being bigger or stronger, but they also could include being smaller and smarter—what counts as fitness depends on the environment and the role that the organism plays within it.
4. Individuals who have more resources because of their particular characteristics tend to produce more offspring that survive. For example, if a bird with a long beak can get more food from holes in trees because of its long beak, it will be more likely to survive and provide more food for its offspring. If beak length is a result of genetics, that bird's offspring are more likely to have long beaks, and the following generations of offspring are more likely to have long beaks, until every bird in the population has a long beak.

The selection for more "adaptive" traits tends to narrow a population of individuals down to those who are best suited for a particular environment. If changes occur in the environment, selection favors individuals best suited to the new environment. The theory of natural selection could explain the differences Darwin observed in species he studied and helps to explain biodiversity in organisms today.

Natural Selection and Variation

Natural selection is the differential survival and reproduction of individuals based on variation in genetically controlled traits. These differing rates of survival and reproduction are due to forces in the environment and/or to forces exhibited by other species. Evolutionary **fitness** is measured by the reproductive success of a species. To understand natural selection, it is necessary to understand variation.

Ultimately, all variation originates in the mutation of DNA in an organism's genome. For a mutation to have an impact on evolution, it must occur in a gamete and be passed on to offspring. If a mutation occurs in a gamete that forms a zygote, the offspring will inherit that mutation and pass it on to its offspring. Genetic variation can also occur during recombination in meiosis I. Most mutations are harmful, but occasionally a mutation exists in viable offspring. These offspring may then exhibit a phenotype that differs from the rest of the population.

> ✔ **AP Expert Note**
>
> **Variation that occurs in a population will have a distribution based upon the kind of natural selection that is taking place in the population.**

The three types of variation that can occur in a population are stabilizing selection, directional selection, and disruptive (diversifying) selection. If a population is subject to **stabilizing selection**, extremes at both ends of a phenotype are eliminated, resulting in less genetic variability. For example, if the variation in color of a bird species ranges from dark gray to white and the population is subject to stabilizing selection, the medium-gray phenotype will be most common. If the population is under **directional selection**, one extreme is preferentially selected against the other (e.g., white birds being more easily spotted by predators leading to the selection of dark gray), so that the average in the population moves in one direction. **Disruptive (diversifying) selection** favors both extremes while selecting against the average, which would mean dark gray and white are selected over medium-gray in the case of the birds.

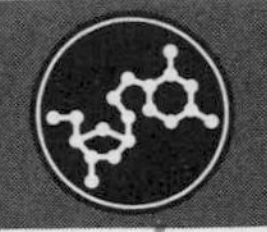

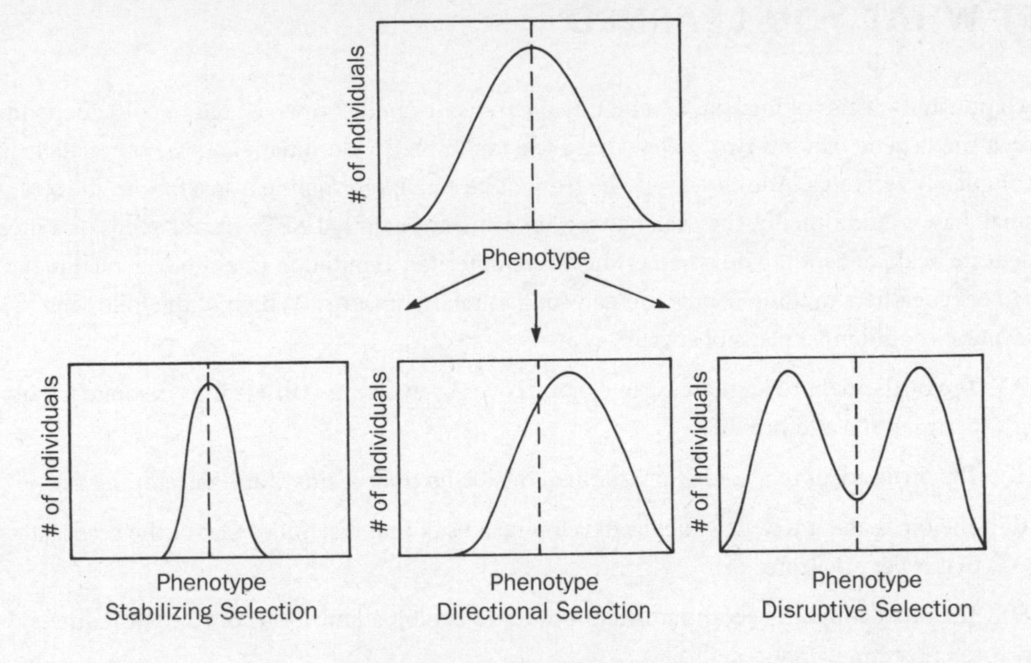

Modes of Natural Selection

RAPID REVIEW

If you take away only 3 things from this chapter:

1. It is thought that the chemical components of life on Earth originated through radiation and storms. These compounds became increasingly complex, forming protobionts and, ultimately, living organisms.
2. According to Darwin, species evolve via natural selection, in which animals with certain traits are more likely to survive and reproduce, passing on those traits.
3. Selection can be stabilizing (median is encouraged), directional (the norm shifts toward one extreme), or disruptive (extremes are favored over the norm).

TEST WHAT YOU LEARNED

1. An unusual species of bird has several notable traits. It can be brown or red in color, depending on a single gene that has two alleles. One allele is completely dominant and the other allele is completely recessive. Adults vary in size from quite small, resembling a sparrow, to the size of a small hawk. Additionally, they can have either a sturdy beak that easily breaks seeds or a more delicate beak, depending on whether they have inherited a mutation in a single gene. Finally, they can have feathers ranging from extremely long to relatively short. Which of the following scenarios could most plausibly occur?

   (A) The birds adapt to an increase in canopy cover by evolving feathers intermediate in color between red and brown.

   (B) The birds adapt to a colder environment by losing their feathers and developing fur.

   (C) The birds adapt to a new diet by developing a beak that resembles that of the seed-eaters but that is very delicate.

   (D) The birds adapt to become smaller in order to exploit a new food source that requires them to enter small crevices.

2. Which of the following statements is most consistent with scientists' understanding of the origin of life on Earth?

   (A) The first cells must have been composed of simple monomers, rather than more complex polymers.

   (B) Early life evolved in an anaerobic environment, meaning that the first cells must have been able to grow without oxygen.

   (C) The first cells grew in a relatively cold environment, requiring enzymes that could function at low temperatures.

   (D) Early life appeared multiple times, meaning that living organisms have evolved from several different original ancestors.

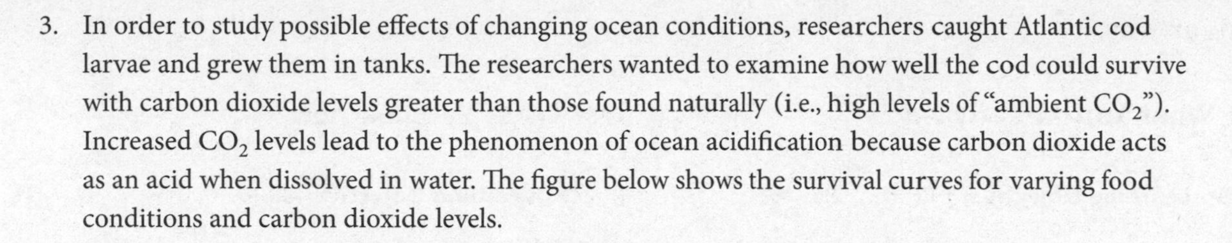

3. In order to study possible effects of changing ocean conditions, researchers caught Atlantic cod larvae and grew them in tanks. The researchers wanted to examine how well the cod could survive with carbon dioxide levels greater than those found naturally (i.e., high levels of "ambient CO_2"). Increased CO_2 levels lead to the phenomenon of ocean acidification because carbon dioxide acts as an acid when dissolved in water. The figure below shows the survival curves for varying food conditions and carbon dioxide levels.

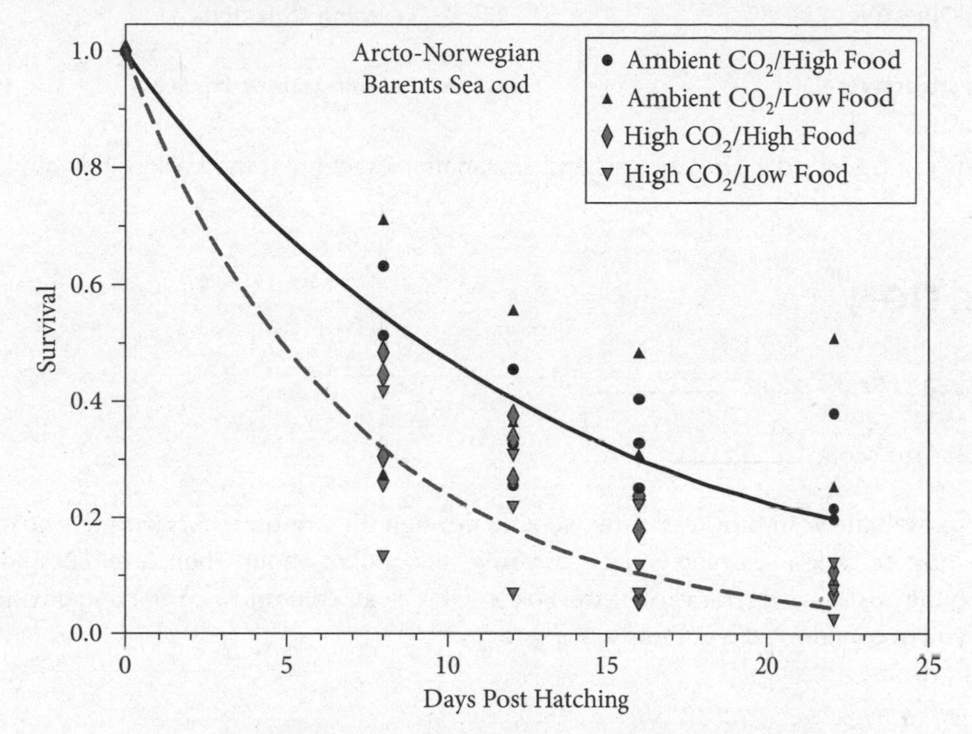

Adapted from Martina H. Stiasny et al., "Ocean Acidification Effects on Atlantic Cod Larval Survival and Recruitment to the Fished Population," *PLoS ONE* 11, no. 8 (August 2016): e0155448.

When conditions change too rapidly, organisms may not be able to adapt and may become extinct. If change is slower, animals may be able to adapt. What would be required for the cod to be able to adapt to the changing conditions described above?

(A) The cod would need genetic variation in genes associated with water acidity.

(B) The cod would need genetic variation in genes allowing them to metabolize carbon dioxide more efficiently.

(C) The cod would need to learn how to consume food more efficiently.

(D) The cod would need to move to cooler locations in order to survive.

Answer Key

Test What You Already Know

1. **B** **Learning Objective:** 4.1
2. **C** **Learning Objective:** 4.2
3. **B** **Learning Objective:** 4.3

Test What You Learned

1. **D** **Learning Objective:** 4.3
2. **B** **Learning Objective:** 4.1
3. **A** **Learning Objective:** 4.2

Detailed solutions can be found in the Answers and Explanations section at the back of this book.

REFLECTION

Test What You Already Know score: __________

Test What You Learned score: __________

Use this section to evaluate your progress. After working through the pre-quiz, check off the boxes in the "Pre" column to indicate which Learning Objectives you feel confident about. Then, after completing the chapter, including the post-quiz, do the same to the boxes in the "Post" column. Keep working on unchecked Objectives until you're confident about them all!

Pre	Post	
❑	❑	**4.1** Evaluate evidence for origin hypotheses
❑	❑	**4.2** Recall Darwin's theory
❑	❑	**4.3** Explain how natural selection impacts evolution

FOR MORE PRACTICE

Complete more practice online at kaptest.com. Haven't registered your book yet? Go to kaptest.com/booksonline to begin.

PART 3

Structures of Life

CHAPTER 5

Cells

LEARNING OBJECTIVES

In this chapter, you will review how to:

5.1 Differentiate between prokaryotic and eukaryotic structures

5.2 Explain the function of common organelles

5.3 Recall membrane composition and function

5.4 Explain transport mechanisms

5.5 Investigate effects of tonicity and gradients

TEST WHAT YOU ALREADY KNOW

1. Biologists isolated a giant cell from the gut of a surgeonfish. The cell is 600 μm long and can be seen with the naked eye. Electron microscope images revealed convoluted membranes and tangles of DNA not surrounded by membranes on the margins of the cell. Ribosomes were visible in the cytoplasm. Which of the following conclusions can the scientists most reasonably draw from their observations?

   (A) The organism is a eukaryote because it is visible to the naked eye and contains loose membranes.

   (B) The organism cannot be yet classified because there are too few observations to make an informed decision.

   (C) The organism is probably a virus because it was found inside the gut.

   (D) The organism is a prokaryote because it does not contain a true nucleus.

2. Scientists measured the density of mitochondria in several human tissues. They estimated the mitochondrial content by measuring the activity of a marker enzyme, citrate synthase, which is present only in mitochondria. Their results are plotted in the bar graph below.

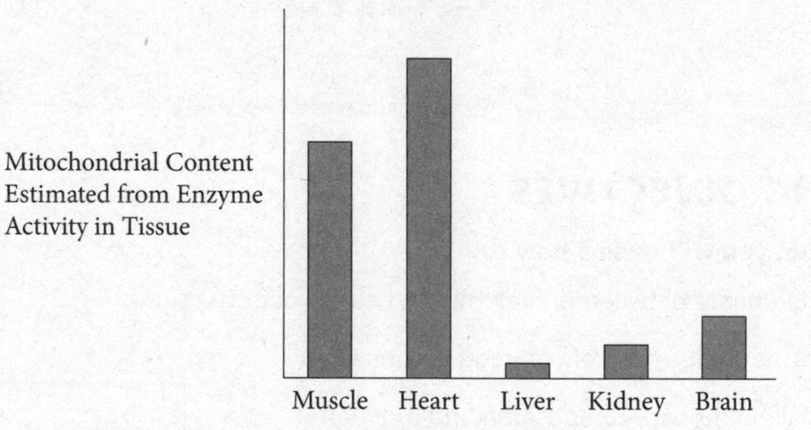

Adapted from G. Benard et al., "Physiological Diversity of Mitochondrial Oxidative Phosphorylation," *Am J Physiol Cell Physiol* 291, no. 6 (June 2006): C1172–C1182.

Which of the following hypotheses would provide the strongest justification for the scientists' observations?

   (A) The heart muscle uses a large supply of ATP because it contracts and relaxes continuously.

   (B) Skeletal muscle needs a large amount of mitochondria to support protein synthesis.

   (C) The liver does not need mitochondria to function.

   (D) The relative amount of mitochondria in each tissue does not have physiological significance.

3. Saturated fats have a high melting point and remain solid at room temperature. Unsaturated fats are liquid at room temperature and solidify at much lower temperatures than saturated fats. In order for the membrane to function properly, membrane fluidity must stay within a certain physiological range. Chemical analysis of membrane lipids in bacteria adapted to different temperature ranges is shown in the table below.

Organism	**Temperature Range (°C)**	**Average Proportion of Unsaturated Fatty Acids**
Bacillus psychrophilus	0–28	Large (17–28%)
Bacillus cereus	10–50	Small (7–12%)
Geobacillus stearothermophilus	35–80	Small to insignificant (~3%)

Adapted from S. E. Diomandé et al., "Role of Fatty Acids in Bacillus Environmental Adaptation," *Frontiers in Microbiology* 6 (August 2015): 813.

Which of the following conclusions can be reasonably drawn from the data provided in the table?

(A) The composition of phospholipids in the membrane is not correlated to ambient temperature.

(B) Bacteria that thrive in cold environments adapt by increasing membrane rigidity.

(C) The membranes of *G. stearothermophilus* are adapted to increase fluidity at high temperatures.

(D) Fatty acids in *B. psychrophilus* allow the membrane to maintain its fluidity at low temperatures.

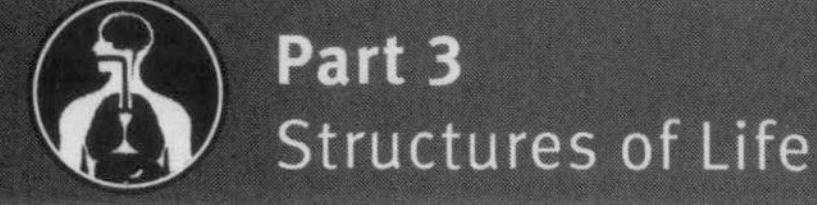

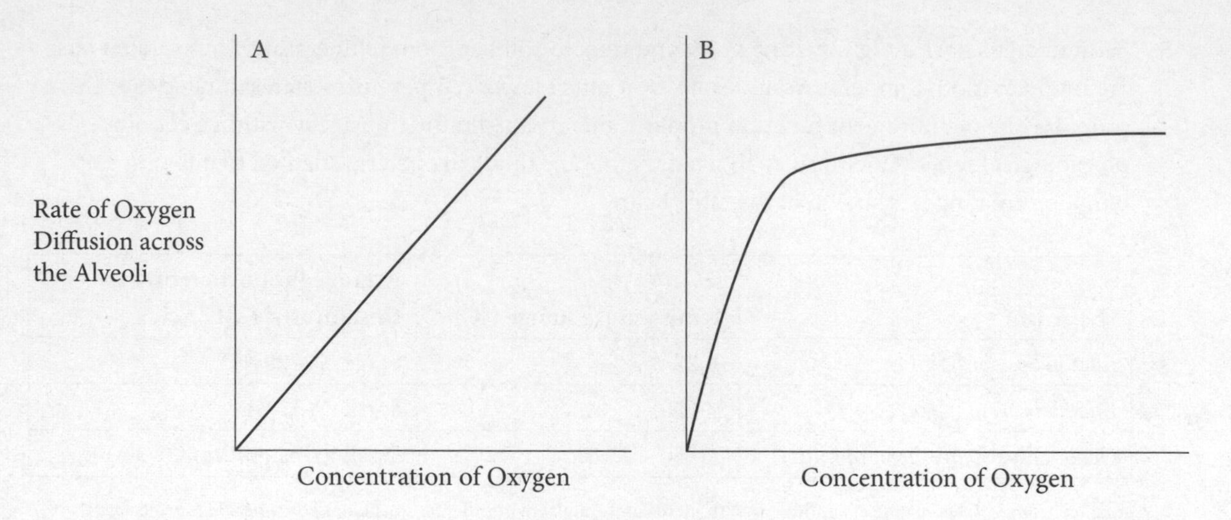

4. The alveoli of the lungs are sac-like structures where gas exchanges take place during respiration. Oxygen and carbon dioxide are small, hydrophobic gases. Two different groups of investigators report different kinetics for oxygen movement across the epithelium. Which of the above graphs is most likely to represent their results correctly?

   (A) Graph A, because the membrane is impermeable to small, hydrophobic molecules; oxygen uses active transport to flow against its concentration gradient

   (B) Graph A, because the membrane is permeable to small, hydrophobic molecules; oxygen diffuses through the membrane down its concentration gradient

   (C) Graph B, because the membrane is impermeable to small, charged molecules; oxygen binds to a carrier that allows it to flow down its concentration gradient

   (D) Graph B, because the membrane is permeable to all gas molecules; oxygen flows by osmosis down its concentration gradient

5. In an experiment to study diffusion across a semi-permeable membrane, students fill a length of dialysis tubing with a solution of 0.9% NaCl. The filled dialysis tube is soaked in a beaker containing 5% albumin (MW 64,500) in distilled water.

The students tested the content of the dialysis tube and the beaker before soaking the dialysis tube in the beaker. After letting the dialysis tube sit in the beaker for half an hour at room temperature, the students recorded that the dialysis tube had swelled and tested the contents of the beaker and the tube for NaCl and protein. Testing results for the experiment are summarized in the table below, in which a + indicates the presence of the compound and a – indicates the absence.

	Start of Experiment		End of Experiment	
Compound	**In Bag**	**In Beaker**	**In Bag**	**In Beaker**
NaCl	+	–	+	+
Albumin	–	+	–	+

Which of the following is a valid comparison between the dialysis tube and the plasma membrane of a red blood cell?

(A) NaCl can never exit a red blood cell because it is composed of ions, but it can cross the dialysis tube membrane.

(B) Albumin can permeate a red blood cell membrane, but it cannot cross the dialysis tube membrane because of its size.

(C) Both membranes are similar in that both are impermeable to albumin but permeable to NaCl.

(D) The membranes are similar in that both are impermeable to albumin, but NaCl can permeate only across the dialysis tube membrane.

Answers to this quiz can be found at the end of this chapter.

PROKARYOTES VERSUS EUKARYOTES

5.1 Differentiate between prokaryotic and eukaryotic structures

There are two main types of cells—**prokaryotes** and **eukaryotes**. Prokaryotes are much simpler than eukaryotes; they are composed of a plasma membrane, cytoplasm, cell wall, DNA, ribosomes, and simple microtubules. Bacteria are an example of prokaryotic cells. Eukaryotes include both plant and animal cells. These cells are much more complex than prokaryotes and contain numerous **organelles**.

Prokaryotes

Prokaryotes are the more basic, simpler of the two major cell types. These cells are considered to be the older form of cells. There are three major regions of prokaryotic cells. The inside region is called the cytoplasmic region. It contains the circular DNA that makes up the genetic material of the cell. Because prokaryotes do not contain a nucleus, DNA is condensed inside an area of the cytoplasm, the nucleoid. Prokaryotes are useful in science because they carry extrachromosomal DNA elements called plasmids. Plasmids are circular bits of DNA that can be added or changed to allow for the addition or suppression of certain functions based on the coding of inserted DNA sections.

The cell envelope usually consists of a cell wall that covers a plasma membrane and may sometimes also include another protective layer called the capsule. The envelope provides structure as well as a protective filter for the cell. Most prokaryotes contain a cell wall that acts as yet another protective barrier from the cell's external environment.

Many prokaryotes have external projections known as flagella and pili. Flagella (singular: flagellum) are long projections or appendages that protrude from the cell body. The primary function of flagella is locomotion, but they can also function as a sensory structure. Pili (singular: pilus) are shorter structures, some of which are used for locomotion and others of which play a role in bacterial conjugation, discussed in the following chapter.

Eukaryotes

All multicellular organisms (such as you, a tree, or a mushroom) and all protoctists (such as amoebas and paramecia) are eukaryotic. The eukaryotes include the protoctists (protists), fungi, animals, and plants. Eukaryotic cells are enclosed within a lipid bilayer cell membrane, as are prokaryotic cells. Unlike prokaryotes, eukaryotic cells contain membrane-bound organelles (see figure). An organelle is a structure within the cell with a specific function that is separated from the rest of the cell by a membrane. The presence of membrane-bound organelles in eukaryotes allows eukaryotic cells to compartmentalize activities in different parts of the cell, making them more efficient. Compartments within a cell allow the cell to carry out activities, such as ATP production and consumption, within the same cell and to control each independently.

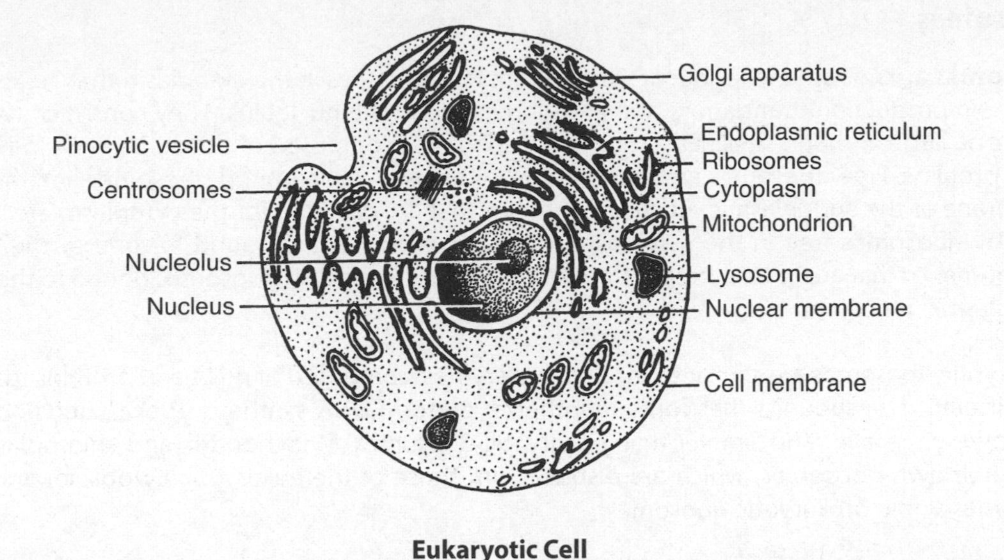

Eukaryotic Cell

ORGANELLES

HIGH YIELD

5.2 Explain the function of common organelles

Nucleus

The genetic material, the DNA genome, is found in the largest organelle of a eukaryotic cell, the **nucleus**. The nucleus is separated from the rest of the cell by the nuclear envelope, a double membrane that has a large number of nuclear pores for communication of material between the interior and exterior of the nucleus. The pores are large enough to allow proteins to pass through but are also selective in the proteins that are transported into the nucleus or excluded from the nucleus. Special sequences in proteins signal a protein to be imported into the nucleus.

While the prokaryotic genome is generally found in a single circular piece of DNA, the eukaryotic genome in each cell is split into chromosomes. Chromosomes contain the DNA genome complexed with structural proteins, called histones, that help package the large strands of DNA in each chromosome within the limited space of the nucleus. Genes in the DNA genome are read (transcribed) to make RNA, which is processed in the nucleus before it is exported to the **cytoplasm**, where the RNA is read in turn (translated) to make proteins. The basic information flow of the cell is DNA to RNA to protein. The DNA genome is replicated in the nucleus when the cell divides. Other metabolic activities, such as energy production, are excluded from the nucleus. The structure and function of the eukaryotic genome will be presented later in more detail.

The dense structure within the nucleus where ribosomal RNA (rRNA) synthesis occurs is known as the nucleolus. The nucleolus is not surrounded by a membrane but is the site of assembly of ribosomal subunits from RNA and protein components. After assembly, the ribosomal subunits are exported from the nucleus to the cytoplasm to carry out protein synthesis.

Ribosomes

Ribosomes are not organelles but are large, complex structures in the cytoplasm that are involved in protein production (translation) and are synthesized in the nucleolus. They consist of two subunits, one large and one small. Each ribosomal subunit is composed of ribosomal RNA (rRNA) and many proteins. Free ribosomes are found in the cytoplasm, while bound ribosomes line the outer membrane of the endoplasmic reticulum. Proteins that are destined for the cytoplasm are synthesized by ribosomes free in the cytoplasm, while proteins that are bound for one of the several membranes or that are to be secreted from the cell are translated on ribosomes bound to the rough endoplasmic reticulum.

Prokaryotic ribosomes are similar to those of eukaryotes, composed of rRNA and proteins that form two different size subunits that come together to perform DNA synthesis. Prokaryotic ribosomes are, however, smaller and simpler than eukaryotic ribosomes. Mitochondria and chloroplasts also have their own ribosomes, which are distinct from those of the eukaryotic cytoplasm and more closely resemble prokaryotic ribosomes.

Endoplasmic Reticulum

The **endoplasmic reticulum** (ER) is an extensive network of membrane-enclosed spaces in the cytoplasm. The interior of the ER between membrane layers is called the lumen and, at points in the ER, the lumen is continuous with the nuclear envelope. If a region of the ER has ribosomes lining its outer surface, it is termed rough endoplasmic reticulum (rough ER); without ribosomes, it is known as smooth endoplasmic reticulum (smooth ER). Smooth ER is involved in lipid synthesis and the detoxification of drugs and poisons, and it has the appearance of a network of tubes, while rough ER is involved in protein synthesis and resembles a series of stacked plates.

Proteins that are secreted or found in the cell membrane, the ER, or the Golgi are made by ribosomes on the rough ER. Proteins synthesized on the rough ER cross into the lumen of the rough ER during synthesis. The presence of a hydrophobic sequence of amino acids at the amino terminus of proteins determines whether the protein will be sorted into the secretory pathway starting at the rough ER or synthesized in the cytoplasm. Proteins that are secreted will have only one hydrophobic signal sequence, the signal peptide, and will be inserted into the ER lumen when they are synthesized, then released from the cell later. Proteins that are destined to be membrane-bound have hydrophobic transmembrane domains that are threaded through the rough ER membrane as the protein is synthesized. When the protein reaches the correct membrane destination along the secretory pathway, additional signals in the protein sequence and structure will cause the protein to stay localized at the current location.

Small regions of ER membrane bud off to form small round membrane-bound vesicles that contain newly synthesized proteins. These cytoplasmic vesicles are then transported to the Golgi apparatus, which is the next stop along the secretory pathway.

Golgi Apparatus

The **Golgi apparatus** is a stack of membrane-enclosed sacs, usually located in the cell between the ER and the plasma membrane (see the figure under the Eukaryotes heading above). The stacks closest to the ER are called the cis Golgi and the stacks farthest from the ER, closer to the plasma membrane, are called the trans Golgi. Vesicles containing newly synthesized proteins bud off of the ER and fuse with the cis Golgi. In the Golgi, these proteins are modified and then repackaged for delivery to other destinations in the cell. For example, the Golgi carries out post-translational modification of proteins through glycosylation, the process of adding sugar groups to the proteins to form glycoproteins. Many proteins destined for the plasma membrane have carbohydrate groups added to the surface of the protein facing the exterior of the cell.

After processing in the cis Golgi, proteins are packaged in vesicles that move to the next layer in the stack, where they fuse and release their contents. Proteins proceed in this manner from one stack to the next until they reach the trans Golgi. In the trans Golgi, proteins are sorted into vesicles based on signals in different proteins that indicate their final destination. The nature of the signal varies but includes the protein's primary sequence, structure, and post-translational modifications. Once the proteins are packaged into vesicles, the vesicles move on to their final destination.

The final destination for a protein may include the lysosome, the plasma membrane, or the exterior of the cell. Some proteins are retained in the Golgi or the ER. Proteins that are destined for the plasma membrane as transmembrane proteins are inserted in the membrane in the ER as they are synthesized, and they maintain their orientation in the membrane as they move from the ER to the Golgi to the vesicle to the plasma membrane. Proteins that are secreted from the cell are inserted in the ER lumen during protein synthesis and remain in the lumen of the ER until they move on to to the Golgi, where they form secretory vesicles. The last step in secretion is the fusion of the secretory vesicle with the plasma membrane, releasing the contents of the vesicle to the cellular exterior.

Lysosomes

Lysosomes contain hydrolytic enzymes involved in intracellular digestion that break down proteins, carbohydrates, and nucleic acids. For white blood cells, the lysosome may degrade bacteria or damaged cells. For a protist, lysosomes may provide food for the cell. They also aid in renewing a cell's own components by breaking them down and releasing their molecular building blocks into the cytosol for reuse. A cell in injured or dying tissue may rupture the lysosome membrane and release its hydrolytic enzymes to digest its own cellular contents.

The lysosome maintains a slightly acidic pH of 5 in its interior, a pH at which lysosomal enzymes are maximally active. The contents of the lysosome are isolated from the cytoplasm by the lysosomal membrane, keeping the pH distinct from the more neutral pH of the cytoplasm. The optimal pH and compartmentalization of lysosomal enzymes prevent the rest of the cellular contents from degrading.

Peroxisomes

Peroxisomes contain oxidative enzymes that catalyze reactions in which hydrogen peroxide is produced and degraded. Peroxisomes break fats down into small molecules that can be used for fuel; they are also used in the liver to detoxify compounds, such as alcohol, that may be harmful to the body. The peroxides produced in the peroxisome would be hazardous to the cell if present in the cytoplasm, because these molecules are highly reactive and could covalently alter molecules such as DNA. Compartmentalization of these activities within the peroxisome reduces this risk.

Mitochondria

Mitochondria are the source of most energy in the eukaryotic cell as the site of aerobic respiration. Mitochondria are bound by an outer and inner phospholipid bilayer membrane. The outer membrane has many pores and acts as a sieve, allowing molecules through on the basis of their size. The area between the inner and outer membranes is known as the intermembrane space. The inner membrane has many convolutions called **cristae**, as well as a high protein content that includes the proteins of the electron transport chain. The area bounded by the inner membrane is known as the mitochondrial matrix and is the site of many of the reactions in cellular respiration, including electron transport, the Krebs cycle, and ATP production.

Mitochondria are somewhat unusual in that they are semiautonomous within the cell. They contain their own circular DNA and ribosomes, which enable them to produce some of their own proteins. The genome and ribosomes of mitochondria resemble those of prokaryotes more than eukaryotes. In addition, they are able to self-replicate through binary fission. Mitochondria are believed to have developed from early prokaryotic cells that began a symbiotic relationship with the ancestors of eukaryotes, with the mitochondria providing energy and the host cell providing nutrients and protection from the exterior environment. This theory of the origin of mitochondria and the modern eukaryotic cell is called the endosymbiotic hypothesis.

Specialized Plant Organelles

Plants also have some organelles that are not found in animal cells. **Chloroplasts** are found only in plant cells and some protists. With the help of one of their primary components, chlorophyll, chloroplasts function as the site of photosynthesis, using the energy of the Sun to produce glucose. Chloroplasts have two membranes: an inner and an outer membrane. Additional membrane sacs called thylakoids inside the chloroplast are derived from the inner membrane and form stacks called grana. The fluid inside the chloroplast surrounding the grana is the stroma. The thylakoid membranes contain the chlorophyll of the cell.

Like mitochondria, chloroplasts contain their own DNA and ribosomes and exhibit the same semiautonomy. They are also believed to have evolved via symbiosis of an early photosynthetic prokaryote that invaded the precursor of the eukaryotic cell. In this arrangement, the chloroplast precursor cell provided food and received protection. Photosynthetic prokaryotes today carry out photosynthesis in a manner similar to the chloroplast.

Vacuoles are membrane-enclosed sacs within the cell. Many types of cells have vacuoles, but plant vacuoles are particularly large, taking up 90 percent of the cell volume in some cases. Plants use the vacuole to store waste products, and the pressure of liquid and solutes in the vacuole helps the plant to maintain stiffness and structure as well.

All plant cells have a cellulose cell wall that distinguishes them from animal cells, which lack a cell wall. The cell wall of plants is also distinct from the peptidoglycan cell wall of bacteria and the chitin cell wall of fungi. The cell wall provides structure and strength to plants.

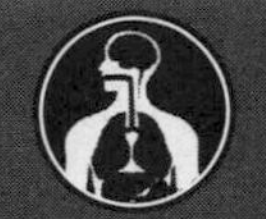

Cilia and Flagella

Cilia and flagella are both anchored into the cell membrane by arrangements of microtubule triplets, which are called basal bodies. Because the microtubules in cilia and flagella must be rebuilt often, tubulin dimers use these basal bodies as the foundation to make new microtubules, which are used to maintain cilia and flagella.

As you can see in the following figure, cilia and flagella are composed of long stabilized microtubules arranged in a "9 + 2" structure (nine pairs of microtubules surrounding two central microtubules for added stability). These nine doublets slide past each other as dynein proteins grab neighboring tubules and pull them. This rapid sliding generates the force needed for the cilia or flagella to quickly beat back and forth and cause movement.

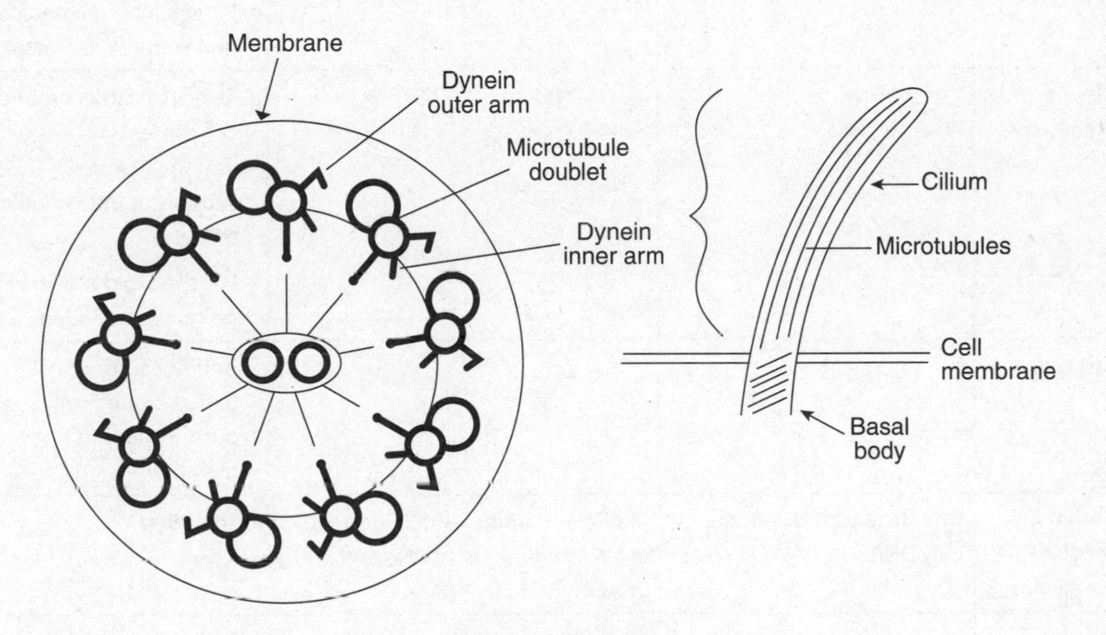

Cilium Cross-Section

CHARACTERISTICS OF CELLS				
		Eukaryotes		
	Prokaryotes	**Plant Cells**	**Animal Cells**	
Size	**0.2–500μm, most 1–10μm**	**Most 30–50μm**	**Most 10–20μm**	
Structure				**Properties**
Cytoplasm	Yes	Yes	Yes	• Intracellular matrix outside of nucleus
Nucleus	No	Yes	Yes	• Contains DNA • Pores allow communication with cellular matrix
Plasma Membrane	Yes	Yes	Yes	• Selective barrier around cell contents allowing the passage of some substances but excluding others • Phospholipid bilayer with proteins embedded
Cell Wall	Most	Yes	No	• Additional structural barrier around cell outside plasma membrane
Chromosomes	One circular chromosome, only DNA	Multiple strands of DNA and protein	Multiple strands of DNA and protein	• The cell's DNA
Ribosomes	Yes	Yes	Yes	• Site of protein synthesis (translation)
Endoplasmic Reticulum (ER)	No	Yes	Yes	• Site of attachment for ribosomes • Protein and membrane synthesis • Formation of vesicles for transport
Golgi Apparatus	No	Yes	Yes	• Synthesis, accumulation, storage, and transport of products

(*Continued*)

CHARACTERISTICS OF CELLS				
		Eukaryotes		
	Prokaryotes	**Plant Cells**	**Animal Cells**	
Size	**0.2–500μm, most 1–10μm**	**Most 30–50μm**	**Most 10–20μm**	
Structure				**Properties**
Lysosomes	No	Some vacuoles function as lysosomes	Usually	• Vesicle containing hydrolytic enzymes
Vacuoles or Vesicles	No	Yes	Some	• Membrane-bound sacs in the cytoplasm
Mitochondria	No	Yes	Yes	• Site of cellular respiration
Plastids	No	Yes	No	• Group of plant organelles that includes chloroplasts • Site of photosynthesis • Carbohydrate storage
Microtubules (Cilia or Flagella)	Simple	On some sperm	Complex (9 + 2 arrangement)	• Tubes of globular protein, tubulins • Provide structural framework for cell • Provide motility
Centrioles	No	No	Yes	• Cell center for microtubule formation

Structures

MEMBRANE TRAFFIC

5.3 Recall membrane composition and function

5.4 Explain transport mechanisms

Membrane Structure

All cells are surrounded by a **plasma membrane**. In eukaryotic cells, the nucleus and most of the organelles are also surrounded by plasma membranes. Membranes are composed mostly of **lipids**, which are full of nonpolar covalent bonds, so they are **hydrophobic** and do not dissolve in water. Because membranes are composed primarily of lipids, most of the material in membranes will not mix well with water.

Most of the lipids in membranes have a phosphate group attached to one end, so they are called **phospholipids**. The charged end (phosphate group) is polar and happy to be in water, which is why it is termed **hydrophilic**. The other end of the phospholipid is a tail that is nonpolar, and it turns in toward the center of the membrane. The attraction between the nonpolar regions of these phospholipids creates the foundation for the bilayer of the membrane. The lipid ends group together like the insides of a sandwich, surrounded by polar barriers.

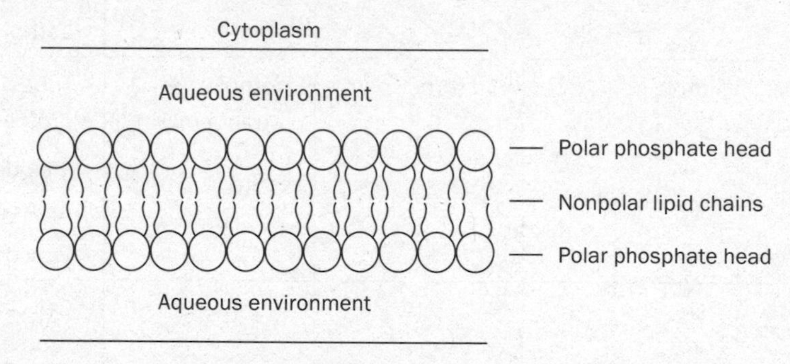

Phospholipid Bilayer

✔ **AP Expert Note**

Make sure you are familiar with the structure and function of macromolecules as they relate to the plasma membrane structure. Be able to identify the *general* structures and functions of phospholipids, proteins, and cholesterol.

Embedded among all of these membrane lipids are **proteins**, **carbohydrates**, and **sterols** (like cholesterol). Some proteins are embedded on the outer surface, some on the inner surface, and some span the entire width of the plasma membrane (these usually function as transport proteins). Some surface proteins have sugar groups attached to them, called glycoproteins.

Each component of a cell membrane contributes to how the membrane functions. Proteins act as transport molecules, receptor sites, attachments to the **cytoskeleton**, and surface enzymes. Carbohydrates on the surface of the cell and glycoproteins contribute to cell recognition, particularly in immune response. Cholesterol helps to maintain the fluidity of the membrane.

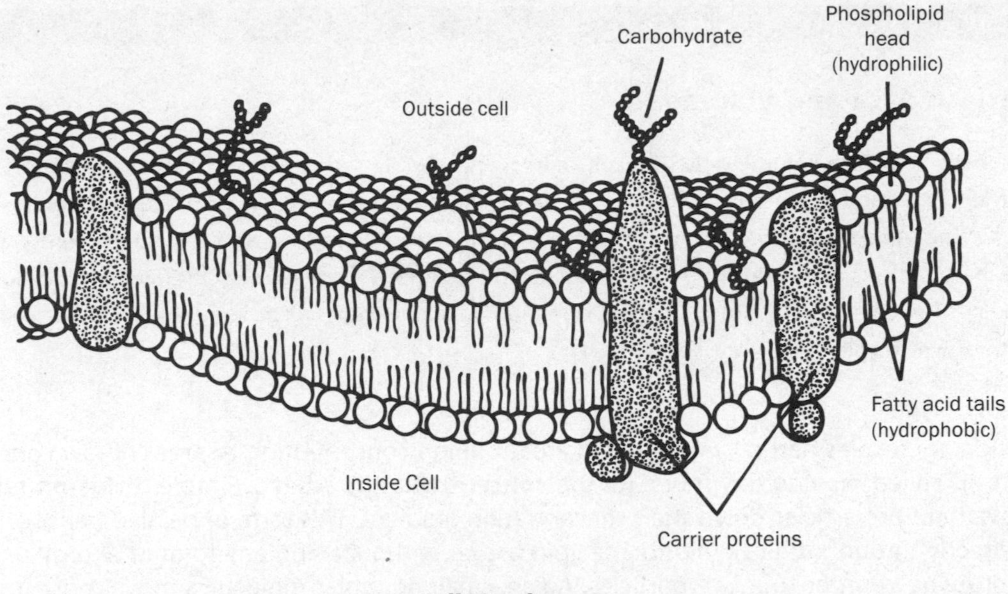

Cell Membrane

Membrane Proteins

Membrane proteins, like the membrane phospholipids, usually have carbohydrate groups attached to them so that the outside surface of the plasma membrane is extremely sugar-rich.

Membrane-spanning proteins have regions that are hydrophobic as well as regions that are hydrophilic, with the nonpolar (hydrophobic) regions passing through the nonpolar interior of the membrane and the polar areas sticking both into the cytoplasm and out into the extracellular space. Other proteins can be located completely intracellularly or extracellularly, anchored to the cell membrane by a variety of special lipids. These proteins are made in the cytosol and bind into the cell membrane only because they subsequently have a lipid molecule attached to their structure.

Transmembrane proteins are involved not only in carrying materials across the membrane, but also in cell recognition, cell adhesion, cell signaling, and enzymatic reactions. Recall that most proteins sticking up from the surface of the membrane are covered in carbohydrates on the extracellular surface. The term used to describe the protein- and carbohydrate-rich coating on the cell surface is glycocalyx. Keep in mind that these sugars reside exclusively on the exterior of the membrane.

Membrane Transport Mechanisms

HIGH YIELD

Membranes are **selectively permeable**, which means that they allow some things to pass through, but not others. The main limiting factors that determine whether a molecule will be able to pass through a cell's membrane are the *size of the particle* and *its charge* (polarity). Simply stated, the molecules quickest to pass through the lipid bilayer are those that are *small* and *nonpolar*, because the interior of the membrane is far too hydrophobic for others to make it through without assistance. This assistance can come in the form of membrane-spanning proteins, which either can bind to extracellular molecules and bring them inside the cell via a conformational change or can open up a temporary tunnel through the membrane lipids so that the molecules can pass through.

✔ **AP Expert Note**

Surface Area and Volume

The surface area-to-volume ratio of a cell limits its size. While surface area affects the transport of nutrients into and waste out of a cell, volume affects the consumption and production of resources and waste. As a cell grows, the surface area-to-volume ratio decreases. At a certain point, the cell will be unable to maintain the rate of transport that is needed to support the increased cellular volume. Thus, the cell will not obtain sufficient nutrients and waste will accumulate within the cell—both of which will lead to cell death.

In solution, molecules naturally move from areas of high concentration to areas of low concentration. This is called moving down or with the **concentration gradient**. Simple **diffusion** refers to the movement of particles down their concentration gradient. This form of passive transport takes place directly through the cell membrane lipid bilayer without using any form of energy or membrane proteins in order to move particles. Again, small nonpolar molecules move most freely by simple diffusion. Examples include water (small but polar), carbon dioxide (nonpolar), and oxygen (nonpolar).

The simple diffusion of water is referred to as **osmosis** and occurs from a region of higher water concentration to a region of lower water concentration. For water to be in high concentration, the amount of dissolved solute (salts, sugars, etc.) must be low, and vice versa for water in low concentration. So, although water diffusion works like any other passive diffusion in terms of movement from high → low concentration, it is generally stated that water moves from an area of lower *solute* concentration to one of higher *solute* concentration.

✔ **AP Expert Note**

Water Potential

The flow of water in a system is determined by its water potential (Ψ), which is a measure of the potential energy in water. Water potential can be described as the sum of pressure potential (Ψ_P) and solute potential (Ψ_S). At atmospheric pressure, the pressure potential of water in an open container is zero and equal to the solute potential. Adding solute decreases solute potential (makes it more negative) and in turn decreases water potential. Negative hydrostatic pressures will also decrease water potential. The gradient of water potential causes water to move (from high to low water potential). In plants, water potential is regulated by osmosis and transpiration to transport water from the roots to the leaves for photosynthesis.

Solutions low in solute concentration relative to other solutions are said to be **hypotonic**, whereas solutions higher in solute than others are **hypertonic**. When two solutions have the same solute concentration as each other, they are said to be **isotonic**. The cell membrane effectively separates two distinct solutions: the extracellular environment and the cytoplasm. If the outside of a cell is hypertonic (e.g., a cell has just been moved from fresh water to salt water), water will move out of the cell into the high solute solution. If possible, some of those solutes will also move into the cell until a balance has been established so that both areas are equivalent in solute concentration. In hypotonic solutions, cells generally take on water, sometimes until they burst.

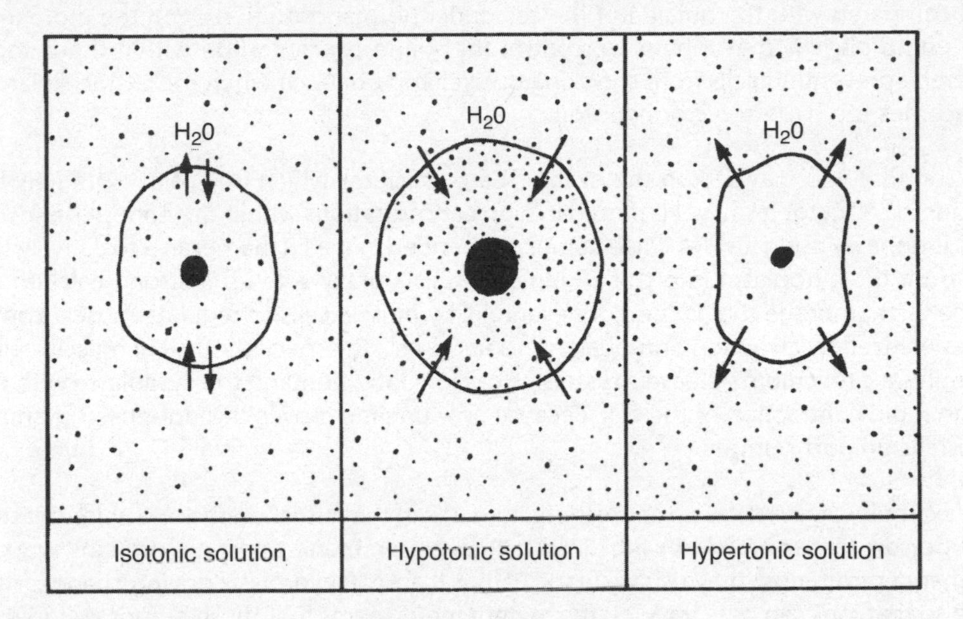

Tonicity and Osmosis

✔ AP Expert Note

In isotonic solutions, water and solutes can and do move across the membrane, yet movement inward is always balanced by reciprocal movement outward. This means that overall concentrations of water and solute on opposite sides of the membrane do not change.

Facilitated diffusion, another type of passive transport, involves the use of channel or carrier proteins embedded in the membrane to allow impermeable molecules to diffuse down a gradient. The structures of the proteins involved in this type of transport are very similar, and amino acid sequences are highly conserved across many species. In some cases, these proteins act as pores for ions; in other cases, they may open and close in response to external signals. Keep in mind that, because cells naturally have a negatively charged cytoplasm, the opening of ion channels favors the movement of positively charged ions into the cytoplasm. The combination of differential solute concentrations and an electrical gradient is called an **electrochemical gradient** and is the key determinant of what moves into and out of membranes when passive transport channels are open.

In **active transport**, membrane proteins use the energy of ATP to change the protein's conformation so that molecules can be brought into and out of the cell against their concentration gradients. These **ATPase** pumps are found in every membrane of cells, and they are extremely important in the maintenance of unequal concentrations of certain ions across the lipid bilayer—something essential for processes like nerve signal conduction.

A commonly cited example of active transport is the Na^+-K^+ ATPase membrane pump, whose conformational change uses the energy of ATP breakdown to pull 2 K^+ (potassium) ions into a cell while kicking out 3 Na^+ (sodium) ions at the same time. This transport of molecules in opposite directions is known as antiport, and it can be contrasted with pumps that pull two different molecules in the same direction (symport). Because the pump, which is present in all cell membranes, pumps out three positive charges for every two it brings in, the inside of the cell remains negatively charged compared with the outside of the cell under normal conditions. Yet, the more important role the pump plays is that it helps to control the solute concentration within the cytoplasm of cells, thereby preventing cells from shrinking or swelling too much when the extracellular environment becomes too hypertonic or hypotonic.

Another ion that you may see on the exam is Ca^{2+} (calcium), which is kept in extremely low concentrations in cell cytoplasm, yet is stored in high concentrations within the endoplasmic reticulum (ER). This is done by using Ca^{2+}-ATPase pumps, embedded in the ER membrane, to actively transport calcium from the cytoplasm into the ER lumen. This naturally sets up a strong calcium gradient across the ER membrane that is used, for example, by muscle cells to regulate muscle contraction. When depolarized by an action potential from a nerve cell, the specialized ER of muscle cells (called the sarcoplasmic reticulum) releases its store of calcium ions, flooding the cytoplasm with Ca^{2+} and leading to rapid contraction of the cell. Because only one ion moves through these channels, they are known as uniport pumps.

As a last example, those ATPases that manufacture ATP in the mitochondria and chloroplasts as part of the **electron transport chain** are simply ATPase membrane-transport proteins working in a reverse manner from how they usually work. Rather than ATP hydrolysis driving changes in protein structure so that ions can pass through the membrane, it seems that these pumps are driven by the flow of H^+ ions moving through them to synthesize ATP from ADP and an inorganic phosphate.

Endocytosis and **exocytosis** are two mechanisms of transport that can move large molecules and even entire cells through the cell membrane. To accomplish this, the cell membrane actually invaginates, or pinches inward, to form a pocket into which the material to be transported can fall. In the case of endocytosis, this invagination pinches off completely, forming a vesicle that contains the transported material and can move freely within the cytoplasm. In exocytosis, a vesicle containing material to be expelled simply merges with the lipid bilayer, and the material is pushed off into the extracellular space.

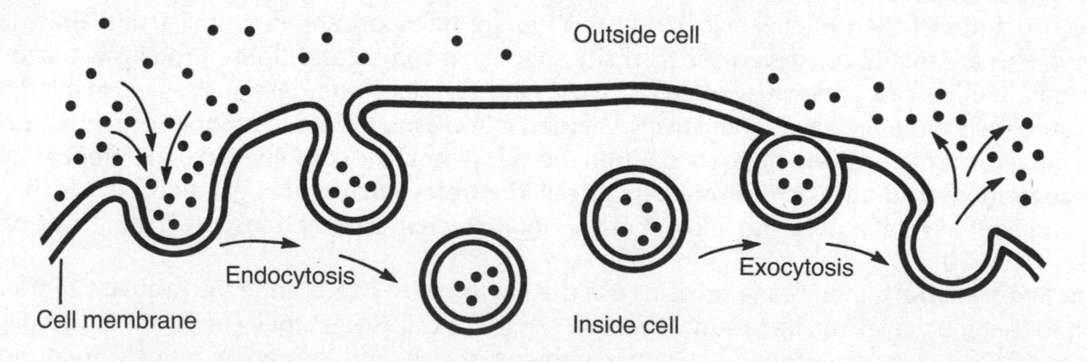

Endocytosis and Exocytosis

There are three distinct types of endocytosis. In phagocytosis, sometimes known as "cellular eating," the membrane invaginates around solid particles. In pinocytosis, or "cellular drinking," the membrane surrounds a mass of liquid. Finally, in receptor-mediated endocytosis, a carrier protein binds to a specific substance, which then triggers the endocytotic response.

✔ **AP Expert Note**

As you prepare for the exam, be familiar with similarities and differences between the different types of transport across the cell membrane. Pay close attention to which ones require energy and which ones do not.

TYPES OF MEMBRANE TRANSPORT

Type of Transport	Requires Energy?	Concentration Gradient
Diffusion	No	Down
Osmosis	No	Down
Facilitated Diffusion	No	Down
Active Transport	Yes	Against
Exocytosis	Yes	N/A
Endocytosis	Yes	N/A

AP BIOLOGY LAB 4: DIFFUSION AND OSMOSIS INVESTIGATION

5.5 Investigate effects of tonicity and gradients

✔ **AP Expert Note**

This and the other lab sections of this book provide supplemental information for the 13 laboratory investigations officially recommended by the College Board for all AP Biology students. Many of these lab sections include high-yield content not covered in other parts of the text, so they are worth reviewing irrespective of whether you completed the labs in class.

The properties of diffusion and osmosis, discussed earlier in this chapter, will be explored in this investigation. Recall that selectively permeable membranes let some molecules through but not others. For this investigation, it is important to know that:

- Isotonic solutions are two solutions that have the same concentration of solute.
- A hypotonic solution has a lower solute concentration than another solution.
- A hypertonic solution has a higher solute concentration than another solution (think *hypo* = "below" and *hyper* = "above").

Keep in mind that water potential is the measure of force a solution has for pulling or drawing water into it. The more negative the water potential, the stronger its pulling force.

Dialysis bags allow for the movement of water but not ions. A common osmosis experiment is to fill dialysis bags with different solutions. These bags are tied at each end and put into hypertonic, hypotonic, and isotonic solutions. The water will move in the direction of the hypertonic solution. Suppose you tied your dialysis bags and put them in solutions to complete the experiment but didn't get the results you expected. Skewed results can be obtained by not tying the knot on your bag tight enough, not getting all the solution washed off the outside of your bag, or by some other slight oversight. Often, one group gets usable results for one part of the lab and another group gets another part right, so everyone shares the "good" results. Occasionally, classes have really good luck and all the groups get "good" results. Either way, you need to know what happened in order to score well on the exam. You can expect to see an osmosis question in one form or another on the exam. It may be in the multiple-choice section or it may be a free-response question.

> ✓ **AP Expert Note**
>
> **Two key skills of the diffusion and osmosis lab are:**
>
> - **Measuring the effects (e.g., weight change) of osmosis**
> - **Determination of the osmotic concentration/water potential of an unknown tissue or solution using solutions of known concentrations**

The first skill uses experimental methods to obtain results. Even if you know how osmosis works, it is important to know how to measure it. You can't really watch water molecules or sugar and starch molecules move because they're too small. In this investigation, you learn how to measure things you don't see by observing things you can. To obtain results, the weight change of a dialysis bag or a piece of potato can be measured. An indicator color will appear if glucose or starch occurs in a solution. If a bag or piece of potato increased in weight, it gained water and had a more negative water potential than (was hypertonic to) the solution it was placed into, and vice versa. If a solution produced an indicator color, the semipermeable membrane allowed the passage of molecules that the dye is an indicator for.

The second key skill deals with using a standard to figure out the "identity" of an unknown and using observed data to interpolate expected data. Part two of the experiment begins with a dialysis bag full of solution (the concentration of which is unknown). Questions to ask during the experiment include: What is the osmolarity/water potential of the unknown, and how do you measure it? The first parts of the investigation teach you to observe the effects of osmosis by measuring the change in weight. Samples are weighed and placed into solutions with known concentrations, usually ranging from distilled water to a high concentration like 1 M. Percentage weight change can be plotted on a graph like this:

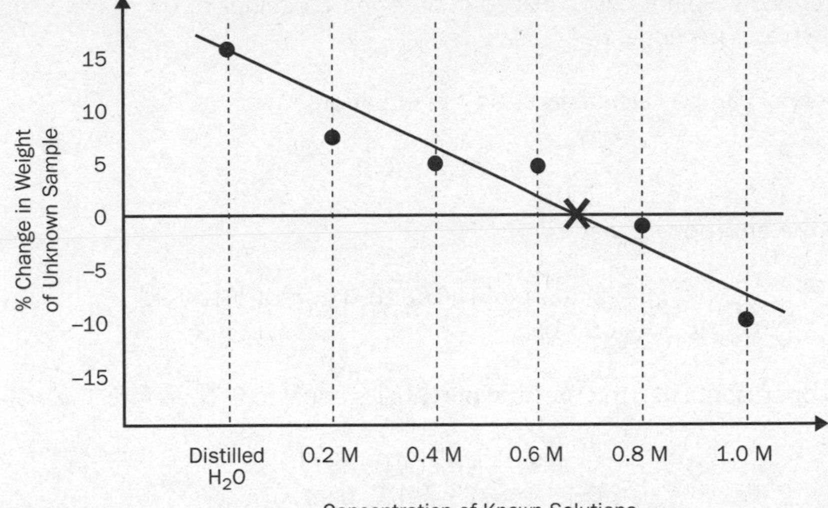

An imaginary line can be drawn that nearly bisects all data points. Note where the line crosses the *x*-axis. At this point, there is no change in weight of the unknown sample, which represents the point at which the osmolarity/water potential of the unknown is the same as a known solution equal to that point on the *x*-axis (on the graph, this is about 0.675 M). By assuming change in weight is linear with respect to change in concentration, the expected concentration of a theoretical solution can be determined. The concentration of the unknown can be estimated by comparing it to the concentration of the theoretical solution (as we saw in the graph above, this is approximately 0.675 M). This type of experimental design and interpolation is very common in biological research. The College Board expects you to be able to design simple experiments like this to test simple hypotheses.

Water potential [represented by the Greek letter psi (Ψ)] predicts which way water diffuses. Water potential is calculated from the solute potential (Ψ_S), which is dependent on solute concentration, and the pressure potential (Ψ_P), which results from the exertion of pressure on a solution. When a solution is open to the atmosphere, the pressure potential is equal to 0 because there is no tension on that solution.

$$\Psi = \Psi_P + \Psi_S$$

$$\text{water potential} = \text{pressure potential} + \text{solute potential}$$

In an open beaker of pure water, the water potential is equal to zero. There is no solute and no tension on the solution, so the solute and pressure potentials are zero. If you add solute in a measured concentration to the beaker, you can calculate the solute potential using the following formula:

$$\Psi_S = -iCRT$$

where i is equal to the number of particles the molecule will dissolve into in water, C is the molar concentration, R is the pressure constant (equal to 0.0831 L·bar/mol · K), and T is the temperature of the solution in Kelvins.

For example, suppose a plant cell is placed in an open container of 0.1 M NaCl solution at 25°C. What would the water potential be?

The solute potential can be calculated using the equation:

$$\Psi_S = -iCRT$$

Substituting in the appropriate values,

$$\Psi_S = -(2)\ (0.1\ \text{mol/L})(0.0831\text{L}\cdot\text{bar/mol}\cdot\text{K})(298\ \text{K})$$
$$\Psi_S = -4.95\ \text{bars}$$

Because it is an open container, the pressure potential is equal to 0. Therefore, the water potential is:

$$\Psi = \Psi_P + \Psi_S$$
$$\Psi = 0 + -4.95\ \text{bars}$$
$$\Psi = -4.95\ \text{bars}$$

A negative water potential means that water is likely to osmose out of the cell from a place of high water potential to a place of low water potential.

RAPID REVIEW

If you take away only 5 things from this chapter:

1. Cells are the basic structural and functional units of all known living organisms. There are two major types: prokaryotic cells and eukaryotic cells.
2. Prokaryotic cells have no real nucleus, have circular DNA, and reproduce using binary fission. Prokaryotes include archaea and bacteria. These cells are small and move via flagella.
3. Eukaryotic cells have a nucleus, linear DNA, highly structured cell membranes, and organelles, and they reproduce via mitosis and meiosis. Eukaryotes include protists, fungi, plants, and animals. These cells are large and move via a variety of methods including flagella and cilia.
4. Cell membranes are made up of a lipid bilayer that includes a hydrophobic and a hydrophilic region. Specific structures embedded within the membrane help to facilitate transport.
5. The cell membrane is selectively permeable, meaning it allows certain things through while keeping others out. Water diffuses across the membrane from areas of lesser to greater solute concentration (osmosis). While certain things can cross the membrane in the processes of diffusion or facilitated diffusion, which do not require energy, others require the expenditure of energy for active transport against the concentration gradient.

TEST WHAT YOU LEARNED

1. A group of investigators isolated a cDNA from the plant *Alonsoa meridionalis* and expressed the protein in yeast cells. They observed that sucrose accumulated in the yeast cells only if the sense cDNA is expressed. The group then followed the accumulation of radioactive sucrose by yeast cells under several conditions as summarized in the table. Sucrose is made of a molecule of fructose linked to a molecule of glucose. Maltose is a disaccharide made of two molecules of glucose. Raffinose is a trisaccharide made of galactose, glucose, and fructose. Chlorophenyl hydrazone is a metabolic poison that prevents synthesis of ATP.

Compound in Solution	**Transport Rate as Percent of Control**
^{14}C-Sucrose (Control)	100
^{14}C-Sucrose + Maltose	45
^{14}C-Sucrose + Raffinose	96
^{14}C-Sucrose + Chlorophenyl Hydrazone	12

Adapted from Christian Knop et al., "AmSUT1, a Sucrose Transporter in Collection and Transport Phloem of the Putative Symplastic Phloem Loader *Alonsoa meridionalis*," Plant Physiology 134, no. 1 (January 2004): 204–214.

What conclusion can the researchers most justifiably draw from this data?

(A) The cloned protein is a carrier protein that facilitates passive movement along a concentration gradient.

(B) The cloned protein is a channel protein that is not specific for sucrose.

(C) Transport of sucrose by the cloned protein does not require energy.

(D) The cloned protein is involved in the active transport of sucrose.

2. In an emergency, a doctor can replace a patient's blood plasma with a solution of 5% human albumin and 0.9% NaCl. A pharmacist does not have ready-made 5% albumin solution and decided to use a 25% albumin solution instead. He mixed 1 part of the 25% solution with 4 parts of sterile distilled water.

The administration of the solution caused severe damage to the patient. How can the consequence of the administration of substitute plasma be explained?

(A) The dilution was incorrect, resulting in an excessive concentration of albumin that led to the shriveling of the red blood cells.

(B) The dilution was incorrect, resulting in a deficient concentration of albumin that led to the bursting of the red blood cells.

(C) The dilution of albumin in distilled water caused the bursting of red blood cells because the solution was hypotonic to the red blood cells.

(D) The dilution of albumin in distilled water caused the shriveling of the red blood cells because the solution was hypertonic to the red blood cells.

3. The energy required for protein segments to span a plasma membrane was calculated and presented in the graphs shown. The higher the free energy required, the less likely the amino acid will be found in that environment. The grayed rectangle shows the region associated with the membrane lipids. Graphs A and B below represent results obtained for different types of amino acids, but are not labeled.

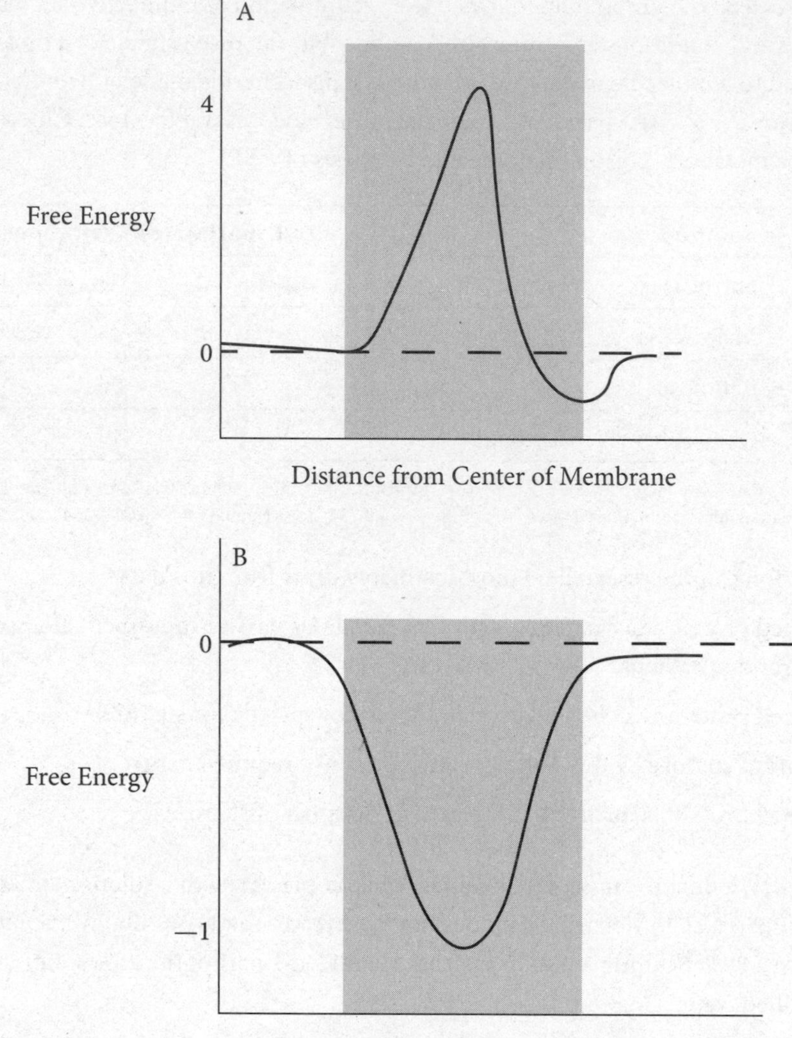

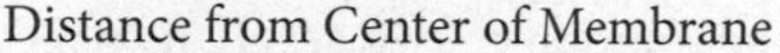

Adapted from Gunnar von Heijne, Membrane-Protein Topology, *Nature Reviews Molecular Cell Biology 7* (December 2006): 909–918.

Which of the following best characterizes how the graphs should be labeled?

(A) Graph A represents nonpolar amino acids and Graph B represents charged amino acids.

(B) Graph A represents amino acids that have short side chains and Graph B represents amino acids that have long side chains.

(C) Graph A represents charged amino acids and Graph B represents nonpolar amino acids.

(D) Graph A represents amino acids that have a long side chains and Graph B represents amino acids that have short side chains.

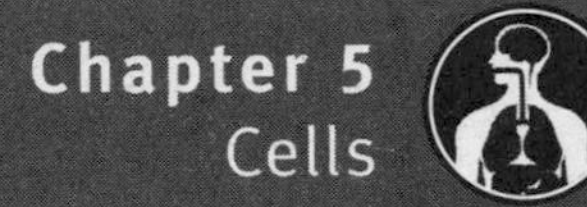

4. A pharmaceutical company is evaluating new types of drugs to treat bacterial infections. They have several candidates with different mechanisms of action. Which of these candidates would most likely be the safest for use in humans?

   (A) A molecule that prevents the formation of pores in the nuclear membrane

   (B) A molecule that inhibits the crosslinking of subunits in the biosynthesis of cell walls

   (C) A molecule that prevents the modification of proteins in the Golgi apparatus

   (D) A molecule that inhibits biosynthesis of the phospholipids in the plasma membrane

5. An investigator supplies radioactive carbon in the form of $^{14}CO_2$ to plants and harvests tissues at regular intervals. The tissue is fractionated and analyzed for radioactive carbon content in phosphoglyceric acid (PGA), a 3-carbon sugar, which is an early product of carbon fixation. The results are plotted in a graph that shows the content of radioactive PGA as a function of time of exposure to radioactive carbon.

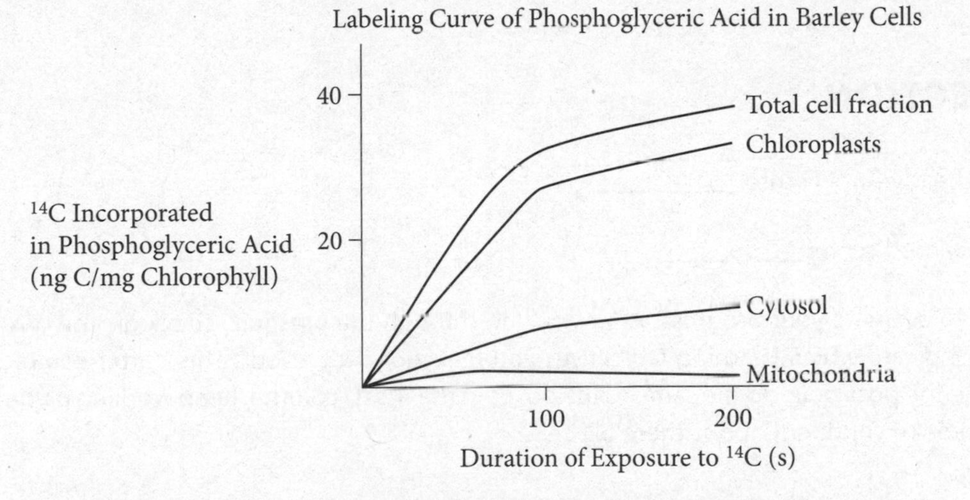

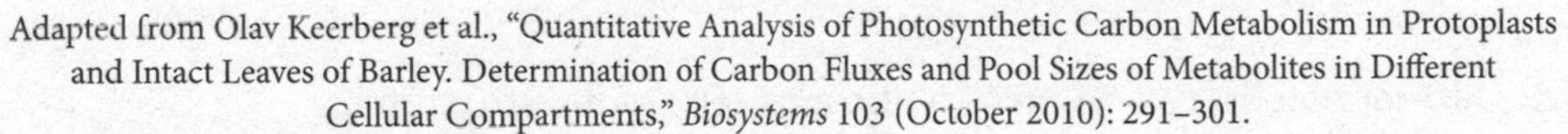

Adapted from Olav Keerberg et al., "Quantitative Analysis of Photosynthetic Carbon Metabolism in Protoplasts and Intact Leaves of Barley. Determination of Carbon Fluxes and Pool Sizes of Metabolites in Different Cellular Compartments," *Biosystems* 103 (October 2010): 291–301.

After analyzing the graphs, the scientist can reasonably conclude that

   (A) carbon fixation occurs in all compartments

   (B) carbon fixation takes place mostly in the cytosol

   (C) PGA produced in the mitochondria comes from CO_2 in the air

   (D) the first steps of carbon fixation take place in the chloroplast

Answer Key

Test What You Already Know

1. **D** **Learning Objective:** 5.1
2. **A** **Learning Objective:** 5.2
3. **D** **Learning Objective:** 5.3
4. **B** **Learning Objective:** 5.4
5. **D** **Learning Objective:** 5.5

Test What You Learned

1. **D** **Learning Objective:** 5.4
2. **C** **Learning Objective:** 5.5
3. **C** **Learning Objective:** 5.3
4. **B** **Learning Objective:** 5.1
5. **D** **Learning Objective:** 5.2

Detailed solutions can be found in the Answers and Explanations section at the back of this book.

REFLECTION

Test What You Already Know score: __________

Test What You Learned score: __________

Use this section to evaluate your progress. After working through the pre-quiz, check off the boxes in the "Pre" column to indicate which Learning Objectives you feel confident about. Then, after completing the chapter, including the post-quiz, do the same to the boxes in the "Post" column. Keep working on unchecked Objectives until you're confident about them all!

Pre	Post	
❑	❑	**5.1** Differentiate between prokaryotic and eukaryotic structures
❑	❑	**5.2** Explain the function of common organelles
❑	❑	**5.3** Recall membrane composition and function
❑	❑	**5.4** Explain transport mechanisms
❑	❑	**5.5** Investigate effects of tonicity and gradients

FOR MORE PRACTICE

Complete more practice online at kaptest.com. Haven't registered your book yet? Go to kaptest.com/booksonline to begin.

CHAPTER 6

Viruses and Bacteria

LEARNING OBJECTIVES

In this chapter, you will review how to:

- **6.1** Recognize common virus types
- **6.2** Recall virus replication pathways
- **6.3** Explain why viruses mutate easily
- **6.4** Explain plasmid function in bacteria
- **6.5** Recall gene transfer processes in bacteria
- **6.6** Investigate bacterial transformation gene transfer

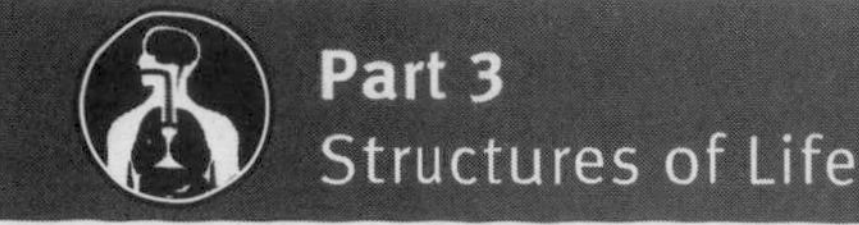

TEST WHAT YOU ALREADY KNOW

1. In the above illustration of an influenza virus, what are structures 2 and 3 and how are their functions related?

    (A) Structure 2 is the nucleoprotein that acts as a protective coating for structure 3, the capsid, which contains the genetic material of the virus.

    (B) Structure 3 is the lipid envelope that acts as a protective coating for structure 2, the capsid, which contains the genetic material of the virus.

    (C) Structure 2 is the lipid envelope that acts as a protective coating for structure 3, the neuraminidase, which contains the genetic material of the virus.

    (D) Structure 2 is the lipid envelope that acts as a protective coating for structure 3, the capsid, which contains the genetic material of the virus.

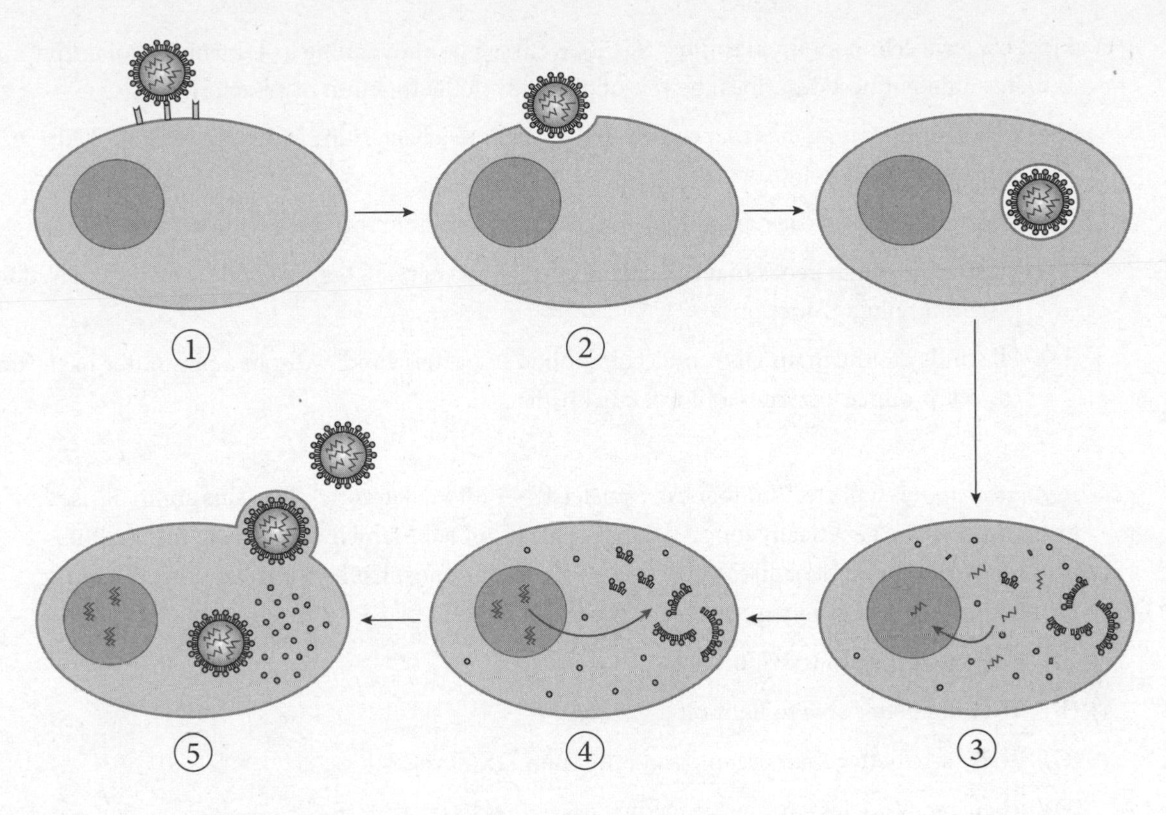

2. The diagram above illustrates the infection of an epithelial cell by an influenza virus. During which step does the viral mRNA start synthesizing new viral proteins?

   (A) 1

   (B) 2

   (C) 3

   (D) 4

3. When a human being is infected with the influenza virus, his or her immune system produces antibodies that help fight off viral antigens. Vaccines for influenza work by stimulating the body to produce the antibodies that recognize the antigens of the virus. Which of the following is the most plausible explanation of why previously inoculated people have to receive the influenza vaccine annually?

   (A) The influenza virus will likely undergo minor mutations in the genes that encode the influenza antigens.

   (B) The influenza virus will likely undergo major mutations that are caused by genetic recombination resulting from two different influenza strains infecting a host cell at the same time.

   (C) Natural selection causes only the most successful viruses from the previous year to survive to the current year.

   (D) Crossing over of the viruses' chromosomes recombines genes, creating the variation required for new strains to emerge.

4. Most bacteria cells contain plasmids. However, these plasmids can be lost during cellular division. Which statement best describes the role of plasmids in the function of a bacterium?

   (A) Plasmids contain genes that control cell growth, so a bacterium without a plasmid will survive but will not grow.

   (B) Plasmids are used for replication, so a bacterium cannot reproduce without plasmids.

   (C) Plasmids contain genes that can enhance bacterial survival, but plasmids are not required for essential cellular functions.

   (D) Plasmids are the main DNA molecules found in bacteria and, without a plasmid, a bacterium cannot produce enzymes and will quickly die.

5. *Agrobacterium* is a bacterium that can transfer DNA into eukaryotic cells. This ability is used by humans to create genetically modified organisms (GMOs). Many industries are using GMOs because of the added benefits. Which of the following is most likely the greatest benefit to the farming industry in growing genetically modified vegetables?

   (A) They are easier to transport.

   (B) They are better able to fight off diseases.

   (C) They taste better than organic and other non-GMO vegetables.

   (D) Their seeds are less expensive for farmers to purchase.

6. A researcher wants to study the effects of penicillin on the growth of *Escherichia coli.* Two Luria broth agar plates, one containing penicillin and the other only nutrient agar, were both inoculated with *E. coli.* Which of the following is the most likely outcome for the experiment?

   (A) The agar plate containing penicillin will show the same amount of bacteria growth as the agar plate without penicillin.

   (B) The agar plate containing penicillin will show more bacteria growth than the agar plate without penicillin.

   (C) The agar plate containing penicillin will show less bacteria growth than the agar plate without penicillin.

   (D) Both agar plates will show reduced bacterial growth.

Answers to this quiz can be found at the end of this chapter.

VIRAL STRUCTURE

6.1 Recognize common virus types

Viruses defy much of the logic with which we approach our study of life on Earth. Their genomes often differ strikingly from those of other living organisms, their life cycles depend upon their ability to enter and replicate within a living host cell, and they have no cell membranes or organelles as part of their structure.

Virus particles are surrounded by a capsid, or protein shell, which can come in a variety of shapes (cones, rods, and polyhedrons). Capsids are built out of proteins, many of which have various sugars attached to them, poking upward from the viral surface. These glycoproteins are used to gain entry into a living cell by binding with surface proteins on the living cell's membrane. Most viral capsids are made up of only one or two different types of protein. In addition, some viruses are able to surround themselves with an envelope of cell membrane as they burst out of a cell they have just infected. This viral envelope can help them avoid detection by the host's immune system, because the viral particles resemble (at least on the outside) the host's own cells.

Bacteriophages

The viruses that infect bacteria are known as bacteriophages, or phages for short. They are a diverse group of organisms that are best characterized by their head and tail structure, which is unique to phages:

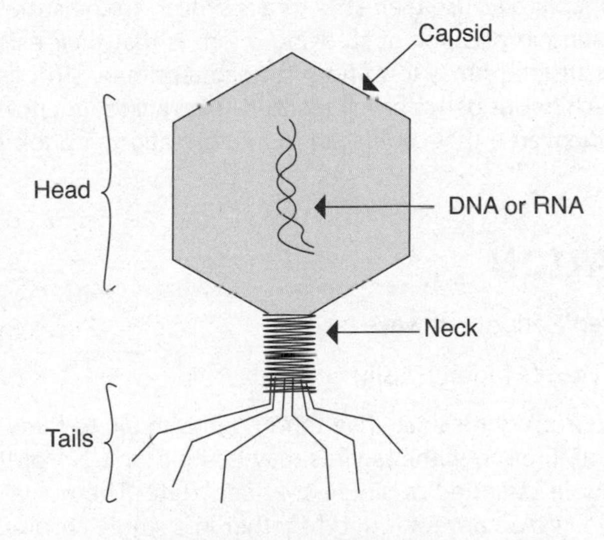

Bacteriophage Structure

The tails are used to latch on to the host cell surface, after which enzymes digest a portion of the cell membrane to insert viral genetic material (either DNA or RNA) into the host cell. The capsid and the tails are left outside of the host cell membrane, differentiating phage infection from that of other viruses. Often, the viral DNA or RNA is rapidly destroyed by powerful bacterial enzymes called restriction enzymes (which are described in detail in the lab section of chapter 16). These enzymes are a primitive type of immune system in bacteria and chew up foreign genetic material.

For the viral genomes that do survive, they will either cause the host cell to produce new viral particles (as described later in the chapter) or will integrate into the bacterial chromosome. Phages that infect bacteria in a lytic way and cause active viral replication are called lytic phages, while those whose DNA gets integrated into the host cell's DNA (in a lysogenic fashion) are called temperate phages. While integrated, the bacteriophage is known as a prophage.

Atypical Virus-like Forms

There are a few exceptions to the typical virus described above, and these particles (which certainly cannot be considered living organisms according to the standard definition of life) are infectious to many living cells:

Viroids are virus-like particles that are composed of a single molecule of circular RNA without any surrounding capsid or envelope. The RNA can replicate using a host's machinery, but it does not seem to code for any specific proteins. Despite this, viroids have been implicated in some plant diseases, though not in diseases in any other organisms.

Prions are simply pieces of protein that are infectious. They have gained fame through the recent spread of mad cow disease (bovine spongiform encephalopathy), and they are connected with diseases that are typically slow to form, taking years before symptoms develop. Despite once being called slow viruses, prions are not viruses at all and do not contain any DNA or RNA. Recent evidence suggests that prions are infectious because they change the structure of an organism's normal proteins, a switch not encoded in genes at all, by coming into contact with the proteins. They are not recognized by the immune system, probably because their structure is similar to the structure of normal proteins. Perhaps the most fascinating implication of studying prions is that their existence shows that more than one tertiary protein structure may form from the same primary structure. Both prions and the normal proteins from which prions derive have identical amino acid sequences, and no traceable DNA mutation has yet been discovered that could lead to the formation of prion particles.

VIRAL REPLICATION

6.2 Recall virus replication pathways

6.3 Explain why viruses mutate easily

Viral genomes can range from quite small (five to ten genes in all) to fairly large (several hundred genes). The genetic material found within a virus may be held on DNA or RNA, both of which can be found in either a double-stranded or single-stranded state. The nucleic acid can be linear or circular, and although the DNA is always found together in a single chromosome, the RNA can be in several pieces. Viruses are considered to be obligate intracellular parasites, meaning that they must live and reproduce within another cell, where they act as a parasite, using the host cell's machinery to copy themselves or to make proteins encoded for by their DNA or RNA. Independent of host cells, viruses conduct no metabolic activity of their own.

Lytic Cycle

Some viruses contain the enzymes they need for replication of their genome, stuffing these proteins into each "baby" virus as it is produced by the host cell. Others do not travel with their proteins

but rather make them only when they are inside a host. The typical viral growth cycle includes the ordered events of attachment and penetration, uncoating, viral mRNA and protein synthesis, replication of the genome, and assembly and release.

Attachment and penetration by parent virion (viral particle): The specificity of the proteins coating the capsid of the virus determines the host range of the particular virus, or how many kinds of cells the virus can infect. Those viruses with a wide range of surface proteins (unusual) or with proteins that can bind to many kinds of cell surface receptors (more common) are said to have a wide host range.

Uncoating of the viral genome: As the viral particle penetrates, the cell traps it in a vesicle, at which point the virus will break open its capsid. When the vesicle breaks due to viral uncoating, the inner core of the virus with its genetic material can dump itself into the cytoplasm.

Viral mRNA and protein synthesis: DNA viruses replicate their DNA in the nucleus of the cell and use the host cell's RNA polymerases to make their proteins. There are a few exceptions to this (e.g., the pox family of viruses that causes smallpox, chicken pox, etc.) that cannot enter the nucleus and actually carry the necessary RNA polymerase with them. RNA viruses replicate in the cytoplasm, sometimes using their RNA as mRNA directly, and sometimes using their RNA as a template for mRNA synthesis. One group of RNA viruses, the retroviruses (e.g., HIV), converts its RNA into DNA first, using the enzyme *reverse transcriptase*. The DNA copy is then transcribed into mRNA by the host's RNA polymerases. Many of the viral proteins that are first synthesized on the host's ribosomes are the ones needed for replication of the genome.

Replication of the genome: All virus particles will first replicate the DNA or RNA they came into the host cell with; the complementary copy that is created is then used as a template for all new copies of the genome. This is analogous to making a negative from a photograph and using the negative for all subsequent copies rather than using the photograph itself. Replication proceeds using host cell DNA and RNA polymerases as cells normally would.

Assembly and release: Viral nucleic acid is packaged within a protein capsid, and "baby" virus particles are released from the cell either en masse or slowly. *En masse* release ruptures and kills the host cell; *slowly* means the particles bud out of the host cell surface and envelop themselves in the host cell membrane as they push their way out. Budding in a slow and deliberate fashion will usually allow the host cell to live.

Lysogenic Cycle

The cycle described in the previous section is typical of most viruses: it is called a lytic cycle because the host cells usually lyse, or burst, in the end. Some viruses, however, use a lysogenic cycle, whereby the viral DNA gets integrated into the host cell's DNA and can remain there indefinitely, allowing viral DNA to be replicated for generations alongside the host cell DNA. The infection continues so that all offspring of the host cell carry the viral DNA. This DNA can simply remain a part of the host cell's genome forever. For example, certain bacteria (e.g., *Corynebacterium diphtheriae*, which causes diphtheria, and *Clostridium botulinum*, which causes botulism) secrete toxins that are coded for by viral genes that were acquired from a bacteriophage (a virus that infects bacteria). In some cases, however, certain environmental events may cause this integrated viral DNA to begin a full replicative cycle (lytic cycle), accompanied by the production of new viruses and the eventual death of the cell.

BACTERIA

6.4 Explain plasmid function in bacteria

6.5 Recall gene transfer processes in bacteria

Bacteria are **prokaryotes**, micrometers in size, that live in diverse environments. Single-celled organisms composed of an outer cell wall and inner cell membrane, bacteria contain a free-floating DNA chromosome (not enclosed in a nucleus). Unlike eukaryotes, bacteria do not have membrane-bound organelles. Bacteria exist in three basic shapes: cocci (round), bacilli (rod), and spirilla (spiral). Some bacteria also contain plasmids and have flagella (whiplike **appendages**) for motility and/or have pili (hair-like appendages) for surface attachment.

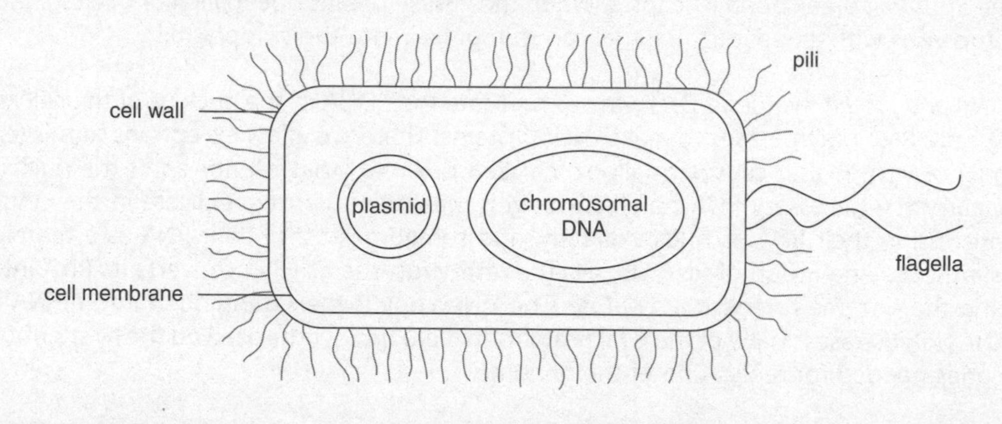

✔ AP Expert Note

The study of bacteria falls under a branch of microbiology known as bacteriology. Bacteriology is a growing field with a rapidly expanding knowledge base, but this section is limited only to material that is highly tested on the AP Biology exam.

In response to changes in their environment, bacteria may undergo chemotaxis and quorum sensing. Chemotaxis involves bacteria responding to a chemical stimulus by moving in a certain direction. For example, bacteria move toward higher concentrations of food molecules like glucose and move away from higher concentrations of toxins like phenol. Quorum sensing involves cell-to-cell communication in which bacteria release signal molecules into the environment to monitor population density. At high cell densities, bacteria regulate gene expression to produce the most beneficial phenotypes.

Bacteria reproduce via **binary fission**, which produces genetically identical copies of the parent bacterium. Aside from random mutations, binary fission does not promote genetic recombination or genetic diversity like sexual reproduction. Genetic variation, which allows organisms to adapt to changes in their environment, is advantageous for evolution and survival. Thus, bacteria have evolved mechanisms to ensure genetic variation and to yield new phenotypes. Transfer of genetic material from one bacterium to another is called horizontal gene transfer and can occur via three mechanisms: 1) transformation, 2) transduction, or 3) conjugation. Bacteria also use transposition to increase genetic variation. In some bacterial species, the generation time is only a few minutes, which allows bacteria to evolve quickly.

Plasmids

HIGH YIELD

Mainly found in bacteria, plasmids also naturally occur in some archaea and eukaryotes such as yeast and plants. Plasmids are small, double-stranded circular DNA molecules that are separate from the chromosomal DNA. Plasmids may carry genes that benefit the survival of the organism by providing a selective advantage. For example, plasmids may encode for **antibiotic** resistance, the production of toxins in a competitive environment, or the utilization of particular organic compounds when nutrients are scarce. Possessing its own origin of replication, a plasmid can replicate independently from its host's chromosomal DNA.

In genetic engineering, recombinant plasmids are used as vectors to clone and amplify or express particular genes. The advantages of using plasmids as vectors include the ability to easily modify and mass-produce them.

Bacterial Gene Transfers

Transformation

Upon cell death, some bacterial species release pieces of their chromosomal DNA and/or plasmids to be taken up by other bacteria. The uptake and incorporation of the exogenous (foreign) DNA into a host (recipient) bacterium is called transformation. For transformation to occur, the host cell must be competent—the cell membrane becomes more permeable to allow the DNA to pass from the environment into the cell. Transformation works best between closely-related species and results in a stable genetic change. In a laboratory setting, artificial competence can be obtained by exposing cells to calcium chloride and heat shock or by using electrical pulses to create pores in the cell membrane.

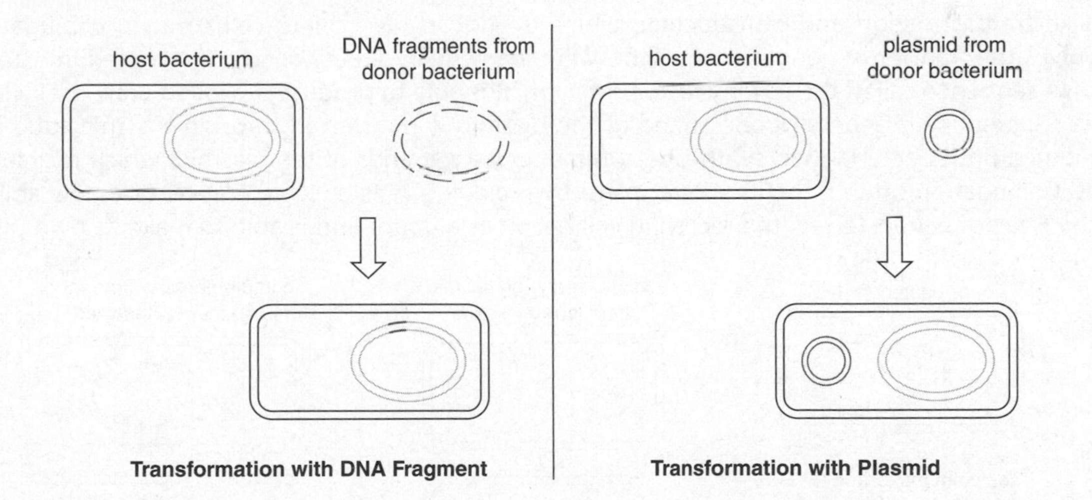

Transformation with DNA Fragment

Transformation with Plasmid

Transduction

Introducing exogenous DNA into a bacterium by a virus is called transduction. The bacteriophage (a virus that infects bacteria) injects viral DNA into a host bacterium and breaks down the host DNA. Infected by the phage, the host cell (donor) creates new phages that may contain host DNA packaged in the viral capsid. Bacterial lysis releases the new phages, which inject the host cell DNA into recipient bacterial cells. In the recipient bacterium, the donor DNA may be recycled for spare parts, it may recircularize and become a plasmid if it was originally one, or it may be exchanged via recombination to create a bacterium with a different genotype.

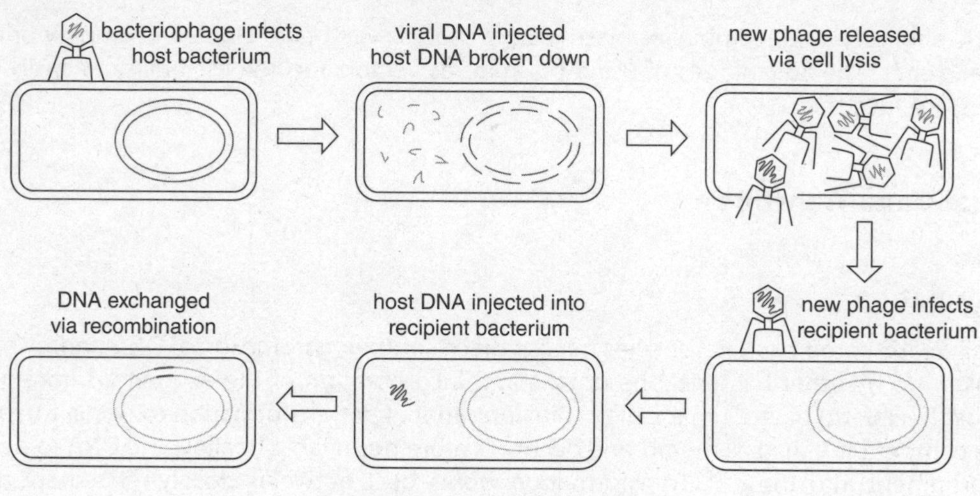

Transduction with DNA Fragment

Conjugation

Unlike transformation and transduction, which do not involve cell-to-cell contact, conjugation involves the transfer of genetic material between bacteria in direct contact. If a bacterium carries a DNA sequence called the fertility factor (F-factor), it is able to produce a pilus to draw itself close to a recipient cell. Generally, one strand of the plasmid is transferred through the thin tube-like structure. Both bacteria then synthesize complementary strands of the plasmid, which recircularizes. Conjugation often benefits the recipient by providing resistance, tolerance, or a new ability. If the F-factor is transferred, the receiving cell becomes a donor and is able to make its own pilus.

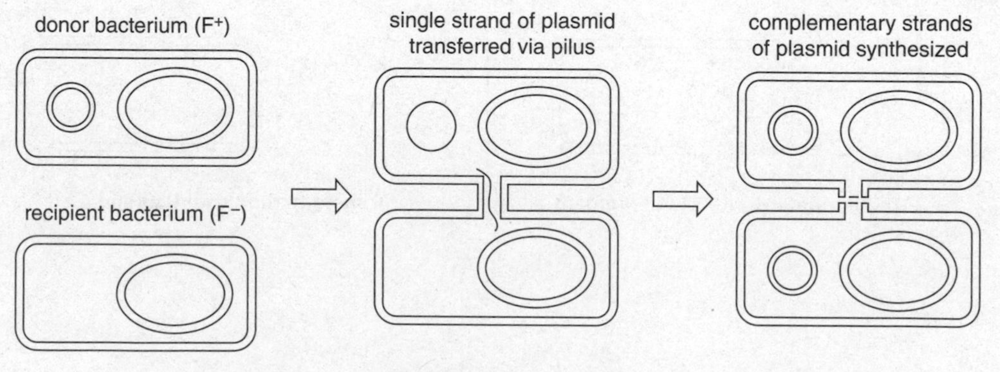

Conjugation with Plasmid

Transposition

Transposition involves the movement of DNA segments within a genome or between the chromosome and plasmid of a bacterium. Transposable elements (transposons), sometimes known as jumping genes, can change their position from one place to another by cutting and inserting themselves in new spots. In bacteria, transposable elements may carry antibiotic resistance or virulence. After transposition, bacteria may then spread the advantageous new genes to the population via some kind of horizontal transfer.

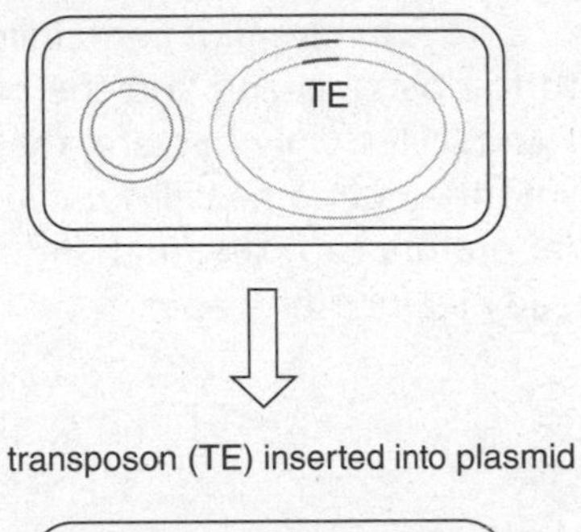

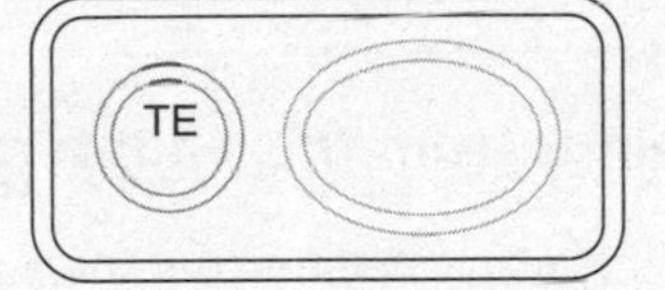

Transposition between Chromsomal DNA and Plasmid

AP BIOLOGY LAB 8: BIOTECHNOLOGY: BACTERIAL TRANSFORMATION INVESTIGATION

6.6 Investigate bacterial transformation gene transfer

This is a difficult investigation to conceptualize. It is also difficult to perform because it requires a considerable amount of specialized materials, expensive equipment, and aseptic conditions. Some companies have even prepared software that replaces the lab with a virtual experiment.

✔ AP Expert Note

You will most likely see questions on the AP Biology exam about the concepts covered in the two biotechnology investigations (the other one, AP Biology Lab 9, is covered in chapter 16)—along with other techniques such as polymerase chain reaction (PCR) and restriction fragment length polymorphism (RFLP) analysis.

Be sure that you familiarize yourself with the general procedures and purposes of using these techniques.

This investigation deals with the transformation of bacteria using plasmids that contain known genes. Bacteria can incorporate plasmids after being shocked by a chemical, temperature, or electrical pulses. Heat shock followed by an ice bath allows the plasmid DNA to penetrate the cell wall more easily. As long as the gene from the plasmid inserts after a promoter region, it will be expressed as part of the new, recombined bacterial genome.

The common example is to take plasmids that contain an ampicillin-resistance gene and transform bacteria that are affected by ampicillin. After applying the treatment, you can test for transformation by looking for bacterial growth on a medium that contains ampicillin. You start with a stock of ampicillin-sensitive bacteria and a stock of plasmids containing the gene for ampicillin resistance. You apply the shock treatment to the bacteria and add the plasmids, then apply the bacteria to growth media with and without ampicillin. Only bacteria that have incorporated the ampicillin-resistant gene from the plasmid into their genome will survive in the medium containing ampicillin. Your goal in this part of the laboratory is to see firsthand the products of recombination and transformation that you have already learned on paper.

RAPID REVIEW

If you take away only 4 things from this chapter:

1. Viruses are obligate intracellular parasites that must replicate inside the cells of living organisms. Viral genomes can range from small to large and be DNA or RNA and double-stranded or single-stranded.
2. There are multiple types of viruses and virus-like structures. Bacteriophages, or phages, are viruses that infect bacteria. Viroids and prions are not viruses in the traditional sense but are still infectious. One of the most famous prions is responsible for causing bovine spongiform encephalopathy, or mad cow disease.
3. Viruses can replicate in several ways, including the lytic and lysogenic cycles.
4. Bacteria use four distinctive mechanisms to increase genetic variation: transformation, transduction, conjugation, and transposition.

TEST WHAT YOU LEARNED

1. It is useful for scientists to study whether genetically transformed organisms can pass on new traits to future offspring. Which of the following organisms would be the most practical to use for the study of genetic transformations and why?

   (A) *Escherichia coli*, because they have short generation times

   (B) Insects, because most use external fertilization

   (C) Laboratory mice, because they reproduce sexually

   (D) Chimpanzees, because they are very genetically similar to humans

2. Which of the following conclusions can most reasonably be drawn if a host cell's function has been impaired following a viral infection?

   (A) The host cell will lyse or undergo apoptosis regardless of the severity of damage.

   (B) The host cell will lyse or undergo apoptosis if the severity of damage exceeds the cell's ability to repair itself.

   (C) The host cell will only lyse.

   (D) The host cell will only undergo apoptosis.

3. Some bacteria, such as *Methicillin-resistant Staphylococcus aureus* (MRSA), are known to carry genes for antibiotic resistance. Which of the following structures could best facilitate the inheritance of antibiotic-resistance genes by a bacterium?

   (A) Flagellum

   (B) Pilus

   (C) Plasmid

   (D) Chlorosome

4. There has been an outbreak of the influenza virus at a school. A student, who happens to work in the cafeteria, has been infected. Which of the following would be the best advice to give to the student to help reduce the spread of infection?

   (A) Stay at home until all of the symptoms are gone.

   (B) Go see the doctor immediately and get a prescription for antibiotics.

   (C) Take over-the-counter flu medication and go back to work.

   (D) Use antibiotics from a previous illness.

5. Horizontal gene transfer between bacteria can occur through several mechanistic pathways as shown in the diagram below.

How does pathway 2 differ from pathway 3?

(A) Pathway 2 is transformation and involves the exchange of DNA using bacteriophages, whereas pathway 3 is conjugation and involves the transfer of genetic material using a pilus.

(B) Pathway 3 is transduction and involves the exchange of genetic material using bacteriophages, whereas pathway 2 is conjugation and involves the transfer of genetic material using a pilus.

(C) Pathway 2 is transduction and involves the exchange of genetic material using a pilus, whereas pathway 3 is conjugation and involves the transfer of genetic material using bacteriophages.

(D) Pathway 2 is transduction and involves the exchange of genetic material using bacteriophages, whereas pathway 3 is conjugation and involves the transfer of genetic material using a pilus.

6. Certain viruses, such as Ebola, have higher mutation rates than other viruses, such as the virus that causes chicken pox. Which of the following statements best explains this difference?

(A) Ebola is a DNA virus, while chicken pox is an RNA virus.

(B) Ebola is an RNA virus, while chicken pox is a DNA virus.

(C) Ebola is a DNA virus, while chicken pox is a retrovirus.

(D) Ebola is an RNA virus, while chicken pox is a retrovirus.

The answer key to this quiz is located on the next page.

Answer Key

Test What You Already Know

1. **D** **Learning Objective:** 6.1
2. **D** **Learning Objective:** 6.2
3. **A** **Learning Objective:** 6.3
4. **C** **Learning Objective:** 6.4
5. **B** **Learning Objective:** 6.5
6. **C** **Learning Objective:** 6.6

Test What You Learned

1. **A** **Learning Objective:** 6.6
2. **B** **Learning Objective:** 6.2
3. **C** **Learning Objective:** 6.4
4. **A** **Learning Objective:** 6.1
5. **D** **Learning Objective:** 6.5
6. **B** **Learning Objective:** 6.3

Detailed solutions can be found in the Answers and Explanations section at the back of this book.

REFLECTION

Test What You Already Know score: __________

Test What You Learned score: __________

Use this section to evaluate your progress. After working through the pre-quiz, check off the boxes in the "Pre" column to indicate which Learning Objectives you feel confident about. Then, after completing the chapter, including the post-quiz, do the same to the boxes in the "Post" column. Keep working on unchecked Objectives until you're confident about them all!

Pre	Post	
❑	❑	**6.1** Recognize common virus types
❑	❑	**6.2** Recall virus replication pathways
❑	❑	**6.3** Explain why viruses mutate easily
❑	❑	**6.4** Explain plasmid function in bacteria
❑	❑	**6.5** Recall gene transfer processes in bacteria
❑	❑	**6.6** Investigate bacterial transformation gene transfer

FOR MORE PRACTICE

Complete more practice online at kaptest.com. Haven't registered your book yet? Go to kaptest.com/booksonline to begin.

CHAPTER 7

Plant Structure and Systems

LEARNING OBJECTIVES

In this chapter, you will review how to:

- **7.1** Describe structures that bring in nutrients
- **7.2** Describe the vascular system in plants
- **7.3** Explain how exposure to light affects plants
- **7.4** Recall methods of plant reproduction
- **7.5** Describe the effect of mutations across generations
- **7.6** Investigate water movement through plants

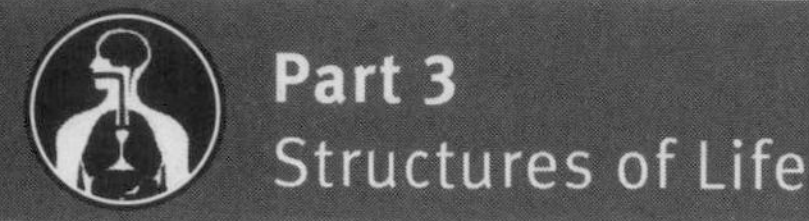

TEST WHAT YOU ALREADY KNOW

1. Mycorrhizae are fungi that live in symbiotic association with plant roots. The figure below shows the results of an experiment that examined the percent of petunia roots colonized by mycorrhizae in environments with different amounts of phosphate (PO_4^{3-}).

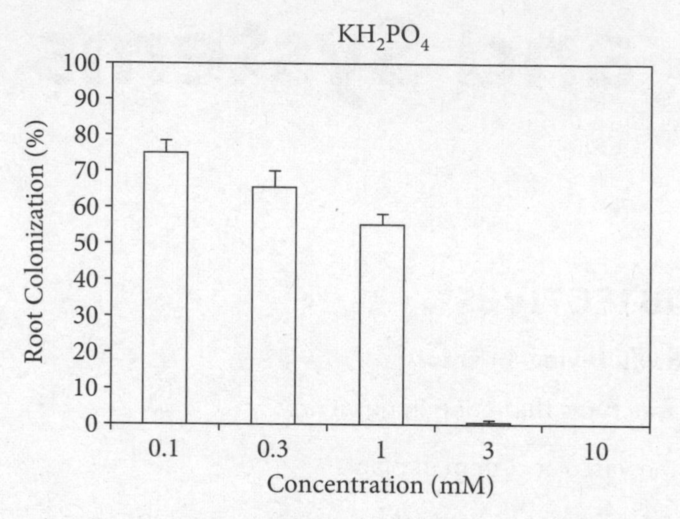

Adapted from Eva Nouri et al., "Phosphorus and Nitrogen Regulate Arbuscular Mycorrhizal Symbiosis in *Petunia hybrid*," *PLoS ONE* 9, no. 3 (March 2014): e90841.

Which of the following provides the most plausible interpretation of the graph?

(A) Petunias can take up more phosphate when they have mycorrhizae, so they are less likely to benefit from mycorrhizal colonization when there is plenty of phosphate available.

(B) Petunias under stress from high or low phosphate levels are less able to form associations with mycorrhizae than those at healthy phosphate levels.

(C) The pattern in the graph is unclear, so another abiotic factor not measured is probably responsible for the change in percent root colonization with changing phosphate concentration.

(D) Increased phosphate levels inhibit mycorrhizal growth, making them less available to colonize petunia roots.

2. Researchers are developing a way to produce clean drinking water using plant tissues. Which of the following strategies would probably be most successful, and why?
   (A) Because periderm forms an outer ring of tissue around a plant stem, it can be used to create a straw-like structure to transport water.
   (B) Because xylem transports water in plants, water could be pushed through it with the natural structure of the xylem acting as a filter.
   (C) Because phloem consists of dead tissue, it forms straw-like structures that can be used to filter water.
   (D) Because the cortex consists of living tissue, it can be used to actively pump water through a membrane.

3. To improve crop yields, a researcher decides to study the effects of light on factors related to photosynthesis in plants. In his study, he examines stomata under the microscope under light and dark conditions and measures changes in sugar concentrations. Which of the following provides the most plausible hypothesis that he could test with this experiment?
   (A) If the plant has receptors on guard cell membranes, it can increase its rate of photosynthesis by closing its stomata in response to light.
   (B) If the plant has receptors on guard cell membranes, it can increase its rate of photosynthesis by opening its stomata in response to light.
   (C) If the plant has receptors on cortical cell membranes, it can increase its rate of photosynthesis by closing its stomata in response to light.
   (D) If the plant has receptors on cortical cell membranes, it can increase its rate of photosynthesis by opening its stomata in response to light.

4. Researchers examined different methods of growing pineapples in Benin, Africa, to understand why some methods yielded larger pineapples than others. The figures below show two experiments. In both cases, plants were either artificially induced to flower (artificial flowering induction) or allowed to flower naturally (natural flowering induction). Additionally, the fruit was either artificially induced to mature (AMI) or allowed to mature naturally (NMI).

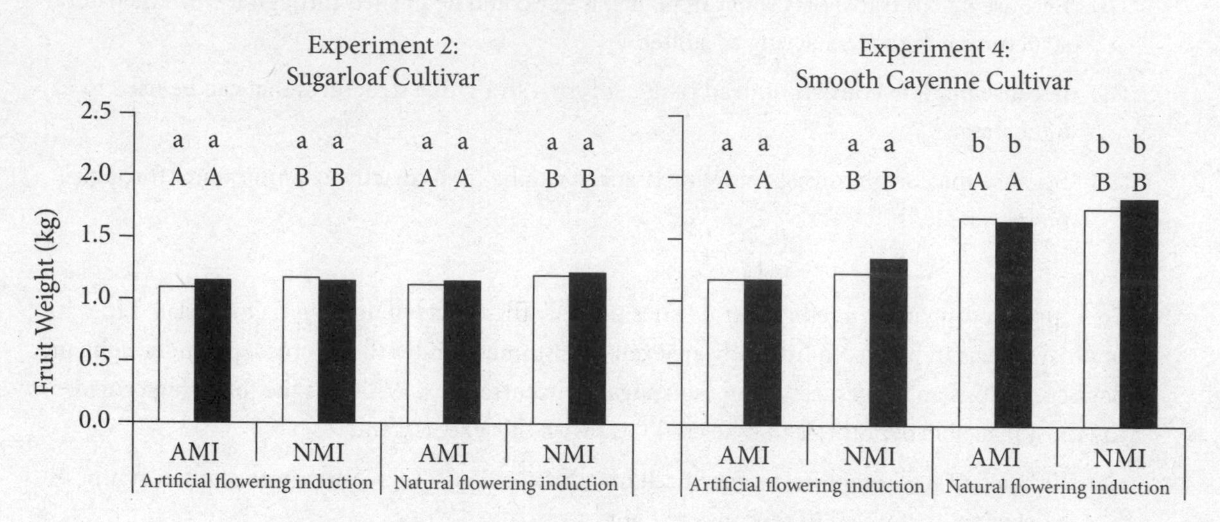

Adapted from V. Nicodème Fassinou Hotegni et al., "Trade-Offs of Flowering and Maturity Synchronisation for Pineapple Quality," *PLoS ONE* 10, no. 11 (November 2015): e0143290.

Which option produces the largest fruit, and why?

(A) The largest fruit is produced by the Sugarloaf cultivar when natural flowering induction is used, regardless of the maturation approach used. This technique allows the plant to invest maximum amounts of energy into fruit growth.

(B) The largest fruit is produced by the Sugarloaf cultivar with artificial flowering induction and by the Smooth Cayenne cultivar with natural flowering induction. This technique allows the plant to save more energy for future fruit production.

(C) The largest fruit is produced by the Smooth Cayenne cultivar when natural flowering induction is used, regardless of the maturation approach used. This technique allows the plant to invest maximum amounts of energy into fruit growth.

(D) The largest fruit is produced by the Sugarloaf cultivar when artificial flowering induction is used, regardless of the maturation approach used. This technique allows the plant to save energy to produce larger numbers of fruit.

5. If an error in meiosis in a diploid plant resulted in the production of diploid gametes, what would be the most likely consequences for the next generation?
   (A) The fusion of three diploid gametes would produce a hexaploid plant with six sets of chromosomes, which would then produce hexaploid gametes.
   (B) If the diploid gametes fused with diploid gametes from another plant, some chromosomes would be lost to produce a new diploid plant.
   (C) There would be no possibility that the individual could reproduce because polyploid organisms are not fertile.
   (D) If the diploid gametes were capable of self-fertilization, then a tetraploid plant with four sets of chromosomes would be produced. If this plant could self-fertilize, more tetraploid plants would be produced.

6. A researcher decides to compare the response of two plants to desiccation by growing them in a hot, dry environmental chamber and then returning the chamber to normal climate conditions. One plant showed a dramatic increase in transpiration relative to its initial transpiration rate as the environment became drier, but then reverted to a normal transpiration rate when the normal climate was restored. The other plant showed a much smaller change in transpiration relative to its initial transpiration rate under the same conditions. What conclusion can most likely be drawn from these results?
   (A) The plant that had a smaller change in transpiration probably normally lives in a moist environment and had water reserves. In contrast, the plant that had a larger change in transpiration is a xerophyte, a plant that is adapted to a dry environment.
   (B) The plant that had a smaller change in transpiration is probably a smaller plant and therefore has less transpiration under any conditions than the plant that had a larger change.
   (C) The plant that responded dramatically does not normally live in a dry environment. In contrast, the plant that had a smaller change in transpiration may be a xerophyte, a plant that is adapted to a dry environment.
   (D) The plant that responded dramatically is probably from a hot climate, whereas the plant that responded only slightly is probably from a cold climate.

Answers to this quiz can be found at the end of this chapter.

PLANT VASCULAR AND LEAF SYSTEMS

7.1 Describe structures that bring in nutrients

7.2 Describe the vascular system in plants

Plant Structure

Plants with vascular tissues usually have three types of structures, or organs: leaves, roots, and branches. The leaves provide most of the photosynthesis of the plant; the roots provide support in the soil, along with water and **minerals**; and the branches hold the leaves up to light and convey nutrients and water between the leaves and the roots. Each of these structures can be specialized in many ways.

Plants with taproots have long roots with a single extension deep into the soil, while other plants have highly branched roots. Cells on the surface of roots often have long extensions called root hairs, which increase the surface area of roots. Some plants without root hairs have a symbiotic relationship with fungi that increase the surface area of the root to absorb water and minerals. In legumes, nitrogen-fixing *Rhizobium* species of bacteria infect roots and form root nodules in symbiosis with the plant. The roots of plants often play an important role in preventing erosion. Tropical rain forest that is cleared is highly vulnerable to erosion of the thin soil if the plants and their root systems are absent.

Leaves can have a variety of shapes. Monocot leaves are usually very narrow with veins that run parallel to the length of the leaf, while dicot leaves are broad with veins that are arranged in a net in the leaf. Modified leaves form thorns in cacti, tendrils in pea plants, and petals in flowers. The leaves produce all of the energy of the plant and are specialized to gather sunlight. The broad shape helps to gather sunlight for themselves and in some cases to block sunlight from getting to competing plants. The shape and arrangement of leaves are key features used to distinguish plant species.

Terrestrial plants have broad leaves that maximize the absorption of sunlight but also tend to increase water loss by evaporation. Leaves of terrestrial plants have a waxy cuticle on top to conserve water. The lower **epidermis** of the leaf is punctuated by **stomata**: openings that allow diffusion of carbon dioxide, water vapor, and oxygen between the leaf interior and the atmosphere. A loosely packed spongy layer of cells inside the leaf contains chloroplasts with air spaces around cells. Another photosynthetic layer in the leaf, the palisade layer, consists of more densely packed elongated cells spread over a large surface area. A moist surface that lines the photosynthetic cells in the spongy layer is necessary for diffusion of gases into and out of cells. Air spaces in leaves increase the surface area available for gas diffusion by the cells. The size of stomata is controlled by guard cells that can open and close the opening. These cells open during the day to admit CO_2 for photosynthesis and close at night to limit the loss of water vapor through **transpiration**, the evaporation of water from leaves that draws water up through the plant's vascular tissues from the soil. The upper surface layer of cells in leaves has no openings, an **adaptation** that reduces water loss from the leaf.

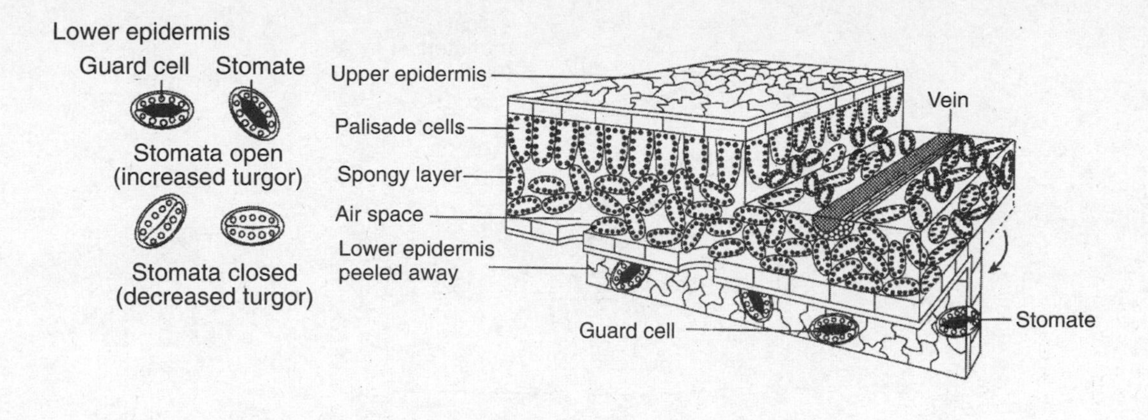

Leaf Structure

Guard cells are kidney-shaped cells in dicots and dumbbell-shaped cells in monocots that change their shape according to the amount of water that exists within them. This water exerts a pressure called turgor pressure. When water is abundant, the guard cells swell, and when water is sparse, they clamp down and shut the stomata. This makes perfect sense, as the guard cells allow transpiration when there is plenty of water in the leaves and water conservation when there is not. Guard cells absorb water in response to potassium ions that are driven into the cells. Water follows these ions due to the osmotic gradient that is created by the presence of extra particles inside the cell as compared with outside the cell.

The presence of blue-light receptors on the guard cell membranes is what drives the movement of K^+ into the guard cell. Sunlight, particularly blue wavelengths, causes K^+ ion channels to open and potassium to flood in. This explains why guard cells are usually open during the day and closed at night, helping photosynthesis to take place as carbon dioxide gas can then enter through the open stomata.

Growth in higher plants is restricted to areas of perpetually embryonic and undifferentiated tissue known as meristems. Meristems are self-renewing populations of cells that divide and cause plant growth either in height or width. Apical meristems exist at the tips of roots and stems, whereas lateral meristems (also known as cambium) are found within the stem between layers of **xylem** and **phloem** on the sides of the plant. The trunk of the plant thickens each year because the embryonic cells of the cambium produce more and more xylem and phloem, supporting the growth of a larger tree with more leaves. Upward growth that occurs as a result of cell division within apical meristems is called primary growth, while outward growth is called secondary growth. It is secondary growth that causes the well-known concentric tree rings that one sees when a large woody tree is cut down.

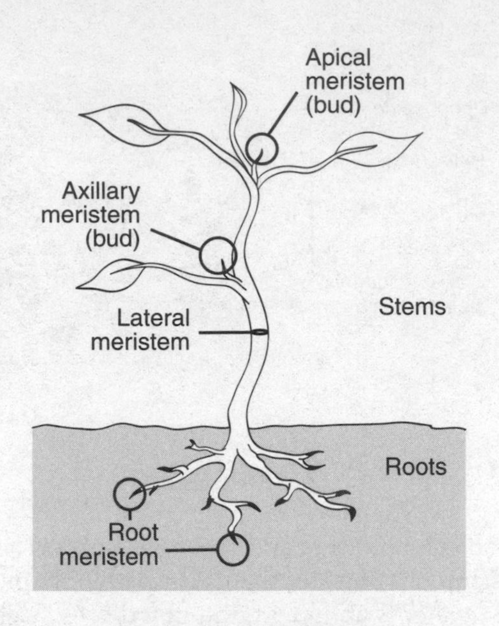

General Plant Anatomy

Plant Vasculature

Terrestrial plants left an aquatic environment during their evolution, and in the process lost some of the benefits that a liquid medium had previously provided. **Cell walls**, which provided simple rigidity for structures in small plants, became aligned, allowing for the growth of trees over 100 meters in height and creating channels for the delivery of important soil nutrients. Water loss became a constant challenge in environments where rainfall could be unpredictable or nearly nonexistent. The sun, formerly relied upon solely for light-giving energy, also became a source of dehydration and overheating.

One of a plant's many challenges is providing all of its structures with the water needed to complete photosynthesis and to maintain an aqueous solution for all biological reactions. Water is heavy and a plant cannot rely on muscular contractions to move materials, as animals do. Instead, a plant has narrow, lifeless channels in its xylem tissue that take advantage of the cohesive properties of water. A plant loses the majority of its water during transpiration while its stomata are open for the exchange of CO_2 and O_2 in photosynthesis. As water is lost through the leaves, it creates a negative pressure in the xylem channels, just like sucking on a straw. The cohesive force of water keeps a steady flow of water moving through the plant, pulled by the negative pressure developed by transpiration occurring in the leaves. The force exerted in the water column due to cohesion is as strong as steel wire of the same diameter.

Plants need a mechanism to deliver the stored energy created during photosynthesis to all of their structures. While water travels upward from the roots through the xylem via the pull of transpiration and the cohesion of water, nutrients flow both up and down through the phloem via the pull of **osmotic pressure**. Osmotic pressure is built up between areas that have high concentrations of nutrients and those that have low concentrations. Translocation of materials in the phloem is like long-range diffusion. Similar to the cohesion of water, lower solute pressure on one end of a phloem vessel is translated along the narrow vessel to an area where the solute pressure is greater.

The adaptive significance of these transport systems is the colonization of a terrestrial environment. Plants also harnessed a physical disadvantage, water loss, and turned it into a benefit, with the force of water loss powering water transport through the xylem.

Xylem

The vascular tissue of the xylem contains a continuous column of water from the roots to the leaves, extending into the veins of the leaves. The leaves regulate the amount of water that is lost through transpiration. Water is transported from the roots to the shoots. In other words, from roots to stems, water is transported through two different mechanisms: root pressure and cohesion-tension.

Osmotic pressure in the roots tends to build up due to water absorption. This pressure pushes water up from the roots into the stem of the plant. Root pressure functions best in extremely humid conditions when there is abundant water in the ground or at night. The drops of water, or dew, that appear on blades of grass or other small plants in the morning result from this root pressure. During the day, transpiration occurs at a high enough rate that dew is generally not seen, because the water that leaves the tips of the grass is evaporating before it can build up.

Transpiration

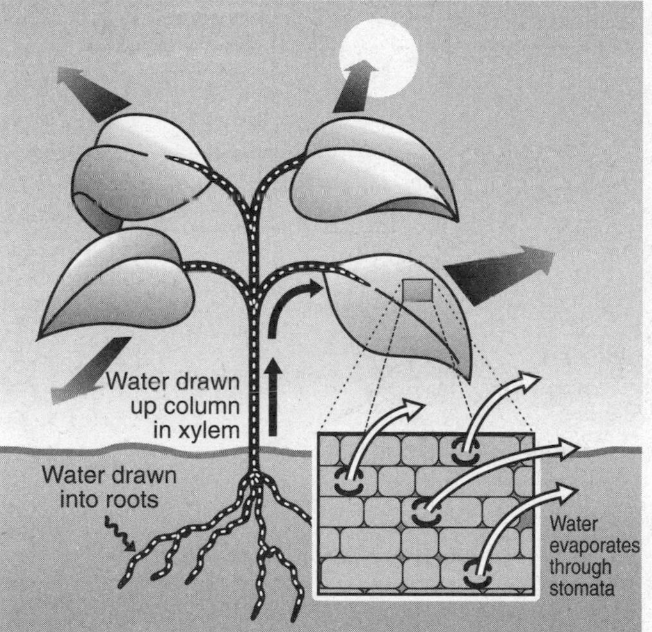

Water Gain and Loss in Plants

Phloem

Phloem tubes are much thinner-walled than xylem and are found toward the outside edges of stems. They begin up in the leaves, where sugars are made by photosynthesis and stored as starch within mesophyll cells. Much of this sugar and starch, though, is transported down phloem tubes from shoots to roots (from leaves and stems to the roots). Active transport between the mesophyll cells, where sugar is made, and the nearby phloem tubes is what allows sugar to move into the phloem's sieve tubes. The active transport pumps are symport pumps that move one H^+ ion along with every sucrose molecule. The H^+ that is used to push the sucrose across into the phloem is rapidly transported back out to the mesophyll cells by an H^+ ion ATPase pump, so that more sucrose can be moved. As sugar is loaded into the phloem, the water potential, or the amount of water pressure, in the phloem is effectively reduced. The influx of sugar into the phloem creates an osmotic potential that pulls more water into the phloem, generating a water pressure that forces sap (essentially sugar-rich water) down the phloem toward the roots. The xylem will recycle this water back from the roots.

As you can see in the following figure, vascular cambium (C) divides laterally to create new phloem (P) on the outer edges of the tree and new xylem (X) near the inner core. This results in the formation of annual tree rings.

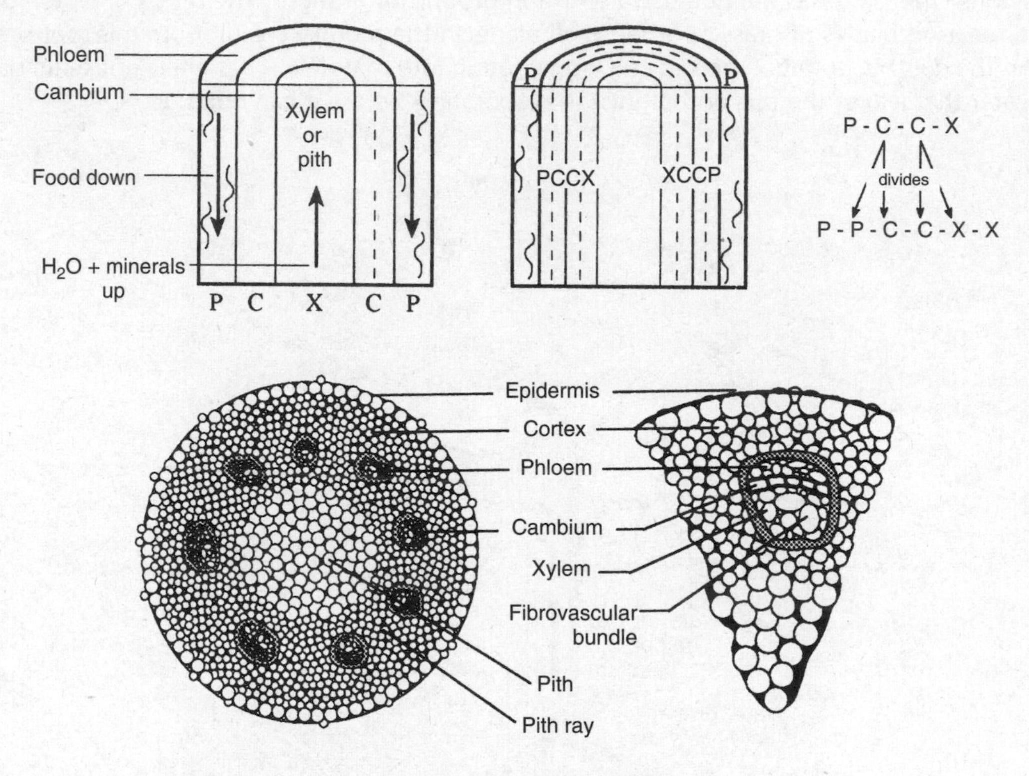

Internal Anatomy of a Plant

Plant Nutrients

Plants require certain nutrients and compounds that are important for normal physiological functions. For example, the photoreceptor phytochrome is an important pigment that plants use to detect light. Plants can use phytochrome to regulate flowering based on day length or to set up other daily rhythms. Large amounts of macronutrients such as potassium, calcium, magnesium, and phosphorous are required for normal plant functions.

PLANT NUTRIENTS			
Element/Compound	**Type**	**Origin**	**Action**
Phytochrome	Photopigment	Systemic	Detection of light to control photoperiodism
Potassium	Nutrient	Root uptake in soil	Protein synthesis, operation of stomata
Calcium	Nutrient	Root uptake in soil	Cell wall stability, enzyme activation
Magnesium	Nutrient	Root uptake in soil	Chlorophyll synthesis, enzyme activation
Phosphorus	Nutrient	Root uptake in soil	Nucleic acid and ATP synthesis

PLANT REPRODUCTION AND DEVELOPMENT

7.3 Explain how exposure to light affects plants

7.4 Recall methods of plant reproduction

7.5 Describe the effect of mutations across generations

Phototropism and Photoperiodism

The ability to respond to external stimuli allows plants to adjust to their environment and optimize growth, which is important for competition and survival. **Phototropism** is the directional growth of plants in response to a light stimulus. For example, if a plant on a window sill detects light coming from the window, it can respond by growing toward the light to obtain more light energy needed for photosynthesis. Shoots generally exhibit positive phototropism or growth toward a light source, while roots generally exhibit negative phototropism or growth away from light.

During phototropism, light is sensed at the coleoptile (the tip of a plant) and detected by blue-light photoreceptors called phototropins, which are composed of a protein bound to a light-absorbing pigment called the chromophore. The absorption of light by the chromophore causes the protein shape to change and activate a signaling pathway. The different levels of phototropin activation, in turn, cause unequal transport of auxin, a plant hormone that promotes cell elongation, down the sides of the coleoptile. More auxin is transported down the side that is away from the light than the illuminated side, which causes the plant to bend in the direction of light.

Photoperiodism is the response to changes in the photoperiod (relative lengths of day and night). Examples of photoperiodism in plants include regulating flowering, tuber formation, bud dormancy, and loss of leaves. To flower, some plants require a certain length of day or night. Long-day plants typically measure the continuous length of night (darkness) and flower when the night length is below a certain threshold (light-dominant). Short-day plants, on the other hand, typically measure the length of day and flower when the day length is below a certain threshold (dark-dominant).

Plants that do not depend on day length are day-neutral. Though it is unclear how plants determine day or night, models suggest the interactions between a plant's **circadian rhythms** (biological clock) and light cues from its environment govern a plant's photoperiodic response.

Plant Reproduction

Plants possess a reproductive advantage in that they can **reproduce vegetatively**, while very few groups of animals can reproduce without some form of sexual reproduction. Plants also live dual lives in that they have specific structures for reproduction (flowers, pollen, and fruit), while the rest of the plant continues life as usual. In contrast, female animals' entire bodies are affected by the reproductive process, so reproduction has to be an entire stage of existence for them. In fact, some animals' adult life stage is solely for the purpose of reproduction; they don't even have mouthparts for the **ingestion** of nutrients. Because of the specialization of certain plant structures, however, plants need not undergo such vast changes to reproduce.

✓AP Expert Note

You do not need to know the details of the sexual reproduction cycles in different plants and animals. Focus instead on the similarities among the processes and how this is important for genetic variation.

Plants can reproduce vegetatively as a function of their **indeterminate growth**. Plant cells can differentiate when isolated from the rest of the plant, so one could place the stem of a rose in the soil and it would grow adventitious roots and become a whole plant. By default, this produces a series of plants that are all genetically identical and reduces variation in the population, but it is advantageous when **sexual reproduction** is not suitable for a given environmental situation.

✓AP Expert Note

All plants exhibit a definite alternation of generations in which the plant life cycle alternates between haploid and diploid stages.

Most terrestrial plants live their lives as diploid **sporophytes** and produce haploid tissue in the form of ovules and **pollen** (**gametophytes**), sometimes within the same flower. Pollen is delivered to the female gametophyte either by an animal, such as a bird, or by the wind, and fertilization takes place. The flower functions as the entire reproductive organ, completely dependent on the sporophyte plant. This is a departure from aquatic plants, in which **spores** travel independently from the plant that produced them. Flowering plants seem to have evolved a specialized region where a plant can become "pregnant." These "seed babies" are housed within a seed coat and often a fruiting body. The seed coat and fruiting body aid in plant propagation by either acting as bait to an animal or providing a source of nutrition to the developing plant embryo.

Variation in Plants

The terrestrial environment provides a great degree of variation, in contrast to the constant temperatures and cycles of aquatic systems. Seasonal changes and the variability of weather requires

plants to synthesize new chemical messengers that allow plant tissues to respond to both acute and chronic environmental cues. **Novel structures**, primarily for reproduction, require new forms of control mechanisms to function in time with a changing environment.

Within the plant kingdom are several phyla, with one of the key distinguishing characteristics between phyla being the presence or absence of vascular tissue. Within the tracheophytes, which have vascular tissue, two of the important modern phyla are the gymnosperms and the angiosperms. The gymnosperms, such as the conifers, have "naked seeds" that do not have endosperm and are not located in true fruits. The angiosperms are the flowering plants. They have flowers, true fruits, and a double fertilization system that creates endosperm to nourish the plant embryo. Angiosperms are the most successful form of plant life on the planet, and the following diagram distinguishes between two main classes of angiosperm: monocots and dicots.

	Monocots	Dicots
Cotyledons in Seed	one cotyledon	two cotyledons
Leaf Shape	narrow leaves	broad leaves
Leaf Veins	parallel	branched
Vascular Bundles in Stem		

**Monocots vs. Dicots:
Two Angiosperm Classes**

Common in flowering plants, **polyploidy** drives adaptation and speciation. Most eukaryotes are diploid, having two sets of chromosomes (2*N*). Polyploids have three or more times the haploid (*N*) chromosome number. For example, triploids (3*N*) have three sets of chromosomes, tetraploids (4*N*) have four sets of chromosomes, and hexaploids (6*N*) have six sets of chromosomes. Polyploids occur spontaneously in nature and result from mutations during cell division, meiotic or mitotic failures, and the fusion of gametes. Combining two diploid cells, for instance, results in a tetraploid, and combining a diploid and haploid cell results in a triploid.

Polyploidy may arise from two closely related species (autopolyploidy) or two different species (allopolyploidy). Normally, **hybrids** are sterile because they do not have homologous pairs of chromosomes needed for gamete formation during meiosis. In some instances, polyploidization results in each chromosome having a homologue from the additional set of chromosomes, so meiosis can occur and the hybrid becomes fertile. Note that sterile polyploids can propagate via **asexual reproduction**. Polyploids like allotetraploids also have increased heterozygosity (varied sets of a chromosome) and thus adaptive advantages to changes in the environment. The different copies of a gene may each express a different phenotype and result in a new trait such as drought tolerance or pathogen resistance that affects growth rate and performance. In addition, some polyploids can interbreed and become a new species.

AP BIOLOGY LAB 11: TRANSPIRATION INVESTIGATION

7.6 Investigate water movement through plants

This investigation is another example of a controlled experiment to evaluate the effects of environmental variables on the rate of transpiration for a plant. It is easy to complete and requires simple equipment. The objectives are threefold: another opportunity to explore the scientific method through controlled experimentation, a chance to make direct observations on the phenomena immediately related to the physical properties of water, and an occasion to observe the environment's effects on an actual organism.

A suitable plant species is obtained and set up with a potometer, a fancy name for a tube that measures the amount of water taken up by a plant during transpiration. The water-filled tube is connected to the bottom of a plant's cut stem so that the water leaving the tube and entering the plant due to transpiration can be measured. The potometer tube takes the place of soil as a source of water. The plant is then subjected to a number of treatment effects that might include increased light, darkness, heat, or air movement. Changes in transpiration, if any, are compared to the rate of transpiration of a control plant that is under stable conditions, usually room temperature and room lighting. The control plant serves as the benchmark with which all other treatment effects are compared.

Transpiration is affected by environmental conditions in a way very similar to human skin. Increasing the amount of heat around skin causes increased sweat and air movement, which increases evaporation on the skin surface; the result is increased water loss. The same is true for the leaf surface of a plant—increased heat causes an increase in water loss. Light or darkness in the absence of changes in temperature only affects a leaf by modifying the rate of photosynthesis. Increasing the light intensity should increase the rate of photosynthesis and the opening of stomata, thereby increasing transpiration and water loss. As light levels decrease in intensity, water loss decreases. Water loss at the leaf surface causes greater uptake of water through the potometer, which can be measured as movement of the meniscus through the tube.

RAPID REVIEW

If you take away only 4 things from this chapter:

1. Plants with vascular tissues usually have three types of structures: leaves, roots, and branches.
2. Plants have specialized structures to deal with water and nutrients. These include stomata controlled by guard cells, a loosely packed spongy layer, the palisade layer, xylem, and phloem.
3. Plants produce energy through photosynthesis and lose water via transpiration. As water evaporates from the leaves, it pulls water up through channels in the xylem. The phloem carries nutrients throughout the plant.
4. Plants can reproduce asexually via vegetative propagation. Sexual reproduction in plants takes place in the flower.

Structures

TEST WHAT YOU LEARNED

1. A researcher uses a potometer to measure transpiration in plants. How is it possible to measure transpiration using a tube attached to the bottom of the plant?
   (A) Transpiration involves the loss of water through the leaves. This water must be replaced and the potometer measures the uptake of replacement water.
   (B) Transpiration involves the uptake of water through the roots. This can be measured directly using a potometer.
   (C) Transpiration involves the loss of water through the leaves. The tube must surround one or more leaves in order to properly measure transpiration.
   (D) Transpiration involves the uptake of water through the roots. This water is actively pushed out through the leaves and a potometer measures this pressure.

2. A researcher wants to understand what factors cause a particular species of plant to grow against buildings. He hypothesizes that this behavior is due to phototropism. Which of the following experimental designs would most effectively test this hypothesis?
   (A) 40 seeds are evenly divided into two groups. Half of the seeds are grown in pots placed against a wall, allowing the plants to grow in close proximity to the wall. The other half of the seeds are grown far from any walls. After two weeks, the average angles of plant growth are compared among the groups.
   (B) 200 seeds are evenly divided into two groups. Half of the seeds are grown with lighting directed from one side only. The other half of the seeds are grown with diffuse lighting from all directions. After two weeks, the average angles of plant growth are compared among the groups.
   (C) 10 seeds are evenly divided into two groups. Half of the seeds are grown with lighting above the pot. The other half of the seeds are grown with lighting below the pot. After two weeks, the average angles of plant growth are compared among the groups.
   (D) 400 seeds are evenly divided into two groups. Half of the seeds are grown in green light and half of the seeds are grown in red light. After two weeks, the average angles of plant growth are compared among the groups.

3. Mycorrhizae are fungi that grow in a symbiotic association with plant roots. Researchers were interested in knowing whether a particular species of fungus, *Tuber melanosporum*, obtained carbon from the soil or from its host plant (the hazel tree). They added ^{13}C to carbon dioxide in the air surrounding a hazel tree, which was enclosed in a cylinder. Which of the following experimental results most strongly supports the hypothesis that *T. melanosporum* obtains carbon primarily from its host rather than from the soil?
   (A) ^{13}C was found in high abundance in the soil surrounding the plant and in *T. melanosporum* cells.
   (B) ^{13}C was found in the leaves of the hazel tree, but not at all in the surrounding soil.
   (C) ^{13}C was found in the leaves of the hazel tree and in the surrounding soil.
   (D) ^{13}C was found in the leaves of the hazel tree and in *T. melanosporum* cells in much greater abundance than in the surrounding soil.

4. To understand the spread of genes associated with herbicide resistance, researchers planted a herbicide-resistant strain of *Lolium rigidum* (annual ryegrass) plant surrounded by herbicide-susceptible plants to examine the spread of resistance genes by pollen. The table below shows the percent of pollen-mediated gene flow at varying distances (0, 1, 5, 10, and 15 meters) from the resistant plant.

	Pollen-Mediated Gene Flow (%)			
Distance (m)	**A**	**B**	**C**	**D**
		2009		
0	2.3	5.9	6.7	5.5
1	3.6	2.5	3.7	2.8
5	3.2	1.8	0.9	3.0
10	1.1	—	—	—
15	0.1	—	—	1.5
		2010		
0	7.1	11.2	7.1	9.0
1	4.9	7.0	6.1	8.4
5	0.9	3.1	4.2	3.6
10	0.7	1.1	1.7	2.5
15	0.5	0.4	0.6	0.6

Adapted from Iñigo Loureiro, María-Concepción Escorial, and María-Cristina Chueca, "Pollen-Mediated Movement of Herbicide Resistance Genes in *Lolium rigidum*," *PLoS ONE* 11, no. 6 (June 2016): e0157892.

Based on these findings, is it reasonable to expect that herbicide resistance may spread from resistant plants?

(A) No, because the percentage of pollen-mediated gene flow is too low to have a significant effect on future generations. This suggests that the amount of resistance will not increase in future years.

(B) No, because it does not matter what genes are in the pollen as long as the resistance genes are not in the seeds. The table would need to show data on resistance in seeds in order to draw conclusions about whether future generations will show resistance.

(C) Yes, because the trials consistently showed pollen-mediated gene flow within 5 meters of the resistant plant and less consistent gene flow within 15 meters. This suggests that more plants will be resistant in the next generation.

(D) Yes, because the trials consistently showed pollen-mediated gene flow within 5 meters of the resistant plant. This suggests that pollen is spreading to plants immediately adjacent to the resistant plant, although it is unlikely to be a problem over longer distances.

5. A farmer growing grape vines is testing out two subspecies of grapes in his vineyard. Some regions of the vineyard have unfertilized soils while other regions of the vineyard are artificially fertilized. In addition, he artificially induces some plants to flower while allowing others to flower naturally. Which of the following observations would be the most likely outcome of this experiment?

(A) Fruit weight is lower for both subspecies when unfertilized because the plants have less energy to invest in fruit production.

(B) Fruit weight is higher for both subspecies when unfertilized because the plants invest more energy in reproduction as they reduce expenditure on their own growth.

(C) The naturally flowering plants in both subspecies have larger fruit because the artificially flowering plants would experience too much stress.

(D) The artificially flowering plants in both subspecies have larger fruit because the naturally flowering plants produce smaller fruit to conserve energy.

6. The diagram below illustrates the movement of water (indicated by the path of the arrows) through vascular tissue in angiosperms and gymnosperms.

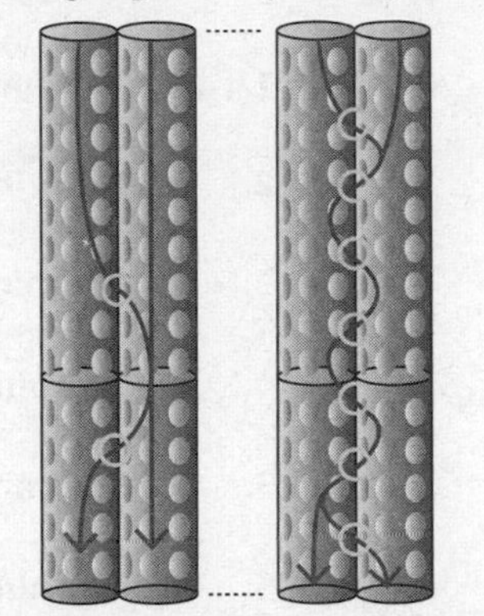

Adapted from Michael S. H. Boutilier et al., "Water Filtration Using Plant Xylem," *PLoS ONE* 9, no. 2 (February 2014): e89934.

Which of these structures would most effectively filter water moving through the plant?

(A) Angiosperm vascular tissue would be more effective because water can travel rapidly through these tissues, making it more efficient to use.

(B) Gymnosperm vascular tissue would be more effective because the water frequently moves from one cell to another through the vessel walls, which act as a filter.

(C) Angiosperm and gymnosperm vascular tissue would be equally effective in filtration because water travels over the same distance in both cases.

(D) It is impossible to determine which would make the better filter without knowing whether the plants are monocots or dicots.

Answer Key

Test What You Already Know

1. **A** **Learning Objective:** 7.1
2. **B** **Learning Objective:** 7.2
3. **B** **Learning Objective:** 7.3
4. **C** **Learning Objective:** 7.4
5. **D** **Learning Objective:** 7.5
6. **C** **Learning Objective:** 7.6

Test What You Learned

1. **A** **Learning Objective:** 7.6
2. **B** **Learning Objective:** 7.3
3. **D** **Learning Objective:** 7.1
4. **C** **Learning Objective:** 7.5
5. **A** **Learning Objective:** 7.4
6. **B** **Learning Objective:** 7.2

Detailed solutions can be found in the Answers and Explanations section at the back of this book.

REFLECTION

Test What You Already Know score: __________

Test What You Learned score: __________

Use this section to evaluate your progress. After working through the pre-quiz, check off the boxes in the "Pre" column to indicate which Learning Objectives you feel confident about. Then, after completing the chapter, including the post-quiz, do the same to the boxes in the "Post" column. Keep working on unchecked Objectives until you're confident about them all!

Pre	Post	
☐	☐	**7.1** Describe structures that bring in nutrients
☐	☐	**7.2** Describe the vascular system in plants
☐	☐	**7.3** Explain how exposure to light affects plants
☐	☐	**7.4** Recall methods of plant reproduction
☐	☐	**7.5** Describe the effect of mutations across generations
☐	☐	**7.6** Investigate water movement through plants

FOR MORE PRACTICE

Complete more practice online at kaptest.com. Haven't registered your book yet? Go to kaptest.com/booksonline to begin.

CHAPTER 8

Immune Systems

LEARNING OBJECTIVES

In this chapter, you will review how to:

8.1 Describe nonspecific immune responses

8.2 Explain why nonspecific immunity exists

8.3 Differentiate between cell-mediated and humoral responses

8.4 Compare the function of cytotoxic T cells and B cells

8.5 Explain antigen and antibody specificity

8.6 Describe the immune response to repeat exposure

TEST WHAT YOU ALREADY KNOW

1. Bacteria that live on and within the human body sometimes produce chemicals called bacteriocins that inhibit the growth of other microbes. There has been interest in the use of these microbes as probiotics. If a researcher wanted to test whether a particular bacteriocin contributed to an organism's nonspecific immune response, which of the following would be the best approach?

   (A) The researchers could measure antibody titers to determine whether an individual had developed antibodies against the bacteriocin.

   (B) The researchers could grow bacteria that produced the bacteriocin on agar plates and expose them to different antibiotics to determine which was most effective in killing them.

   (C) The researchers could determine the sequence of the gene for the bacteriocin to compare its sequence to other known bacterial proteins.

   (D) The researchers could take bacteria that produced the bacteriocin from an organism's gut and culture them with other bacteria to observe if they inhibited the growth of the other species.

2. Opsonization is a process in which microbes are coated with chemicals that make them better targets for phagocytosis by immune cells. Although these chemicals can be antibodies, they can also be host proteins such as C3b. Researchers wanted to understand how some *Staphylococcus aureus* appear to evade the immune system. The figure below shows the amount of C3b bound to red blood cells both as an isolated protein and when it is bound to a Staphylococcal protein, Ecb.

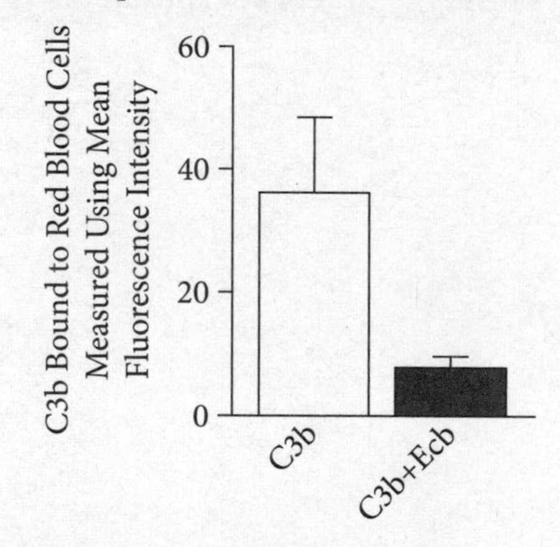

Adapted from Hanne Amdahl et al., "Staphylococcal Protein Ecb Impairs Complement Receptor-1 Mediated Recognition of Opsonized Bacteria," *PLoS ONE* 12, no. 3 (March 2017): e0172675.

   Based on the results, what is the likely role of Ecb?

   (A) Ecb helps *S. aureus* evade the immune system and establish an infection because it interferes with the ability of C3b to bind to cells.

   (B) Ecb helps *S. aureus* evade the immune system to establish an infection because it increases the activity of C3b.

   (C) Ecb helps prevent *S. aureus* infections because it increases the likelihood of *S. aureus* opsonization.

   (D) Ecb helps prevent *S. aureus* infection because it interferes with the ability of C3b to bind to red blood cells.

3. The figure below shows the levels of Anti-Arp antibody in a study of the mouse immune response to the pathogen *Borrelia burgdorferi* (Bb). The *x*-axis shows the number of weeks since initial infection and the *y*-axis shows the levels of antibody. The solid line is the control condition (untreated) and the dashed line represents mice treated with antibiotics beginning at the time indicated with an arrow.

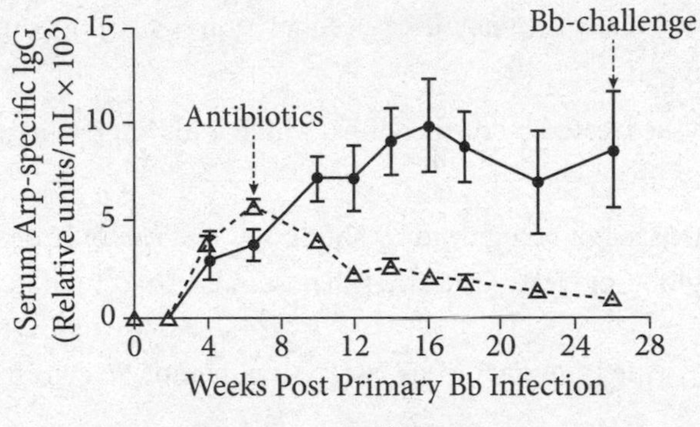

Anti-Arp

Adapted from Rebecca A. Elsner et al., "Suppression of Long-Lived Humoral Immunity Following *Borrelia burgdorferi* Infection," *PLoS Pathog* 11, no. 7 (July 2015): e1004976.

Which of the following best describes these results?

(A) The cell-mediated immune response was more substantial in the control group than in the antibiotic-treated group at day 20, suggesting that the immune response did not persist once treatment began.

(B) The cell-mediated immune response was more substantial in the group treated with antibiotics than in the control group at day 20, suggesting that the immune response was enhanced by antibiotic treatment.

(C) The humoral immune response was more substantial in the control group than in the antibiotic-treated group at day 20, suggesting that the immune response was enhanced by exposure to the pathogen.

(D) The humoral immune response was more substantial in the group treated with antibiotics than in the control group at day 20, suggesting that the immune response was enhanced by antibiotic treatment.

4. B cells in the rabbit intestine can be stimulated by exposure to *Bacillus* spore surface molecules. What effect is this most likely to have?

(A) B cells increase in number in the gut, so there is an abundance of phagocytic cells.

(B) B cells increase in number in the gut, resulting in an increased ability to produce antibodies.

(C) B cells mature more rapidly in the gut, meaning that more cells are available to target and directly kill cancer cells.

(D) B cells mature more rapidly in the gut after developing in the thymus.

5. There has been considerable interest in trying to develop a universal influenza vaccine to replace the annual vaccine currently in use. Unfortunately, this has been difficult to accomplish. Why has it been so challenging to develop a universal vaccine?
   (A) The influenza virus changes rapidly, undergoing antigenic drift and antigenic shift. Existing antibodies do not recognize the new strains.
   (B) Influenza is an unusually virulent disease, so it requires a stronger immune response than most other pathogens.
   (C) Influenza cannot be treated by medications, which is an important prerequisite in vaccine development.
   (D) The influenza virus is not recognized by antibodies and can only be treated with shorter-acting immunoglobulin treatment rather than standard vaccination.

6. The figure below shows the frequency of IgG-secreting memory B cells in a study of immune response to malaria.

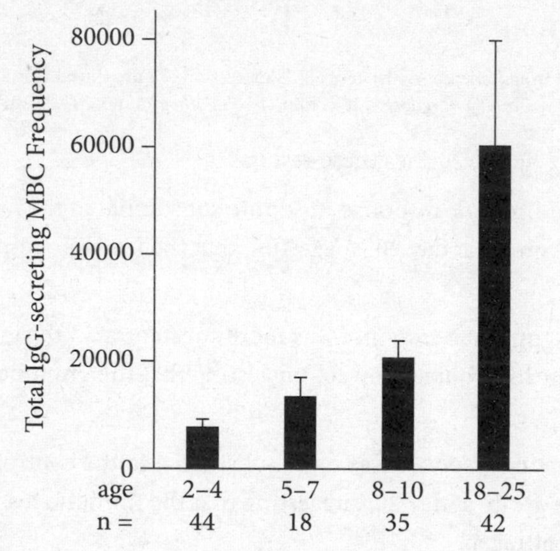

Adapted from Greta E. Weiss et al., "The *Plasmodium falciparum*-Specific Human Memory B Cell Compartment Expands Gradually with Repeated Malaria Infections," *PLoS Pathog* 6, no. 5 (May 2010): e1000912.

What is the best explanation for these results?
   (A) Younger individuals acquire substantial amounts of immunity from their mothers, but this decreases with age.
   (B) Older individuals have a greater likelihood of developing malaria in any particular year and therefore have a higher probability of an active infection when tested.
   (C) Individuals have an increased chance of being exposed to malaria as they age. Because memory B cells persist after exposure, older individuals are more likely to have them.
   (D) The total number of memory B cells increases with age, irrespective of exposure to specific pathogens.

Answers to this quiz can be found at the end of this chapter.

NONSPECIFIC DEFENSES

8.1 Describe nonspecific immune responses

8.2 Explain why nonspecific immunity exists

The task of fighting infections is quite demanding: The human immune system must be able to respond to disease-causing organisms that are as small as a few nanometers in diameter (e.g., the polio virus) and as large as 10 meters long (e.g., a tapeworm). Because the major cells of the immune system are all approximately 10–30 micrometers in diameter, the cells must be able to fight off a range of organisms from 1,000 times smaller than they are to a million times larger! Because of this, the immune system is made up of a network of cells and organs that perform a variety of simultaneous tasks.

The immune system can be divided into two major divisions: nonspecific and specific. The nonspecific immune system is composed of defenses that are used to fight off infection in general and are not targeted to specific pathogens. The specific immune system is able to attack very specific disease-causing organisms by protein-to-protein interaction and is responsible for our ability to become immune to future infections from pathogens we have already fought off.

Structures of Nonspecific Defense

The skin and mucous membranes form one part of the nonspecific defenses that our body uses against foreign cells or viruses. Intact skin cannot normally be penetrated by bacteria or viruses. Additionally, oil and sweat secretions give the skin a pH that ranges from 4.5 to 6, which is acidic enough to discourage microbes that would proliferate at the normal pH in the body (7.35–7.45) from being there. In addition, saliva, tears, and mucus all contain the enzyme lysozyme, which can destroy bacterial cell walls (causing bacteria to rupture due to osmotic pressure) and some viral capsids. Mucus is able to trap foreign particles and microbes and transport them to the stomach (through swallowing) or to the outside (by coughing or blowing the nose).

Certain white blood cells are also part of the nonspecific defenses. Macrophages are large white blood cells that circulate, looking for foreign material or cells to engulf, which they do through phagocytosis. Macrophages circulate through the blood and are able to transport themselves through capillary walls and into tissues that have been infected or wounded. Once in the tissues, macrophages use their pseudopodia (like amoebas) to pull in foreign particles and destroy them within lysosomes.

Macrophages are called antigen-presenting cells (APCs) because of their ability to "display" on their own cell surface the proteins that were on the surface of the cell or viral particle they have just digested. Because macrophages and other APCs do not distinguish between "self-proteins" destined for their cell membrane and "non-self" proteins previously on another organism's membrane, both types of proteins get shipped to the macrophage's cell surface. The advantage of this is that macrophages are able to display to other, more specific immune system cells the **antigens** (foreign proteins) they have just encountered. That, in turn, often results in a more intensive immune response from these more specific cells (see the discussions of B and T cells later in this chapter).

Neutrophils are white blood cells that are actively phagocytic like macrophages, but are not APCs. Our bodies normally produce approximately 1 million neutrophils per second, and they can be found anywhere in the body. They usually destroy themselves as they fight off pathogens. People who have decreased numbers of neutrophils circulating through their blood are extremely susceptible to bacterial and fungal infections. Other white blood cells that secrete toxic substances without fine-tuned specificity include eosinophils, basophils, and mast cells.

Inflammatory Response

Both basophils and mast cells release large quantities of a molecule called histamine, responsible for dilating the walls of capillaries nearby and making those capillaries "leaky." For this reason, histamine is considered a potent vasodilator, which can lower blood pressure across the whole body if enough is released at once (blood pressure is maintained by the diameter and integrity of the capillary walls).

Leaky capillary walls allow macrophages and neutrophils to more easily reach the site of an injury. This nonspecific defense increases overall blood flow to areas of tissue injury and is responsible for the characteristic redness and heat felt in areas of injury. Basophils and mast cells have large secretory vesicles filled with histamine molecules, but they also release cytokines, chemicals that cause specific immune defenses to activate. Other responses to injury that are more systemic (body-wide), rather than local (at the injury site), include fever and increased production of all types of white blood cells. Fever is thought to both decrease the replication of infectious microbes and enhance immune system functioning.

✔ **AP Expert Note**

Antihistamines

Antihistamines are compounds that can limit immunologic reactions. They bind to histamine receptors and shut down their ability to cause leakage at the site of injury. Antihistamines can be lifesaving for certain reactions, particularly allergic ones in which the immune system overreacts to a seemingly harmless substance and leaky capillaries in the lungs and face make breathing difficult or impossible.

SPECIFIC DEFENSES

8.3 Differentiate between cell-mediated and humoral responses

8.4 Compare the function of cytotoxic T cells and B cells

8.5 Explain antigen and antibody specificity

8.6 Describe the immune response to repeat exposure

Lymphocytes

The major specific defense of the immune system is the use of specialized white blood cells known as **lymphocytes**. These lymphocytes come in two varieties, B cells and T cells. Both are produced by stem cells in the bone marrow after embryologic development has finished. Although T cells mature in the thymus, B cells do not.

The thymus is essential for "educating" T cells through the processes of positive and negative selection. In positive selection, only T cells that are able to react with a specific set of glycoproteins, called MHC (major histocompatibility complex) proteins, are kept while those that do not are killed off. This is followed by negative selection during which T cells that react too strongly to "self" antigens (proteins found on one's own cell surfaces) are killed off so that autoimmune reactions are less likely to occur. The remaining T cells are highly capable of bonding to an MHC molecule and a foreign antigen simultaneously while only reacting to the latter, which is essential for T cells to work properly.

There are three types of T cells: helper (T_H), cytotoxic (T_C), and suppressor (T_S). While T_H cells are mediators between macrophages and B cells, T_C cells are able to kill virally infected cells directly. Because virally infected cells display some viral proteins on their cell surfaces, T_C cells can bind to self-MHC proteins and viral proteins on the cell surfaces and secrete enzymes that perforate the cell membrane and kill the cell. Cytotoxic T cells are an essential part of the body's defenses against viruses. T_S cells are involved in controlling the immune response so that it does not run out of control. They accomplish this by suppressing the production of antibodies by B cells. It seems likely that these T_S cells are not a separate class of T cells altogether, but rather certain T_H cells that secrete inhibitory chemical messengers (cytokines).

✔ **AP Expert Note**

Helper T Cells

Recall the antigen-presenting macrophages that dot their surfaces with foreign proteins they have digested. A certain class of T cells, known as helper T cells, is able to bind simultaneously to self-MHC proteins on the macrophage surface and the displayed foreign proteins. This combination of signals is needed to activate the helper T cells.

Keep in mind that T cells cannot detect free antigens; they can only respond to displayed antigens and MHC on the surfaces of cells. When they do recognize a displayed antigen, it is always in combination with a self-MHC protein displayed along with the antigen on the host cell surface. Interactions between T cells and APCs are enhanced by certain proteins that hold the T cell to the APC as it recognizes the antigen/MHC combination. One of these "holding" proteins is the CD4 protein.

B cells make up about 30 percent of one's lymphocytes, and the average B cell lives for only days or weeks. About 1 billion are made each day in the bone marrow. Every B cell has surface receptors that are identical in structure to a certain class of **antibody** protein that the B cell is able to produce and secrete into the bloodstream. Essentially, the receptors on B cell surfaces are bound Y-shaped antibody proteins that can recognize a specific set of foreign antigens (proteins found on the surfaces of foreign cells and viruses). B cells can be "activated" in one of two ways: either they can come into contact with a foreign antigen that can bind to the B cell surface receptors, or they can engulf a pathogen, displaying its antigens on the B cell surface much as a macrophage would. Then they can get stimulated to divide by chemicals released by a helper T cell (T_H) that recognizes the foreign proteins sitting on the B cell surface.

Both B and T cells each have unique cell receptors. That means that *almost every one of the several billion B and T cells in the body is capable of responding to a slightly different foreign antigen.* When a particular B or T cell gets activated, it begins to divide rapidly to produce identical clones. In the case of B cells, these clones will all produce antibodies of the same structure, capable of responding to the same invading antigens. B cell clones are known as plasma B cells and can produce thousands of antibody molecules per second as long as they live.

Humoral versus Cell-Mediated Defenses

Within the immune system, the specific arm, which uses B and T cells to target individual microbes, can work either by secreting antibodies into the bloodstream (considered one of the body's "humors") or by directly killing cells. While antibody secretion (humoral immunity) is the job of the B cells, direct killing of cells and overall activation of the immune system (cell-mediated immunity) is the job of T cells.

Humoral Response

When the body is exposed to an antigenic stimulus for the first time, many events must take place before a protective antibody response is produced. The period after exposure to a pathogen, but before helpful levels of antibodies have been made by B cells, is called the lag period. During this period, APCs such as macrophages and neutrophils process and display antigens to T_H cells, which rapidly grow and divide. These T cells activate B cells, some of which have also contacted antigens, and these B cells grow and divide as well.

After the first exposure to a microbe, the lag period lasts seven to ten days before enough antibodies are present in the blood to noticeably slow the infection. Within a week or so, the infection will have likely subsided. If the person is exposed to the same antigen a second time, there is a very quick secondary response (with a lag period of one to four days), and levels of antibody in the bloodstream typically reach much higher levels than they did in the primary response.

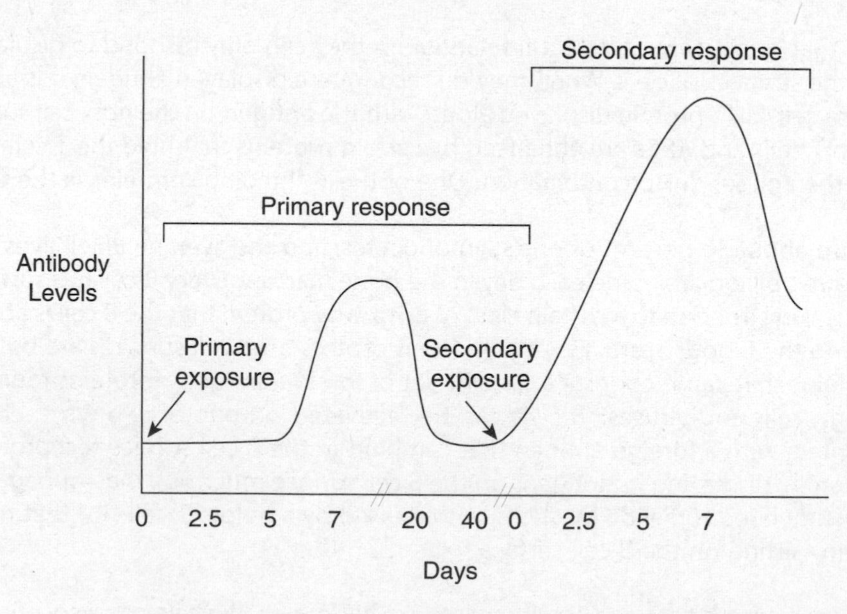

Antibody Levels during Exposure and Response

This is because certain antigen-specific memory cells remain after a primary infection. Both B and T cells form memory cells. While a typical B or T lymphocyte may live for only days or weeks, memory B and T cells can live for decades. The reasons for this are unclear, but the formation of memory cells is the basis of **immunity** and forms the concept behind vaccination. Small doses of antigen given as an injection or swallowed allow the body to recognize and form a primary response against the antigen so that upon actual exposure to the pathogen carrying that antigen, the body will mount a quick and sufficient response.

Because the lag period can be unacceptably long for some pathogens, doctors often provide preformed antibodies (usually made in horses or chickens) to people who have been exposed to a microbe or who face probable exposure. This passive immunity, to be distinguished from the **active immunity** that forms when B cells make their own antibodies, is temporary. Individuals who travel abroad are often given a shot of gamma globulin before leaving the country. This mixture of preformed antibodies to several tropical diseases is like a soup of temporary protection and will disintegrate within a week or two.

Cell-Mediated Response

The timing of the T cell response follows almost the same pattern as shown in the graph of the humoral response. After primary exposure to an antigen, specific T cells rapidly divide and form clones of identical cells. Upon secondary exposure, memory T cells left over from these clones will divide much more rapidly than they did at the first exposure. The cell-mediated response severely limits the ability of viruses to proliferate within cells, because cytotoxic (killer) T cells will destroy any cells harboring viruses. Helper T cells participate in antigen recognition and B cell activation, and natural killer cells (related to cytotoxic T cells) destroy infected and cancerous cells directly.

Altogether, nonspecific defenses such as skin, fever, and macrophages along with specific defenses such as B and T lymphocytes and the complement system make up a highly efficient and adaptable machine that serves to protect the body from a wide assortment of invaders.

Antibodies

HIGH YIELD

Antibodies are also known as immunoglobulins and are *produced solely by B cells*. Each immature B cell has specific antibodies stuck to its surface so that the B cell can be activated if it comes into contact with an antigen that its antibodies are specific for. A single B cell can produce billions of antibodies in its short lifetime.

Antibody Structure and Class

There are five types or classes of antibody: IgM, IgG, IgD, IgE, and IgA (Ig stands for immunoglobulin). Each class serves a different purpose in the body, and the same B cell can produce each type at different times during an infection. B cells always produce IgM class antibodies first, then *class-switch* to another type, usually IgG, depending upon where the B cell is in the body and the type of antigen it is responding to.

The structure of all five classes is based on the same four polypeptide chains. Two of the chains are called heavy chains and two are light chains. Both light chains are identical to each other in amino acid sequence, as are the two heavy chains:

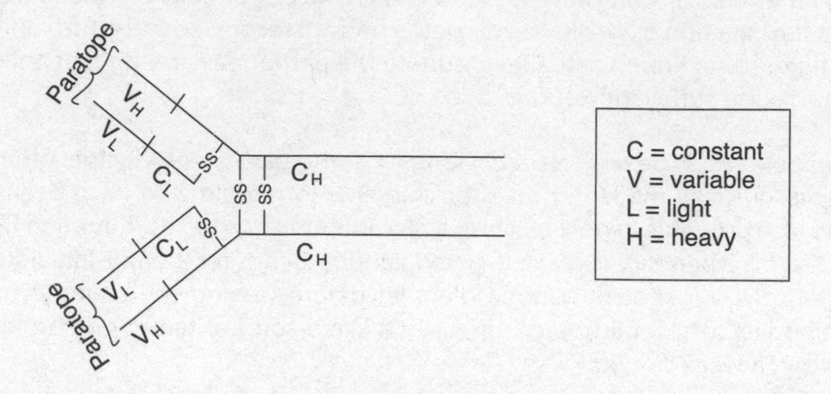

Basic Antibody Structure

Notice in the above diagram that both the heavy chains and the light chains have constant regions and variable regions. The variable regions are responsible for the specificity of a particular antibody for a particular antigen. The antigen-binding site, where antibodies can bind to foreign proteins, is also known as the paratope. The paratopes are each formed from a combination of variable amino acids in one heavy chain and one light chain. It should be clear from the diagram that each Y-shaped antibody has *two antigen-binding sites.*

Keep in mind that, while the B cells may have individual antibodies bound to their surface for use as receptors, the vast majority of antibodies that are produced are sent out to float freely through the bloodstream. IgG is the simplest type of antibody, with a structure approximated in the previous diagram. IgA, however, is a dimer made of two Y-shaped antibodies placed back to back, and IgM is a pentamer of five Y-shaped antibodies placed outward-facing in a circle:

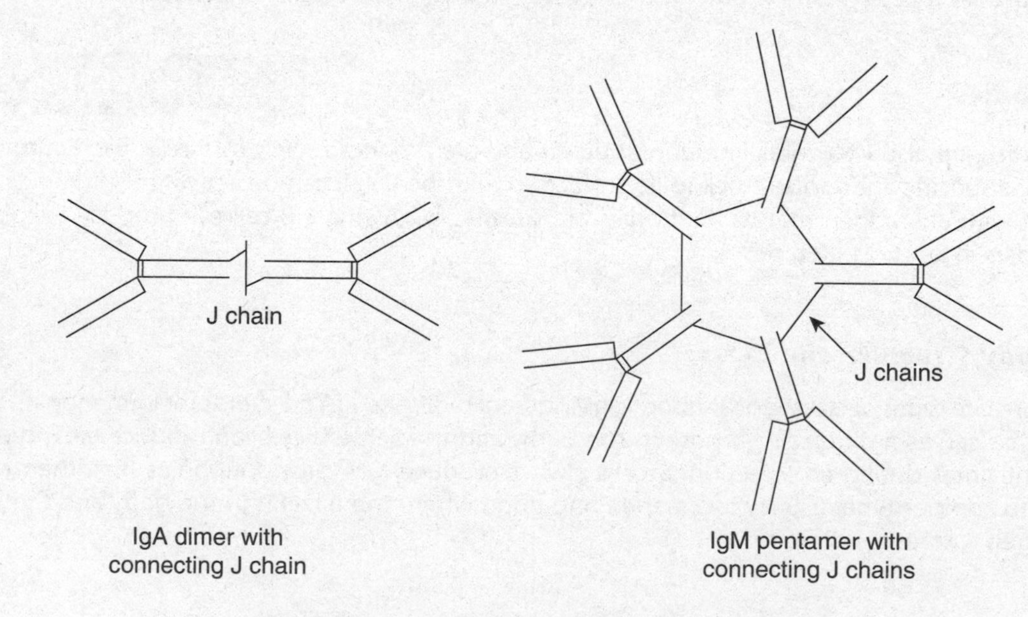

Antibody Structure with Antigen Attachment Sites

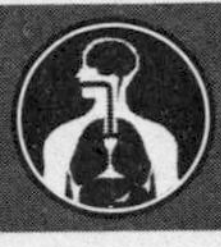

Each IgA antibody can attach to 4 antigens, while each IgM can attach to 10 antigens. The ability of the antibodies to attach to antigens is central to their function.

How Antibodies Work

Agglutination/neutralization occurs when antibodies cross-link adjacent antigen molecules (on bacteria and other organisms) so that these invaders literally get stuck together by the antibodies circulating in the bloodstream. Because each Y-shaped antibody can stick to two different organisms, antibodies can cause the clumping together of many pathogens in a short time. These agglutinated bunches of bacteria or virus particles form large, insoluble masses that are no longer able to invade cells and can be easily engulfed by circulating macrophages.

Precipitation is similar to agglutination, but is used for soluble antigen molecules such as small bacterial toxins, which dissolve in the bloodstream. Antibody binding allows rapid phagocytosis and destruction of small proteins by macrophages.

Complement activation occurs when antibodies bound to the surfaces of foreign cells activate a system of 20 different complement proteins that circulate in the bloodstream. These proteins are turned on in a cascade-like fashion with each one activating the next, allowing for a great deal of control over the process.

- *Classical pathway*: This requires antibodies bound to antigens. Complement proteins bridge the gap between two adjacent antibody molecules and use a protein complex called the membrane-attack complex to lyse the cell membrane of the invader. Complement proteins also activate mast cells to release histamine, which brings more blood cells to the area.
- *Alternative pathway*: This occurs independently of antigen/antibody binding. Cell surface molecules of many bacteria, yeasts, viruses, and protozoan parasites can cause membrane-attack complexes to form without the help of antibodies. Cells of our body present complement regulatory proteins that prevent the formation of membrane attack complexes. This nonspecific reaction to foreign cells makes the alternative pathway a form of innate immunity.

Hybridomas and Monoclonal Antibodies

Antibodies that arise in the natural course of fighting many pathogens are considered polyclonal—that is, they are produced by several different clones of plasma B cells and cover a wide range of specificity. Antibodies arising from a single clone, rapidly divided into identical B cells, are called monoclonal, and this single B cell has important scientific uses. The specific nature of antibody binding makes them attractive as research targets for disease cures.

Imagine, for example, being able to target specific cancer cells with an injection of antibodies that seek out and destroy only those cancer cells. Such antibodies can be made in the lab by fusing a myeloma cell (a cancerous, always-dividing B cell) with an antibody-producing cell from a mouse. The resulting cell, called a hybridoma because it is a hybrid cell from two different species, can produce almost unlimited quantities of a particular monoclonal antibody, which can be used in research.

RAPID REVIEW

If you take away only 4 things from this chapter:

1. The immune system can be divided into two major divisions: nonspecific (targets general infections) and specific (attacks specific disease-causing organisms by protein-to-protein interaction). Specific immunity is how we become immune to future infections from pathogens we have already fought off.
2. Specialized white blood cells, lymphocytes, come in two varieties: B cells and T cells. They are produced by stem cells in the bone marrow. There are three types of T cells: helper (T_H), cytotoxic (T_C), and suppressor (T_S).
3. There are five types of antibodies and their structures are based on the same four polypeptide chains.
4. Antibodies work through agglutination/neutralization, precipitation, and complement activation.

TEST WHAT YOU LEARNED

1. Researchers have been working to develop a universal influenza vaccine. In one experiment, recombinant technology was used to add an influenza protein (M2e) to the spore coats of several strains of *Bacillus subtilis*. The figure below shows end-point mouse antibody titers for several different recombinant *B. subtilis* strains.

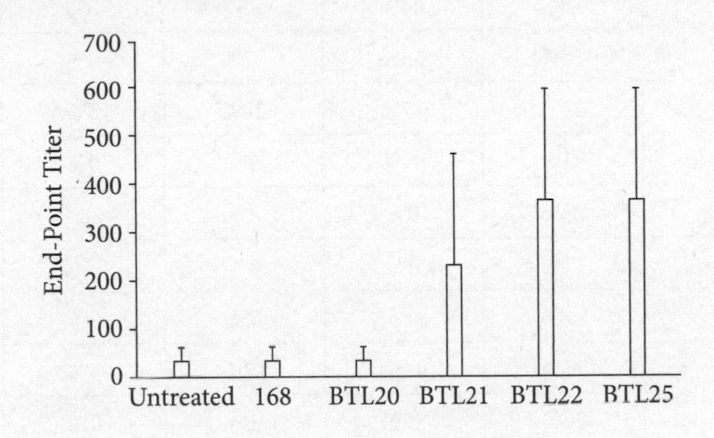

Adapted from Tomasz Łęga et al., "Presenting Influenza A M2e Antigen on Recombinant Spores of *Bacillus subtilis*," *PLoS ONE* 11, no. 11 (November 2016): e0167225.

Which of the following can most reasonably be concluded from the results?

(A) There was minimal variability in the amount of antibody produced in all of the strains tested, making this a consistent method for producing an immune response.

(B) The M2e protein will not likely be an effective universal vaccination because it is only targeted by the immune system when combined with *B. subtilis* endospores.

(C) The M2e protein causes a significant amount of antibody production with or without the addition of *B. subtilis* endospores.

(D) The M2e protein can lead to the production of influenza-specific antibodies when expressed on the spore coats of some strains of *B. subtilis*.

Structures

2. A study investigated the population of activated CD8 T cells in a group of patients with acute HBV infection. The figure below shows the percentage of activated CD8 T cells that were specific for HBV, HCMV, EBV, and FLU viruses respectively.

HBV Acute Patients	Percentage of Activated T Cells				
	Total CD8	**HBV**	**HCMV**	**EBV**	**FLU**
Patient 1	18%	81%	5.5%	0%	0%
Patient 2	18%	43%	13%	0%	0%
Patient 3	25%	77%	20%	0%	0%
Patient 4	21%	81%	0%	0%	0.9%
Patient 5	28%	0%	8%	41%	0%
Patient 6	17%	0%	0%	31%	0%
Patient 7	17%	0%	0%	0%	2%
Patient 8	15%	0%	12%	0%	1.3%
Patient 9	26%	89%	20%	0%	1.4%
Patient 10	12%	0%	9%	26%	0%

Adapted from Elena Sandalova et al., "Contribution of Herpesvirus Specific CD8 T Cells to Anti-Viral T Cell Response in Humans," *PLoS Pathog* 6, no. 8 (August 2010): e1001051.

Which of the following best describes these results?

(A) Even though patients had an acute HBV infection, they often showed elevated humoral immunity toward other viruses.

(B) All patients with HBV had activated T cells specific for HBV.

(C) Even though they exhibited an acute HBV infection, patients often showed elevated cell-mediated immunity toward other viruses.

(D) Patients who have been vaccinated show residual levels of activated T cells to the viruses for which they have received vaccinations.

3. Researchers have observed that certain gut microbiota may increase antibody diversity. Would such an increase in antibody diversity most directly affect the humoral or the cytotoxic immune response?

(A) An increase in antibody diversity would affect the cytotoxic immune response because the body would be able to directly kill a wider array of virally infected cells.

(B) An increase in antibody diversity would affect the humoral immune response because the body would be able to respond to a wider range of invading microbes.

(C) An increase in antibody diversity would affect the cytotoxic immune response because antibodies are secreted by helper T cells.

(D) An increase in antibody diversity would affect both responses equally because both the humoral and cytotoxic systems cannot function without antibodies.

4. Bacteriocins are chemicals produced by bacteria that can inhibit the growth of other bacteria. The plates below show colonies of bacteria taken from mouse fecal samples. The plate labeled "SakA(+)" was made using bacteria that produce the bacteriocin SakA. The plate labeled "SakA(−)" was made using non-SakA-producing bacteria.

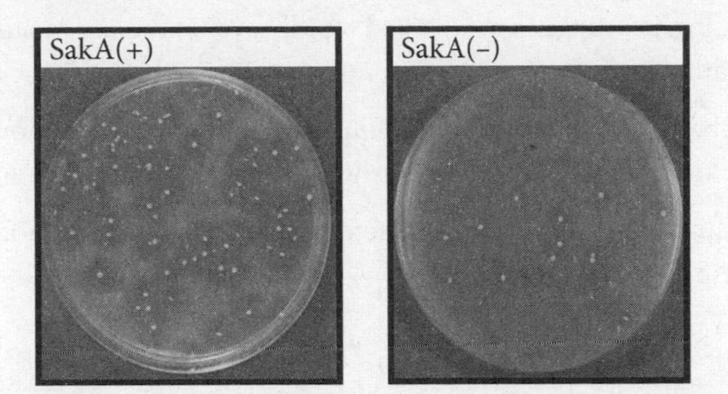

Adapted from Özgün C. O. Umu et al., "The Potential of Class II Bacteriocins to Modify Gut Microbiota to Improve Host Health," *PLoS ONE* 11, no. 10 (October 2016): e0164036.

Which of the following can most reasonably be concluded based on the experimental results?

(A) SakA may contribute to the nonspecific immune systems of mice by helping to kill microbes in the gut, as shown by the clear zones around bacterial colonies in the SakA(+) plate that are not present in the SakA(−) plate.

(B) SakA may contribute to the nonspecific immune systems of mice by helping to kill microbes in the gut, as shown by the clear zones around bacterial colonies in the SakA(−) plate that are not present in the SakA(+) plate.

(C) SakA does not contribute to the nonspecific immune systems of mice because only the SakA(+) plate shows any inhibition of bacterial growth.

(D) SakA does not contribute to the nonspecific immune systems of mice because the SakA(−) plate does not show inhibited bacterial growth around the colonies.

5. Influenza vaccination is recommended annually for people in certain risk categories. If a woman received a vaccination against a particular influenza strain one year and then was exposed to that strain many years later, how would she most likely respond?

(A) Innate immunity would cause her to resist infection without delay despite significant exposure.

(B) She would have a slower than normal immune response due to the time elapsed.

(C) She would not have an effective immune response due to the time elapsed.

(D) Memory cells would allow her to respond rapidly to the virus.

6. The acidic environment of the stomach serves as a barrier to many pathogens that may enter the body. Which of the following best describes the main advantage of this type of barrier?

   (A) This barrier allows pathogens to be killed immediately upon entry to the body, preventing any possibility of infection.

   (B) This barrier allows pathogens to be killed rapidly even if the organism has never encountered the specific pathogen before, preventing unknown infections.

   (C) This barrier produces a memory effect, so pathogens are targeted more rapidly by the stomach acid and mucus if they previously entered the body.

   (D) This barrier allows selection of which bacteria to kill, targeting pathogens and allowing only beneficial bacteria to pass through.

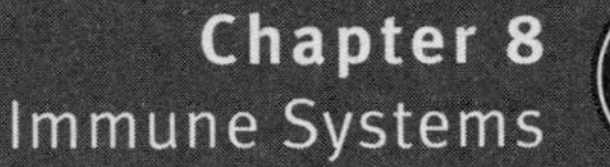

The answer key to this quiz is located on the next page.

Answer Key

Test What You Already Know

1. **D Learning Objective:** 8.1
2. **A Learning Objective:** 8.2
3. **C Learning Objective:** 8.3
4. **B Learning Objective:** 8.4
5. **A Learning Objective:** 8.5
6. **C Learning Objective:** 8.6

Test What You Learned

1. **D Learning Objective:** 8.5
2. **C Learning Objective:** 8.4
3. **B Learning Objective:** 8.3
4. **A Learning Objective:** 8.1
5. **D Learning Objective:** 8.6
6. **B Learning Objective:** 8.2

Detailed solutions can be found in the Answers and Explanations section at the back of this book.

REFLECTION

Test What You Already Know score: __________

Test What You Learned score: __________

Use this section to evaluate your progress. After working through the pre-quiz, check off the boxes in the "Pre" column to indicate which Learning Objectives you feel confident about. Then, after completing the chapter, including the post-quiz, do the same to the boxes in the "Post" column. Keep working on unchecked Objectives until you're confident about them all!

Pre	Post	
❑	❑	**8.1** Describe nonspecific immune responses
❑	❑	**8.2** Explain why nonspecific immunity exists
❑	❑	**8.3** Differentiate between cell-mediated and humoral responses
❑	❑	**8.4** Compare the function of cytotoxic T cells and B cells
❑	❑	**8.5** Explain antigen and antibody specificity
❑	❑	**8.6** Describe the immune response to repeat exposure

FOR MORE PRACTICE

Complete more practice online at kaptest.com. Haven't registered your book yet? Go to kaptest.com/booksonline to begin.

CHAPTER 9

Endocrine Systems

LEARNING OBJECTIVES

In this chapter, you will review how to:

9.1 Differentiate between stimulatory and inhibitory transduction signaling

9.2 Explain how selective pressure impacts pathway choice

9.3 Describe receptor recognition of ligands and protein shape change

9.4 Explain how blocking transduction or reception affects cells

9.5 Contrast contact and distance forms of cell communication

9.6 Explain the endocrine system as long-distance communication

TEST WHAT YOU ALREADY KNOW

1. Physiologists observe that epinephrine can have opposite effects on skeletal and smooth muscles. Epinephrine stimulates contraction in skeletal muscles, whereas it slows down the contraction of the smooth muscles found in bronchial and intestinal walls. Researchers have proposed the following signaling cascade to explain these differential effects.

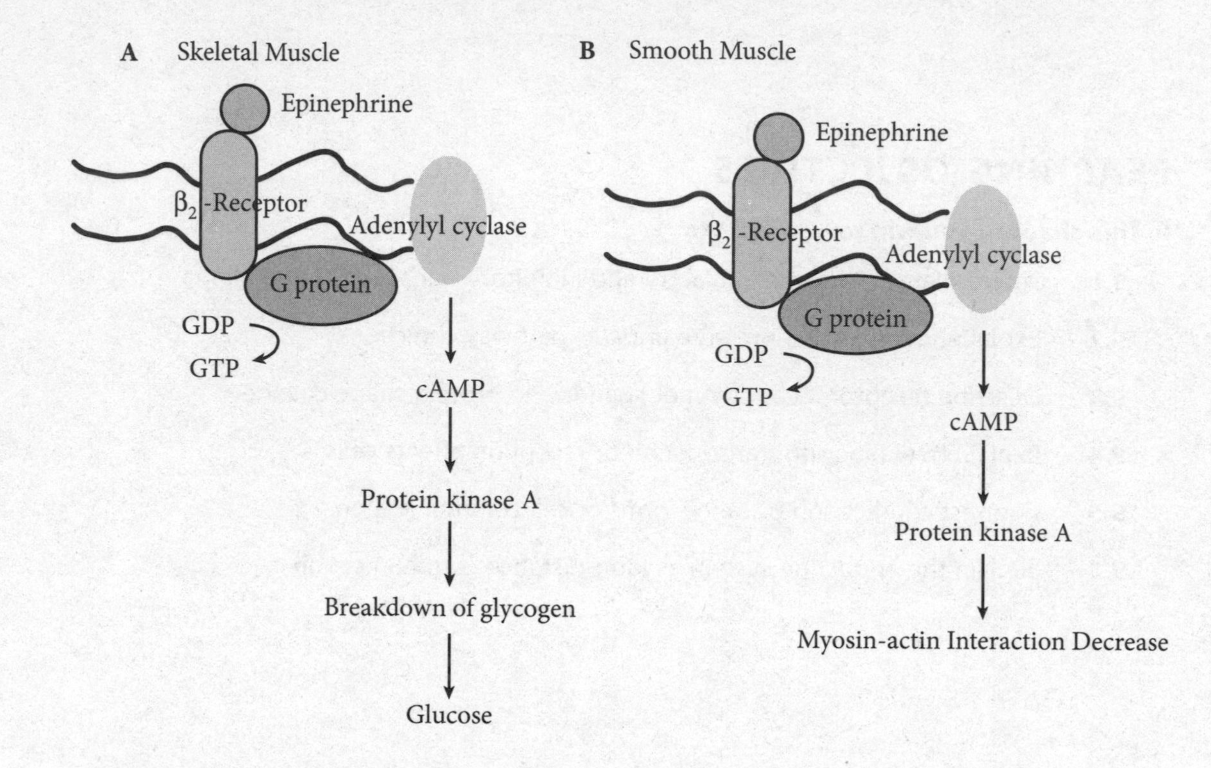

Based on the researchers' proposal, which statement best explains the differential effects of epinephrine on skeletal and smooth muscles?

(A) Epinephrine attaches to different types of receptors in skeletal and smooth muscles.

(B) Different second messengers transduce the signal in skeletal and smooth muscles.

(C) Glycogen metabolism is inhibited in smooth muscle and enhanced in skeletal muscle by epinephrine.

(D) The target of protein kinase A is different in skeletal and smooth muscles.

2. Infection by the bacterium *Clostridium difficile* (*C. diff*) results in extensive tissue damage caused by the secretion of a toxin. Microbiologists investigate secretion of the toxin by measuring its accumulation in the nutrient broth in which the bacteria grow.

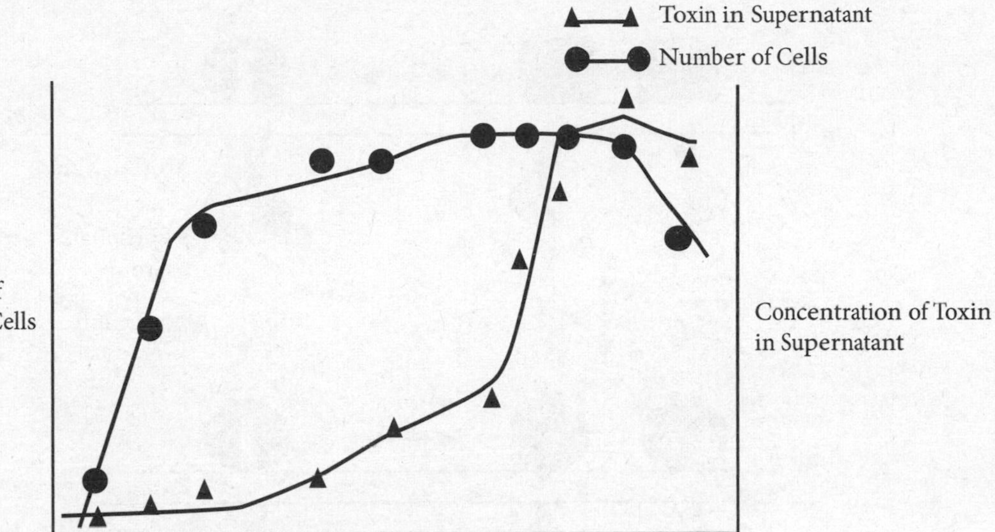

Adapted from Charles Darkoh et al., "Toxin Synthesis by *Clostridium difficile* Is Regulated through Quorum Signaling," *mBio* 6, no. 2 (February 2015): e02569–14.

They hypothesize that bacteria secrete the toxin only when the population reaches a certain cell density and that they coordinate expression of the protein through quorum sensing mediated by a diffusible inducer molecule.

Which of the following experiments would most effectively test their hypothesis?

(A) Resuspend cells in the late stationary phase of growth in the supernatant conditioned by cells in early stages of growth.

(B) Resuspend cells from the early stages of growth in the supernatant conditioned by cells in the late phase of growth.

(C) Resuspend cells in the late phase of growth in fresh medium.

(D) Resuspend cells from the early stages of growth in fresh medium.

3. Receptor tyrosine kinases (RTKs) are found mostly as monomers but also as dimers in the membrane. Signal transduction inside the cell is activated when the intracellular tyrosine kinase domain of the receptor is modified with four phosphate groups. Researchers studied ligand-receptor interactions and summarized their data in the diagram.

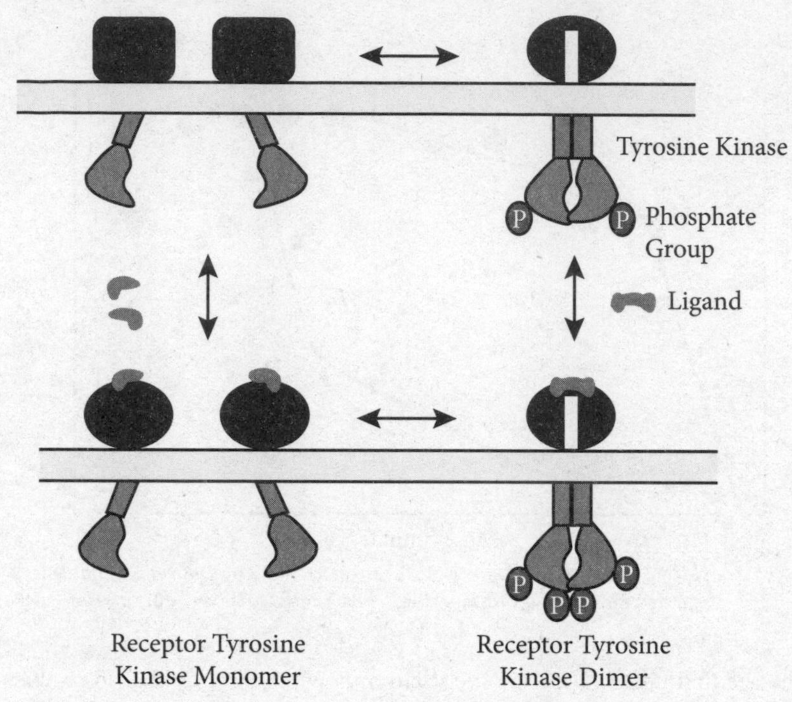

Adapted from Lijuan He and Kalina Hristova, "Physical-Chemical Principles Underlying RTK Activation, and Their Implications for Human Disease," *Biochim Biophys Acta* 1818, no. 4 (April 2012): 995–1005.

Based on the diagram, which of the following statements best describes the activation of an RTK?

(A) Ligands can bind either to monomers or dimers and, in each case, fully activate tyrosine kinase phosphorylation.

(B) Ligands can bind to monomers or dimers, but only the ligand-dimer complex can be phosphorylated.

(C) Only dimers bound to the ligand undergo extensive phosphorylation, initiating the signaling cascade.

(D) Monomers can form dimers without the ligand bound, can become phosphorylated, and can start the signaling cascade in the absence of a signal.

4. The cholera toxin secreted by the bacterium *Vibrio cholerae* enters the cells lining the intestinal tract and locks a G-protein in its GTP-bound form, which stimulates the continuous production of cAMP by adenylyl cyclase. In turn, high cAMP levels activate the cystic fibrosis transmembrane conductance regulator (CFTR), which pumps Cl^- out of the cell.

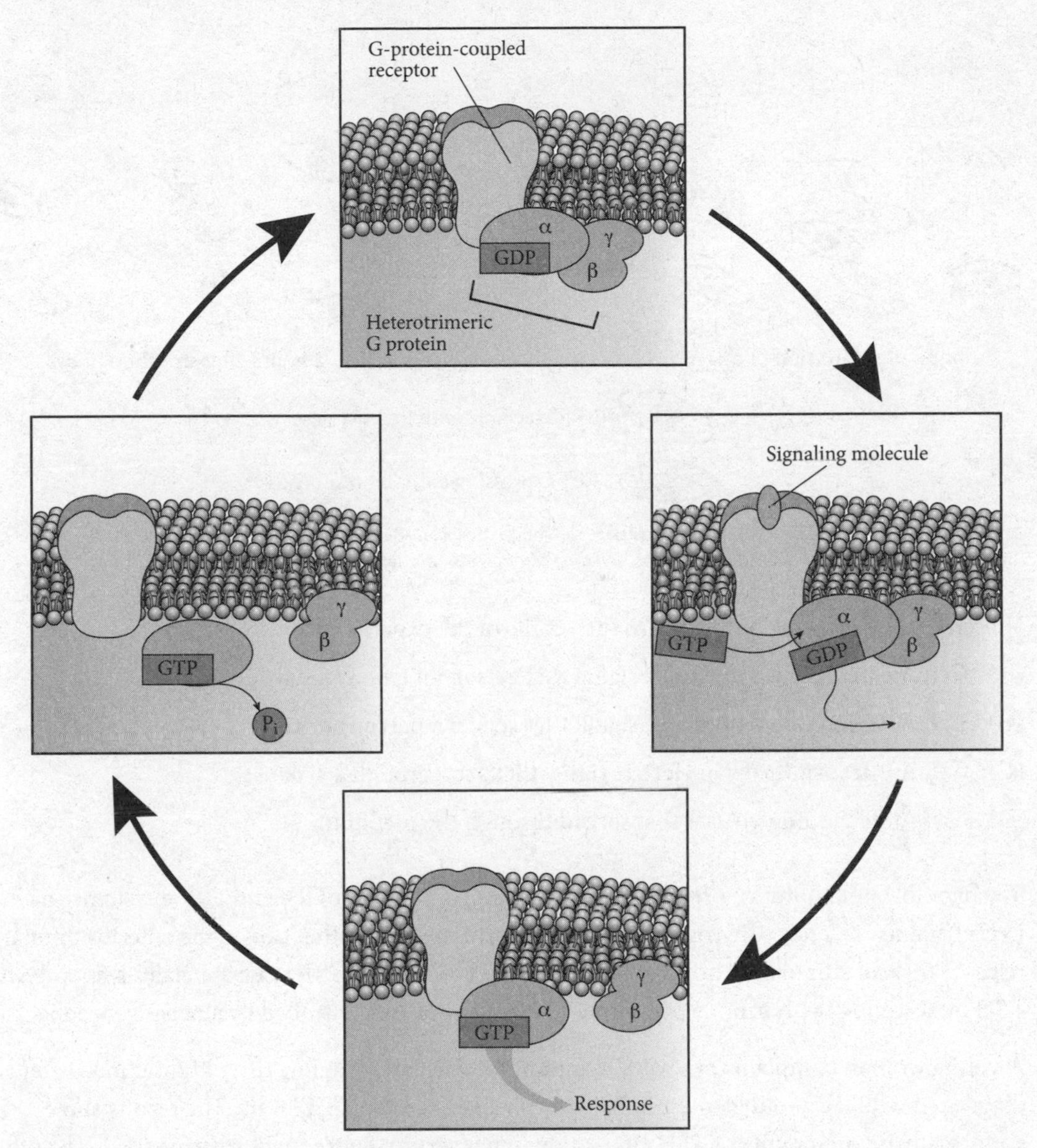

Which stage of the signaling cycle by a G-protein-coupled receptor is affected by the action of cholera toxin?

(A) Cholera toxin interferes with the binding of the signal molecule with its cognate receptor.

(B) The G-protein-coupled receptor activates dissociation of the G-protein complex into subunits.

(C) Activation of the enzyme adenylyl cyclase by a G-protein subunit leads to increased levels of cAMP.

(D) G-protein subunits reassemble to form an inactive G-protein that can reattach to the receptor.

5. Many viruses elicit the biosynthesis of the protein interferons in the cells they infect. The following diagrams show the synthesis of antiviral proteins in cells immobilized on a solid substrate in growth medium. Cell A is infected with a nonreplicating virus at time zero. The synthesis of mRNA encoding antiviral proteins is measured in all cells at the 1 and 2 hour time points.

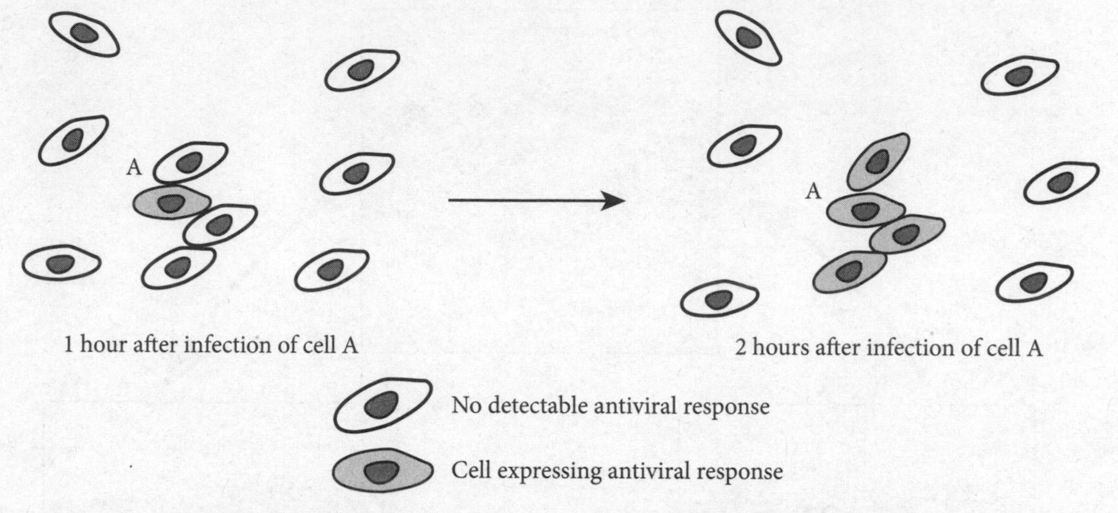

Adapted from A. K. Shalek et al., "Single-Cell RNA-Seq Reveals Dynamic Paracrine Control of Cellular Variation," *Nature* 510, no. 7505 (June 2014): 363–369.

Which of the following best explains the results of the experiment?

(A) Cell A is secreting a diffusible signal that acts in an endocrine way.

(B) Cell A is secreting a diffusible signal that acts in a paracrine way.

(C) Cell A is transmitting an electric signal that acts through a synapse.

(D) Cell A is shedding viruses that spread through the medium.

6. The hypothalamus-pituitary-thyroid axis is a major component of the endocrine system. The hypothalamus secretes a thyrotropin-releasing hormone (TRH) that causes the anterior pituitary to releases thyroid stimulating hormone (TSH) in the bloodstream. The thyroid gland responds to TSH by secreting the hormone thyroxine. TRH and TSH are controlled by a negative feedback loop.

A young woman complains of several symptoms associated with low thyroid function. After she is diagnosed with a thyroid gland insufficiency, she is tested for TSH levels. The results show abnormally high amounts of TSH circulating in her serum. Subsequent treatment with levothyroxine, a synthetic thyroid hormone, relieved her symptoms. Based on her medical history, which of the following is the most likely explanation for her condition?

(A) Her thyroid gland does not produce enough hormone.

(B) Her body cells lack a receptor for thyroxine.

(C) Her pituitary gland does not respond to stimulation by the hypothalamus.

(D) The feedback loop in her system is not responsive.

Answers to this quiz can be found at the end of this chapter.

CELL COMMUNICATION

9.1 Differentiate between stimulatory and inhibitory transduction signaling

9.2 Explain how selective pressure impacts pathway choice

9.3 Describe receptor recognition of ligands and protein shape change

9.4 Explain how blocking transduction or reception affects cells

9.5 Contrast contact and distance forms of cell communication

General Principles of Signaling

HIGH YIELD

Two main ways that animal cells communicate with one another are (1) via signaling molecules that are secreted by the cells and (2) through molecules that rest on the cells' surfaces and remain attached to the cell even as they signal other cells. The target cell receives information through receptors on its surface, which are generally membrane-spanning proteins with binding domains on the outside of the plasma membrane.

Although some signaling molecules may act far away from the cell that secreted them, many bind only to receptors on cells in the immediate vicinity. *Paracrine signaling* refers to the process of signaling only nearby cells, with signaling molecules quickly pulled out of the **extracellular matrix**. *Synaptic signaling* occurs between a nerve cell and either another nerve cell, a gland cell, or a muscle cell. Synaptic signaling itself occurs over extremely short distances, but the combined effect of multiple synaptic signals plus action potential transmission along lengthy axons can result in long distance communication. As electric signals reach axon terminals, they cause the release of chemicals called neurotransmitters. These chemical messengers bind to nearby receptors in post-synaptic cells and cause an electrical signal to be propagated or continued. By sending information from one neuron to another, nervous systems can produce rapid communication over long distances without a diminished effect (which can happen when sending individual molecules over long distances, such as through the bloodstream). *Endocrine signaling* refers to the secretion of chemical messengers into the bloodstream (or other liquid medium, such as is the case with plants) for widespread distribution throughout the entire organism.

Cells communicate with each other using a wide variety of molecules. Some can pass through the cell membranes of the target cells and act directly within the cell's cytoplasm. Others bind to plasma membrane receptors and act through second messengers. All of these molecules can be considered hormones, circulating signals that are released by specialized cells and travel throughout the bloodstream.

Responses to signal transduction (the conversion of an extracellular signal to a change in an intra-cellular process) may be stimulatory or inhibitory. Stimulatory responses typically involve activators or growth factors that promote gene expression, cell growth, and cell proliferation, whereas inhibitory responses involve inhibitors that may block signals from one molecule to another and/or lower enzyme activity. Epinephrine, also known as **adrenaline**, is an example that elicits both stimulatory and inhibitory responses. Epinephrine not only activates glycogen breakdown, but it also inhibits the enzyme that catalyzes glycogen synthesis. Another example is apoptosis. In a normal cell, apoptosis is inhibited by inhibiting the expression of pro-apoptotic factors and by promoting the expression of anti-apoptotic factors. However, when DNA damage is beyond repair, the cell signals apoptosis to occur.

Selective pressures such as predators, diseases, and environmental threats also impact which signaling pathway cells within an organism choose. If the selective pressure is neutral, the present signal transduction pathway persists because the pathway results in a stable population that is able to survive and reproduce. Otherwise, the selective pressure will cause the population to undergo natural selection and evolve. For instance, a mutation that alters the phenotype of an organism will spread in a population if the mutation enhances the behavior, fitness, and reproduction of the organism.

Signal transduction pathways of hormones (specifically nonsteroid hormones) generally involve the binding of a hormone to a cell-surface receptor, which induces a conformational change in the receptor protein and sets off an intracellular cascade to alter the cell's behavior in some way. Cell-surface receptors come in three different forms: ion-channel-linked, G-protein-linked, and enzyme-linked. All of these receptors can bind to the hormone or chemical signal (also referred to as the ligand) very accurately and very tightly and can turn on intracellular signals.

Ion-channel-linked receptors are also called ligand-gated channels. These membrane-spanning proteins undergo a conformational change when a ligand binds to them so that a "tunnel" is opened through the membrane to allow the passage of a specific molecule. These ligands can be neurotransmitters or peptide hormones, and the molecules that pass through are often ions, such as sodium (Na^+) or potassium (K^+), which can alter the charge across the membrane. The ion channels, or pores, are opened only for a short time, after which the ligand dissociates from the receptor and the receptor is available once again for a new ligand to bind.

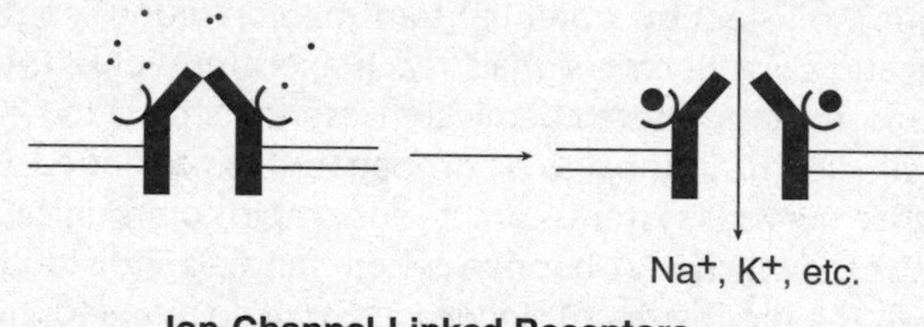

Ion-Channel-Linked Receptors

G-protein-linked channels cause G-proteins to dissociate from the cytoplasmic side of the receptor protein and bind to a nearby enzyme. This enzyme continues the signaling cascade by inducing changes in other intracellular molecules. In addition, it can also cause other membrane channels to open in areas some distance from the originating receptor.

Most G-proteins activate what are known as second messengers, small intracellular molecules like cyclic AMP (cAMP), calcium, and phosphates, which in turn activate key enzymes or transcription factors involved in essential reactions. Signaling cascades involving G-proteins can be very complex and involve many different enzymes and conversions, which prevents the reactions from running out of control.

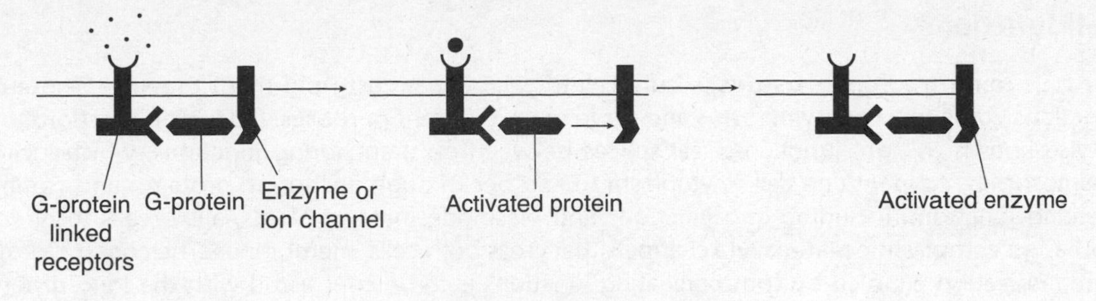

G-Protein-Linked Receptors

Enzyme-linked receptors can act directly as enzymes, catalyzing a reaction inside the cell, or they can be associated with enzymes that they activate within the cell. Most enzyme-linked receptors turn on a special class of enzymes called protein kinases, which add free-floating phosphate groups to proteins, regulating their activity. Protein phosphorylation is an essential means of intracellular signaling and control, as proteins become activated or deactivated simply by the addition or removal of phosphates. Protein kinases often target other protein kinases, initiating a cascade of kinase activity. Protein phosphatases reverse the action of protein kinases.

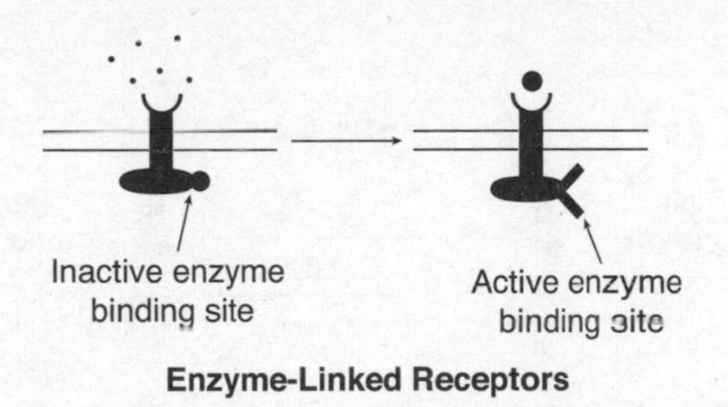

Enzyme-Linked Receptors

G-protein-linked and enzyme-linked receptors use complex relays of signal proteins to amplify and/or regulate their signal transduction. In some cases, a measure of safety requires that two different receptors on the cell surface become activated to turn on a particular intracellular protein. In other cases, signals at different receptors lead to the phosphorylation of different proteins that activate together only when *both* proteins have been phosphorylated. This signal integration leads to a measure of control over reactions and the ability to use multiple inputs to cause a certain effect that can vary in degree.

Signal transduction pathways that are blocked or defective can be deleterious (harmful) or prophylactic (preventative). For example, type I and type II diabetes arise from blocked signal transduction and blocked signal reception, respectively. Normally, the pancreas releases insulin, which signals liver, muscle, and fat cells to store sugar. In type I diabetes, pancreatic cells that produce insulin are destroyed by the immune system, so little to no insulin is produced. As a result, the effect is deleterious as sugar accumulates to toxic levels in the blood. In type II diabetes, cells are unable to respond to insulin, resulting in insulin resistance as well as toxic sugar levels in the blood. Some drugs such as antihistamines are prophylactic. Used to prevent and treat allergies, antihistamines inhibit the action of histamine (a compound released in allergic reactions) by suppressing the expression of histamine receptors in nasal mucosa.

Cell Junctions

For cells to form complex **tissues**, secure cell-to-cell bonds must hold them together. These cell junctions come in a variety of forms and serve many different purposes. Occluding junctions, otherwise known as tight junctions, seal spaces between cells; anchoring junctions, which include desmosomes, connect one cell's cytoplasm to another through anchoring proteins; and communicating junctions, including gap junctions and plasmodesmata in plants, allow cells to directly exchange cytoplasmic material via channels that cross both cells' membranes. The contact form of communication enabled by communicating junctions is to be contrasted with the long-distance communication allowed by the endocrine system.

Occluding (Tight) Junctions

These connections are thought to be formed by proteins that tightly wind between the adjacent plasma membranes of neighboring cells, binding the cells together at those points so tightly that *nothing can diffuse between cells or past the junction*. In such a way, tight junctions form a total barrier to transport and diffusion where they exist.

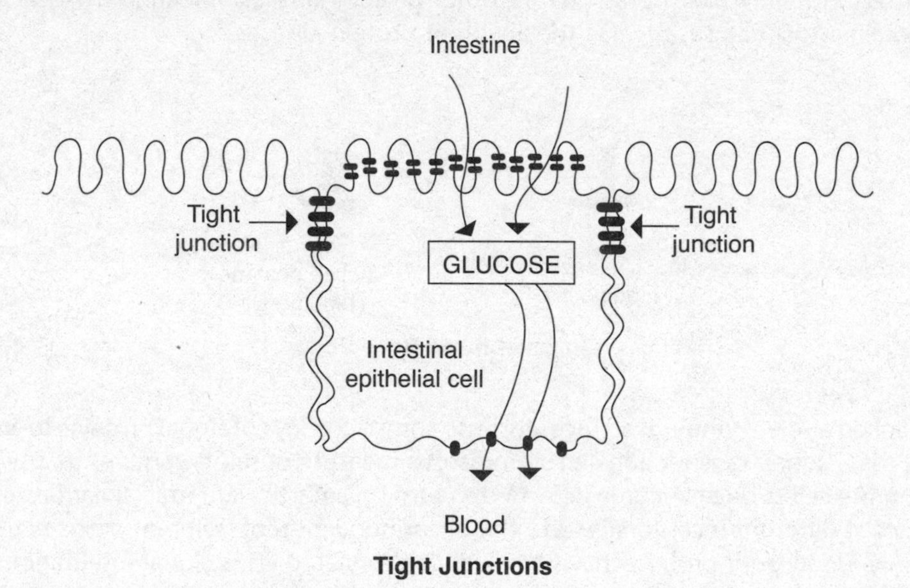

Tight Junctions

Where they are most useful, as shown in the above diagram, is in places such as the **intestines**, where specialized cells absorb nutrients from one side of the cell (the intestinal side) and transport them through the cell and out the other end (into the bloodstream). Certain transporters (e.g., sodium ion-driven glucose transporters) exist on the intestinal side of the cell but not on the bloodstream side, and glucose that enters the bloodstream is prevented from diffusing back into the intestinal tract by tight junctions.

Anchoring Junctions

Found between cells subjected to fair amounts of stress, either from shearing forces or contacting forces, these junctions connect one cell's cytoplasm to another's via a series of proteins. Regardless of classification, anchoring junctions not only allow cells to adhere to neighbors, but also may allow them to contract themselves into large tubelike tissues. This occurs when the fibers holding the junctions together contract across the cells' cytoplasms.

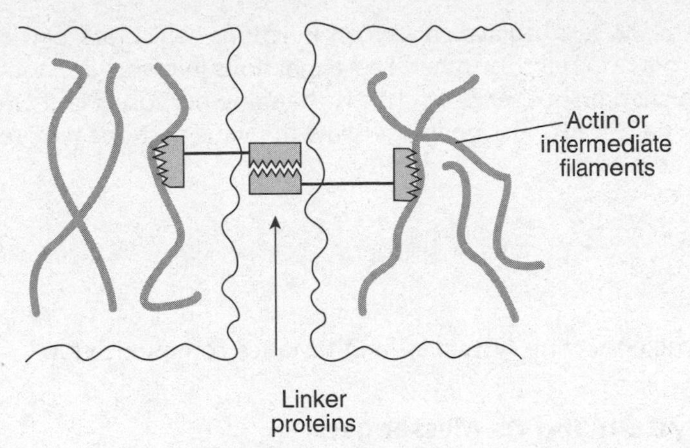

Anchoring Junctions

Notice in the above diagram that anchoring junctions involve proteins that attach to actin filaments within the cell cytoplasm and also attach to linker proteins across the intercellular space. Thus, the junction is not an actual linking of cytoplasm, where material can be freely exchanged, but rather a *physical joining so that the cells do not shear away from each other.*

The most often mentioned of these attachments are the desmosomes, found in heart cells and between epithelial cells in the skin. Although they function in the same way as other anchoring junctions, they attach two cells using intermediate filaments within the cytoplasm rather than actin filaments.

Communicating Junctions

The best known of these cell-to-cell connections are the gap junctions, formed by proteins called connexins, which build tubes or pores between two adjacent cells' cytoplasms. It is through these pores that ions and other material can pass from one cell to the other. In cells that rapidly transmit chemical or electrical signals across tissues, gap junctions are everywhere. Because chemical and electrical transmission is mediated through the movement of ions and other messengers, gap junctions *allow for undisrupted and very fast signal transmission* across wide areas of tissue. In heart cells, gap junctions allow for rhythmic contractions of large sections of the heart all at once. In the digestive tract, gap junctions allow for waves of muscle contraction such as those found in the esophagus. These junctions also allow for coordination of *rapid and complex movements,* such as a fish's tail-flip to escape an oncoming predator. The following diagram shows a generic gap junction with many connexin proteins that form a connexon.

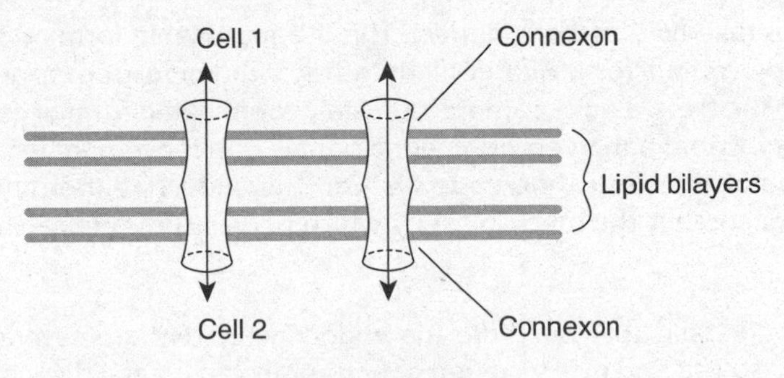

Gap Junctions

Plasmodesmata are plant cells' equivalent of gap junctions, which are particularly useful for the free flow of nuclei from one cell to another. These junctions in plants are not nearly as complex in structure as those in animals, but serve essentially the same purpose. Plant viruses, however, often exploit plasmodesmata because the openings allow the virus particles to spread rapidly from one section of the plant to others.

HORMONES

9.6 Explain the endocrine system as long-distance communication

The Endocrine System and Its Messengers

The **endocrine system** acts along with the nervous system as a means of internal communication, coordinating the activities of organ systems around the body. **Endocrine glands** synthesize and secrete chemical hormones that are dumped directly into the bloodstream and affect specific target organs or tissues. Compared with the nervous system, hormones take much longer to communicate with cells of the body. After all, they can travel only as fast as the blood flows. Yet hormonal signaling can cause behavioral effects that last for days, far longer than any nervous signal can last.

If an animal were injured, it would release the hormone epinephrine (adrenaline) from the adrenal glands. The epinephrine would travel in the bloodstream throughout the entire body. Any cell in the body that is accessible by blood flow and has **receptor proteins** for epinephrine responds to the "message" of the injury. Blood vessels respond by constricting blood flow, the heart and lungs increase their rate of activity, and the liver releases sugar to the bloodstream. The key issues of any endocrine response are the widespread release of hormones, specificity controlled by the receptor's ability to recognize the signal, and the location of the cells that contain the receptor.

✓ AP Expert Note

Hormones are defined as chemical signals that:

- **Are synthesized by specialized cells**
- **Travel throughout a multicellular organism by some kind of bodily fluid**
- **Coordinate systemic (total body) responses by activating specific receptor cells**

Hormonal responses take time and are indirect. This is a reasonable form of communication for a plant that doesn't move, but for a bird in flight, a fish swimming from a predator, or a gibbon brachiating through the trees, a quicker, more intricately coordinated form of communication inside the body is necessary. Coordinating environmental signals with movement responses and complex processing is the job of the **central nervous system**. The endocrine (hormone) system interacts with the nervous system, but the mechanism by which nerves communicate messages is entirely different.

It's important to understand that built into the endocrine system are several controls based on **negative feedback**, so that the oversecretion of hormones can be avoided. In general, high levels of a particular hormone in the bloodstream inhibit further production of that hormone. In some

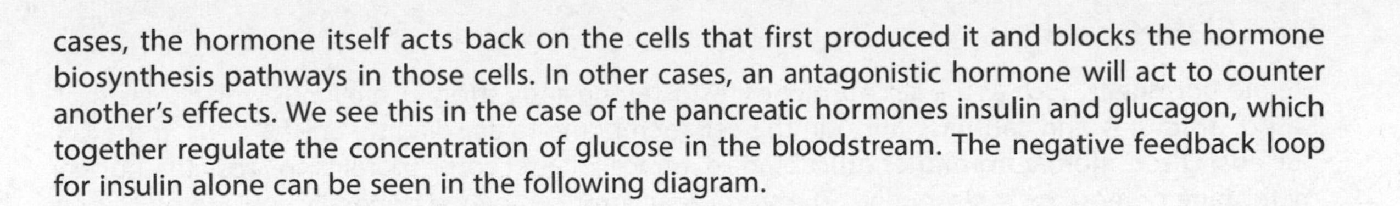

cases, the hormone itself acts back on the cells that first produced it and blocks the hormone biosynthesis pathways in those cells. In other cases, an antagonistic hormone will act to counter another's effects. We see this in the case of the pancreatic hormones insulin and glucagon, which together regulate the concentration of glucose in the bloodstream. The negative feedback loop for insulin alone can be seen in the following diagram.

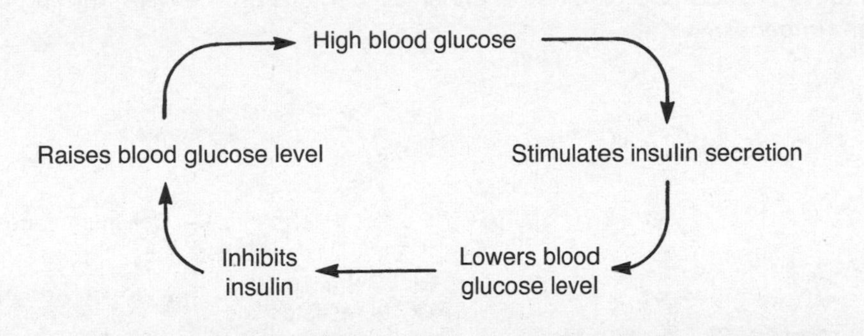

✔ AP Expert Note

You will not need to memorize the names, molecular structures, or specific effects of any hormones. These are not tested on the AP Biology exam.

IMPORTANT HORMONES IN VERTEBRATES

Compound	Type	Origin	Action
Androgens (e.g., testosterone)	Steroid	Testis	Secondary sex characteristics
Epinephrine (adrenaline)	Catecholamine	Adrenal gland	Fight-or-flight response
Estrogen	Steroid	Ovary	Secondary sex characteristics
Follicle-stimulating hormone (FSH)	Glycoprotein	Anterior pituitary	Regulate gonads
Glucagon	Polypeptide	Pancreas	Increase blood glucose
Insulin	Polypeptide	Pancreas	Decrease blood glucose
Melatonin	Catecholamine	Pineal gland	Circadian rhythm
T_3 and T_4	Amino acids	Thyroid	Stabilize metabolic rate
Thyroid-stimulating hormone (TSH)	Glycoprotein	Anterior pituitary	Regulate thyroid

Steroid Hormones

Steroid hormones, such as the sex hormones testosterone and estrogen, are lipids with cholesterol-based structures and can pass through the cell membrane to act directly on the DNA in the cell nucleus. Often, steroid hormones must bind to intracellular receptor proteins to cross the nuclear membrane or regulate transcription. The steroid-receptor complex is generally considered to be a **transcription factor**, as it is able to bind to "enhancer" regions on the DNA, turning on the transcription of certain genes.

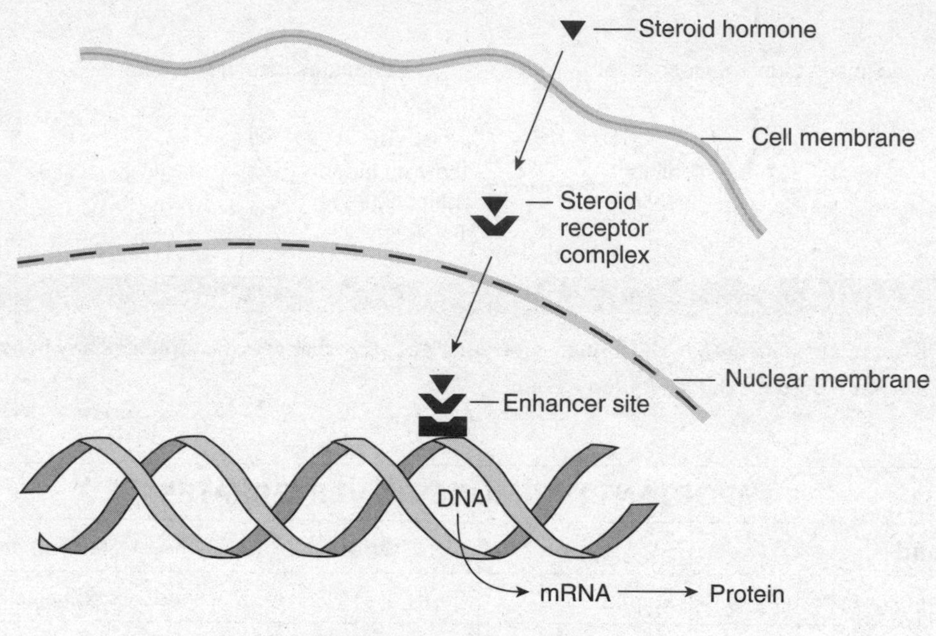

Steroid Hormone Action

The basis for the production of steroid hormones, as well as compounds such as vitamin D, is cholesterol. Steroid hormones travel through the bloodstream bound to carrier molecules, from which they dissociate when entering a cell. *Because these hormones are lipids and do not dissolve in water, their effects can last for hours or days* in the bloodstream after being released, a much longer period of time than water-soluble nonsteroid hormones can last.

Nonsteroid Hormones

Many compounds that are unable to cross the plasma membrane and enter the cytoplasm still act as powerful signaling molecules. Examples include peptides, such as the atrial natriuretic peptide, released by the heart to influence water absorption in the kidneys, calcitonin, involved in calcium regulation, and glucagon, involved in blood sugar regulation. Nonsteroid hormones act via signal-transduction pathways, as described earlier in the chapter.

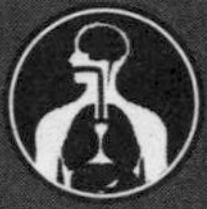

Other Signaling Factors

Another molecule worth mentioning here is not exactly a hormone, but it can easily pass through a cell's lipid bilayer (like a steroid hormone) because it is so small. Nitric oxide (NO), recently recognized as an important signaling molecule, passes into the cell's cytoplasm, acting on the enzyme guanylyl cyclase, which in turn produces cyclic-GMP. Cyclic-GMP is an important molecule that, in the case of NO, causes smooth muscle cells in blood vessels to relax when stimulated by acetylcholine. Keep in mind, too, that different cells can respond to the same hormone or neurotransmitter in different ways. For example, although acetylcholine can cause smooth muscle cells to relax, it causes skeletal muscle cells to contract. It is postulated that these differences are caused by the variety of receptor complexes that can respond to the same hormone.

RAPID REVIEW

If you take away only 5 things from this chapter:

1. There are two major ways animal cells communicate: through signaling molecules secreted by cells and through molecules that rest on the cell surface. Signaling molecules can be further divided into paracrine, synaptic, and endocrine.
2. Some cell communication occurs through cell junctions and is important for complex tissues to be able to function properly. There are three major types of cell junctions: tight junctions, desmosomes, and gap junctions.
3. The endocrine system works with the nervous system to produce certain reactions as needed in response to changes in the body's organ systems.
4. Hormones are one of the body's methods of internal communication. Hormones are secreted by endocrine glands and travel throughout the body. They activate receptor cells that produce receptor proteins, thereby creating a reaction.
5. There are two major types of hormones: steroid hormones and nonsteroid hormones. Nonsteroid hormones are usually made up of proteins or modified amino acids.

TEST WHAT YOU LEARNED

1. The sympathetic branch of the autonomic nervous system transmits signals through the release of the neurotransmitter acetylcholine. The muscarinic acetylcholine receptors associated with the sympathetic system are G-protein-coupled receptors usually found in the membrane of the target organ. The graph shows the response of a target organ to acetylcholine with or without atropine. The researchers observed that adding atropine shifted the dose-response curve to the right. An increased dose of acetylcholine was required to overcome the effect of atropine on the target organ.

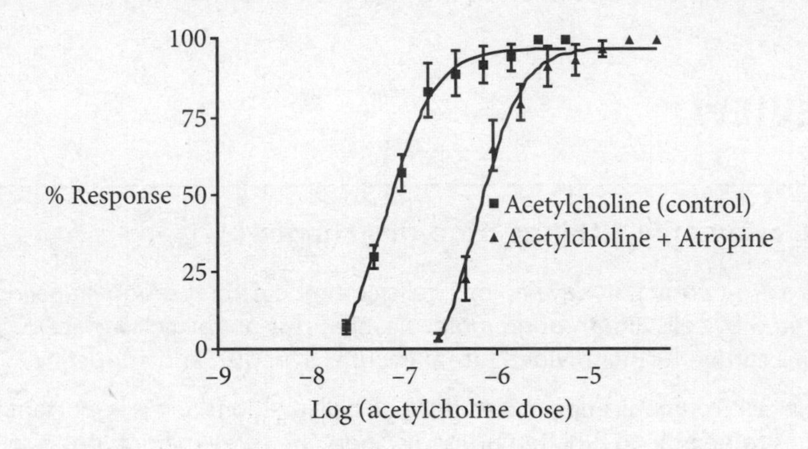

Which mechanism of action of atropine would produce the effects recorded in the experiment?

(A) Atropine and acetylcholine bind to the same site on the receptor.

(B) Atropine increases the effectiveness of acetylcholine.

(C) Atropine is an irreversible inhibitor of the muscarinic receptor.

(D) Atropine has no substantial effect on the response of the target organ to acetylcholine.

2. Acetylcholine is a known neurotransmitter that initiates muscle contraction when it binds to ion-gated channels. Epinephrine is a neurotransmitter of the sympathetic system, which decreases smooth muscle contraction. It does so by binding to a G-protein-coupled receptor that triggers adenylyl cyclase to release cAMP, which activates protein kinase A, causing less interaction between myosin and actin. In an experiment on smooth muscle tissue, a baseline of contraction in the absence of chemicals was recorded for 1 minute before introducing acetylcholine to the system. Epinephrine was applied 3 minutes after acetylcholine. The graph shows the response of the smooth muscle tissue to the addition of acetylcholine followed by epinephrine.

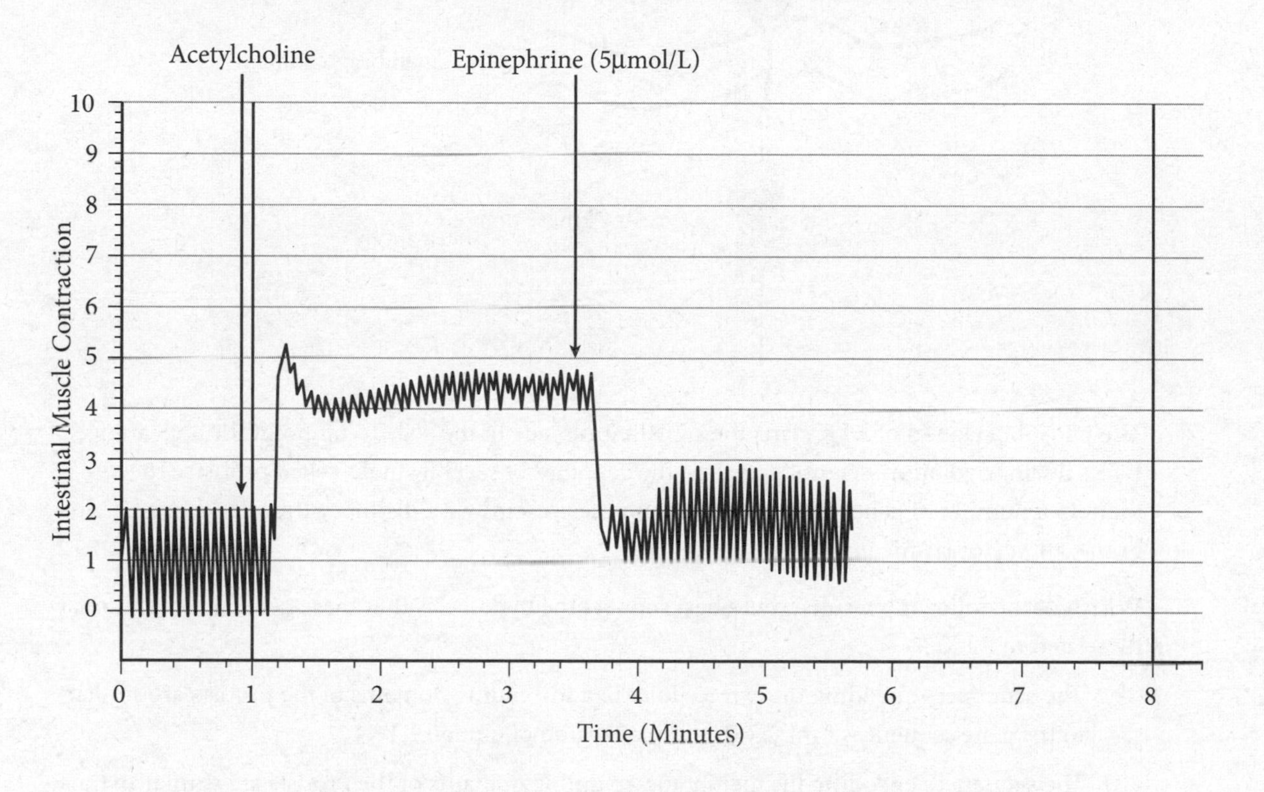

Which of the following hypotheses is best supported by the results shown in the graph?

(A) The contractions of intestinal muscle decrease after the addition of epinephrine because epinephrine competes with acetylcholine for a binding receptor in the membrane.

(B) The activation of the transducing cascade triggers activation of protein kinase A and inhibition of glucose synthesis.

(C) Epinephrine initiates a signaling cascade that activates protein kinase A, thereby indirectly decreasing muscle contraction.

(D) Epinephrine activates the membrane receptor, which inhibits the activation of adenylyl cyclase, preventing production of a second messenger.

Structures

3. Receptor tyrosine kinases (RTKs) are signaling receptors that consist of an extracellular region that binds the ligand, a hydrophobic region that spans the membrane, and a cytoplasmic region that acts as a tyrosine kinase enzyme.

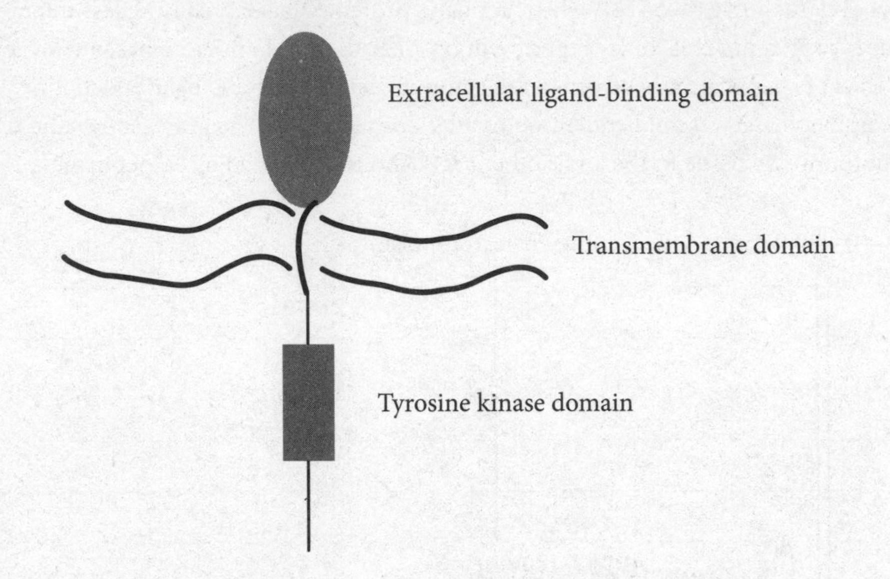

The phosphorylation of RTK starts the signaling cascade in the cell. To pinpoint the appearance of RTKs during evolution, scientists analyze the genomes of several single-celled protists. They identify a number of genes that encode membrane proteins with distinct extracellular, transmembrane, and cytoplasmic domains.

Which of the following results would best support the hypothesis that these genes are evolutionarily related to RTKs?

(A) The sequences encoding the extracellular ligand-binding domains of the protists are similar to the gene sequences for the extracellular region of human RTKs.

(B) The sequences encoding the membrane-spanning domains of the protists are similar to the gene sequences for the membrane-spanning region of human RTKs.

(C) The sequences encoding the tyrosine kinase domain of the protists are similar to the gene sequences for the tyrosine kinase regions of human RTKs.

(D) The entire gene sequences for the proteins of the protists are similar to the gene sequences for the entire coding region of human RTKs.

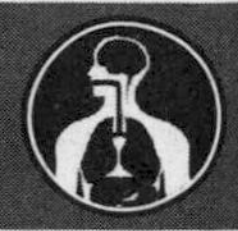

4. The hormone insulin is produced by the β cells of the pancreas. When the levels of glucose rise above a set threshold, insulin is released in the blood and promotes the uptake of glucose by cells. In type I diabetes, the body is unable to produce insulin. In type II diabetes, cells are insulin-resistant and do not respond to insulin stimulus. The graph shows levels of glucose and insulin in the blood of three patients after eating.

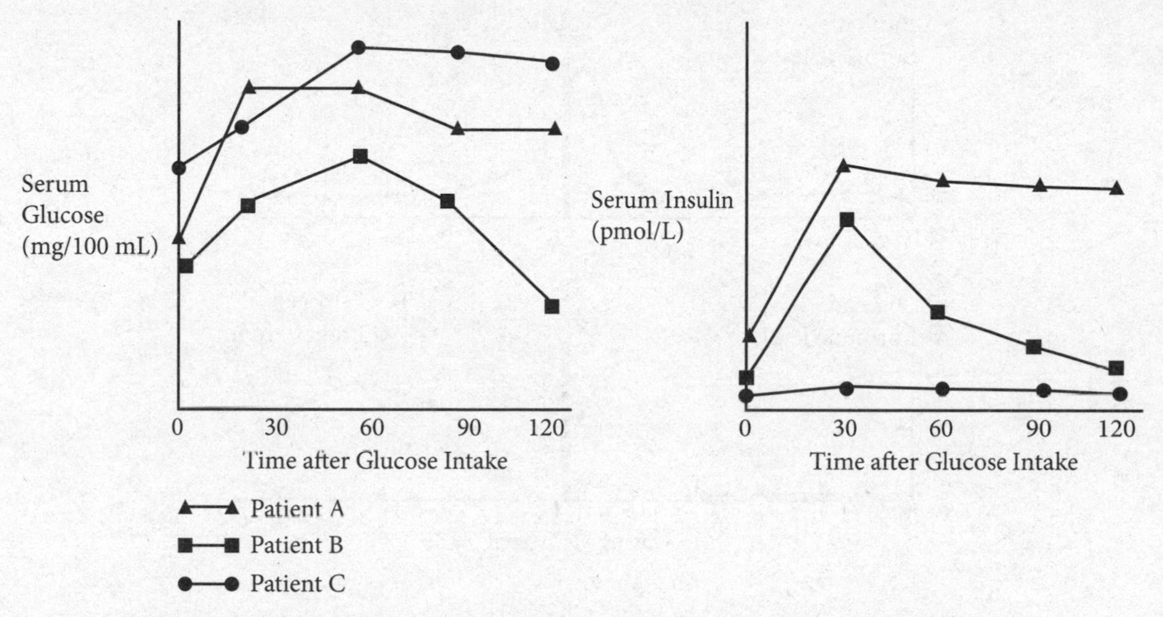

Adapted from N. Geidenstam et al., "Metabolite Profile Deviations in an Oral Glucose Tolerance Test—A Comparison between Lean and Obese Individuals," *Obesity* 22, no. 11 (November 2014). 2388–2395.

Based on the curves, which patient is the most likely to be diagnosed with type II diabetes?

(A) Patient A, because his insulin and glucose levels remain high after 2 hours

(B) Patient B, because her insulin levels show a strong peak after ingestion of glucose

(C) Patient C, because his insulin levels never increase

(D) All three patients, because the glucose levels in their blood increase sharply during the glucose tolerance test

5. Luteinizing hormone (LH) and follicle stimulating hormone (FSH) are released by the pituitary gland during menstruation. The pituitary hormones influence the production of hormones in the ovaries, including estrogen and progesterone. A biotechnology company is testing an ovulation kit to plan pregnancy. The window of fertilization is restricted by the viability of the ovule, which is fertile up to 12 hours after ovulation.

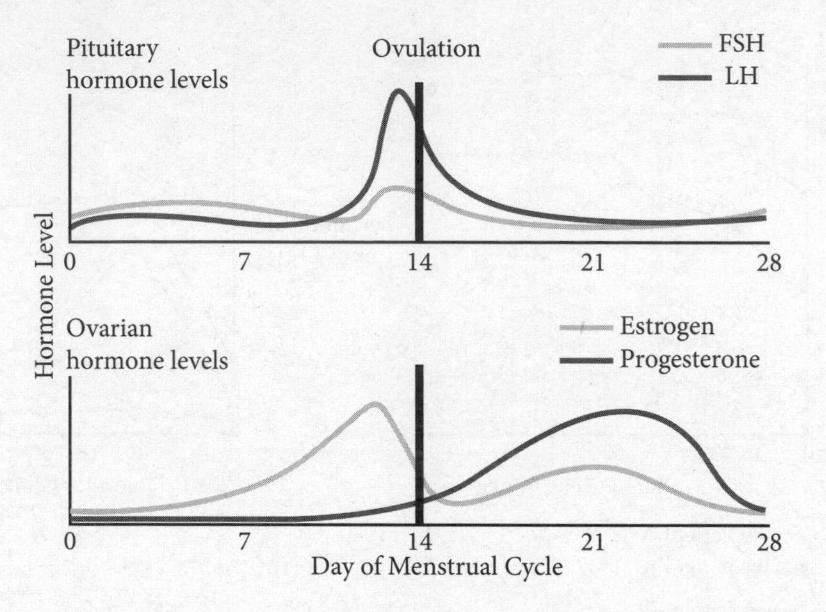

According to the diagram, which hormone would be the most effective target for the ovulation kit to measure?

(A) FSH, because it peaks early enough to "catch" the ovule

(B) LH, because it displays a single peak just before ovulation

(C) Estrogen, because it is produced by mature follicles, which indicate that an egg is ready for release

(D) Progesterone, because it promotes readiness of the uterus for implantation of an embryo

6. Tubocurarine was first purified from curare, a poison obtained from the bark of a South American tree. Poisoning by tubocurarine can be reversed by the administration of a cholinesterase inhibitor, which results in an increased concentration of acetylcholine in the synaptic cleft. The inhibition of muscle contraction by tubocurarine is most likely explained by which of the following statements?

(A) Tubocurarine prevents the release of acetylcholine from the nerve ending, causing paralysis of the muscle.

(B) Tubocurarine binds to the same site on the ion-gated channel as acetylcholine in a competitive inhibition reaction and causes paralysis of the muscle.

(C) Tubocurarine blocks the influx of Na^+ ions, which triggers the release of calcium ions and contraction of the muscle.

(D) Tubocurarine interferes with the contraction mechanism and causes paralysis of the muscle.

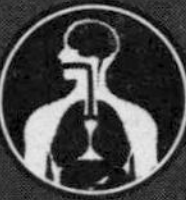

The answer key to this quiz is located on the next page.

Answer Key

Test What You Already Know

1. **D** **Learning Objective:** 9.1
2. **B** **Learning Objective:** 9.2
3. **C** **Learning Objective:** 9.3
4. **D** **Learning Objective:** 9.4
5. **B** **Learning Objective:** 9.5
6. **A** **Learning Objective:** 9.6

Test What You Learned

1. **A** **Learning Objective:** 9.3
2. **C** **Learning Objective:** 9.1
3. **C** **Learning Objective:** 9.2
4. **A** **Learning Objective:** 9.5
5. **B** **Learning Objective:** 9.6
6. **B** **Learning Objective:** 9.4

Detailed solutions can be found in the Answers and Explanations section at the back of this book.

REFLECTION

Test What You Already Know score: __________

Test What You Learned score: __________

Use this section to evaluate your progress. After working through the pre-quiz, check off the boxes in the "Pre" column to indicate which Learning Objectives you feel confident about. Then, after completing the chapter, including the post-quiz, do the same to the boxes in the "Post" column. Keep working on unchecked Objectives until you're confident about them all!

Pre	Post	
❑	❑	**9.1** Differentiate between stimulatory and inhibitory transduction signaling
❑	❑	**9.2** Explain how selective pressure impacts pathway choice
❑	❑	**9.3** Describe receptor recognition of ligands and protein shape change
❑	❑	**9.4** Explain how blocking transduction or reception affects cells
❑	❑	**9.5** Contrast contact and distance forms of cell communication
❑	❑	**9.6** Explain the endocrine system as long-distance communication

FOR MORE PRACTICE

Complete more practice online at kaptest.com. Haven't registered your book yet? Go to kaptest.com/booksonline to begin.

CHAPTER 10

Nervous Systems

LEARNING OBJECTIVES

In this chapter, you will review how to:

10.1 Explain the function of neuron structures

10.2 Describe synapse transmission and neurotransmitters

10.3 Describe electrical potential and ion requirements for an impulse

10.4 Describe how the nervous system transmits information

10.5 Explain why different brain regions have unique functions

10.6 Describe senses as signal detection, processing, and response

TEST WHAT YOU ALREADY KNOW

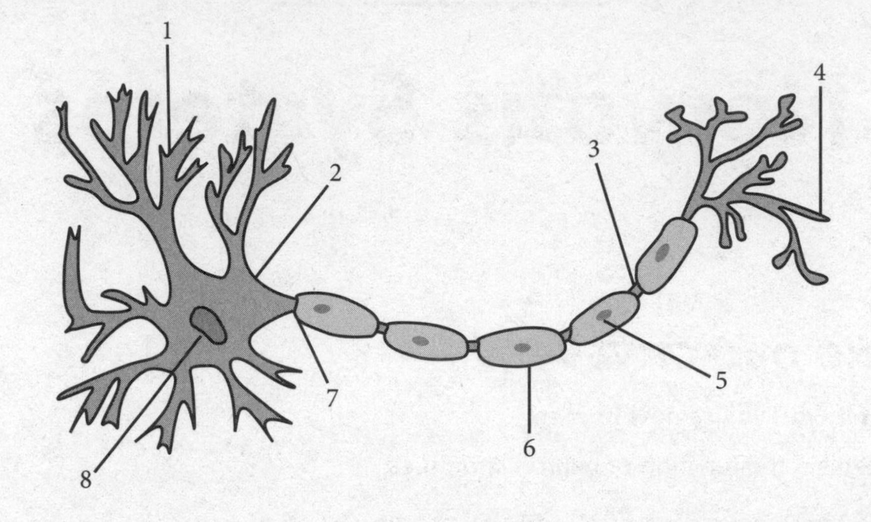

1. Neurons communicate using special chemical messengers called neurotransmitters, which travel across the synaptic gap between two neurons. In the above diagram of a neuron, how do structures 1 and 4 contribute to this chemical signal transmission?

   (A) Structure 1 is a dendrite and structure 4 is an axon. An axon on the end of one neuron transmits a chemical signal to a dendrite of another neuron.

   (B) Structure 1 is a dendrite and structure 4 is an axon. A dendrite on the end of one neuron transmits a chemical signal to an axon of another neuron.

   (C) Structure 1 is an axon and structure 4 is a dendrite. An axon on the end of one neuron transmits a chemical signal to a dendrite of another neuron.

   (D) Structure 1 is an axon and structure 4 is a dendrite. A dendrite on the end of one neuron transmits a chemical signal to an axon of another neuron.

2. Neurotransmitters are chemical messengers needed for impulse transmission between neurons. In response to an impulse, neurotransmitters are released at the presynaptic terminal into the synaptic cleft. The released neurotransmitters diffuse across the synaptic cleft and bind with receptors on the postsynaptic neuron. What normally happens to excess neurotransmitters produced by the presynaptic terminal in a healthy individual?

   (A) They are absorbed by other neurons in the region.

   (B) They are taken up by the postsynaptic neuron.

   (C) They are either enzymatically destroyed or taken back up by the presynaptic terminal.

   (D) They quickly degrade within the synaptic cleft without the need for additional enzymes.

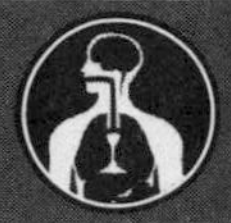

3. A neuron is in the absolute refractory period of transmitting an action potential and then receives a very strong stimulating signal. Can this neuron create a second impulse? Why or why not?
   (A) A second impulse can occur because the potassium channels remain active during the refractory period.
   (B) A second impulse can occur because the sodium channels remain active during the refractory period.
   (C) A second impulse cannot occur because the potassium channels are inactive for a short period.
   (D) A second impulse cannot occur because the sodium channels are inactive for a short period.

4. The human nervous system is a highly specialized, complex network of neurons and fibers that transmit impulses within the body. Some basic nervous system functions include sensory, motor, and integration functions. How would the human sensation of feeling warm best be classified within the nervous system?
   (A) It is classified as a sensory function because it causes muscles to generate heat when they move.
   (B) It is classified as a sensory function because it is associated with sensory receptors.
   (C) It is classified as a motor function because it leads to the movement of skin cells.
   (D) It is classified as a motor function because it involves communication between skin cells.

5. The cerebral cortex is composed of four different lobes: frontal, parietal, occipital, and temporal. The frontal lobe is in the foremost portion of the brain, the parietal lobe to the rear of the frontal lobe, the occipital lobe at the very rear, and the temporal lobe is nested beneath them. The frontal lobe supervises processes associated with perception, memory, emotion, impulse control, and long-term planning. The parietal lobe is involved in somatosensory information processing and spatial processing and manipulation. The occipital lobe is responsible for visual processing. The temporal lobe controls auditory processing and language comprehension.

   A patient was brought into an emergency room with a suspected head injury after falling in a bathtub. An initial physical exam revealed that he experienced a loss in spatial perception and the ability to distinguish left from right. In addition, an MRI scan of the patient uncovered a possible cancerous lesion on the brain in the location shown below. (Note that the top of the figure corresponds to the front.)

   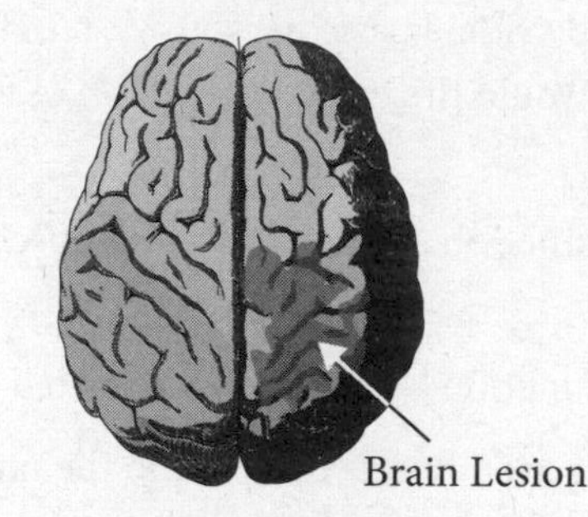

   Which of the following provides the best interpretation of this data?

   (A) The brain lesion likely caused the fall and the patient's loss of spatial perception, since these functions are controlled by the frontal lobe.

   (B) The patient's loss of spatial perception likely resulted from the fall, since the lesion is in the occipital lobe, which is not the area of the brain that controls these abilities.

   (C) The brain lesion likely caused the fall and the patient's loss of spatial perception, since these functions are controlled by the parietal lobe.

   (D) The patient's loss of spatial perception likely resulted from the fall, since the lesion is in an area of the brain that controls these abilities.

6. The proximal cause of a process within the body describes how that process occurs, in terms of the actions that cells, tissues, and organs perform to make that process occur. The ultimate cause of a body process describes why that process evolved within the organism. Which of the following best describes the proximal and ultimate causes of the sensation of pain in human bodies?

   (A) Proximal: Pain arises from the stimulation of internal sensory receptors that help monitor aspects of the internal environment such as blood pressure and blood pH. Ultimate: The perception of pain allows the conscious mind to perceive when the body is being injured.

   (B) Proximal: Pain arises from the stimulation of external sensory receptors that help organisms to sense their external environment. Ultimate: The perception of pain allows the conscious mind to perceive when the body is being injured.

   (C) Proximal: The perception of pain allows the conscious mind to perceive when the body is being injured. Ultimate: Pain arises from the stimulation of internal sensory receptors that help monitor aspects of the internal environment, such as blood pressure and blood pH.

   (D) Proximal: The perception of pain allows the conscious mind to perceive when the body is being injured. Ultimate: Pain arises from the stimulation of external sensory receptors that help organisms to sense their external environment.

Answers to this quiz can be found at the end of this chapter.

NEURONS

10.1 Explain the function of neuron structures

The nerve cell (or **neuron**), is the basic functional unit of the mammalian nervous system. Each neuron contains a cell body, dendrites, Schwann cells, nodes of Ranvier, an axon, and synaptic knobs. The cell body (or **soma**) contains the nucleus and organelles and is the site of protein synthesis. **Dendrites** are extensions of cytoplasm from the cell body that receive chemical signals from nearby neurons and can initiate a new electrical signal. Schwann cells are separate cells from the neurons. They secrete myelin, an insulation for nerve cells that helps signals move faster down the axon. Nodes of Ranvier are spaces along myelinated axons where Schwann cells have not laid down myelin covering. The **axon** is the main elongated extension of the cell body through which the electrical signal travels; the signal is in one direction only, from the cell body toward the synaptic knobs. Synaptic knobs release neurotransmitters, chemicals that communicate with surrounding nerve cells.

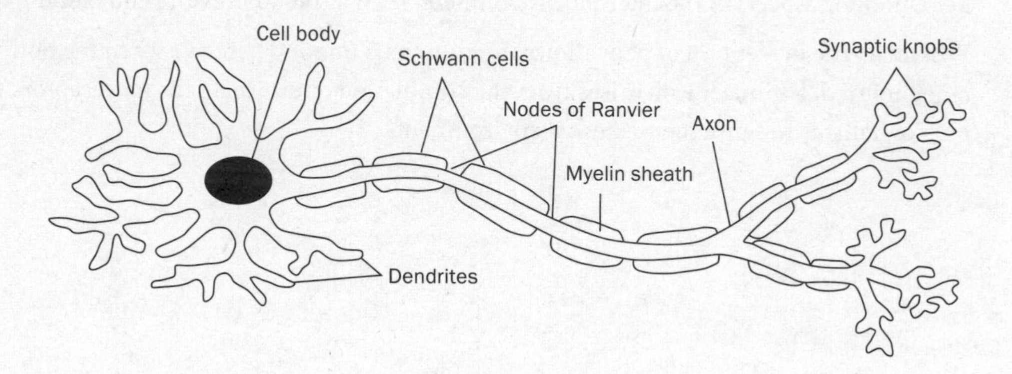

Neuron Structure

Dendrites are multi-branched and can receive signals from many different neurons at the same time. In fact, some specialized neurons such as the Purkinje cells of the brain can receive signals from over 1,000 different neurons at once into a single cell body. The amount of integration that takes place in a cell such as this, filtering out and summing together all of this input, is enormous.

Most neurons have a single long axon enclosed in layers of myelin. **Myelin** insulates the axon, much like electrical tape insulates bare wire, and neurons that lose their myelin sheaths cannot transmit signals fast enough to give appropriate stimuli to muscles or organs. Each Schwann cell can contribute a single internodal segment of myelin, so many Schwann cells are needed to coat an entire axon with myelin. In the central nervous system (brain and spinal cord), the equivalent cell to a Schwann cell is known as an oligodendrocyte. Whereas a single Schwann cell can myelinate only a single axon, oligodendrocytes can send off myelin sheaths in several directions at once.

Schwann cells and oligodendrocytes are not the only "support cells" present around neurons. Astrocytes in the central nervous system (CNS) far outnumber the neurons of the CNS. The extensions of astrocytes stick to various parts of neurons and help to break down and remove certain neurotransmitter chemicals as well as engulf debris. Ependymal cells line the fluid-filled cavities of the brain and spinal cord and secrete cerebrospinal fluid (CSF), which helps cushion the CNS.

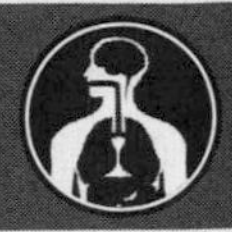

ACTION POTENTIALS

10.2 Describe synapse transmission and neurotransmitters

10.3 Describe electrical potential and ion requirements for an impulse

10.4 Describe how the nervous system transmits information

Resting Potentials and Action Potentials

Signals along the axon are electrical in nature and depend upon the flow of ions across the axon membrane and cell body membrane. How does electrical voltage arise in a cell and how does a signal arise from ion flow? All living cells have an electrical charge difference across their membranes, the inside of the cells being more negatively charged than the outside. The reasons for this are:

- DNA is a negatively charged molecule due to copious negative phosphate groups (remember the basic units of nucleotides).
- Many proteins (amino acid side chains) in the cell are negatively charged.
- The Na^+/K^+ (sodium/potassium) pumps in the cell membrane kick out three positive sodium (Na^+) ions for every two positive potassium (K^+) ions they move into the cell. Overall, that means that one positive charge is leaving the cell every time one Na^+/K^+ pump "turns."
- While sodium is kicked out of the cell and potassium is brought in due to the action of the Na^+/K^+ ATPase pump (ATPase means that active transport is involved here—the pump uses ATP to work), some potassium leaks out. This passive diffusion of potassium through leakage channels works along with the sodium-potassium pump to create an overall charge difference across the axon or cell body membrane.

An **action potential** is a temporary discharging of the battery power stored in neurons. The up-and-down signal transmitted on a heart monitor, as commonly seen in movies and on television shows, tracks the progression of these electrical signals through the heart. Almost all cells in the body constantly pump Na^+ ions out and K^+ ions in, creating an electrical potential between the more positively charged outside matrix and the more negatively charged cytoplasm. The **resting membrane potential** of most neurons is about –70 mV (the minus sign indicates a negative charge).

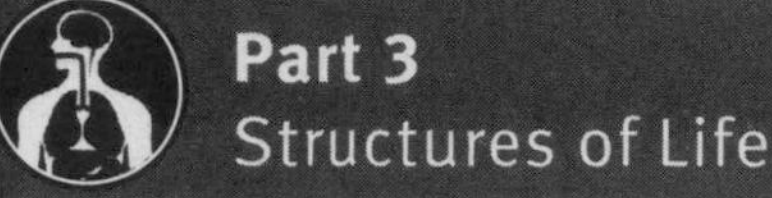

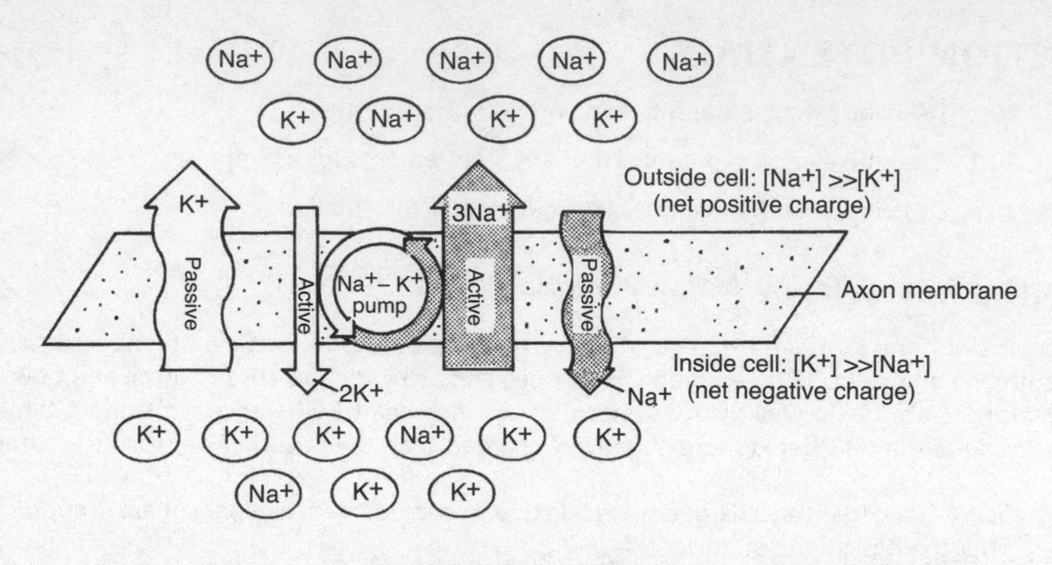

Resting Potential of a Neuron

When some kind of stimulus depolarizes the membrane potential to a specific, predetermined threshold, gates open that allow Na^+ into the cell, down the **electrochemical gradient** (a combination of the attraction of unlike charges and the diffusion gradient). After the neuron completes its depolarization, the Na^+ gates close and the K^+ gates open, repolarizing the membrane as positive charge leaves the cell. K^+ gates slowly close as the membrane returns to its resting potential, causing an overshoot known as hyperpolarization, which prevents action potentials from flowing in the reverse direction. The action potential takes place locally across the cell membrane, but propagates down the axon as adjacent parts of the membrane depolarize as well. The depolarization travels down the length of the axon like the current in an electrical wire.

The following diagrams and graph sum up this action potential, or nerve impulse:

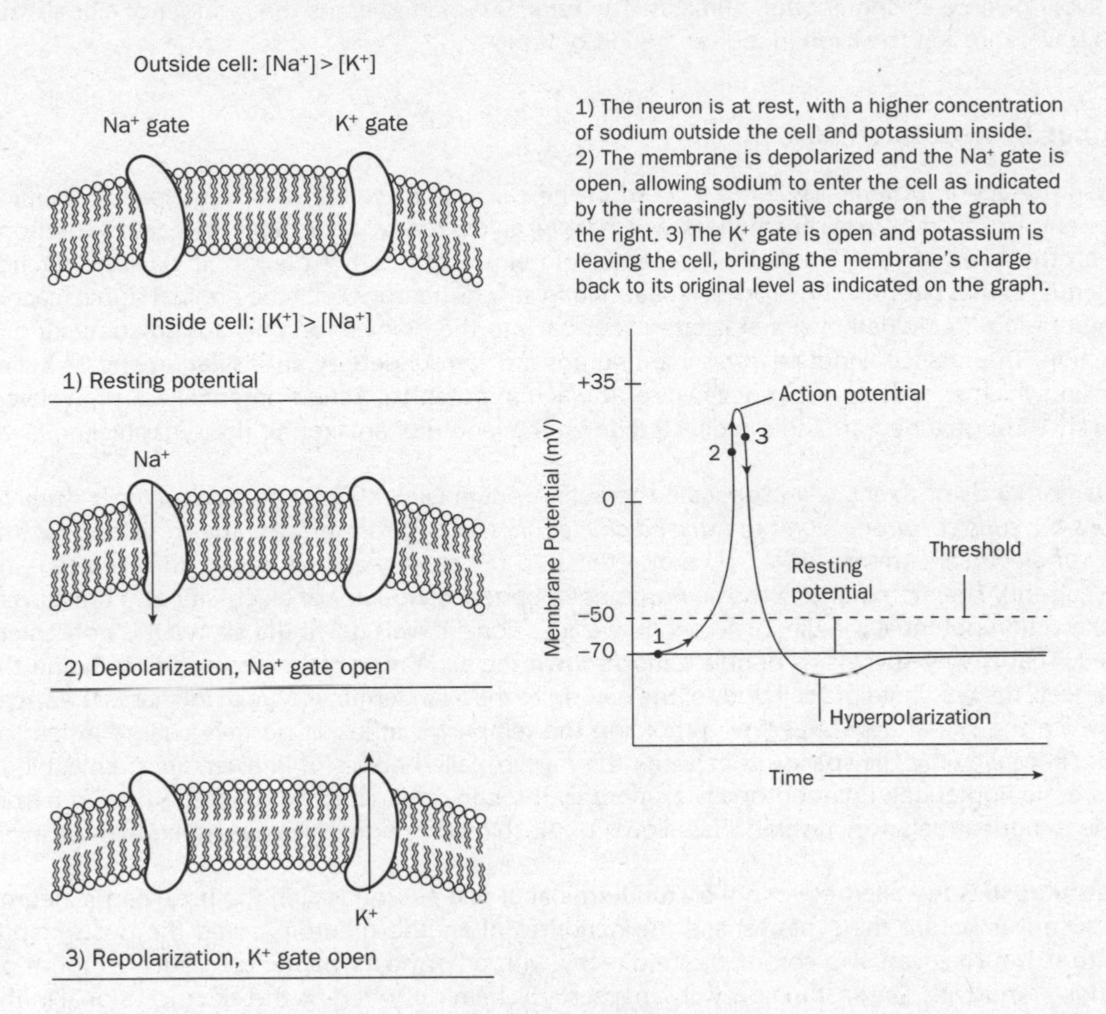

Action Potential

Ions do *not* readily travel across cell membranes because of their charges, and they must be carried across by either membrane-spanning proteins or specific protein channels that let only certain types of ions through. Some channels let only Na^+ through (sodium channels), some only K^+, some only Cl^-, and so on. The depolarization that occurs across the membrane of a neuron is controlled by special gates that control the diffusion of Na^+ and K^+ ions. The above figure shows the timing of the ion gates during the course of an action potential.

Sodium and potassium channels are open for a very short time (milliseconds), yet the rush of positive charge into one region of a neuron can set off a cascade of channel opening along the entire length of the axon. Voltage-gated sodium channels adjacent to where the axon first depolarizes (usually at the axon hillock) will open, causing sodium channels farther down the axon membrane to do the same. In such a way, the action potential is propagated down the entire length of the axon. Once opened, sodium channels inactivate for a brief period of time, which means that the action potential (flow of positive charge down the axon) occurs in one direction only, down toward the axon terminal. The inability of channels to open again creates an absolute refractory period in which another impulse cannot travel along the axon. This is followed by a relative refractory period during which it is more difficult than normal to send an action potential, but possible given a strong enough stimulus. The refractory period limits the number of signals that can travel through the axon in a given period of time.

Signal Transmission

When the action potential reaches the end of the cell, it initiates the release of special chemical messengers called **neurotransmitters**, which travel across the synaptic gap between two neurons. When the neurotransmitter binds to receptor proteins on the other neuron, it causes an action potential in that neuron. The process repeats along several neurons until the original signal reaches a processing area where a signal is often sent back to the point of origin, causing some kind of reaction. To enhance conductivity, some neurons are surrounded by an insulating sheath called myelin, which speeds up the movement of the action potential. Action potentials can travel very quickly through a neuron and are directed down the length of an axon to the synaptic knob.

Different kinds of axons will propagate the action potential at different speeds. *Larger diameter axons have faster transmission than small diameter ones* (recall from physics that resistance to the flow of electrical current is inversely proportional to the cross-sectional area of the wire carrying the current). The membrane of the neuron is not a perfect conductor of electricity, and the current of the action potential will diminish over time as the signal travels down the axon if it is not replenished. That is why successive depolarizations down the axon membrane succeed in bringing the charge all the way from the cell body of the neuron to the axon terminal. Myelin increases the speed at which the signal travels, as well as holding the temporary influx of positive charge inside the axon. Because only the spaces in between the myelin, called nodes of Ranvier, are permeable to ions, action potentials do not propagate smoothly through myelinated axons, but rather jump from node to node in saltatory fashion. This allows for much quicker signal transmission down the axon.

The **synapse** is the gap between the axon terminal of one neuron (called the presynaptic neuron because it is before the synapse) and the dendrites of another neuron (called the postsynaptic neuron). Neurons can also communicate directly with other postsynaptic cells, such as those on glands or muscles. The vast majority of synapses are chemical, whereby the electrical signal in the presynaptic neuron is converted into a chemical signal in the synapse, which in turn incites a new electrical signal in the postsynaptic neuron. Electrical synapses can occur where gap junctions join the cytoplasms of neighboring cells, directly transmitting ionic signals.

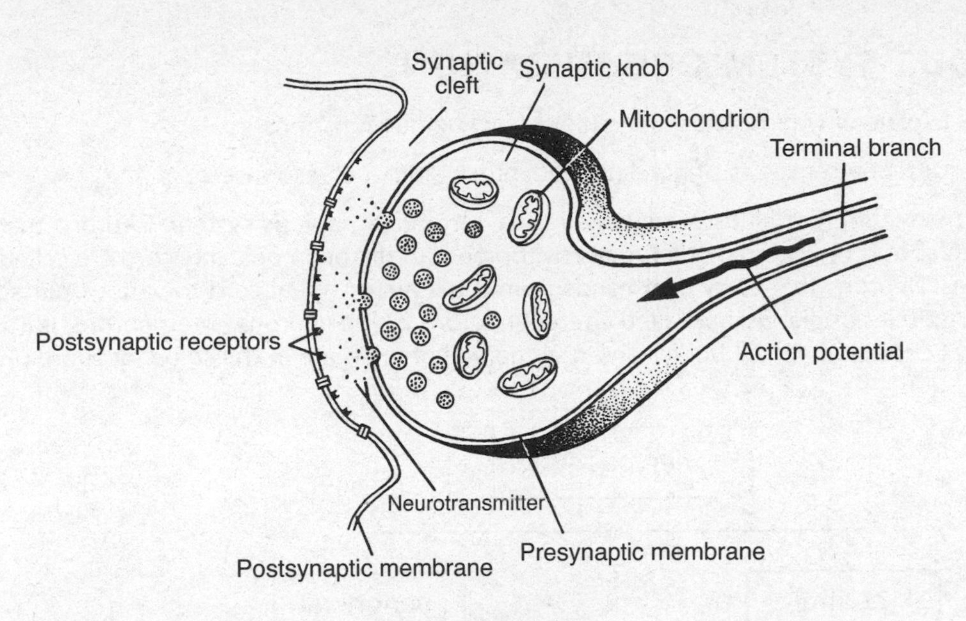

Synaptic Knob

As the action potential reaches the synaptic knob (terminal), voltage-gated calcium (Ca^{2+}) channels at the terminal end open up, and the rapid influx of calcium ions causes membrane-bound vesicles filled with neurotransmitters to merge with the presynaptic membrane, releasing their contents into the synapse. The synapse is a very small space, comprising at most a few micrometers from the presynaptic neuron to the postsynaptic one. Excitatory neurotransmitters, such as acetylcholine (ACh), will bind to membrane receptors on the postsynaptic dendrites or cell body and cause the opening of sodium channels (ligand-gated channels) on the postsynaptic cell, starting the process of an action potential all over again.

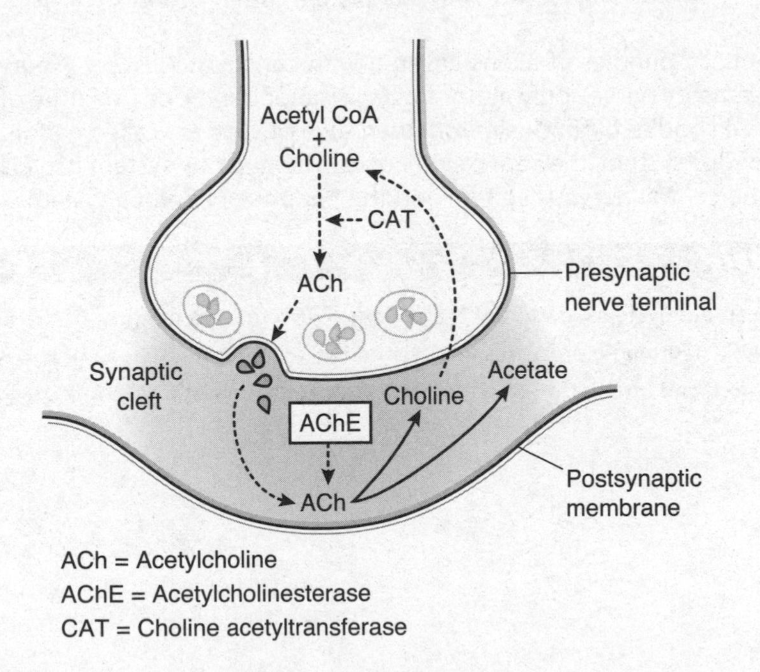

Synapse Ion Transmission

NERVOUS SYSTEM ORGANIZATION

10.5 Explain why different brain regions have unique functions

10.6 Describe senses as signal detection, processing, and response

There are many different kinds of neurons in the vertebrate nervous system. Neurons that carry information about the external or internal environment to the brain or spinal cord are called afferent neurons. Neurons that carry commands from the brain or spinal cord to various parts of the body (e.g., muscles or glands) are called efferent neurons. Some neurons (interneurons) participate only in local circuits; their cell bodies and their nerve terminals are in the same location.

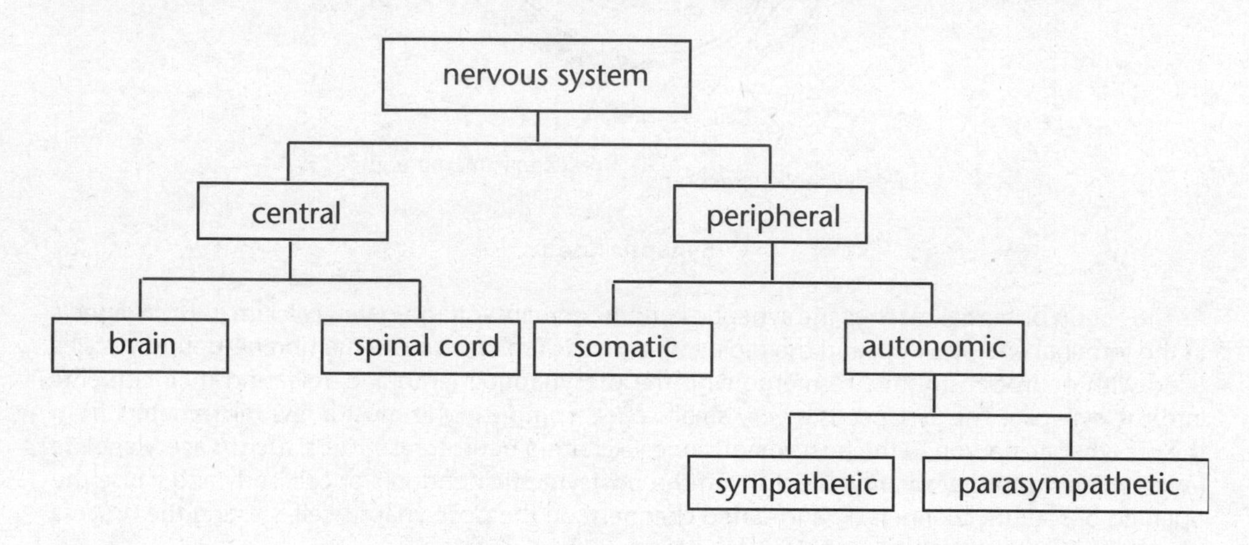

Vertebrate Nervous System Organization

Nerves are essentially bundles of axons covered with connective tissue. A nerve may carry only sensory fibers (a sensory nerve), only motor fibers (a motor nerve), or a mixture of the two (a mixed nerve). Neuronal cell bodies often cluster together; such clusters are called ganglia in the periphery; in the central nervous system, they are called nuclei. The nervous system itself is divided into two major systems, the central nervous system and the peripheral nervous system.

AP Expert Note

The types of nervous systems, details of the various structures and features of the brain parts, and details of specific neurologic processes are beyond the scope of the AP exam. Focus on the detection, transmission, and integration of stimuli and the production of a response.

Central Nervous System

The **central nervous system** (CNS) consists of the brain and spinal cord.

Brain

The brain is a jellylike mass of neurons that resides in the skull. Its functions include interpreting sensory information, forming motor plans, and cognitive function (thinking). The brain consists of gray matter (cell bodies) and white matter (myelinated axons). The brain can be divided into the forebrain, midbrain, and hindbrain.

Forebrain

The forebrain consists of a number of different structures, including the thalamus, hypothalamus, posterior pituitary gland, pineal gland, basal ganglia, limbic system, and cerebral cortex. The cerebral cortex is divided into two halves, called cerebral hemispheres. The surface of the cortex is divided into four lobes—the frontal lobe, parietal lobe, occipital lobe, and temporal lobe. The cortex processes and integrates sensory input and motor responses and is important for memory and creative thought. Right and left cerebral cortices communicate with each other through the corpus callosum.

Midbrain

The midbrain is a relay center for visual and auditory impulses. It also plays an important role in motor control.

Hindbrain

The hindbrain is the posterior part of the brain and consists of the cerebellum, the pons, and the medulla. The cerebellum helps to modulate motor impulses initiated by the motor cortex and is important in the maintenance of balance, hand-eye coordination, and the timing of rapid movements. One function of the pons is to act as a relay center to allow the cortex to communicate with the cerebellum. The medulla (also called the medulla oblongata) controls many vital functions such as breathing, heart rate, and gastrointestinal activity. Together, the midbrain, pons, and medulla constitute the brain stem.

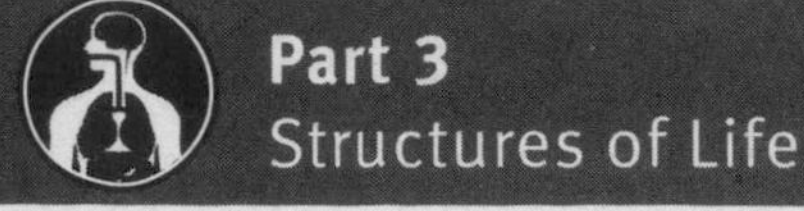

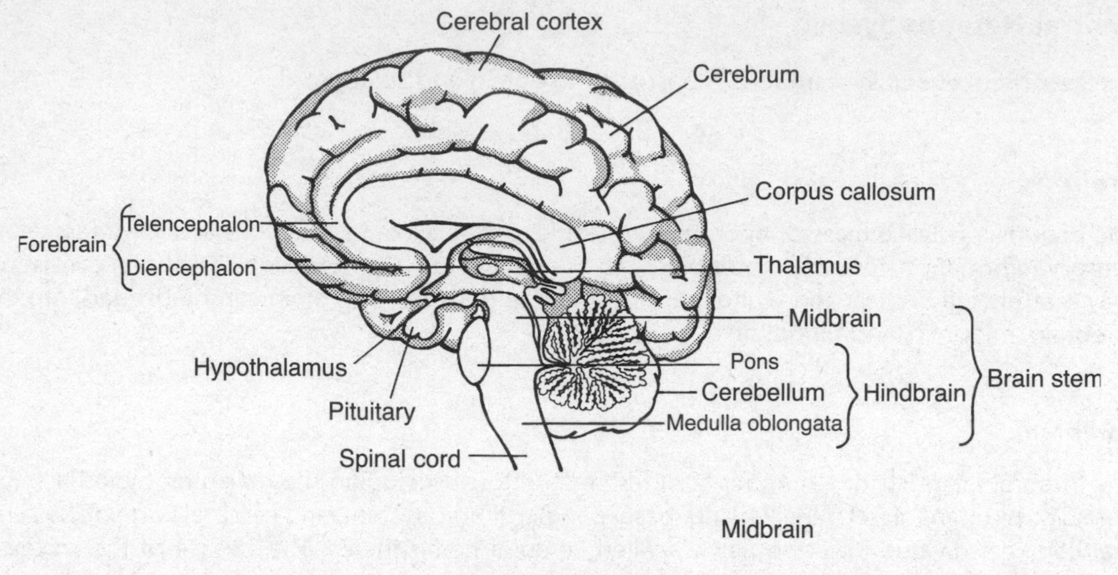

The Human Brain

Spinal Cord

The spinal cord is an elongated structure continuous with the brain stem that extends down the dorsal side of vertebrates. Nearly all nerves that innervate the viscera or muscles below the head pass through the spinal cord, and nearly all sensory information from below the head passes through the spinal cord on the way to the brain. The spinal cord can also integrate simple motor responses (e.g., reflexes) by itself.

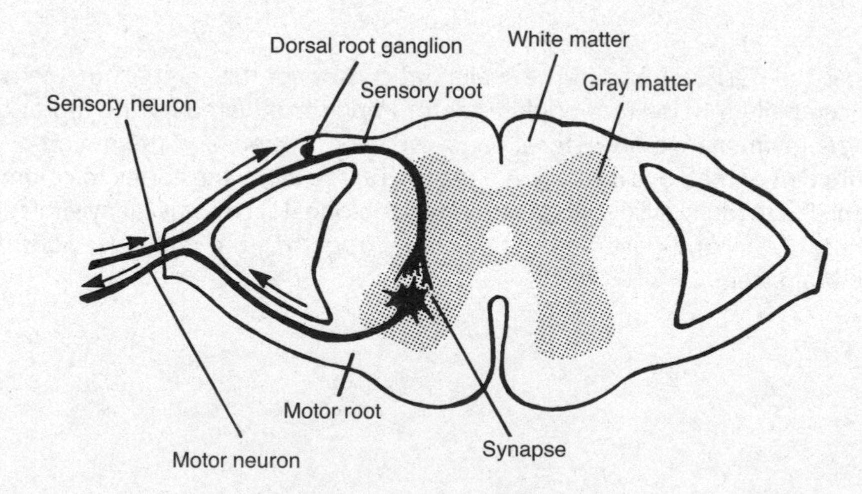

Cross-Section of Spinal Cord

A cross-section of the spinal cord reveals an outer white matter area containing motor and sensory axons and an inner gray matter area containing nerve cell bodies. Sensory information enters the spinal cord dorsally; the cell bodies of these sensory neurons are located in the dorsal root ganglia.

All motor information exits the spinal cord ventrally. Nerve branches entering and leaving the cord are called roots. The spinal cord is divided into four regions (in order from the brain stem to the tail): cervical, thoracic, lumbar, and sacral.

Peripheral Nervous System

The peripheral nervous system (PNS) consists of 12 pairs of cranial nerves, which primarily innervate the head and shoulders, and 31 pairs of spinal nerves, which innervate the rest of the body. Most of the cranial nerves exit from the brain stem, while spinal nerves exit from the spinal cord.

The PNS has two primary divisions: the somatic and the autonomic nervous systems, each of which has both motor and sensory components.

Somatic Nervous System

The somatic nervous system (SNS) innervates skeletal muscles and is responsible for voluntary movement. Motor neurons release the neurotransmitter acetylcholine (ACh) onto ACh receptors located on skeletal muscle. This causes depolarization of the skeletal muscle, leading to muscle contraction. In addition to voluntary movement, the somatic nervous system is also important for reflex action. There are both monosynaptic and polysynaptic reflexes.

Monosynaptic reflex pathways have only one synapse between the sensory neuron and the motor neuron. The classic example is the knee-jerk reflex. When the tendon beneath the patella (kneecap) is hit, stretch receptors sense this and action potentials are sent up the sensory neuron and into the spinal cord. The sensory neuron synapses with a motor neuron in the spinal cord, which in turn stimulates the quadriceps muscle to contract, causing the lower leg to kick forward.

In polysynaptic reflexes, sensory neurons synapse with more than one neuron. A classic example of this is the withdrawal reflex. When a person steps on a nail, the injured leg withdraws in pain, while the other leg extends to retain balance.

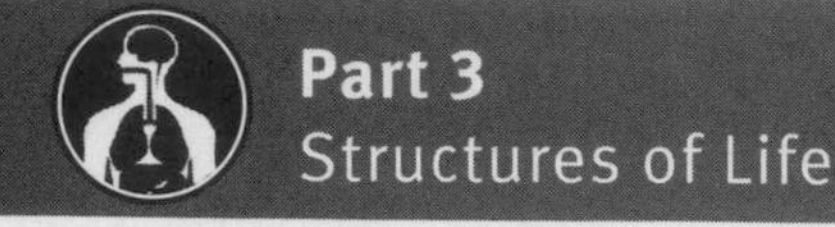

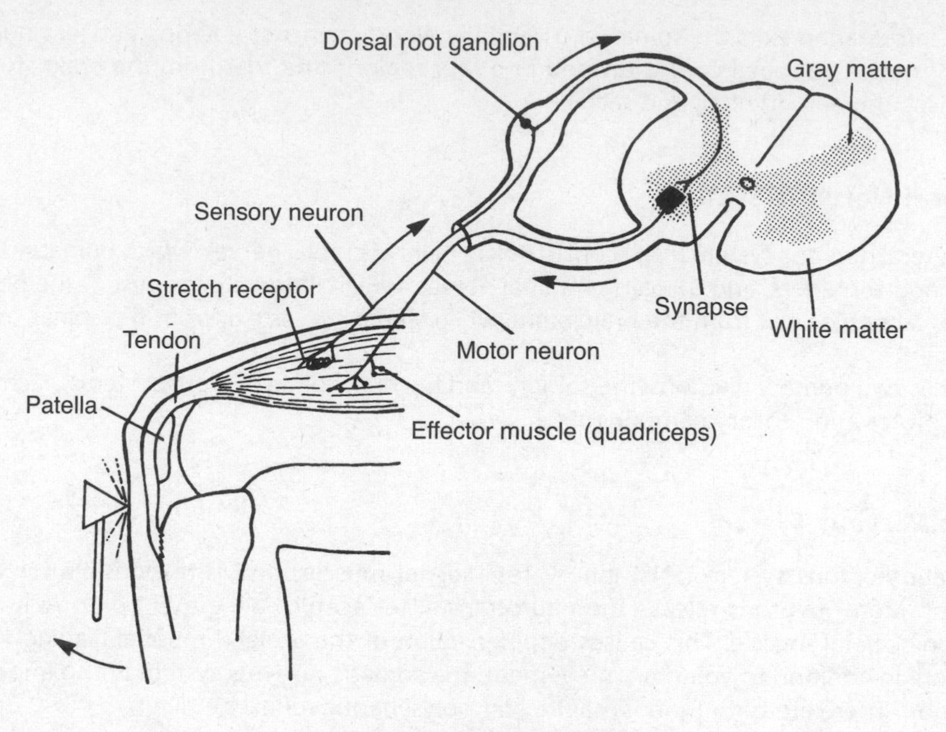

Reflex Arc for Knee Jerk

Autonomic Nervous System

The **autonomic nervous system** (ANS) is sometimes also called the involuntary nervous system because it regulates the body's internal environment without the aid of conscious control. Whereas the somatic nervous system innervates skeletal muscle, the ANS innervates cardiac and smooth muscle. Smooth muscle is located in areas such as blood vessels, the digestive tract, the bladder, and bronchi, so it isn't surprising that the ANS is important in blood pressure control, gastrointestinal motility, excretory processes, respiration, and reproductive processes. ANS pathways are characterized by a two-neuron system. The first neuron (preganglionic neuron) has a cell body located within the CNS and its axon synapses in peripheral ganglia. The second neuron (postganglionic neuron) has its cell body in the ganglia and then synapses on cardiac or smooth muscle.

The ANS is composed of two subdivisions, the sympathetic and the parasympathetic nervous systems, which generally act in opposition to one another.

Sympathetic Nervous System

The sympathetic division is responsible for the fight-or-flight responses that ready the body for action. It basically does everything you would want it to do in an emergency situation. It increases blood pressure and heart rate, increases blood flow to skeletal muscles, and decreases gut motility. The preganglionic neurons emerge from the thoracic and lumbar regions of the spinal cord and use acetylcholine as their neurotransmitter; the postganglionic neurons typically release norepinephrine. The action of preganglionic sympathetic neurons also causes the adrenal medulla to release **adrenaline** (epinephrine) into the bloodstream.

Parasympathetic Nervous System

The parasympathetic division acts to conserve energy and restore the body to resting activity levels following exertion (rest and digest). It acts to lower heart rate and to increase gut motility. One very important parasympathetic nerve that innervates many of the thoracic and abdominal viscera is the vagus nerve. Parasympathetic neurons originate in the brain stem (cranial nerves) and the sacral part of the spinal cord. Both the preganglionic and postganglionic neurons release acetylcholine.

The Senses

The body has three types of sensory receptors to monitor its internal and external environments: interoceptors, proprioceptors, and exteroceptors. Interoceptors monitor aspects of the internal environment such as blood pressure, the partial pressure of CO_2 in the blood, and blood pH. Proprioceptors transmit information regarding the position of the body in space. These receptors are located in muscles and tendons to tell the brain where the limbs are in space, and they are also located in the inner ear to tell the brain where the head is in space. Exteroceptors sense things in the external environment such as light, sound, taste, pain, touch, and temperature.

The Eye

The eye detects light energy (as photons) and transmits information about intensity, color, and shape to the brain. The eyeball is covered by a thick, opaque layer known as the sclera, which is also known as the white of the eye. Beneath the sclera is the choroid layer, which helps to supply the retina with blood. The innermost layer of the eye is the retina, which contains the photoreceptors that sense the light.

The transparent cornea at the front of the eye bends and focuses light rays. The rays then travel through an opening called the pupil, whose diameter is controlled by the pigmented, muscular iris. The iris responds to the intensity of light in the surroundings (light makes the pupil constrict). The light continues through the lens, which is suspended behind the pupil. The lens, the shape of which is controlled by the ciliary muscles, focuses the image onto the retina. In the retina are photoreceptors that transduce light into action potentials. There are two main types of photoreceptors: cones and rods. Cones respond to high-intensity illumination and are sensitive to color, while rods detect low-intensity illumination and are important in night vision. The cones and rods contain various pigments that absorb specific wavelengths of light.

The cones contain three different pigments that absorb red, green, and blue wavelengths; the rod pigment, rhodopsin, absorbs one wavelength. The photoreceptor cells synapse onto bipolar cells, which in turn synapse onto ganglion cells. Axons of the ganglion cells bundle to form the right and left optic nerves, which conduct visual information to the brain. The point at which the optic nerve exits the eye is called the blind spot because photoreceptors are not present there. There is also a small area of the retina called the fovea, which is densely packed with cones and is important for high acuity vision.

The eye also has its own circulation system. Near the base of the iris, the eye secretes aqueous humor, which travels to the anterior chamber of the eye, from which it exits and eventually joins venous blood.

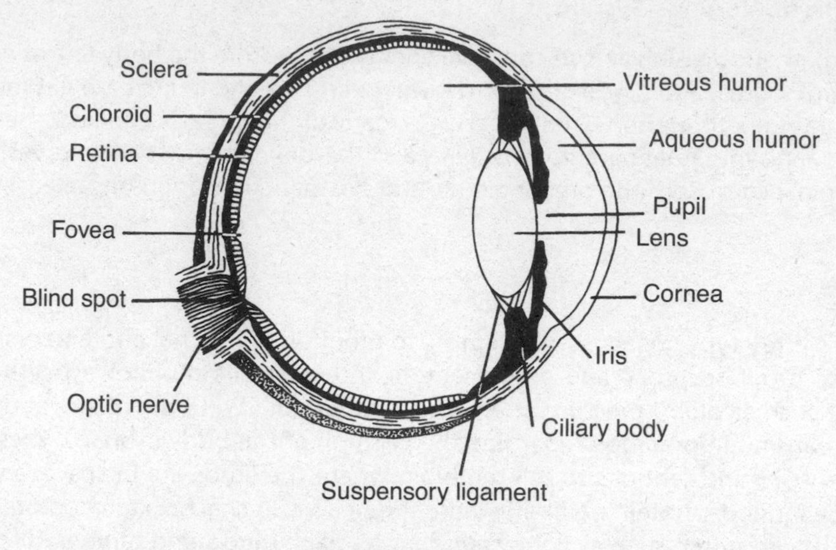

The Human Eye

The Ear

The ear transduces sound energy (pressure waves) into impulses perceived by the brain as sound. The ear is also responsible for maintaining equilibrium (balance) in the body.

Sound waves pass through three regions as they enter the ear. First, they enter the outer ear, which consists of the auricle (pinna) and the auditory canal. At the end of the auditory canal is the tympanic membrane (eardrum) of the middle ear, which vibrates at the same frequency as the incoming sound. Next, the three bones, or ossicles (malleus, incus, and stapes), amplify the stimulus and transmit it through the oval window, which leads to the fluid-filled inner ear. The inner ear consists of the cochlea and the semicircular canals. The cochlea contains the organ of Corti, which has specialized sensory cells called hair cells. Vibration of the ossicles exerts pressure on the fluid in the cochlea, stimulating the hair cells to transduce the pressure into action potentials, which travel via the auditory (cochlear) nerve to the brain for processing.

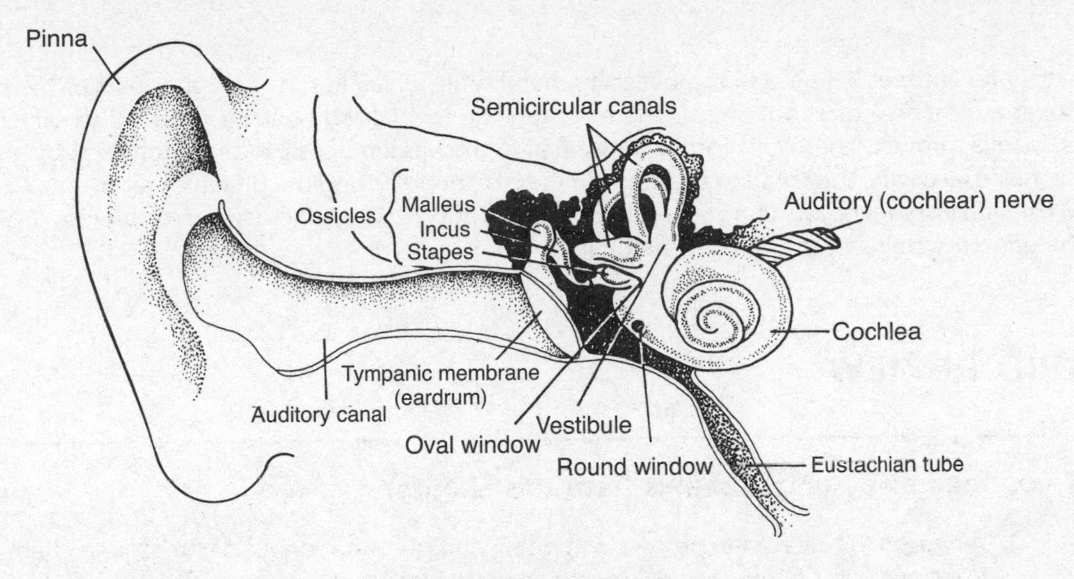

The Human Ear

The three semicircular canals are each perpendicular to the other two and filled with a fluid called endolymph. At the base of each canal is a chamber with sensory hair cells; movement of the head displaces endolymph in one or more of the canals, putting pressure on the hair cells within them. This changes the nature of impulses sent by the vestibular nerve to the brain. The brain interprets this information to determine the position of the head.

The Chemical Senses

The chemical senses are taste and smell. These senses transduce chemical changes in the environment, specifically in the mouth and nose, into gustatory and olfactory sensory impulses, which are interpreted by the nervous system.

Taste

Taste receptors, bundled into clusters known as taste buds, are located on the tongue, the soft palate, and the epiglottis. Taste buds are composed of approximately 50–100 epithelial cells. The outer surface of a taste bud contains a taste pore, from which microvilli, or taste hairs, protrude. The receptor surfaces for taste are on the taste hairs. Interwoven around the taste buds is a network of nerve fibers that are stimulated by the taste receptors. These neurons transmit gustatory information to the brainstem via three cranial nerves. There are five kinds of taste sensations: sour, salty, sweet, bitter, and umami (savory). Although most taste buds will respond to all five stimuli, they respond preferentially (i.e., at a lower threshold) to one or two of them.

Smell

Olfactory receptors are found in the olfactory membrane, which lies in the upper part of the nostrils over a total area of about 5 cm^2. The receptors are specialized neurons from which olfactory hairs, or cilia, project. These cilia form a dense mat in the nasal mucosa. When odorous substances enter the nasal cavity, they bind to receptors in the cilia, depolarizing the olfactory receptors. Axons from the olfactory receptors join to form the olfactory nerves. The olfactory nerves project directly to the olfactory bulbs in the base of the brain.

RAPID REVIEW

If you take away only 4 things from this chapter:

1. A neuron (or nerve cell) processes and transmits information via electrical and chemical signaling. Chemical signals occur across a synapse via neurotransmitters. Electrical signals along the axon depend upon the flow of ions across the axon membrane and cell body membrane. The opening and closing of sodium and potassium voltage-gated channels are responsible for the transmission of electrical signals.
2. The nervous system is broken down into two main parts: the central nervous system and the peripheral nervous system. The central nervous system contains the brain and spinal column and is used to control animal behavior based on input from the environment. The peripheral nervous system connects the central nervous system to all other parts of the body.
3. The autonomic nervous system regulates the body's internal environment without the aid of conscious control. It is composed of two subdivisions, the sympathetic and the parasympathetic nervous systems. The sympathetic division is responsible for the "fight-or-flight" responses that ready the body for action. The parasympathetic division acts to conserve energy and restore the body to resting activity levels following exertion.
4. The body has several types of special sensors and sensory receptors to monitor its internal and external environments: interoceptors, proprioceptors, and exteroceptors. The chemical senses are taste and smell. The ears transduce sound energy (pressure waves) into impulses and are responsible for maintaining equilibrium, while the eyes use light input for vision.

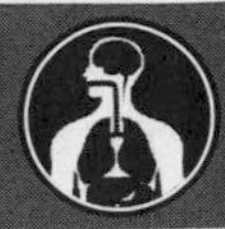

TEST WHAT YOU LEARNED

1. The feeling of fear is a perceived danger that humans experience, which can lead to a behavioral response, such as hiding or running away. Fear can be an irrational response, as observed with certain phobias, or it can arise from some former traumatic event. For example, if a person was involved in a serious car accident as a child, then as an adult a feeling of fear might emerge every time he or she goes for a long car ride. Which of the following statements best describes the neurological basis of fear?

   (A) Fear is a sensory function because it is associated with perception.

   (B) Fear is an integration function because it interprets the sense of touch.

   (C) Fear is a motor response to a given sensory stimuli.

   (D) Fear is a higher function involving multiple regions of the brain.

2. Human vision is a very complex system. The human eye detects a variety of light energy at different wavelengths, all reflected from a multitude of objects. These signals are then sent to the brain, which interprets the inputs and creates the visual field.

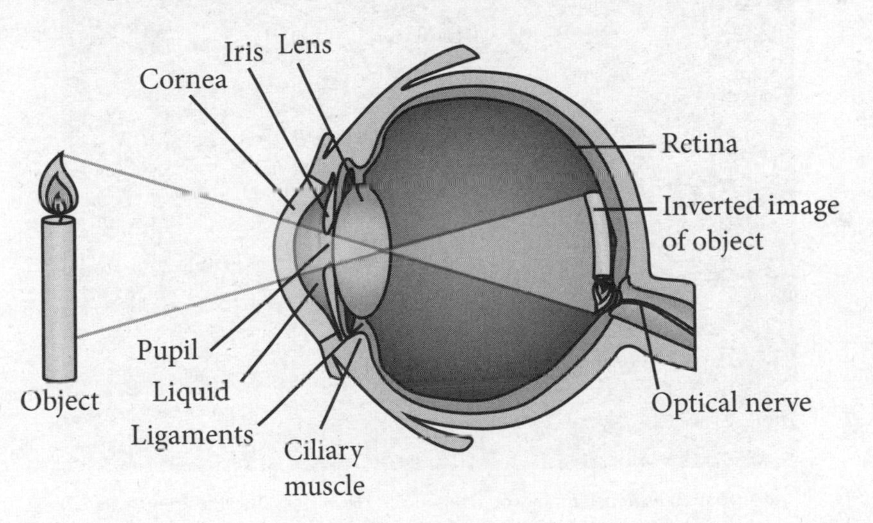

   Based on the cross-section of the human eye above, which of the following best describes how the human eye functions?

   (A) The light from the object is focused on the iris, where it will then be converted to nerve impulses that will travel from the iris to the brain for processing.

   (B) The light from the object is focused on the lens from the retina. The lens will then convert the light energy into nerve impulses that will travel from the iris to the brain for processing.

   (C) The light from the object is focused on the retina by the lens. The retina will then convert the light energy into nerve impulses that will travel to the brain for processing.

   (D) The light from the object is focused on the optical nerve by the retina. The optical nerve will then convert the light energy into nerve impulses that will travel to the brain for processing.

3. Alzheimer's disease is a neurodegenerative disorder that progresses over time. It leads to nerve cell damage and death throughout the brain. Changes in the forebrain are typically observed in Alzheimer's patients. How would these changes affect the memory of patients with Alzheimer's disease?
   (A) Patients would lose the ability to form new memories because the forebrain contains the hippocampus, which regulates memory development.
   (B) Patients' memory would be unaffected because the forebrain is only responsible for motor functions.
   (C) Patients would lose the ability to form new memories because the forebrain contains the hypothalamus, which is responsible for coordinating the autonomic and homeostatic nervous systems.
   (D) Patients' memory would be unaffected because the forebrain is only responsible for auditory and visual processing.

4. Cocaine is an illegal psychostimulant drug that can cause a sense of increased alertness. Scientists have proposed the following model of how cocaine exerts its effects in the nervous system.

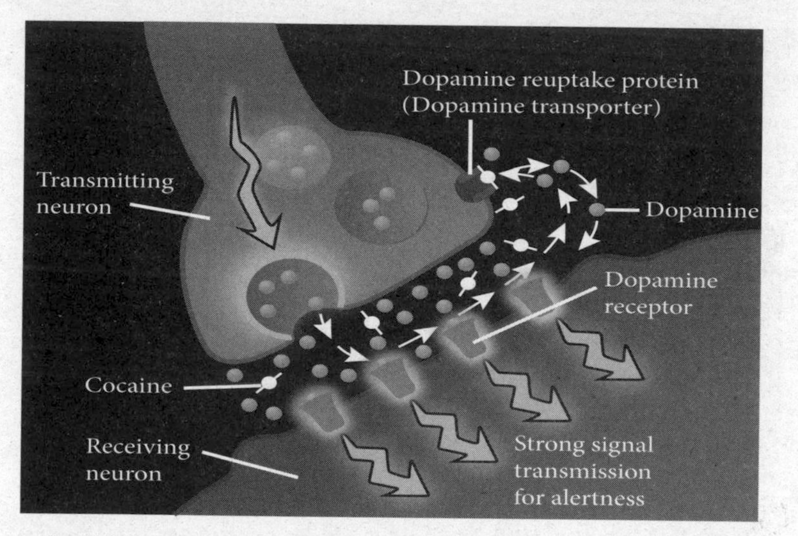

Adapted from National Institute on Drug Abuse, "How Does Cocaine Produce Its Effects?" https://www.drugabuse.gov/publications/research-reports/cocaine/how-does-cocaine-produce-its-effects.

Which of the following statements best summarizes cocaine's mechanism of action?
   (A) Cocaine binds to the dopamine reuptake proteins in the presynaptic cells, which increases dopamine concentration in synaptic clefts.
   (B) Cocaine binds to the dopamine receptors in the presynaptic cells, which inhibits production of dopamine.
   (C) Cocaine reacts with dopamine and increases its concentration in the brain.
   (D) Cocaine causes the formation of plaque buildup in the presynaptic cells.

5. Local anesthesia is preferred over general anesthesia in patients who undergo minor surgical procedures and are at high risk for complications. Local anesthetics block the depolarization of neurons containing pain receptors. How is this most likely accomplished by local anesthetics?
   (A) They prevent the reuptake of neurotransmitters within synapses.
   (B) They increase the rate at which potassium moves across nerve cell membranes.
   (C) They inhibit voltage-gated sodium channels.
   (D) They stimulate voltage-gated sodium channels.

6. Many neurons have axons that are insulated in a myelin sheath. Which statement best describes the effect on signal transmission if the myelin sheath deteriorates on a neuron?
   (A) Loss of the myelin sheath would have no adverse effect on signal transmission.
   (B) Loss of the myelin sheath would result in diminished signal transmission.
   (C) Loss of the myelin sheath would increase signal transmission.
   (D) Loss of the myelin sheath would completely stop signal transmission.

Answer Key

Test What You Already Know

1. **A** **Learning Objective:** 10.1
2. **C** **Learning Objective:** 10.2
3. **D** **Learning Objective:** 10.3
4. **B** **Learning Objective:** 10.4
5. **C** **Learning Objective:** 10.5
6. **B** **Learning Objective:** 10.6

Test What You Learned

1. **D** **Learning Objective:** 10.4
2. **C** **Learning Objective:** 10.6
3. **A** **Learning Objective:** 10.5
4. **A** **Learning Objective:** 10.2
5. **C** **Learning Objective:** 10.3
6. **B** **Learning Objective:** 10.1

Detailed solutions can be found in the Answers and Explanations section at the back of this book.

REFLECTION

Test What You Already Know score: __________

Test What You Learned score: __________

Use this section to evaluate your progress. After working through the pre-quiz, check off the boxes in the "Pre" column to indicate which Learning Objectives you feel confident about. Then, after completing the chapter, including the post-quiz, do the same to the boxes in the "Post" column. Keep working on unchecked Objectives until you're confident about them all!

Pre	Post	
☐	☐	**10.1** Explain the function of neuron structures
☐	☐	**10.2** Describe synapse transmission and neurotransmitters
☐	☐	**10.3** Describe electrical potential and ion requirements for an impulse
☐	☐	**10.4** Describe how the nervous system transmits information
☐	☐	**10.5** Explain why different brain regions have unique functions
☐	☐	**10.6** Describe senses as signal detection, processing, and response

FOR MORE PRACTICE

Complete more practice online at kaptest.com. Haven't registered your book yet? Go to kaptest.com/booksonline to begin.

CHAPTER 11

Other Organ Systems

LEARNING OBJECTIVES

In this chapter, you will review how to:

11.1	Explain how the respiratory system supports the organism
11.2	Describe circulation as nutrient and waste transfer
11.3	Differentiate between myoglobin and hemoglobin
11.4	Describe the cooperation of organs required for digestion
11.5	Explain the excretory systems as waste transfer

TEST WHAT YOU ALREADY KNOW

1. The bicarbonate buffer system in the blood can be represented by the following equation.

$$CO_2 + H_2O \leftrightarrow H_2CO_3 \leftrightarrow HCO_3^- + H^+$$

An increase in CO_2 concentration will shift the reaction to the right, causing an increase in H^+ ions and a drop in pH. The dissociation curve shows the relative amount of O_2 bound to hemoglobin as a function of the partial pressure of O_2.

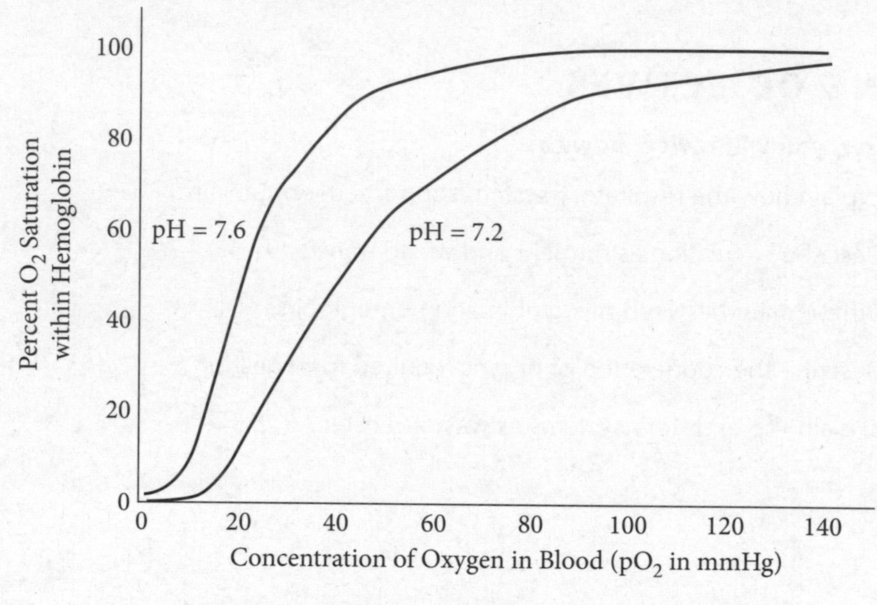

Adapted from R. J. Gillies et al., "MRI of the Tumor *Microenvironment*," *J Magn Reson Imaging* 16, no. 4 (December 2002): 430–450

According to the diagram and the bicarbonate equation, which of the following statements best explains the movement of O_2 between red blood cells and muscle tissue during exercise?

(A) Oxygen binds more tightly to hemoglobin when CO_2 is released by exercising muscle tissue.

(B) Oxygen is released from hemoglobin because the pH increases in exercising muscle tissue.

(C) Oxygen is released from hemoglobin because the pH decreases in exercising muscle tissue.

(D) Oxygen binds more tightly to hemoglobin because CO_2 is taken up by exercising muscle.

2. After a meal rich in fatty food, blood will appear milky. The appearance is caused by the transportation of hydrophobic lipids in the hydrophilic blood. Which statement correctly describes the transport of cholesterol, triglycerides, and other lipids in blood?

(A) Lipids form minuscule fat droplets in the plasma, which are transported through the body.

(B) Lipids circulate in the form of bilayers arranged in micelles.

(C) Lipids bind to carbohydrates and are transported as carbohydrate-lipid particles.

(D) Lipids are transported as lipoprotein particles made of phospholipids and apolipoprotein.

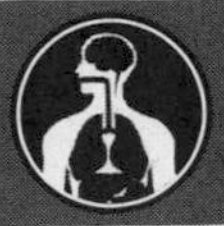

3. Carbon dioxide reacts with water as shown in the following equation.

$$CO_2 + H_2O \leftrightarrow H_2CO_3 \leftrightarrow HCO_3^- + H^+$$

When a tissue is more active, carbon dioxide production and ATP consumption increase. The graph below shows the oxygen-binding affinities of hemoglobin and myoglobin.

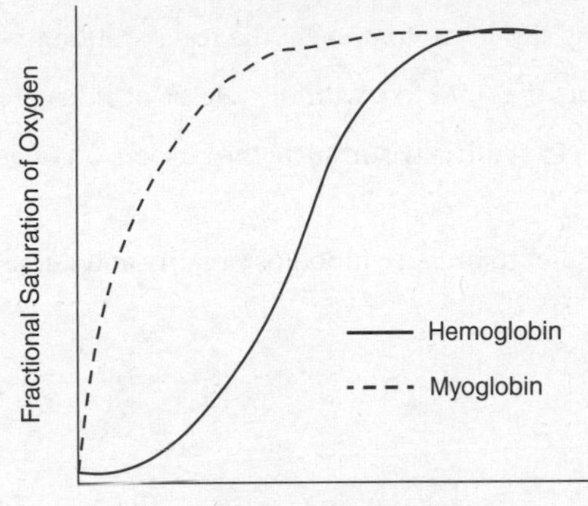

Which of the following best explains the responses of hemoglobin and myoglobin in a tissue with a high metabolic rate?

(A) Hemoglobin is more inclined to release oxygen because its oxygen-binding affinity is lower than myoglobin's.

(B) Myoglobin is more inclined to release oxygen because its oxygen-binding affinity is lower than hemoglobin's.

(C) Hemoglobin is more inclined to bind oxygen because its oxygen-binding affinity is higher than myoglobin's.

(D) Myoglobin is more inclined to bind oxygen because its oxygen-binding affinity is higher than hemoglobin's.

4. Many digestive enzymes that operate in the duodenum can only function under basic conditions. The duodenum increases the production of the hormone secretin when chyme (a mixture of partially digested food and stomach acid) is released from the stomach. The target organ for secretin is the pancreas, which secretes bicarbonate (HCO_3^-) ions into the duodenum in response to the secretin signal. Which of the following statements best explains the release of bicarbonate ions from the pancreas during digestion?

(A) Bicarbonate ions act as cofactors of digestive enzymes secreted in the duodenum.

(B) Bicarbonate ions help in the emulsification of fats into droplets for digestion.

(C) Bicarbonate ions neutralize chyme and raise the pH in the duodenum.

(D) Bicarbonate ions stimulate smooth muscles in the walls of the duodenum.

5. A drop in blood osmolarity stimulates the secretion of renin by the kidneys. Renin is an enzyme that cleaves the precursor hormone angiotensinogen to angiotensin I. Angiotensin-converting enzyme (ACE) then converts angiotensin I to angiotensin II. Angiotensin II promotes the release of the hormone aldosterone, which increases reabsorption of sodium from nephrons. Based on this feedback mechanism, which of the following treatments would be beneficial to patients who have high blood pressure?

   (A) Taking Na^+/K^+ pump inhibitors to prevent the reabsorption of sodium ions in the nephrons

   (B) Taking ACE inhibitors that interfere with the release of aldosterone

   (C) Taking renin analogs to artificially stimulate the conversion of angiotensinogen to angiotensin I

   (D) Going on a high salt diet to increase blood osmolarity and decrease the release of renin

Answers to this quiz can be found at the end of this chapter.

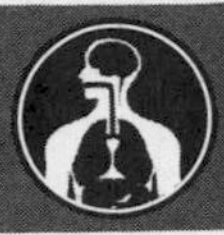

RESPIRATORY AND CIRCULATORY SYSTEMS

11.1 Explain how the respiratory system supports the organism

11.2 Describe circulation as nutrient and waste transfer

11.3 Differentiate between myoglobin and hemoglobin

It is a wonder that organisms function at all. Most organisms are complex machines with many intricate parts and processes. It is perhaps easier to fathom the workings of a simple prokaryotic cell—coordinating activity with other organisms and the liquid matrix in which it lives—than to study complex organisms. The synchronicity of the cells, tissues, and **organs** in a multicellular organism involves a highly complex set of processes that are not easy to simplify.

✔ **AP Expert Note**

The specific structures of the four organ systems discussed in this chapter are not required knowledge for the AP Biology exam. The information contained here is useful, however, because it can be used as examples for free response questions and because it may provide helpful background for some multiple choice questions. The most important content to focus on in this chapter is how organs cooperate to form organ systems and how organ systems cooperate to maintain dynamic homeostasis within organisms.

Respiratory System

Gas exchange and transport depend upon the movement of gases or fluids around an organism's body and the passage of these gases or materials contained within the fluids into and out of cells. All gas exchange is based upon simple diffusion across cell membranes. The key characteristic that has evolved across almost all species for maximal gas exchange is utilizing moist membranes that cover a *large amount of surface area.*

✔ **AP Expert Note**

Positive vs. Negative Pressure Breathing

Frogs and other amphibians can ventilate themselves by literally pushing air down their windpipes. They open and lower the floor of their mouths and gulp in air. This is positive pressure breathing. Mammals have a muscular band of tissue called the diaphragm, which separates the chest cavity from the abdominal cavity. As this muscle contracts and pulls down, the pressure in the chest cavity decreases, sucking air in from outside. This drop in air pressure from within is what allows negative pressure breathing to take place.

While organisms such as amphibians, reptiles, and even some fishes (e.g., lungfish) may possess **lungs** to aid in breathing and gas exchange, these organisms all use other structures as their primary means of oxygen and carbon dioxide exchange. In mammals, the lungs are the primary structure. Because the lungs are restricted to one location in the body, a **circulatory system** is needed to bridge the gap between the lungs and the cells around the organism's body. All mammalian lungs end in blind sacs called alveoli, which are surrounded by a dense net of capillaries into which oxygen diffuses and from which carbon dioxide is excreted.

In mammals, the path that air takes once inhaled is as follows: once brought into the nose or mouth, air enters the pharynx (throat) and passes across the larynx (voice box), which lies within the trachea (windpipe). The trachea branches into two bronchi (singular: bronchus), which further branch into bronchioles and then alveoli. Cells that line the bronchioles and alveoli keep the lungs moist, secreting mucus and other watery fluids into the lung openings. There are also a great many cells covered with short hairlike cilia to keep mucus and other particles flowing across the inside epithelial surface. The millions of alveoli where the gas exchange actually takes place are separated from capillaries by only a thin layer or two of cells.

Deoxygenated blood enters the pulmonary (lung) capillaries from the systemic circulation having a low partial pressure of oxygen (it has lost its store of O_2 at the tissues). Inhaled air in the alveoli has a much higher partial pressure of oxygen (there is much more oxygen relative to other gases in the alveoli than within the capillaries). Therefore, oxygen diffuses down its concentration gradient into the capillaries, where it binds to hemoglobin molecules in red blood cells and returns to the heart to be pumped out to the body. In contrast, the partial pressure of CO_2 in the capillaries is greater than that of the inhaled alveolar air; thus, CO_2 diffuses from the capillaries into the alveoli, where it is exhaled.

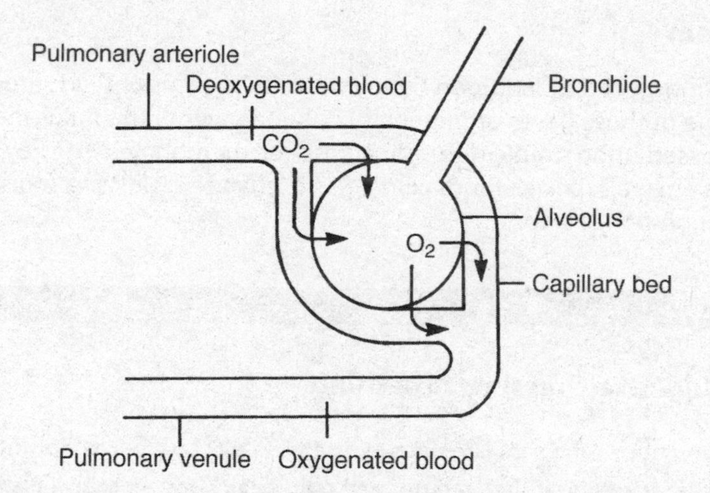

Gas Exchange in the Alveoli

At high altitudes, the partial pressure of O_2 in the atmosphere declines, making it more difficult to get sufficient oxygen to diffuse into the capillaries. The body often compensates for this by increasing the rate of breathing (hyperventilation) and by increasing the number of red blood cells available to carry oxygen. Erythropoietin, a hormone released by the kidneys, is involved in the increased production of red blood cells: it travels to the bone marrow to stimulate red blood cell production there.

Circulatory System

In mammals and birds, blood follows a double circuit. As the diagram below shows, the right and left sides of the heart are each their own separate pumps: the right side pumps deoxygenated blood to the lungs, while the left side pumps oxygenated blood to the aorta and out to the body. The two upper (receiving) chambers are atria, and the two lower chambers are ventricles. While the atria are fairly thin-walled and not very muscular, the ventricles, particularly the left ventricle, are thick with muscle tissue for sustained, forceful pumping.

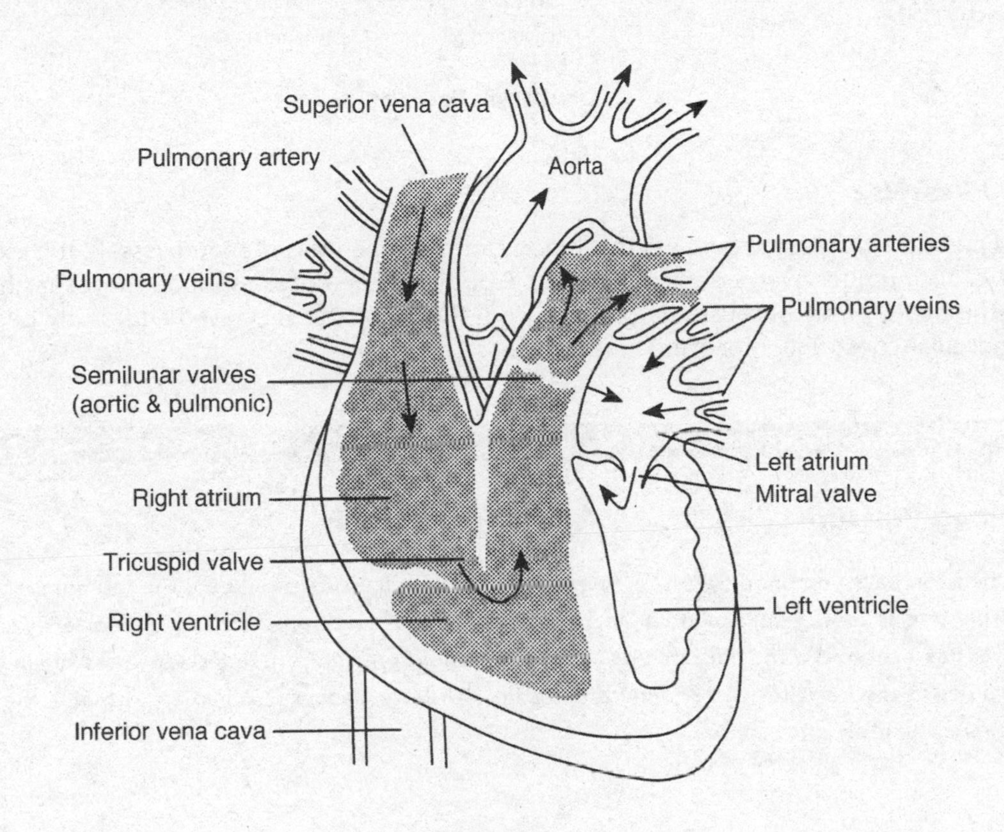

The Heart

The atrioventricular (AV) valves, located between the atria and ventricles on either side of the heart, prevent backflow of blood into the atria when the ventricles are contracting. The semilunar valves in the pulmonary artery and the aorta serve a similar function: to prevent the backflow of blood into the ventricles from the lungs or aorta once the blood has been pumped out of the heart. The heart's pumping cycle is divided into two alternating phases: systole and diastole, which together make up the heartbeat. Systole is the period during which the ventricles contract, forcing blood into the lungs and aorta, and diastole is when the ventricles relax and fill with blood.

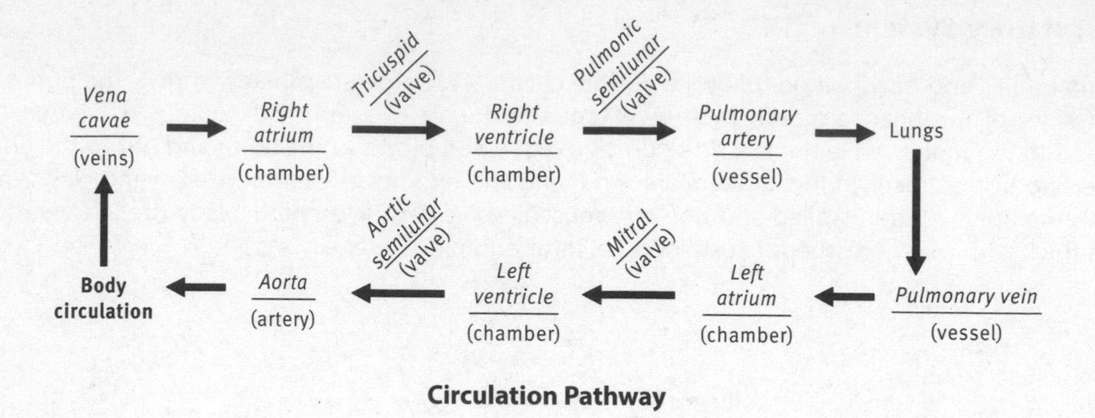

Circulation Pathway

Blood Vessels

Blood pressure is the force per area that blood exerts on the walls of blood vessels. It is expressed as a systolic number over a diastolic number. Blood pressure drops as blood moves farther away from the heart, partly because the distance from the pump has increased and partly because of the muscular construction of arteries versus veins.

> ✔ **AP Expert Note**
>
> **Blood Pressure**
>
> **When you have your blood pressure taken, you get a number such as 120/80. The top number, the larger one, is your systolic blood pressure and reflects how much pressure is in your arteries when your heart contracts. The bottom number is your diastolic pressure and reflects the pressure in your arteries as the heart relaxes. As your arteries fill with fatty plaques and harden with age, the top number typically goes up.**

Arteries are thick-walled, muscular, elastic vessels that transport oxygenated blood away from the heart, while veins are relatively thin-walled inelastic vessels that bring deoxygenated blood back to the heart. When veins are not filled with blood, they often collapse on themselves, only to swell open when blood is pushed through. In fact, because veins have little contractile ability of their own, blood is often pushed back to the heart from the extremities when the skeletal muscles near the veins in those areas contract. As you walk, for instance, contractions of your leg muscles propel blood sitting in leg veins back toward the heart. In contrast, arteries maintain an open **lumen** at all times and can contract against the blood within, helping to propel it away from the heart. Capillaries are the tiniest blood vessels in organisms, usually wide enough to allow blood cells to pass through only in single file. It is here, and not in arteries or veins, that diffusion of gases between the blood cells and tissue cells takes place. Keep in mind that the predominant muscle type found in all blood vessels is smooth muscle, under autonomic nervous system and hormonal control.

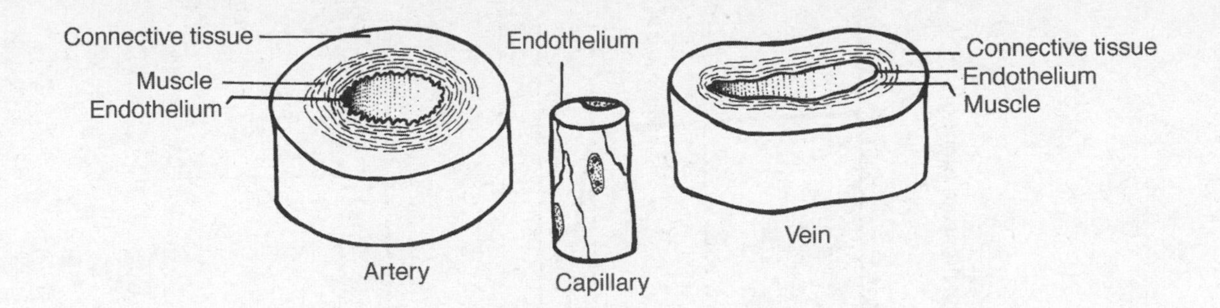

Blood Vessel and Wall Width

The path that blood takes from the heart to the rest of the body begins in the aorta, the major artery leading away from the left ventricle. The aorta splits into other arteries, smaller arterioles, and finally capillaries. As you can see from the following diagram, the increase in *overall surface area* from arteries to capillaries is huge, and this surface area is immediately decreased as capillaries coalesce into venules and then veins for the trip back to the heart.

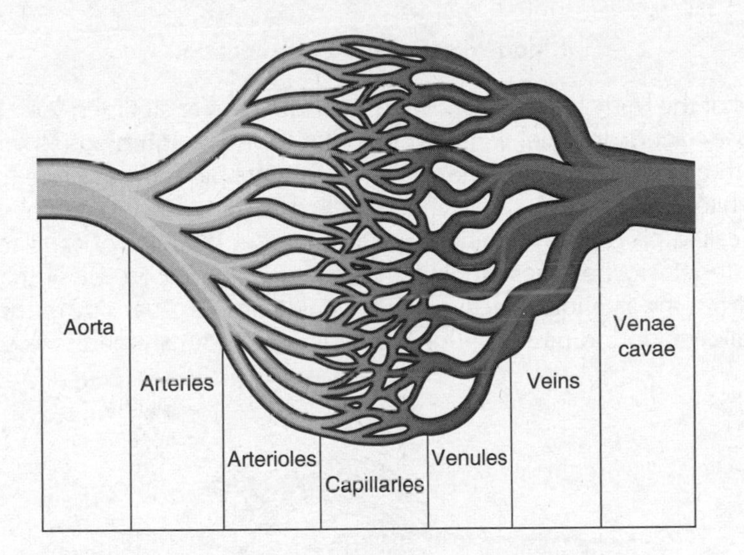

Relative Size of Blood Vessels

Because of this increase in surface area and friction between the blood cells and the blood vessel walls, average blood pressure decreases the farther away blood travels from the heart. In the veins, there is virtually no blood pressure at all.

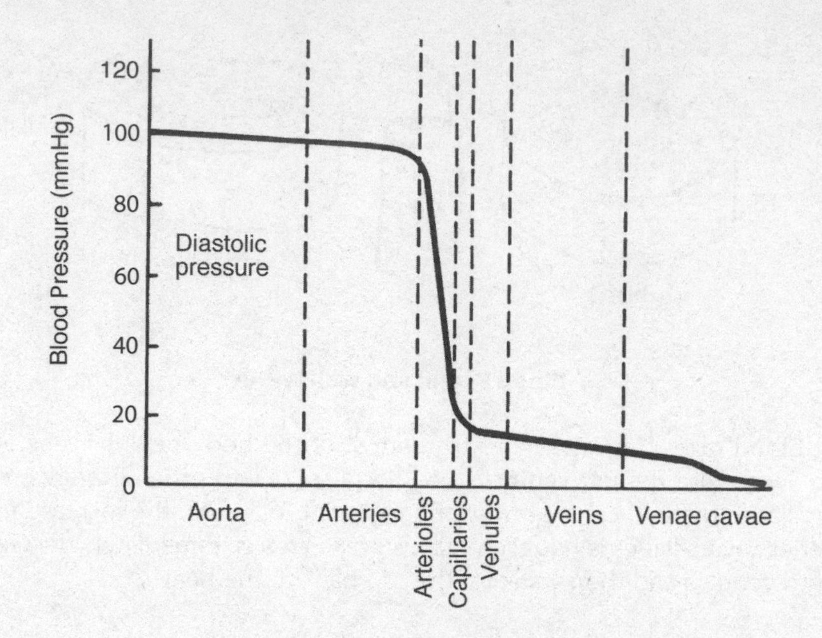

Blood Pressure and Body Location

If all the capillaries of the body had blood flowing through them at all times, your overall blood pressure would be close to zero. In other words, you do not have enough blood in your body to supply every capillary with blood all the time. Thus, the body shunts blood from place to place, closing off some capillaries while opening others. This selective dispersal of blood is accomplished by using small bands of muscle, called precapillary sphincters, which exist at the base of capillaries as they branch off from nearby arterioles. When these sphincters are closed, the blood will simply bypass the capillary bed, moving from the arteriole directly into the venule on the other side of the bed. When open, these sphincters allow blood to pass into the capillary bed to nourish the tissue cells there.

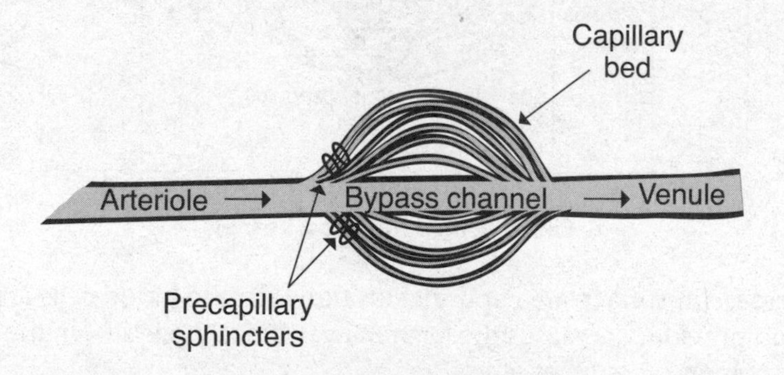

Capillary Sphincters

Another consequence of regulating blood flow using precapillary sphincters is temperature regulation. Have you ever noticed how your skin turns paler when it is cold outside and redder when it is hot? Exercising also causes the skin to become hotter and redder in color. The reason for this is that when it is cold outside, autonomic nerves in your skin cause as many precapillary sphincters as possible to close, to conserve heat (the more blood near the surface of your body, the faster you lose heat). When it's hot or when you're exercising, many sphincters near the surface of the skin will be open, allowing more efficient cooling of the body. Vasodilation, or keeping lots of blood vessels open in the extremities, and sweating are the two main mechanisms the body uses to cool itself.

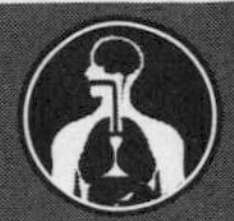

Hemoglobin and Myoglobin

Within each erythrocyte (red blood cell), there are approximately one million **hemoglobin (Hb)** molecules. Each hemoglobin molecule is made of four subunits capable of binding to four different oxygen molecules. The binding of the first oxygen molecule *induces a conformational change in the hemoglobin* molecule as a whole so that the binding of the other three molecules becomes progressively easier. Similarly, the unloading of one oxygen molecule facilitates the unloading of the other three.

This allosteric effect, known as cooperative binding, is reflected in the S-shaped oxygen dissociation curve for hemoglobin. Another useful aspect of this system is that it allows the greatest unloading of oxygen to occur at active tissues. Hemoglobin is able to hold on to its oxygen molecules with great affinity until partial pressures reach about 40 mm Hg (roughly three-quarters of the hemoglobin remain saturated until that point). Then, within a relatively small pressure change, they quickly unload large amounts of oxygen. This is quite useful because it makes sure that the oxygen in red blood cells is not completely grabbed up by other cells before reaching the tissues that need it the most. Since active tissues utilize oxygen at a greater rate, they have a lower partial pressure of O_2, triggering the release of greater amounts of oxygen from passing hemoglobin.

Myoglobin (Mb), like hemoglobin, is a cytoplasmic protein that reversibly binds oxygen molecules, but unlike hemoglobin consists of a single subunit instead of four. Although the heme group of myoglobin is identical to those in hemoglobin, myoglobin does not exhibit allostery and thus has a higher affinity for oxygen and is less inclined to release oxygen. Accordingly, the function of myoglobin is to store rather than transport oxygen, and it is found mainly in the muscle tissue of most mammals. On an oxygen-binding curve, which shows fractional saturation (bound oxygen) versus partial pressure of oxygen (oxygen concentration), the dissociation curve of myoglobin is a rectangular hyperbola as compared to S-shaped for hemoglobin. When oxygen is consumed faster than it is being transported, tissue cells experience hypoxia (oxygen deficiency) and myoglobin releases an emergency supply of oxygen for use by the tissue.

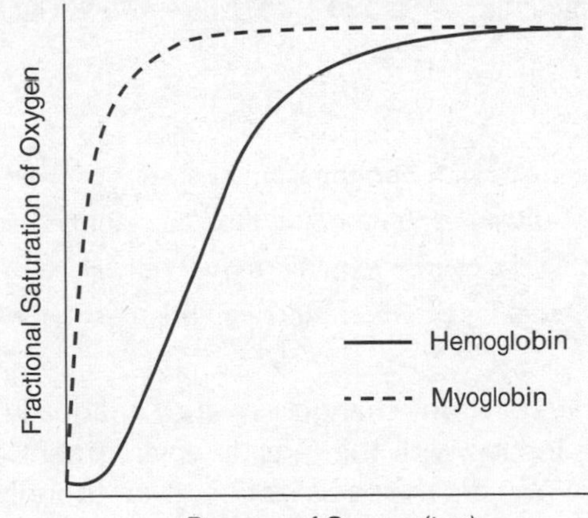

DIGESTIVE AND EXCRETORY SYSTEMS

11.4 Describe the cooperation of organs required for digestion

11.5 Explain the excretory systems as waste transfer

Digestion involves the degradation of large molecules into smaller molecules that are then absorbed and used directly by cells. In complex animals (such as humans), branching digestive canals and **absorption** into the bloodstream are necessary for the effective breakdown of nutrients and their transport to all cells of the body. Mammalian digestive tracts are organized into regions specialized for the digestion and absorption of specific nutrients.

Excretion refers to the removal of metabolic wastes produced in the body. It is distinguished from elimination, the removal of indigestible material. Most of the body's activities produce metabolic wastes that must be removed. For example, aerobic respiration leads to the production of carbon dioxide waste. Deamination of amino acids leads to the production of nitrogenous wastes, such as urea or ammonia. All metabolic processes lead to the production of mineral salts, which must also be excreted.

Digestive system

The **digestive tract** is simply a long tube that travels through a mammal, allowing the exchange of nutrients. The lumen, the space inside the tube, is filled with a liquid matrix that carries particles along like a conveyor belt. By the time the digestive tract reaches the end of the animal through the anus, the tissues have taken what they want and eliminated what they don't, leaving only waste.

The different tissues of the digestive tract are localized in organs such as the **stomach** and **intestines** that coordinate with each other in the uptake of carbohydrates, proteins, fats, vitamins, and any other nutrients the body needs. Just as with gas exchange and the circulatory system, the digestive tract simply provides a mode of transportation for these specialized tissues to transport goods. Don't think of the digestive tract as being inside the body. Think of it as an external surface that has been moved inside so it can imitate the liquid environment from which all mammals originated.

✔ **AP Expert Note**

Nutrients

After a mammal regulates its body with hormones and nervous coordination and supplies its cells with the gases necessary for cellular respiration, the next most immediate concern is getting nutrients. Nutrients include not only the energy required to fuel cellular activity, but the micronutrients (vitamins and minerals) necessary for chemical function and, most importantly, water.

Animals' digestion, like their respiration, changed as they gradually evolved from single-celled organisms that exchanged molecules with the outside environment across their entire bodies to complex organisms with tissue and organ specialization. Even though the exchange of molecules occurs inside the body cavity of these organisms, the actual surface where this exchange takes place in the respiratory and digestive systems is the interface between the inside and outside of the animal. This means that the contents of the stomach, although inside the abdomen of an animal, are actually part of the outside environment (think of the entire **digestive system** as being a hole that passes through an organism). The internal compartments of the alveoli are also part of the outside environment. These systems occur externally in more primitive animals.

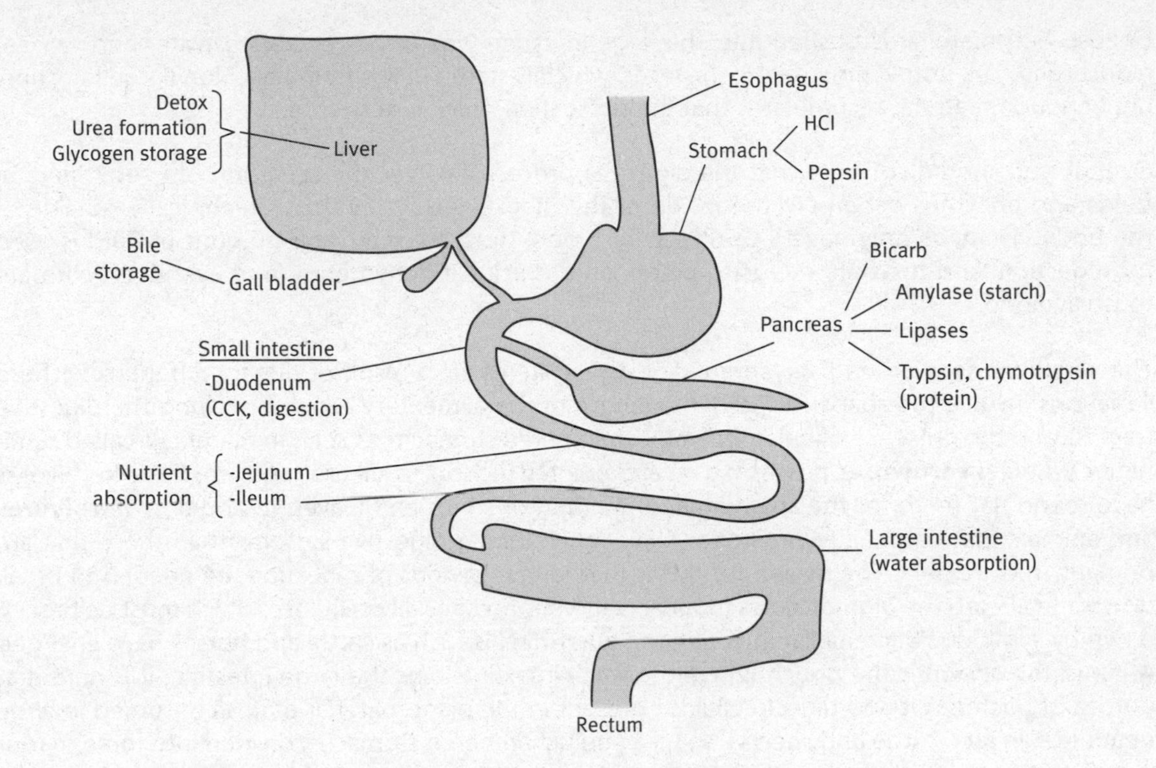

Components of the Digestive Tract

Ingested food is first subjected to mechanical digestion by shearing and grinding in the mouth to reduce the size of food particles. Saliva begins chemical digestion of carbohydrates and lubricates the ball of food that is known as a bolus. Stretch receptors in the stomach cause chief cells and parietal cells to secrete gastric juices, a highly acidic ionic solution that denatures proteins by disrupting hydrogen bonds and secondary structure. Gastric juice, along with the mechanical churning of the stomach, causes connective tissues of animal meat to break apart. Unfolding of proteins provides increased surface area for proteases active *only* at low pH levels.

As the food mass breaks up, it stratifies in the stomach according to size, density, and water solubility. Fat cells and lipids float to the top of the stomach, carbohydrates and smaller proteins remain in the middle, and the larger proteins sink to the bottom. This is very important for the next step, because the contractions of the stomach force proteins and carbohydrates out first and the lipids next. Evolutionarily speaking, this design favors early digestion and absorption of the most readily available sources of chemical energy (polysaccharides); protein and fat digestion come afterward.

The partially digested mass of food, known as chyme, enters the small intestine. Pancreatic fluid, rich in bicarbonate ions to neutralize stomach acid and full of enzymes to break down proteins and lipids, is secreted. **Bile** is secreted by the liver to help fat absorption by emulsifying the chyme. The proper mixture (depending on what's in the meal that was eaten) of chyme, pancreatic fluid, and bile is obtained by adjusting the rates of gastric emptying, pancreatic secretion, bile secretion, and intestinal motility (muscle movement). The smooth muscle of the small intestine, like an elevator, continuously moves the chyme down the intestine. Chemical breakdown continues, and products are absorbed by cells lining the inside of the small intestine. By the time the chyme gets to the end of the small intestine—the terminal ileum—all of the digestible food will have been digested and absorbed. Bile salts are secreted by the liver to be used again for the next meal.

Unabsorbed material is pushed into the large intestine where the remaining water and salt are reabsorbed. The leftover material is the feces, which is stored in the terminal colon (large intestine) until enough material accumulates that the defecation reflex is activated.

Overall, you should consider that the digestive process involves the secretion of over 7 liters of fluids and enzymes *per day* to the inside of the digestive tract (which is technically outside of the body). Humans only have 5 or 6 liters of blood; thus, an enormous amount of fluid is used for digestion, and the fluid needs to be restored quickly if dehydration and eventual death are to be avoided.

The variations seen among vertebrate digestive systems are a result of dietary differences. These variations involve the shape of teeth, the length of the alimentary canal (how long the digestive tract is), and the presence of enlarged, multichambered stomachs as seen in mammals called ruminants. Whereas **carnivores** possess sharp and pointed incisors to kill prey and tear flesh, herbivores have round flat teeth for the crushing and grinding of stems and leaves. In addition, **herbivores** and animals that eat mainly vegetation have much longer digestive systems than those that are primarily meat eaters. The reason for this is that longer periods of digestion are needed to break down and absorb the biomolecules found in vegetation, especially cellulose, which must be broken down by symbiotic bacteria living within the gut. Animals such as cattle and horses have enlarged cecums; the cecum is the pouch where the small intestine joins the large intestine. It is here that hordes of bacteria sit and digest cellulose present in the plant material that is consumed in large quantities. In fact, cattle and sheep also have multichambered stomachs, divided into three or four sections (chambers) that allow slow and repeated digestion of food before the food actually proceeds into the small intestine for absorption.

Excretory System

The principal organs of excretion in humans are the lungs, liver, skin, and kidneys. In the lungs, carbon dioxide and water vapor diffuse from the blood and are continually exhaled. Sweat glands in the skin excrete water and dissolved salts (and small quantities of urea). Perspiration serves to regulate body temperature because the evaporation of sweat produces a cooling effect. The liver processes blood pigment wastes and other chemicals for excretion. **Urea** is produced by the deamination of amino acids in the liver, and it diffuses into the blood for excretion in the kidneys. The kidneys function to maintain the osmolarity of the blood and to conserve glucose, salts, and water.

On an average day-to-day basis, an adult human maintains more or less the same weight, a relatively constant blood pressure, and a certain amount of salts, metabolic wastes, and water volume in his or her tissues. This is due to the fact that renal (kidney) output varies around a set point for each individual, whereby renal output of salt, water, and urea goes up as dietary intake increases and goes down as dietary intake decreases. The mechanism by which the kidney achieves this is through filtration of the blood through many small capillary bundles called glomeruli (singular: glomerulus) and reabsorption or rejection of this filtrate as it passes down long, twisting tubes known as nephrons. Within each kidney, there are perhaps one million glomeruli, each attached to a nephron tube. The filtrate that makes it through the nephron tubes without being reabsorbed back into the bloodstream is called the final filtrate, or urine. It empties from the ends of the nephrons into the renal pelvis and is then carried via the ureter to the bladder for storage and eventual release during urination.

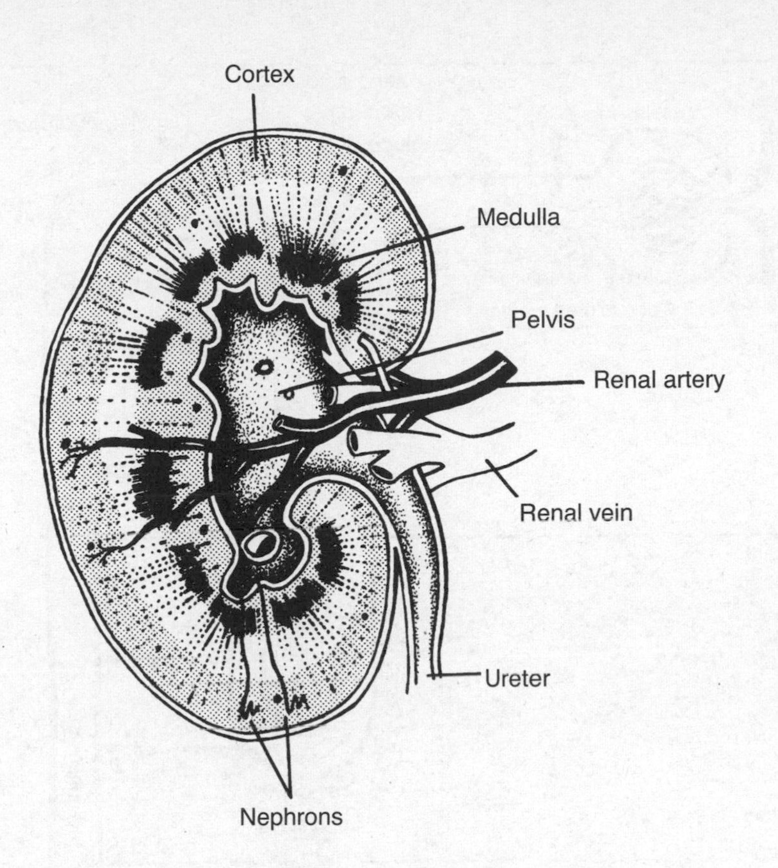

Structure of the Kidney

Blood enters the kidney via the renal artery. From there, it goes into the afferent arterioles and into the glomeruli. Blood leaves the kidney via the efferent arterioles and out the renal vein.

Approximately 1–1.2 liters of blood pass through the two kidneys every minute, entering via the renal artery and exiting via the renal vein. The average human body has only 6 or 7 liters of blood, which means that *the entire blood volume is swept through the kidneys at least 10 times every hour.* As blood enters, it is quickly disseminated into tiny capillaries that end at a glomerulus, a tiny ball of capillaries contained within a pouch called Bowman's capsule. Bowman's capsule leads into a long and coiled nephron that is divided into functionally separate units: the proximal convoluted tubule, the loop of Henle, the distal convoluted tubule, and the collecting duct. The loop of Henle of most nephrons dips down through the medulla of the kidney, while the glomeruli and convoluted tubules remain positioned in the cortex.

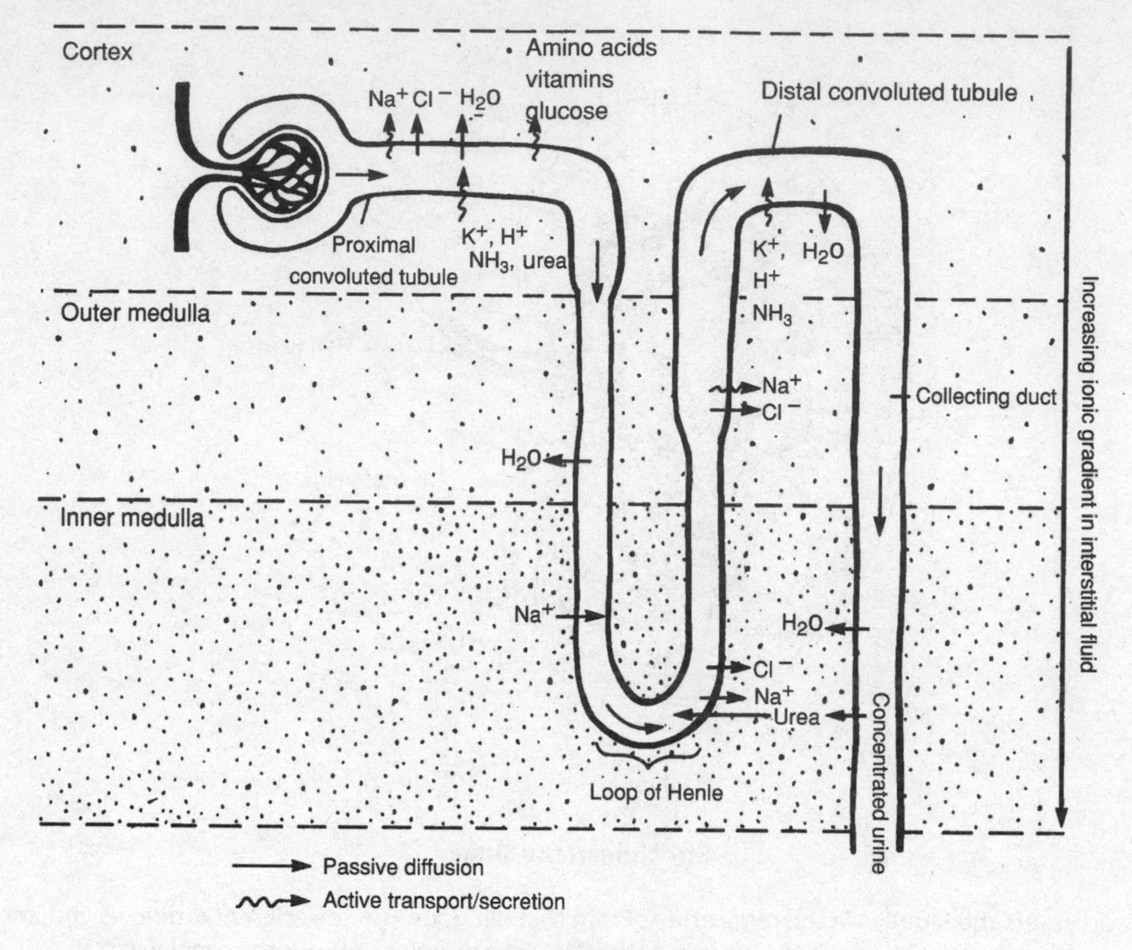

Structure of the Nephron

The kidneys filter, secrete, and reabsorb a variety of molecules and ions to maintain the body's homeostasis.

Filtration

Filtration occurs at the glomerulus in a fashion similar to pouring a pot of boiling pasta into a colander. The glomerulus has small holes that allow smaller molecules and water through, but exclude cells and large proteins (the pasta).

Secretion

Some solutes are actively secreted directly from blood into the nephron by transporters in the membranes of epithelial cells. Examples include H^+ and NH_4^+ for maintaining pH homeostasis and organic molecules such as *p*-aminohippuric acid and penicillin.

Reabsorption

Some molecules are reabsorbed from the filtrate back into the blood via a network of capillaries that surround the nephron. Examples are water, electrolytes, glucose, and amino acids. These may require a transporter or they may be reabsorbed by passive diffusion.

Nephron Flow

In the glomerulus, blood is filtered out of capillaries into Bowman's capsule and the nephron. Many substances do not pass through the glomerulus into the nephron; instead, they simply exit the kidneys through the renal vein without ever passing through the nephron tubes. Large proteins, such as albumin, and proteins with lots of negative charge, are repelled due to the structure of the glomerulus, with its triple-layered capillary cells. Cells remain within the bloodstream. Water and smaller molecules such as salts, sugars, and urea enter the nephron, however, and this fluid is called the filtrate. After filtration, the next step is the selective reabsorption of most of the filtrate. After exiting the glomerulus via the efferent arteriole, the bloodstream doubles back to wind very closely around the proximal and distal tubules as well as the loop of Henle. The closeness of the bloodstream to the filtrate in the nephron sets up a concentration gradient whereby much of the material in the filtrate will diffuse back into the nearby blood or be moved there by active transport.

Upon leaving Bowman's capsule, the filtrate finds itself within the proximal tubule. On the surface of the cells that line this tubule are Na^+/H^+ ion exchange pumps. While kicking out hydrogen ions into the tubule, these membrane proteins absorb sodium from the filtrate so that it travels into the cells that line the tubule. This sodium moves through to the opposing end of the cells where it is extruded into the bloodstream by a Na^+/K^+ ATPase pump. The end result is the movement of sodium ions from the filtrate into the bloodstream. Water naturally follows due to the hypertonic conditions that occur as salt concentration goes up in the nearby bloodstream, and it is here in the proximal tubule that most of the water and salt that were originally filtered out of the blood are returned back into the blood. The proximal tubule is also responsible for the reabsorption of glucose using active transport. In a healthy individual, there is no glucose excreted in the urine because all glucose is efficiently reabsorbed in the proximal tubule.

In the loop of Henle, either water or salt is absorbed independently. The descending part of the loop absorbs water only, and the ascending loop absorbs salt ions only. More water and salt reabsorption takes place in the distal tubule, and in the collecting duct, certain hormones regulate the concentration of the urine.

The selective permeability of the tubules establishes an osmolarity gradient in the surrounding interstitial fluid. By exiting and reentering at different segments of the nephron tube, solutes create a situation in which tissue osmolarity increases from the cortex into the inner medulla. While most of the urea that ends up in the collecting duct is excreted, some diffuses into the nearby interstitial fluid and reenters the ascending loop of Henle. Overall, the gradient that is established by the concentration of solutes in the kidney medulla helps to maximize water conservation and make the urine as hypertonic (concentrated) as possible.

Hormonal Regulation

Principal cells in the collecting duct respond to various hormones, such as aldosterone, to absorb more water or salt depending on how concentrated the filtrate is at that point. Aldosterone is produced in the adrenal cortex and stimulates the reabsorption of Na^+ from the collecting duct. Na^+ stimulates water reabsorption as well, so aldosterone is released when blood pressure falls in the body. People with Addison's disease produce insufficient aldosterone and have urine that has too high Na^+ concentrations. This causes dehydration and a drop in blood pressure, as water is pulled out of the bloodstream to follow the excess salt in the urine.

ADH, or antidiuretic hormone, also controls the concentration of the urine and acts on cells lining the collecting duct of the nephron. As its name implies, release of this hormone, which takes place in the posterior pituitary, causes water reabsorption and a decrease in urine output. ADH is also known as vasopressin, and it works mainly by opening up active transport water channels called aquaporins on the luminal surfaces of principal cells in the collecting duct. This causes water to leave the filtrate and move into the medulla, where it is eventually reabsorbed into the bloodstream, leaving the filtrate more concentrated and the eventual urine much less diluted.

Blood is mostly water, so anything that causes water loss, such as vomiting, diarrhea, blood loss, or heavy sweating, can cause a drop in blood pressure. The body responds by releasing a cascade of hormones known as the renin-angiotensin-aldosterone system, which constricts the blood vessels and increases reabsorption of salt and water. This extra fluid adds volume, and therefore pressure, back to the blood.

RAPID REVIEW

If you take away only 5 things from this chapter:

1. Lungs are the primary structures for respiration in mammals. The circulatory system connects the lungs and the cells around an organism's body. All mammalian lungs end in blind sacs called alveoli, which are surrounded by a dense net of capillaries into which oxygen diffuses and from which carbon dioxide is excreted.
2. Blood pressure is the force per area that blood exerts on the walls of blood vessels and is expressed as a systolic number over a diastolic number. Blood pressure drops as blood moves farther away from the heart because the distance from the pump has increased and because of the muscular construction of arteries versus veins.
3. Mechanical digestion is the physical breakdown of food, by chewing in the mouth and churning in the stomach. Chemical digestion starts in the mouth with the breakdown of carbohydrates and continues through the stomach and small intestine. Nutrients, water, and minerals are absorbed into the body in the colon.
4. Excretion is the removal of metabolic wastes, and all metabolic processes lead to the production of mineral salts that must be excreted.
5. Kidneys regulate the concentration of water, ions, and other dissolved substances in the blood through the formation and excretion of urine.

TEST WHAT YOU LEARNED

1. Bile salts are produced by the liver and are structurally related to cholic acid, which is shown in the diagram. Bile salts are stored in the gall bladder and are released in the duodenum.

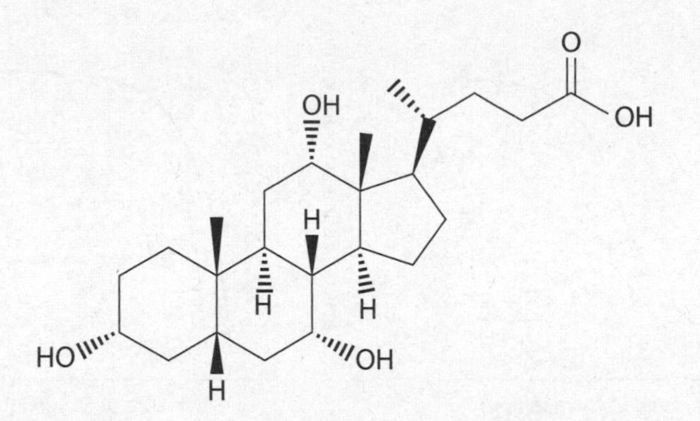

Which of the following statements best explains the function of bile salts?

(A) Bile salts combine with lipids to form emulsions, facilitating further digestion by lipases.

(B) Bile salts combine with proteins to denature them, facilitating further digestion by proteolytic enzymes.

(C) Bile salts combine with fat soluble vitamins and facilitate their elimination from the body.

(D) Bile salts combine with carbohydrate chains and facilitate their digestion by intestinal bacteria.

2. A sample of urine from a patient is positive for sodium and potassium ions, urea, uric acid, and the blood protein albumin. The urine is negative for glucose, white blood cells, red blood cells, bacteria, and yeast. Which of the results in the analysis indicates that the patient has a medical problem?

(A) Urine should not contain uric acid.

(B) Urine should not contain sodium and potassium ions.

(C) Urine should not contain albumin.

(D) Urine should contain white blood cells and some bacteria.

3. The Bohr effect describes the inverse relationship between oxygen binding affinity and acidity. A tetramer with allostery (a phenomenon in which binding at one receptor site affects the binding affinity at a second site) exhibits the Bohr effect while a monomer with no allostery does not. Which of the following pairs of graphs properly represents the effect of decreased pH in tissue on oxygen affinity for hemoglobin and myoglobin, respectively?

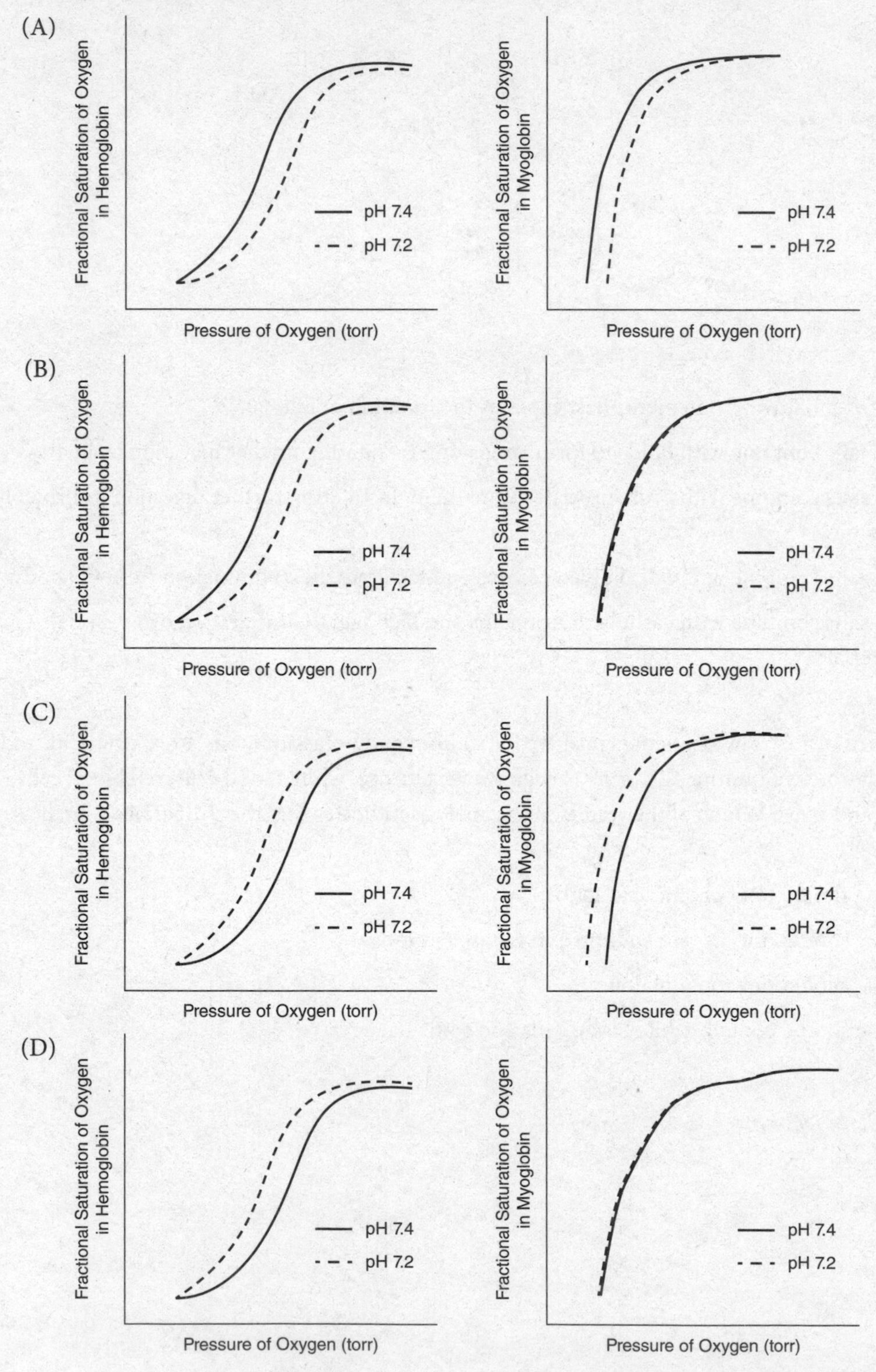

4. The partial pressure of O_2 decreases with altitude. Athletes who plan to compete at high altitudes usually live in a similar environment for several weeks before the sports events to undergo acclimatization. If the physiological response of such athletes is examined, which of the following changes is most likely to have occurred?

   (A) Hemoglobin in the athlete has mutated so that its dissociation curve is now saturated at a lower partial pressure of O_2.

   (B) More red blood cells are produced by the athlete so that the capacity to transport O_2 and deliver it to tissues has increased.

   (C) The maximum volume of air that can be inspired and expired by the lungs in a single breath has increased so that gas exchange is more efficient.

   (D) The pH of blood plasma has increased so that hemoglobin saturates at a lower partial pressure of O_2.

5. The diagram below shows body cells exchanging materials with the blood.

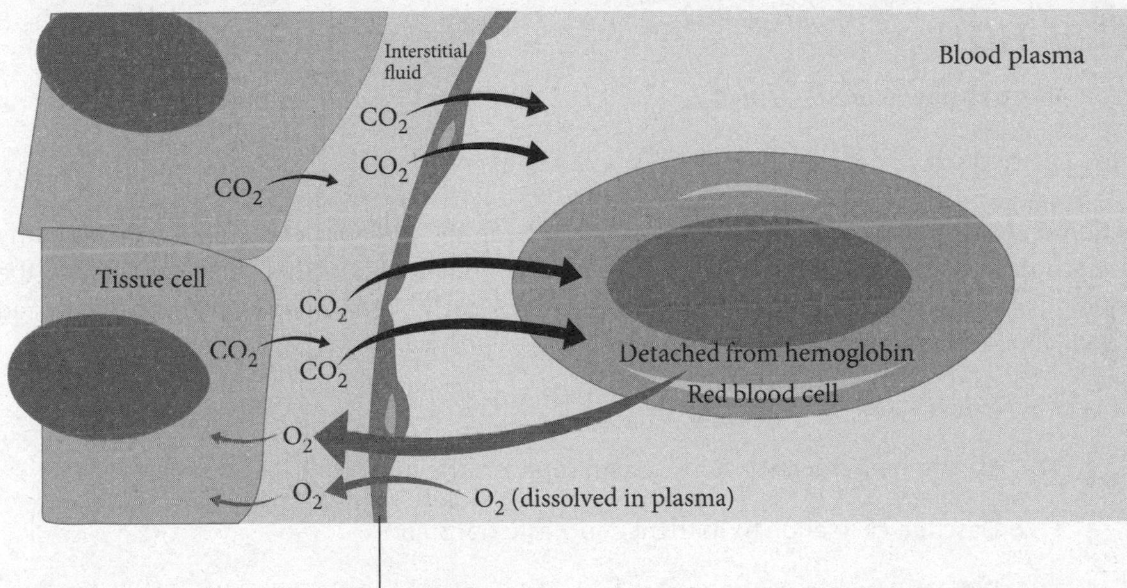

Which of the following scenarios in the human body is best modeled by the diagram?

   (A) Retrieving recovered water and nitrogenous wastes from a kidney cell

   (B) Delivering chemical messengers from the brain to their target cells

   (C) Retrieving nutrients from the liver and delivering wastes to be detoxified

   (D) Delivering nutrients to muscle cells and removing wastes they produce

Answer Key

Test What You Already Know

1. **C** **Learning Objective:** 11.1
2. **D** **Learning Objective:** 11.2
3. **A** **Learning Objective:** 11.3
4. **C** **Learning Objective:** 11.4
5. **B** **Learning Objective:** 11.5

Test What You Learned

1. **A** **Learning Objective:** 11.4
2. **C** **Learning Objective:** 11.5
3. **B** **Learning Objective:** 11.3
4. **B** **Learning Objective:** 11.1
5. **D** **Learning Objective:** 11.2

Detailed solutions can be found in the Answers and Explanations section at the back of this book.

REFLECTION

Test What You Already Know score: __________

Test What You Learned score: __________

Use this section to evaluate your progress. After working through the pre-quiz, check off the boxes in the "Pre" column to indicate which Learning Objectives you feel confident about. Then, after completing the chapter, including the post-quiz, do the same to the boxes in the "Post" column. Keep working on unchecked Objectives until you're confident about them all!

Pre	Post	
☐	☐	**11.1** Explain how the respiratory system supports the organism
☐	☐	**11.2** Describe circulation as nutrient and waste transfer
☐	☐	**11.3** Differentiate between myoglobin and hemoglobin
☐	☐	**11.4** Describe the cooperation of organs required for digestion
☐	☐	**11.5** Explain the excretory systems as waste transfer

FOR MORE PRACTICE

Complete more practice online at kaptest.com. Haven't registered your book yet? Go to kaptest.com/booksonline to begin.

PART 4

Processes of Life

CHAPTER 12

Enzymes

LEARNING OBJECTIVES

In this chapter, you will review how to:

12.1 Explain how enzymes and substrates interact

12.2 Explain how cofactors or coenzymes affect enzyme function

12.3 Contrast competitive and noncompetitive inhibition

12.4 Explain how concentration affects enzyme function

12.5 Investigate how enzyme reactivity is affected by concentration

TEST WHAT YOU ALREADY KNOW

1. Amplex red (AR) is an indicator used to monitor the activity of peroxidases. In the presence of peroxidases, AR reacts with hydrogen peroxide to produce a fluorescent red product, resorufin, which can be measured with a fluorometer. The figure shows an experiment that examined the rate of a reaction at different concentrations of H_2O_2 and AR.

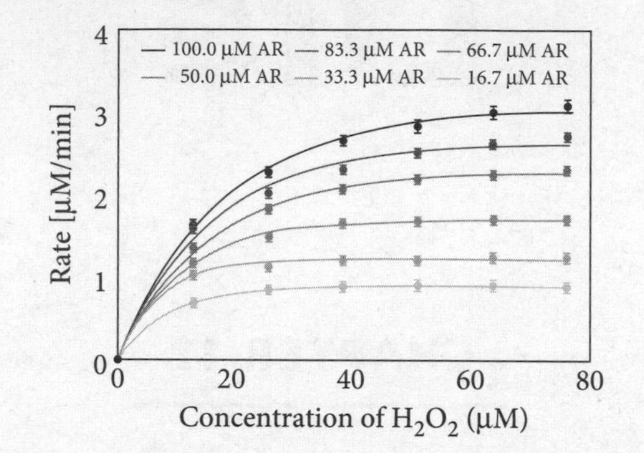

Adapted from Hoon Suk Rho et al., "Mapping of Enzyme Kinetics on a Microfluidic Device," *PLoS ONE* 11, no. 4 (April 2016): e0153437..

Why do the curves level off at a concentration slightly above 20 µM of H_2O_2?

(A) The reaction rate levels off as the temperature increases because most enzymes are proteins and denature at high temperatures.

(B) The reaction rate levels off as the substrate concentration decreases; as the H_2O_2 concentration increases towards the right of the graph, the concentration of AR decreases simultaneously.

(C) The induced fit model states that the active site changes shape as the substrate binds. This is not possible when the H_2O_2 concentration becomes too high.

(D) During an enzymatic reaction, the substrate binds to the active site of an enzyme. This means that the enzymes can only catalyze one reaction at a time, which limits the reaction rate.

2. Glucose-6-phosphate dehydrogenase (G6PDH) is an enzyme that can use NAD^+ or $NADP^+$ as a cofactor. The table shows the K_m and k_{cat} of this enzyme with the two cofactors. The K_m represents the substrate concentration at which the reaction rate is half of the maximum rate. k_{cat} is known as the turnover number and represents how quickly the enzyme processes substrate molecules.

Cofactor	K_M (μM)	k_{cat} (s^{-1})
$NADP^+$	243 ± 5.3	282 ± 1
NAD^+	8352 ± 307	298 ± 4

Adapted from M. Fuentealba et al., "Determinants of Cofactor Specificity for the Glucose-6-Phosphate Dehydrogenase from *Escherichia coli*: Simulation, Kinetics and Evolutionary Studies," *PLoS ONE* 11, no. 3 (March 2016): e0152403.

Based on the data in the table, which cofactor yields the more efficient enzyme, and why?

(A) $NADP^+$, because the reaction reaches maximum velocity at a lower substrate concentration and the turnover rate is higher

(B) $NADP^+$, because the reaction reaches maximum velocity at a lower substrate concentration and the turnover rate is lower

(C) NAD^+, because the reaction reaches maximum velocity at a lower substrate concentration and the turnover rate is higher

(D) NAD^+, because the reaction reaches maximum velocity at a lower substrate concentration and the turnover rate is lower

3. In a recent study, researchers examined the effects of an inhibitor to determine how it affected the activity of angiotensin I-converting enzyme. On the graph below, the maximum velocity of the reaction is represented by the *y*-intercept. The substrate concentration at which the reaction proceeds at half-maximum velocity is represented by the *x*-intercept.

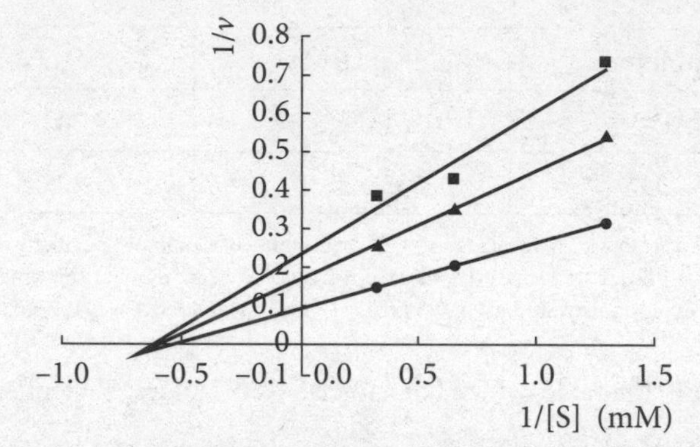

Adapted from He Ni et al., "Inhibition Mechanism and Model of an Angiotensin I-Converting Enzyme (ACE)-Inhibitory Hexapeptide from Yeast (*Saccharomyces cerevisiae*)," *PLoS ONE* 7, no. 5 (May 2012): e37077.

Based on this data, is this a competitive inhibitor or a noncompetitive inhibitor, and why?

(A) The *x*-intercept is the same for all of the reactions, so the maximum velocity must be the same with and without the inhibitor. This is competitive inhibition because high concentrations of substrate can outcompete a competitive inhibitor.

(B) The *y*-intercept is different for all of the reactions, so the maximum velocity must be lower with the inhibitor. This is noncompetitive inhibition because high concentrations of substrate can outcompete a competitive inhibitor and reach the same maximum velocity.

(C) The *y*-intercept is the same for all of the reactions, so the maximum velocity must be the same with and without the inhibitor. This is competitive inhibition because high concentrations of substrate can outcompete a competitive inhibitor.

(D) The *x*-intercept is different for all of the reactions, so the maximum velocity must be less with the inhibitor. This is noncompetitive inhibition because high concentrations of substrate can outcompete a competitive inhibitor and reach the same maximum velocity.

4. Researchers are studying the reaction rate of mevalonate-5-phosphate (Mev-P), an enzyme. The graph below shows the rate of the enzyme-catalyzed reaction as the initial concentration of Mev-P is increased.

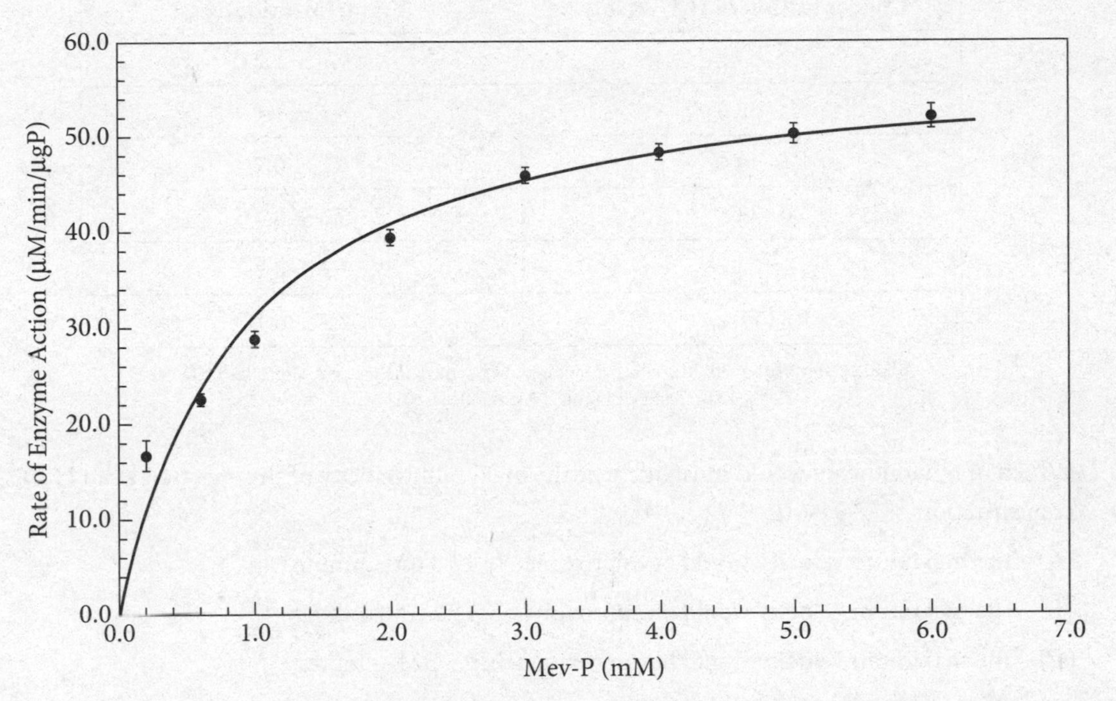

Adapted from David E. Garcia and Jay D. Keasling, "Kinetics of Phosphomevalonate Kinase from *Saccharomyces cerevisiae*," *PLoS ONE* 9, no. 1 (January 2014): e87112.

The graph is plotted using the initial concentration and reaction rate. Which of the following provides the most likely reason why the researchers used these initial values instead of values determined after the reaction had been allowed to progress?

(A) At the beginning of the reaction, the enzyme works more slowly. Over time, the reaction increases in rate as the concentration of the substrate goes down.

(B) The reaction will generate heat, which will increase the reaction rate as the reaction progresses. This will make the reaction appear to proceed faster than expected.

(C) At the start of the reaction, the researchers knew the exact concentrations of all species in the reaction. As the reaction proceeded, it would be more difficult to determine exact concentrations.

(D) To save time, the researchers measured the initial rate of reaction so they do not need to wait for the reaction to progress.

5. The table shows an experiment that examined the reaction rate of peroxidase with hydrogen peroxide. V_{max} refers to the maximum velocity of the reaction.

Concentration of H_2O_2 (μM)	V_{max} (μM/minute)
66.7	9.7 ± 1.2
55.6	8.3 ± 1.1
44.4	6.7 ± 0.7
33.3	5.3 ± 0.4
22.2	3.9 ± 0.2
11.1	2.3 ± 0.1

Adapted from Hoon Suk Rho et al., "Mapping of Enzyme Kinetics on a Microfluidic Device," *PLoS ONE* 11, no. 4 (April 2016): e0153437.

Which of the following would most likely be the maximum velocity of this reaction at an H_2O_2 concentration of 77.8 μM?

(A) The maximum velocity would be approximately 12.8 μM/minute.

(B) The maximum velocity would be approximately 11.2 μM/minute.

(C) The maximum velocity would be approximately 9.7 μM/minute.

(D) The maximum velocity would be approximately 8.9 μM/minute.

Answers to this quiz can be found at the end of this chapter.

ENZYME STRUCTURE

12.1 Explain how enzymes and substrates interact

12.2 Explain how cofactors or coenzymes affect enzyme function

Binding

HIGH YIELD

Most **enzymes** are proteins with specific 3-D structures that allow them to bind to very particular molecules (called **substrate** molecules) and increase the rate of reactions between these molecules. In many cases, enzymes bind to larger molecules and break them into smaller ones. Enzymes can synthesize or break down molecules at a rate of thousands or millions per second—they are extremely fast-acting! They allow reactions to occur that either would not take place or would take place far too slowly under normal conditions to be useful.

Enzymes are very specific for the molecules they bind to and the reactions they catalyze. Each enzyme has a name that usually indicates exactly what it does, and the name often ends in *-ase*. For example, the lactase enzyme breaks the complex sugar lactose into the simple sugars glucose and galactose. The enzyme pyruvate decarboxylase removes a carbon from the three-carbon molecule pyruvate. Thinking about enzyme names in this way may be helpful on the AP Biology exam.

For the exam, you should be familiar with the term *catalyst*, which refers to any chemical agent that accelerates a reaction without being permanently changed in the reaction. Enzymes are biological catalysts, which can be used over and over again.

Enzyme specificity means that an enzyme will only catalyze one specific reaction. Molecules upon which an enzyme acts are called substrates, and the substrate binds to the active site on the enzyme, speeding up the conversion from substrate to product.

All enzymes possess an active site, a 3-D pocket within their structure in which substrate molecules can be held in a certain orientation to facilitate a reaction. The two models of enzyme-substrate interaction are shown below. In the *lock-and-key model*, the spatial structure of an enzyme's active site is exactly complementary to the spatial structure of the substrate so that the enzyme and substrate fit together like a lock and key. In the *induced fit model*, the active site has a flexibility that allows the 3-D shape of the enzyme to shift to accommodate the incoming substrate molecule.

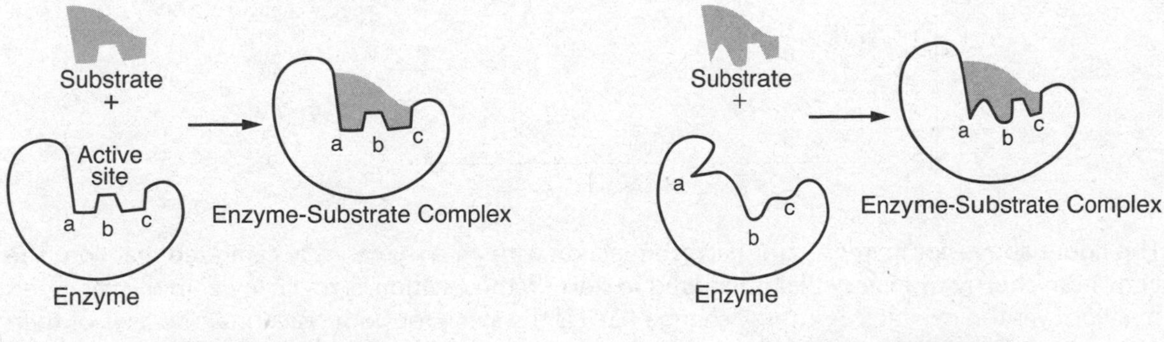

Processes

Cofactors and Coenzymes

Most enzymes require cofactors to become active. **Cofactors** are nonprotein (inorganic) species that either play a role in binding to the substrate or stabilize the enzyme's active conformation. Two examples of cofactors are zinc and the iron in hemoglobin.

Coenzymes are other organic molecules that play a similar role. Most coenzymes cannot be synthesized by the body but are obtained from the diet as **vitamins**.

Without the appropriate cofactor or coenzyme, an enzyme will be less capable of catalyzing its reaction and may even lose its function entirely. As a consequence, compounds that must be synthesized to maintain homeostasis and growth could become deficient or waste products that, should they be broken down, could accumulate to toxic levels. This is why vitamins and minerals are such essential aspects of nutrition.

ENZYME REGULATION

HIGH YIELD

12.3 Contrast competitive and noncompetitive inhibition

12.4 Explain how concentration affects enzyme function

Thermodynamics and Kinetics

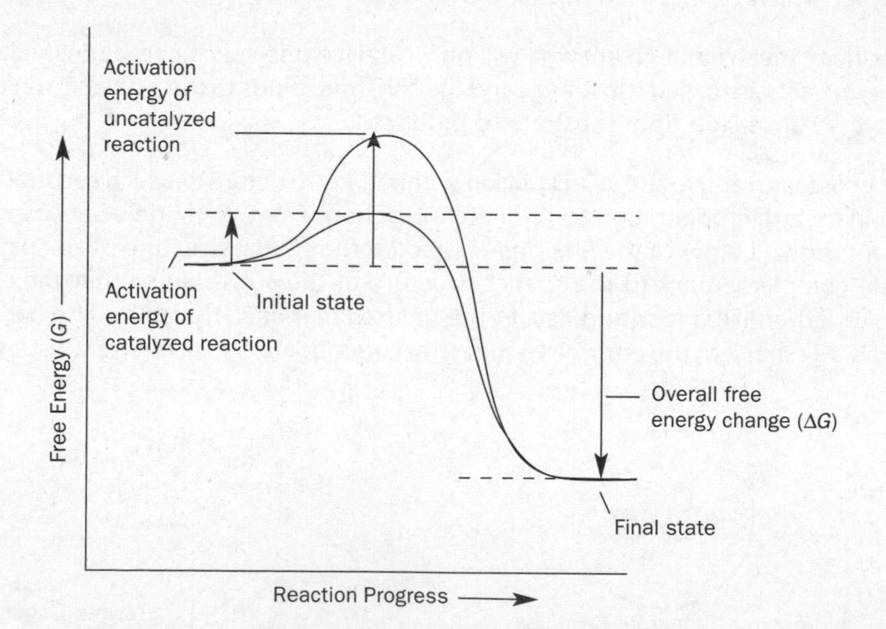

The figure above compares an uncatalyzed reaction with an enzymatically catalyzed reaction. The activation energy, or energy that is required to start up the reaction, is much lower in the catalyzed reaction, yet the overall free energy change (ΔG) is the same for both reactions. The laws of thermodynamics can be used to predict if a reaction will occur or not. If products have less free energy (G) than the reactants, the reaction has an overall negative change in G ($\Delta G < 0$) and will occur spontaneously. If products have more free energy than reactants, the reaction is an uphill one, needing a great deal of supplied energy to occur. Free energy is a measure of the potential energy of the molecules in a reaction. Those starting out with high potential energy, or higher G, are more likely to react and lower their G through the reaction than vice-versa. What that means is that reactions having a $\Delta G < 0$ are deemed favorable. Keep in mind that most biosynthetic reactions have $\Delta G > 0$ and will not occur spontaneously without the help of both enzymes and ATP.

Even though thermodynamics and ΔG alone may predict that a reaction is favorable or can occur spontaneously, the kinetics, or rate, of the reaction may be so slow that these reactions are not feasible for living systems. Sure, a hamburger will break down eventually if exposed to enough acid in your stomach, but without digestive enzymes to help speed up this breakdown, the hamburger might take months to break down sufficiently to be useful to you. Thus, although hamburger breakdown may be spontaneous, the limiting factor in the reaction is the reaction rate, which is dependent on energy being provided to start this breakdown. This energy is the activation energy, and enzymes provide a foundation on which molecules can react so that the energy needed to start a reaction is not as great as it would have to be without the enzyme's presence.

Inhibition

Cells must regulate enzyme action to keep these rapid reactions under control. Most enzymes are inactive most of the time, and enzyme pathways are regulated in complex fashions to ensure efficiency and safety. Enzymes can be regulated by inhibitors, molecules that bind to the enzyme either at the active site or the allosteric (regulatory) site.

Competitive inhibition occurs when the inhibitor and the substrate compete for the active site. This type of inhibition can be overcome by increasing the concentration of substrate.

Noncompetitive inhibition occurs when the inhibitor binds to the allosteric site, inducing a conformation change in the enzyme, rendering the active site inactive. This type of inhibition cannot be overcome by adding more substrate because the enzyme's shape has been altered. Therefore, noncompetitive inhibition may or may not be irreversible. Keep in mind that enzymes do not alter reaction equilibrium or affect the free energy of a reaction. They accelerate the forward and reverse reactions by the same factor.

Feedback inhibition occurs when the end product of an enzyme-catalyzed reaction works to block the activity of the enzymes that started the reaction in the first place. (This is an example of negative feedback, as distinct from positive feedback, which amplifies a particular bodily response.) In the following diagram, feedback inhibition would occur if product C could bind to the active site of enzyme 1 (this would be a kind of competitive inhibition), thereby preventing any more of compound A from becoming compound B.

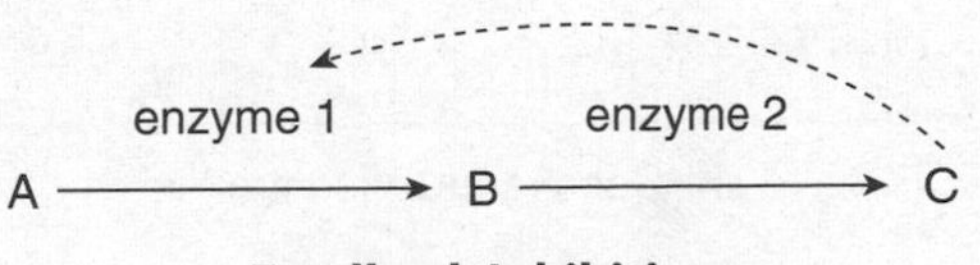

Feedback Inhibition

Physiological reactions can take place without enzymes, but they would take much longer to proceed. Though enzymes are neither changed nor consumed during the reaction, reaction conditions such as high temperatures, detergents, or acidic/basic conditions can cause enzymes to denature (lose their 3-D structure) and thereby lose their activity.

As a general rule, the rate of an enzyme will increase with increasing temperature but only up to a point. If the temperature increases too much, then the enzyme will become denatured. Moreover, enzymes are active only within a specific **pH** range. In the human body, most enzymes work best around neutral (pH = 7).

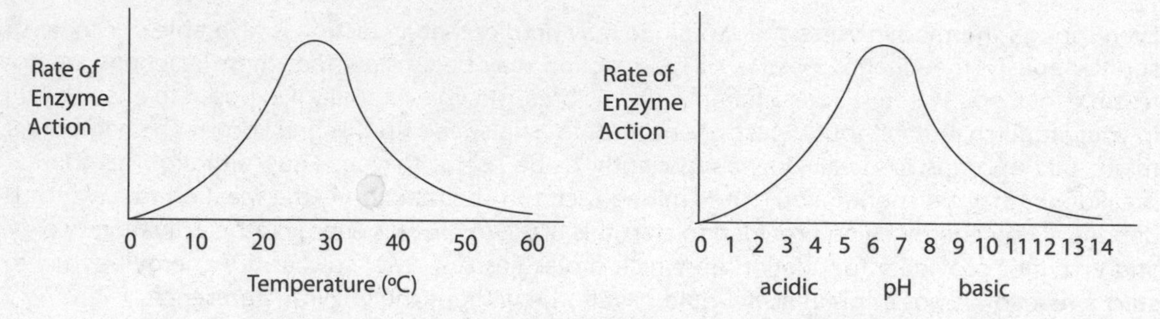

Temperature- and pH-Dependent Enzyme Reaction Rates

Effects of Concentration

Reaction rates increase as more and more enzyme is added to a particular environment. If the enzyme concentration is kept constant, the reaction rate will plateau at a maximum speed even as substrate concentration increases, because the enzymes can only work so fast. This maximum reaction rate is termed V_{max} and is illustrated in the following graph as the flat part of the rate versus substrate concentration curve for an enzymatic reaction. This point occurs when the enzymes become saturated with substrate. Enzymes become saturated at high substrate concentrations because substrate must bind to enzymes at a particular place: the active site. While the entire process is somewhat complex, the main idea is that increasing the substrate can produce the same V_{max} if competitive inhibitors are present, but noncompetitive inhibitors lower V_{max} regardless of the amount of substrate.

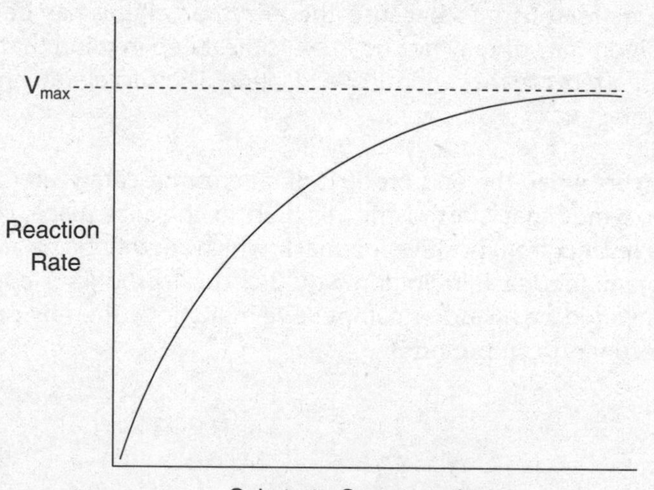

Concentration and Enzyme Reaction Rates

Increasing the concentration of the enzyme itself will also speed up the rate of the reaction. However, note that the synthesis of enzymes requires energy and resources that could be deployed elsewhere in an organism, so there is a trade-off to producing more copies of a particular enzyme.

AP BIOLOGY LAB 13: ENZYME ACTIVITY INVESTIGATION

12.5 Investigate how enzyme reactivity is affected by concentration

The enzyme catalysis lab is designed to allow you to explore structural proteins and environmental conditions that affect chemical activity in the natural world. In the lab, you will measure the chemical activity of the enzyme catalase as it breaks down hydrogen peroxide (H_2O_2) into water and oxygen gas. Specific experimental conditions—temperature, substrate and enzyme concentrations, pH, and so on—will be altered to help you understand the effects of environmental conditions on a specific chemical reaction. This experiment helps you to learn the experimental method through manipulation of experimental treatments compared with a control treatment.

✔ **AP Expert Note**

Graphing

For labs requiring graph construction and calculations, make sure that your graph is labeled thoroughly and accurately. Make sure you identify appropriate axes, show units, plot points accurately, and assign your graph a title. You should also follow these steps when answering AP Biology free response questions that require you to create graphs.

Understanding of the experiment itself can be encapsulated in just a few graphs. One or more of these graphs may be on the exam. The first graph is about free energy and enzyme activity.

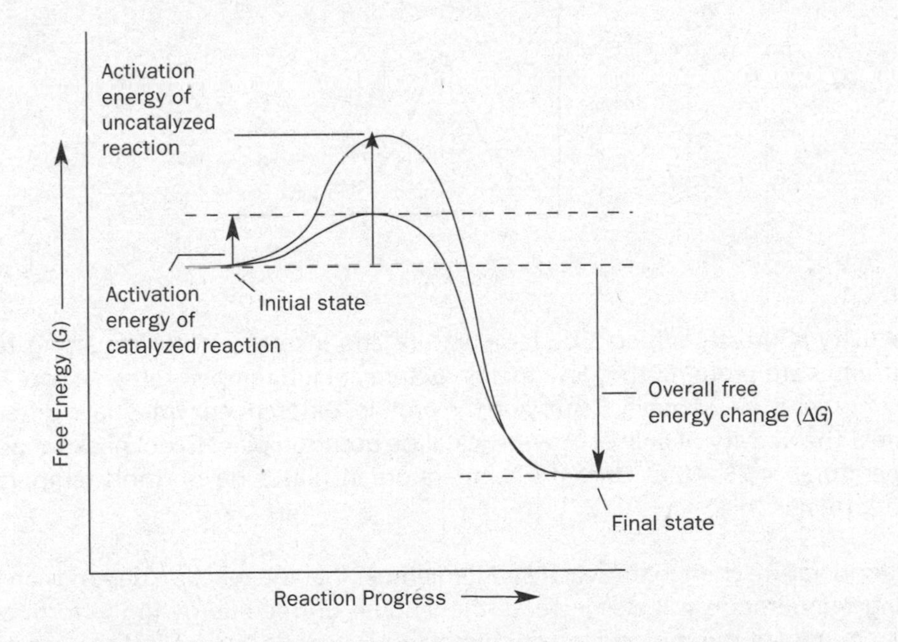

Before a reaction can take place, it must reach a point called its activation energy by receiving enough energy from the environment, termed free energy, in the form of heat or kinetic energy. An enzyme catalyzes a reaction by lowering the activation energy needed (it requires less free energy) to allow the reaction to take place. Hydrogen peroxide breaks down into water and oxygen gas on its own, but at an incredibly slow rate. The enzyme catalase lowers the activation energy of the reaction and the reaction happens very quickly.

✔ AP Expert Note

Many reactions that occur in biology can be sped up through the addition of enzymes that lower activation energy. Natural compounds readily metabolize and synthesize in the presence of certain enzymes, and biological systems control chemical activity largely through synchronizing the presence or absence of enzymes. Enzymes also control the release of energy so that cells don't combust during certain reactions from a sudden influx of energy.

Notice on the graph that the amount of free energy in the system changes over the progress of the reaction. Reactants of the reaction have an initial amount of energy, receive energy from the environment, and then energy is released with the products. Remember that the energy of the reactants and the energy of the products remain at the same values, whether the reaction is enzyme mediated or not. All the enzyme does is lower the activation energy for the molecule to do "what comes naturally" and make the reaction happen. When the reaction is complete, the enzyme is free and helps to speed up the reaction for another H_2O_2 molecule. This lab experiment demonstrates firsthand what occurs in all living systems on a regular basis and points out the importance of these specialized proteins.

The environment can also have a profound effect on the reaction. In this experiment, you measure the experimental effect of changing the reaction temperature and pH, as well as the effect of changing concentrations of the enzyme and substrate.

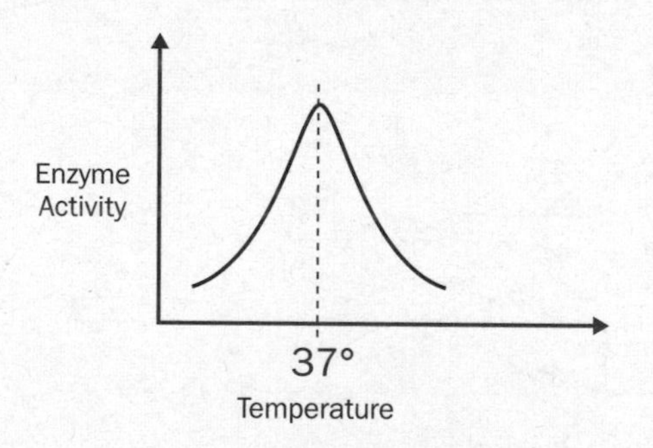

Catalytic activity is greatly affected by temperature and increases with increasing temperature. Because enzymes are proteins, they lose their structure at high temperatures, not only eliminating catalytic properties but essentially destroying the protein. Different enzymes have different specific temperatures. The activity of animal catalase (catalase occurs in plants, too) peaks at about normal body temperature, or 35–40°C. Once the temperature increases beyond this temperature range, the catalase proteins "die."

How does temperature change the reactivity? Remember that the reaction depends on free energy, so increasing temperature will also increase the amount of free energy in the form of both heat and kinetic energy (all the molecules will be moving faster). Because all of the molecules in the system will be moving faster at a higher temperature, there is an increased rate of collision between enzyme and substrate molecules. In essence, there are more chance meetings.

The pH of the environment affects the reactivity of enzymes in the same way that temperature does, in that there is a level of peak activity.

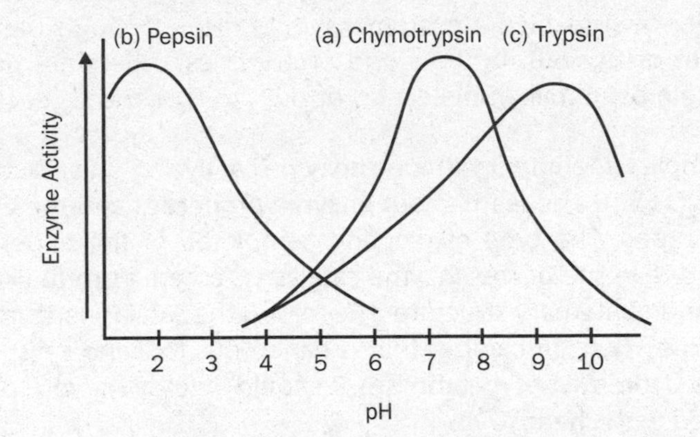

In the above graph, you can see the peak activity for different common digestive enzymes. Pepsin, for example, has a peak activity at a very low pH because it catalyzes reactions in the stomach, which is very acidic. The peak activity for chymotrypsin occurs at a pH of 7 to 8, which is fairly neutral. The pH of a system affects protein activity by altering the structure of the protein. Proteins have a tertiary structure based on the electrochemical properties of the amino acids in the chain. The charges on these amino acids can change as the pH changes because of the available H^+ and OH^- ions.

The last things you need to understand from this lab are the effects of changing the amount of enzyme and substrate available.

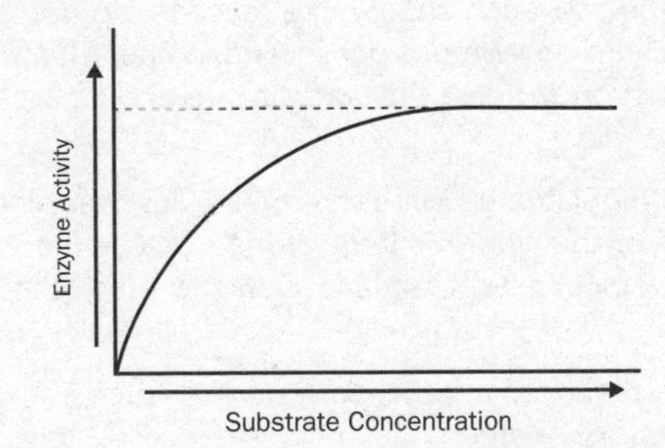

You may think that as you add more H_2O_2, you get more reaction. You would be right ... to a point. Eventually, no matter how much substrate you add, the reaction maxes out. This is because there are a limited number of enzyme molecules in the system. These enzymes can handle only one molecule of H_2O_2 at a time. Think of enzymes as cashiers at the grocery store. If several customers want to check out at the same time, they can only be helped one at a time by any single cashier. Once all of the cashiers are busy, a customer has to wait for the next available cashier. Thus, H_2O_2 has to wait for the next available enzyme to catalyze the reaction.

Changing the enzyme concentration affects the reaction in a similar way, except the reaction becomes almost instantaneous as the number of enzyme molecules becomes equal to the number of substrate molecules. An increase in enzyme concentration is like every customer in the grocery store having his or her own cashier. A customer would move through the line as quickly as the cashier can check him or her out. In the world of enzymes, orders are processed very quickly. Catalase can convert almost 6 million molecules of H_2O_2 to H_2O and O_2 each minute.

By completing this simple experiment with one enzyme catalyzing one reaction, you not only learn about catalase and H_2O_2, you also learn about enzyme properties and the effect the environment has on chemical reactions. The type of reaction completed in the experiment was oxidation-reduction, where the active site of the enzyme catalase is a heme group like in hemoglobin, so it has a similar tertiary and quaternary structure. The reason the catalase enzyme exists in animal and plant tissue is to remove H_2O_2, an oxidizer that is dangerous to living cells and must be removed when it is created as a byproduct of metabolism. It should be easy for you to see now how biology is intricately connected with chemistry.

RAPID REVIEW

If you take away only 4 things from this chapter:

1. All enzymes possess an active site, a 3-D pocket within their structures, in which substrate molecules can be held in a certain orientation to facilitate a reaction. The two models of enzyme-substrate interaction are lock-and-key and induced fit.
2. Enzymes lower the activation energy of a reaction, do not get used up in the reaction, catalyze millions of reactions per second, do not affect the overall free-energy change of the reaction, increase the reaction rate, and do not change the equilibrium of reactions.
3. Enzymes can be regulated by inhibitors, molecules that bind to the enzyme either at the active site or the allosteric (regulatory) site. Feedback inhibition is when the end product of a biochemical reaction works to block the activity of the original enzyme.
4. Enzymes can be influenced by reaction conditions such as high temperatures, detergents, or acidic/basic conditions.

TEST WHAT YOU LEARNED

1. Researchers are studying the reaction rate of mevalonate-5-phosphate (Mev-P), an enzyme. The graph below shows the rate of the enzyme-catalyzed reaction as the initial concentration of Mev-P is increased.

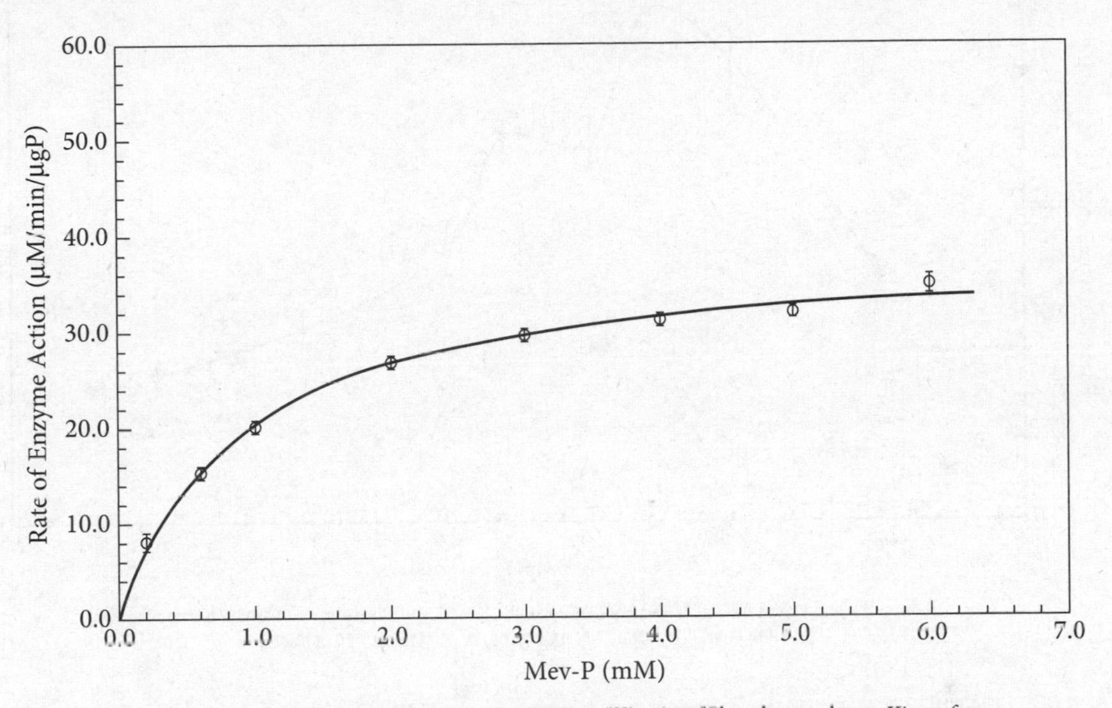

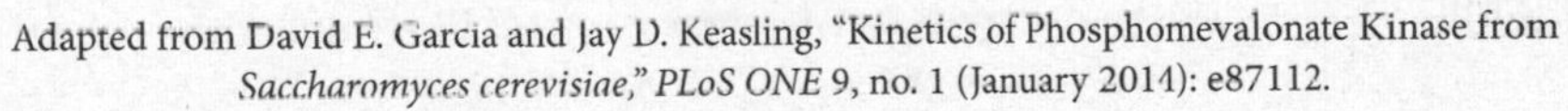
Adapted from David E. Garcia and Jay D. Keasling, "Kinetics of Phosphomevalonate Kinase from *Saccharomyces cerevisiae*," *PLoS ONE* 9, no. 1 (January 2014): e87112.

What would happen if the concentration of Mev-P was doubled from 6.0 mM to 12.0 mM?

(A) The reaction rate would increase significantly.

(B) The reaction rate would decrease sharply.

(C) The reaction rate would not change much.

(D) The reaction rate would change in an unpredictable way.

2. The figure below shows how the rate of an enzymatic reaction changed as the pH of the reaction changed.

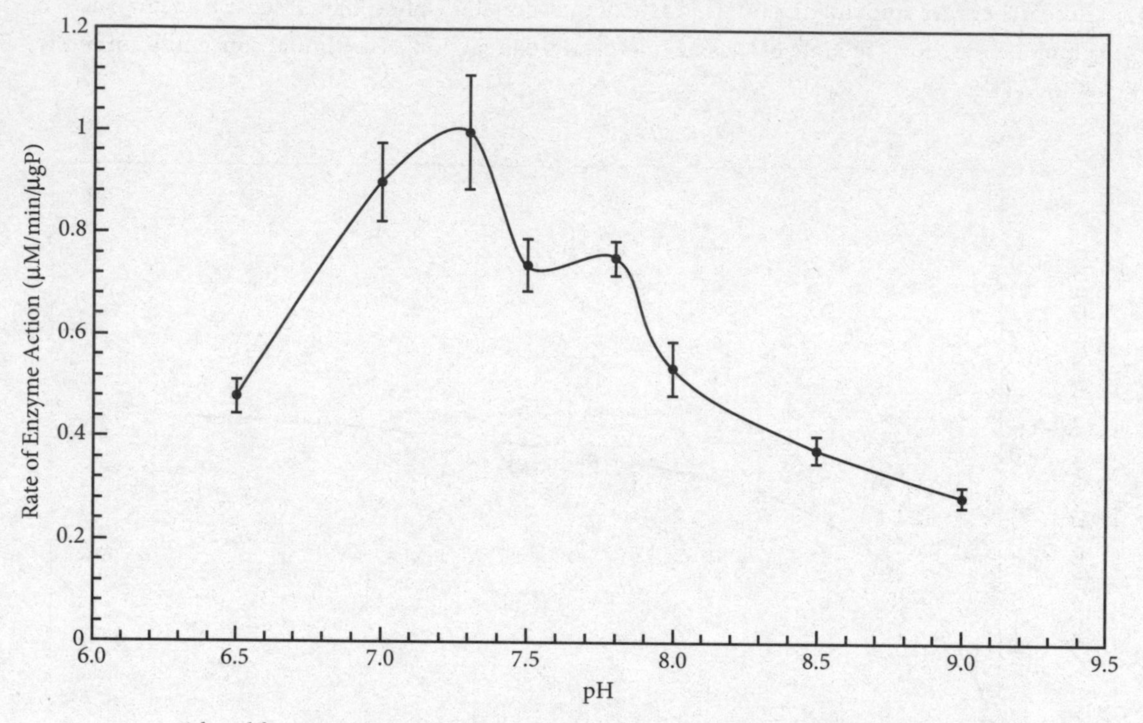

Adapted from David E. Garcia and Jay D. Keasling, "Kinetics of Phosphomevalonate Kinase from *Saccharomyces cerevisiae*," *PLoS ONE* 9, no. 1 (January 2014): e87112.

How would increasing the concentration of substrate most likely affect the rate of reaction at a pH of 7.75?

(A) The reaction rate would be lower, regardless of how the reaction rate compared with the maximum possible rate.

(B) The reaction rate would be higher unless the reaction was already at the maximum rate for that pH.

(C) Because the reaction rate at a pH of 7.75 appears to be unusually favorable, the reaction rate probably cannot increase further if substrate is added.

(D) Because the reaction rate at a pH of 7.75 is lower than the maximum rate shown at a pH of 7.25, the reaction rate will increase back to the maximum of 1.

Processes

3. Glucose-6-phosphate dehydrogenase (G6PDH) is an enzyme that can use NAD^+ or $NADP^+$ as a cofactor. The table shows the K_m and k_{cat}/K_m of a wild type and mutant version of G6PDH. The K_m represents the substrate concentration at which the reaction rate is half of the maximum rate. k_{cat}/K_m is a measure of substrate specificity that indicates the strength of preference for a particular enzyme. Data for both cofactors are shown below.

	$NADP^+$		NAD^+	
G6PDH Type	**K_M (μM)**	**k_{cat}/K_M ($\mu M^{-1}s^{-1}$)**	**K_M (μM)**	**k_{cat}/K_M ($\mu M^{-1}s^{-1}$)**
Wild type	7.5 ± 0.8	23.2 ± 2.4	5090 ± 400	$0.06 \pm 4 \times 10^{-3}$
Mutant	17696 ± 1453	$0.01 \pm 9 \times 10^{-4}$	11736 ± 804	$0.01 \pm 7 \times 10^{-4}$

Adapted from M. Fuentealba et al., "Determinants of Cofactor Specificity for the Glucose-6-Phosphate Dehydrogenase from *Escherichia coli*: Simulation, Kinetics and Evolutionary Studies," *PLoS ONE* 11, no. 3 (March 2016): e0152403.

Based on the data in the table, which of the following statements is true?

(A) The wild type enzyme works less efficiently with NAD^+ than the mutant enzyme. With this cofactor, it reaches the maximum velocity at a much higher substrate concentration than the mutant enzyme and also has higher substrate specificity.

(B) The wild type enzyme works more efficiently with NAD^+ than the mutant enzyme. With this cofactor, it reaches the maximum velocity at a much higher substrate concentration than the mutant enzyme and also has lower substrate specificity.

(C) The wild type enzyme works less efficiently with $NADP^+$ than the mutant enzyme. With this cofactor, it reaches the maximum velocity at a much lower substrate concentration and also has lower substrate specificity.

(D) The wild type enzyme works more efficiently with $NADP^+$ than the mutant enzyme. With this cofactor, it reaches the maximum velocity at a much lower substrate concentration than the mutant enzyme and also has higher substrate specificity.

4. Protein kinase C (PKC) can be inhibited by multiple inhibitors. Some are competitive inhibitors, but others are not. Which of the findings below would suggest competitive inhibition?

(A) The reaction reaches the same maximum velocity with or without the inhibitor, although at different substrate concentrations.

(B) The inhibitor is found to bind to the enzyme relatively far from its active site.

(C) Increasing the concentration of the substrate without increasing the concentration of the enzyme has no effect on the reaction rate.

(D) Raising the temperature increases the reaction rate until an optimal temperature is reached, at which point the reaction rate decreases again.

5. Researchers studied the activity of mevalonate kinase, an enzyme that may be important for biofuel production. The graph below shows the rate of a reaction as the concentration of ATP is increased while substrate concentration remains constant. The open circles represent data points collected at 30°C and the closed circles represent data points collected at 37°C.

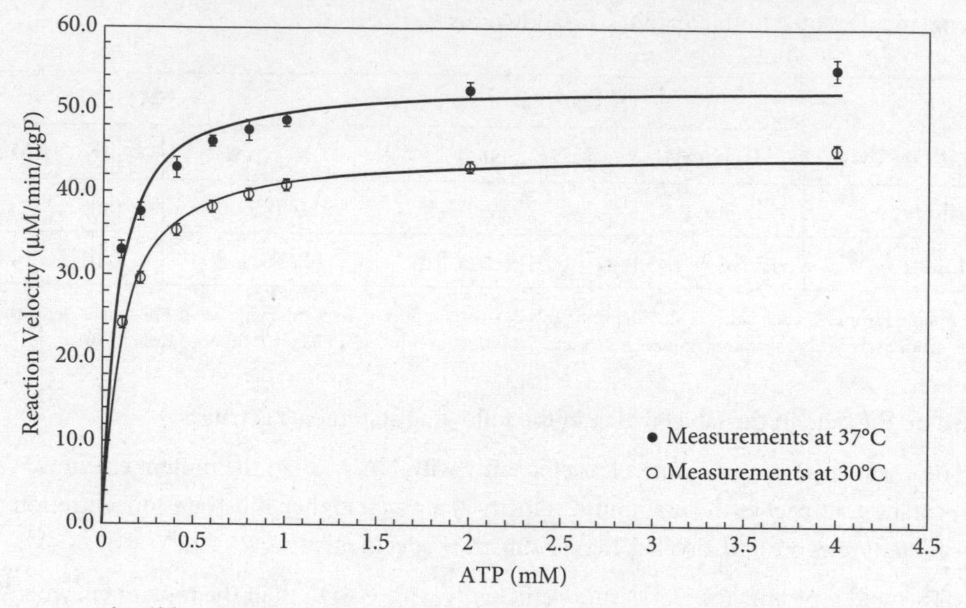

Adapted from David E. Garcia and Jay D. Keasling, "Kinetics of Phosphomevalonate Kinase from *Saccharomyces cerevisiae*," *PLoS ONE* 9, no. 1 (January 2014): e87112.

How is the reaction between enzyme and substrate modified by the different temperatures?

(A) At 30°C, the substrate molecules have less kinetic energy than at 37°C. This reduces the rate of random collisions and therefore the rate of the reaction.

(B) At 30°C, the substrate molecules have more kinetic energy than at 37°C. The excess kinetic energy makes it harder for the substrate to bind to the enzyme.

(C) At 37°C, the enzymes have formed the correct shape to maximize the reaction rate. At 30°C, the enzymes have denatured due to excess heat and the substrates have more difficulty binding to the active site.

(D) At 37°C, the concentration of the substrates is higher than at 30°C. This allows the reaction to proceed more rapidly.

The answer key to this quiz is located on the next page.

Answer Key

Test What You Already Know

1. **D** **Learning Objective:** 12.1
2. **B** **Learning Objective:** 12.2
3. **B** **Learning Objective:** 12.3
4. **C** **Learning Objective:** 12.4
5. **B** **Learning Objective:** 12.5

Test What You Learned

1. **C** **Learning Objective:** 12.4
2. **B** **Learning Objective:** 12.5
3. **D** **Learning Objective:** 12.2
4. **A** **Learning Objective:** 12.3
5. **A** **Learning Objective:** 12.1

Detailed solutions can be found in the Answers and Explanations section at the back of this book.

REFLECTION

Test What You Already Know score: __________

Test What You Learned score: __________

Use this section to evaluate your progress. After working through the pre-quiz, check off the boxes in the "Pre" column to indicate which Learning Objectives you feel confident about. Then, after completing the chapter, including the post-quiz, do the same to the boxes in the "Post" column. Keep working on unchecked Objectives until you're confident about them all!

Pre	Post	
☐	☐	**12.1** Explain how enzymes and substrates interact
☐	☐	**12.2** Explain how cofactors or coenzymes affect enzyme function
☐	☐	**12.3** Contrast competitive and noncompetitive inhibition
☐	☐	**12.4** Explain how concentration affects enzyme function
☐	☐	**12.5** Investigate how enzyme reactivity is affected by concentration

FOR MORE PRACTICE

Complete more practice online at kaptest.com. Haven't registered your book yet? Go to kaptest.com/booksonline to begin.

CHAPTER 13

Metabolism

LEARNING OBJECTIVES

In this chapter, you will review how to:

- **13.1** Describe structures required for photosynthesis
- **13.2** Explain the light-dependent reactions of photosynthesis
- **13.3** Explain energy pathways for glycolysis and fermentation
- **13.4** Explain the function of the electron transport chain
- **13.5** Compare input and output molecules and energy
- **13.6** Investigate effects on the rate of photosynthesis
- **13.7** Investigate effects on the rate of cellular respiration

TEST WHAT YOU ALREADY KNOW

1. In chloroplasts, electron transport chain (ETC) reactions establish an electrochemical gradient of protons across the thylakoid membrane. Synthesized ATP and NADPH are then used to produce carbohydrates in the chloroplast stroma. Which of the following figures most accurately illustrates ETC reactions and ATP synthesis in chloroplasts?

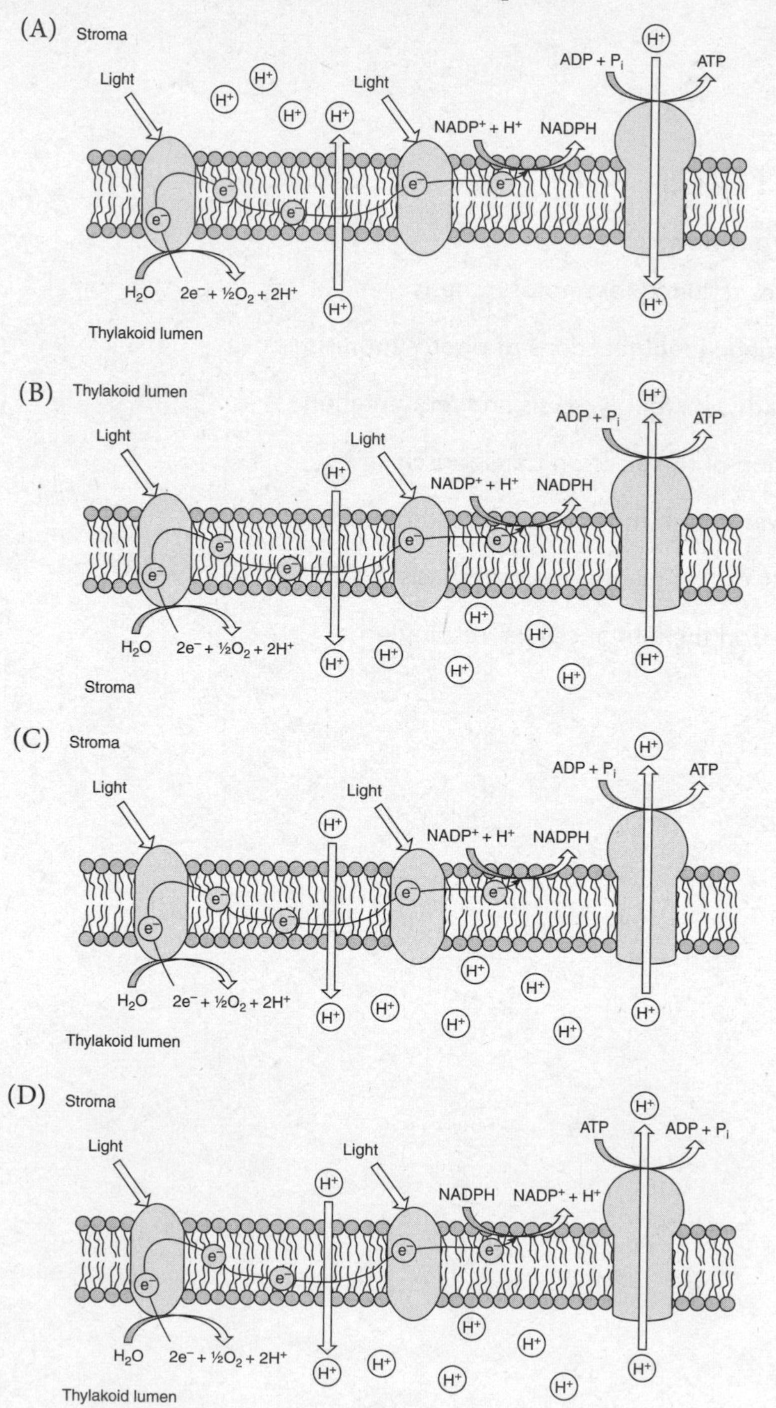

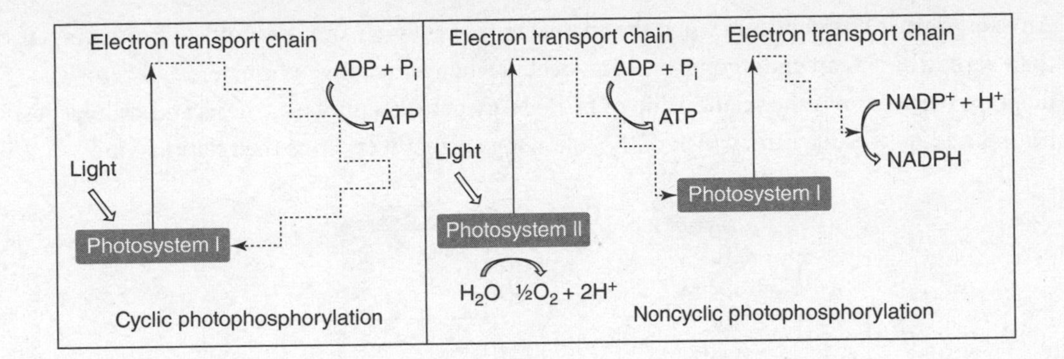

2. The diagram above illustrates cyclic and noncyclic photophosphorylation. High concentrations of NADPH in the chloroplast of a plant cell will cause a shift from noncyclic photophosphorylation to cyclic photophosphorylation. Which of the following best describes a likely result of this shift?

   (A) ATP levels will increase to meet cell energy demands.

   (B) ATP levels will become depleted as oxygen is produced.

   (C) NADPH and ATP will be used in the Krebs cycle to synthesize carbohydrates.

   (D) The reaction center in photosystem I will produce increased levels of oxygen and NADPH.

3. Human red blood cells lack mitochondria. Which of the following correctly explains the primary pathway that red blood cells use to produce energy?

   (A) Red blood cells generate ATP and NADH via aerobic respiration.

   (B) Red blood cells generate ATP and NADH via anaerobic fermentation.

   (C) Red blood cells metabolize glucose via glycolysis followed by carbon dioxide and ethanol production to produce ATP.

   (D) Red blood cells produce ATP via glycolysis followed by lactic acid production.

4. Thermogenin, an uncoupling protein found in brown adipose tissue, increases the permeability of the inner mitochondrial membrane. Which of the following accurately describes how the uncoupler affects the electron transport chain?

   (A) Thermogenin decreases the proton gradient and thus decreases ATP production.

   (B) Thermogenin decreases the proton gradient and thus increases ATP production.

   (C) Thermogenin increases the proton gradient and thus decreases ATP production.

   (D) Thermogenin increases the proton gradient and thus increases ATP production.

5. An exergonic reaction occurs when the change in free energy from reactants to products is less than zero, whereas an endergonic reaction occurs when the change is greater than zero. The diagram below shows the relationship of free energy between photosynthesis and cellular respiration. Based on the diagram, which of the following correctly explains the relationship?

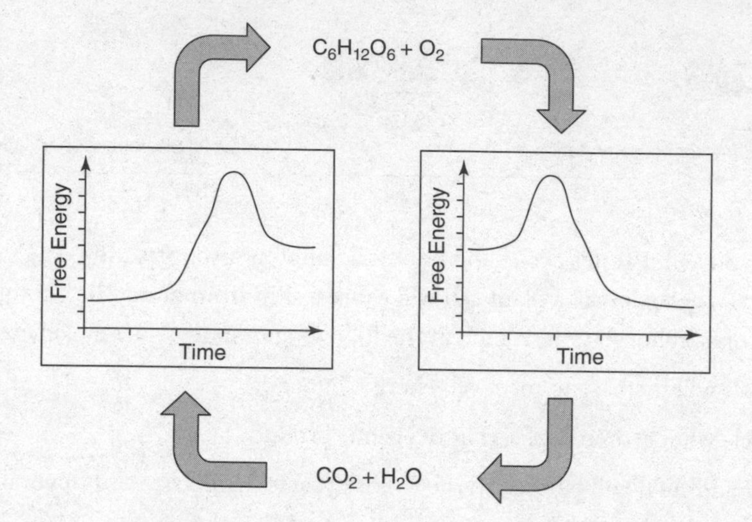

(A) Photosynthesis is exergonic, producing ATP and CO_2; cellular respiration is endergonic, consuming glucose and H_2O.

(B) Photosynthesis is exergonic, absorbing light energy to make ATP; cellular respiration is endergonic, releasing heat and CO_2.

(C) Cellular respiration is exergonic, consuming ATP to make glucose; photosynthesis is endergonic, absorbing light energy and releasing heat.

(D) Cellular respiration is exergonic, making ATP and releasing heat; photosynthesis is endergonic, converting light energy to chemical energy and glucose.

6. Dichlorophenolindophenol (DPIP), an electron acceptor that changes from blue to clear when reduced, can be used to visually determine the rate of photosynthesis. In an experiment, a solution of chloroplasts containing DPIP was divided among 4 tubes. The samples were then exposed to light (1,500 lumens) and/or heat (a temperature of 85°C), and light transmittance was measured over time using a spectrophotometer. Higher transmittance is correlated to lighter color. The results of the experiment are provided below.

	Transmittance (%)			
Time (min)	**No Light and No Heat**	**Light Only**	**Heat Only**	**Both Light and Heat**
0	23.2	20.8	20.4	20.4
5	25.3	37.8	22.9	21.1
10	26.5	48.7	20.6	21.8
15	24.2	63.4	21.6	22.7
20	25.7	77.5	22.1	23.0

Which of the following can most reasonably be concluded from the experimental results?

(A) The onset of photosynthesis is visible when DPIP is oxidized and changes from clear to blue.

(B) Chloroplasts exposed to heat had the highest rate of photosynthesis.

(C) Photosynthesis is stimulated when chloroplasts are exposed to light only.

(D) The solution in all four tubes was clear at time 0.

7. In a respiration experiment, transgenic mice (those possessing a particular genetic mutation) were compared to wild type (normal) mice. The ADP levels in the transgenic mice were found to be higher than those in the wild type. Which of the following is the most plausible explanation for this finding?

(A) The rate of oxygen consumption in mitochondria was higher in transgenic mice.

(B) The transgenic mice have impaired mitochondria, reducing their ability to convert ADP to ATP.

(C) The transgenic mice are unable to utilize ATP in the cytoplasm for metabolism.

(D) The rate of carbon dioxide consumption in cells was lower in transgenic mice.

Answers to this quiz can be found at the end of this chapter.

PHOTOSYNTHESIS

HIGH YIELD

13.1 Describe structures required for photosynthesis

13.2 Explain the light-dependent reactions of photosynthesis

Cellular **metabolism** is the sum total of all chemical reactions that take place in a cell. These reactions can be generally categorized as either *anabolic* or *catabolic*. Anabolic processes are energy-requiring, involving the biosynthesis of complex organic compounds from simpler molecules. Catabolic processes release energy as they break down complex organic compounds into smaller molecules. The metabolic reactions of cells are coupled so that energy released from catabolic reactions can be harnessed to fuel anabolic reactions.

To survive, all organisms need energy. ATP is an energy intermediary used to drive biosynthesis and other processes. ATP is generated in mitochondria using the chemical energy of glucose and other nutrients. The energy foundation of almost all ecosystems is **photosynthesis**. Plants are **autotrophs**, or self-feeders, that generate their own chemical energy from the energy of the Sun through photosynthesis. The chemical energy that plants get from the Sun is used to produce glucose. This glucose can then be burned in plant mitochondria to make ATP, which is used to drive all of the energy-requiring processes in the plant, including the production of proteins, lipids, carbohydrates, and nucleic acids. Similarly, animals can eat plants to extract the energy for their own metabolic needs. In this way, photosynthesis is the energy foundation of most living systems.

> **✔ AP Expert Note**
>
> Not all autotrophs use photosynthesis. Some, such as bacteria that live near hydrothermal vents at the ocean floor, use chemosynthesis, a process that extracts energy through the oxidation of inorganic compounds.

Processes

Photosynthesis Overview

Whereas cellular respiration takes place in all living organisms, only certain organisms contain the necessary pigments to perform photosynthesis, a process of capturing the energy of light and storing it in the chemical bonds of carbohydrates for later use. Plant cells have both **chloroplasts** and mitochondria to perform photosynthesis and cellular respiration, respectively. Photosynthesis is actually the reverse reaction of cellular respiration.

$$6\,CO_2 + 12\,H_2O + \text{light energy} \rightarrow C_6H_{12}O_6 + 6\,O_2 + 6\,H_2O$$

Perhaps the most surprising aspect of photosynthesis is the ability of photosynthetic proteins to split water molecules (H_2O). Once the water molecules have been split, the oxygen atoms are immediately released as O_2 and the hydrogen (H) atoms donate their electrons, which are used to form ATP and NADPH. The H atoms combine with the C and O atoms from CO_2 to form carbohydrates and more water. Sugars are used for energy storage, and O_2 is a waste product that other organisms use for respiration.

> **✔ AP Expert Note**
>
> Photosynthetic organisms are **primary producers**, providing food that supplies the rest of the food web. This "food" starts out in the form of glucose.

Photosynthesis takes place in two stages: the **light** (or light-dependent) **reactions** and the **dark** (or Calvin or light-independent) **reactions**. In the light reactions, light energy is harnessed to produce chemical energy in the form of ATP and NADPH in a process called **photophosphorylation**. The dark reactions complete **carbon fixation,** the process by which CO_2 from the environment is incorporated into sugars with the help of energy released from the **oxidation** of ATP and the NADPH. In short, light reactions produce energy and dark reactions make sugars.

Photosynthesis will be explored further in the first lab section of this chapter. Comparisons of C3, C4, and CAM plants may show up within the data provided for a question. Most plants are C3. This means that the initial products of C fixation are two three-carbon molecules (phosphoglycerate, or PGA), synthesized through the intermediate enzyme **rubisco**. In C4 plants, CO_2 is initially fixed into a four-carbon molecule (oxaloacetate, also found in the Krebs cycle) by the intermediate enzyme phosphoenolpyruvic acid (PEP) in a mesophyll cell. This four-carbon molecule later releases a CO_2 molecule when it enters a bundle sheath cell. The enzyme PEP is much more likely to bind to CO_2 because it has a higher affinity for CO_2 than rubisco. C4 plants have a physiological advantage in hot, arid environments where they often have to limit the opening of stomata during the day.

✔ AP Expert Note

You do not need to memorize the differences between C3, C4, and CAM plants for Test Day.

Plants that go through C4 photosynthesis are grasses, which include semi-arid to arid crops like corn and sorghum. Many succulent plants, such as cacti, use an alternative method of limiting water loss in arid environments and are called CAM (crassulacean acid metabolism) plants because they collect CO_2 at night when it is cooler. CO_2 is then stored in the form of organic acids. C3, C4, and CAM plants all carry out the dark reactions in the Calvin cycle. However, C4 plants complete a carbon fixation step in separate *parts* of the plant, and CAM plants complete a carbon fixation step at separate *times*.

Photophosphorylation is driven by light energy absorbed by pigments (such as **chlorophylls**) in chloroplasts. White light is composed of many different wavelengths. Plants in particular have developed ways to use light of more than one wavelength. Chloroplasts can only use light energy if the energy is absorbed. Visible color is caused by reflected light. Plants reflect green light, so green light is not very useful in photosynthesis.

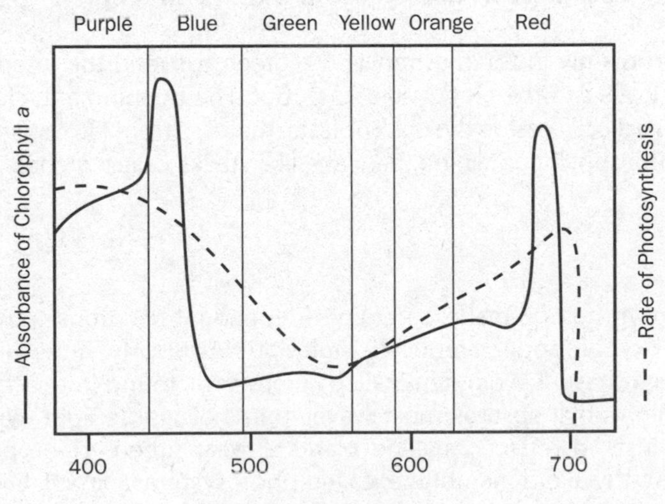

The previous image shows action spectra for the rate of photosynthesis (dotted line) and the absorbance of chlorophyll *a*. Note that the two action spectra do not correlate identically, indicating that the rate of photosynthesis depends on the presence of other photopigments as well. Chlorophyll *a* is medium green, chlorophyll *b* is yellow-green, and the carotenoids range in color from yellow to orange. Chlorophyll *a* is the only photopigment that participates in light reactions, but chlorophyll *b* and the carotenoids indirectly supplement photosynthesis by providing energy to chlorophyll *a*. (The carotenoids absorb light the chlorophyll cannot and transfer the energy to the chlorophyll.) The point to keep in mind is that many different photopigments absorb light energy from different wavelengths during photosynthesis. In order for plants to be healthy, they need blue and red light, not green.

✔ AP Expert Note

You may feel overwhelmed by the amount of information on photosynthesis, but don't panic. Only the major points are required for Test Day.

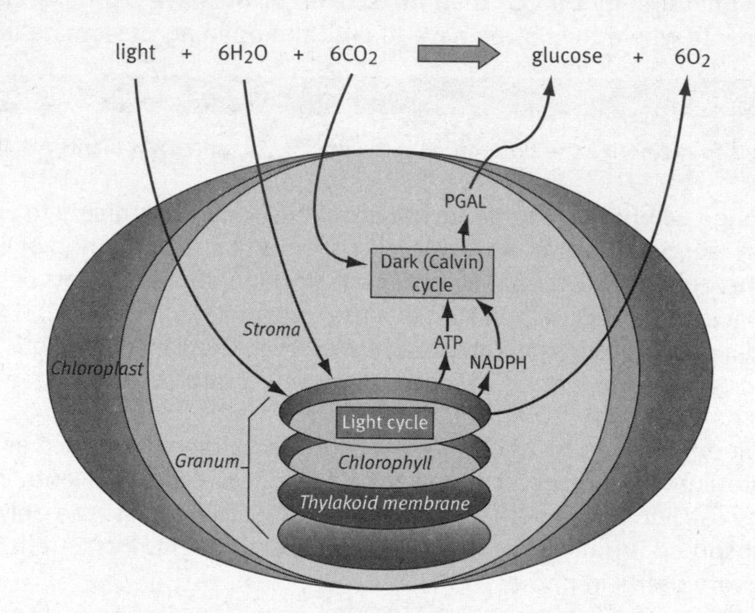

Locations of Photosynthesis within a Chloroplast

Chloroplasts are found mainly in the mesophyll, the green tissue in the interior of the plant. The stomata, pores in the leaf's surface, let CO_2 in and O_2 out. The opening and closing of the stomata is controlled by the guard cells. Inside the chloroplasts, there is stroma (a dense fluid) and thylakoid sacs (arranged into chlorophyll-containing, pancake-like stacks called grana).

Light Reactions

The first part of photosynthesis is made up of light-dependent reactions, in which light energy is used to generate ATP, oxygen, and the reducing molecule NADPH. The molecule that captures light energy to start photosynthesis is a pigment called chlorophyll, found in the thylakoid membranes of the chloroplast. Chlorophyll absorbs most wavelengths of visible light, with the exception of green light, which it instead reflects, making plants appear green. Chlorophyll is used by two complex systems in the thylakoid membrane, called photosystems I and II. Each photosystem is a complex assembly of protein and pigments in the membrane. When photons strike chlorophyll, electrons are excited and transferred through the photosystems to a reaction center. When electrons reach the reaction center, the reaction center gives up excited electrons that enter an electron transport chain, where they are used to generate chemical energy as either reduced NADPH or ATP.

Two different processes occur in the photosystems, *cyclic photophosphorylation* and *noncyclic photophosphorylation*. Both are used to generate ATP but in different ways. The ATP, in turn, is used to generate glucose in the dark reactions. Cyclic photophosphorylation occurs in photosystem I to produce ATP. In the cyclic method, electrons move from the reaction center, through an electron transport chain, then back to the same reaction center again (see figure below). The reaction center in photosystem I includes a chlorophyll called P700 because its maximal light absorbance occurs at 700 nm. This process does not produce oxygen and does not produce NADPH.

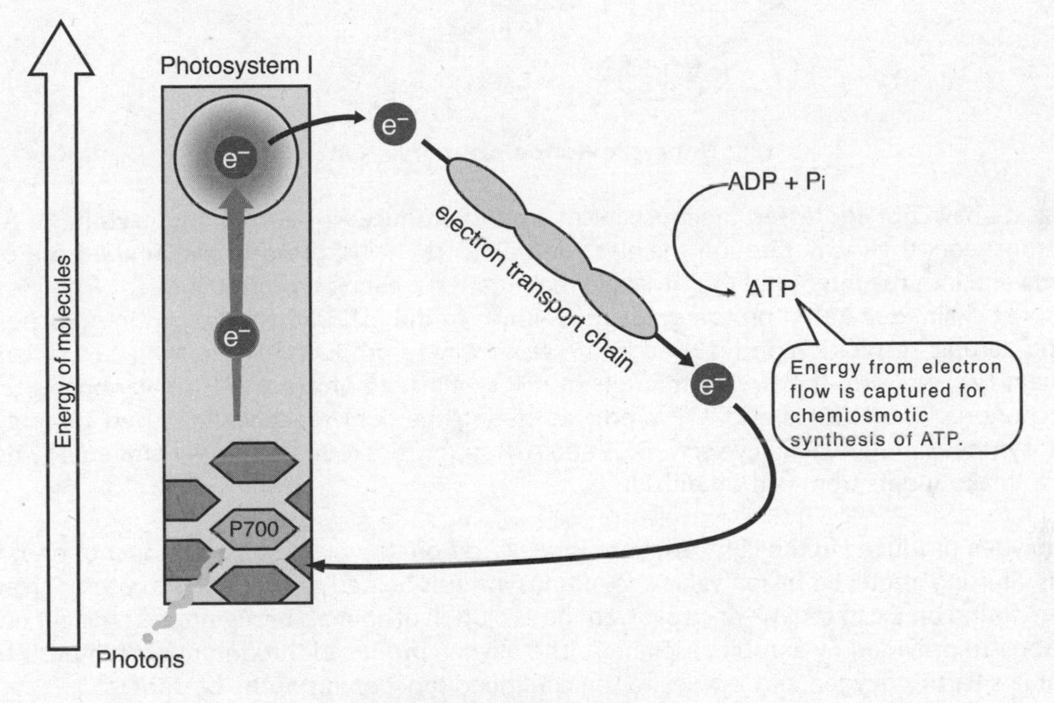

Cyclic Photophosphorylation

Noncyclic photophosphorylation starts in photosystem II (see figure below). In noncyclic photophosphorylation, chlorophyll pigment absorbs light and passes excited electrons to a reaction center, a process equivalent to cyclic photophosphorylation. The photosystem II reaction center contains a P680 chlorophyll, distinct from photosystem I. From the photosystem II reaction center, the electrons are passed to an electron transport chain. In this case, however, the electrons are not returned to the reaction center at the end of the electron transport chain but are passed to photosystem I. Photosystem II replaces the electrons it lost by getting them from water, producing oxygen in the process. In this case, the electrons that enter photosystem I are used to produce NADPH.

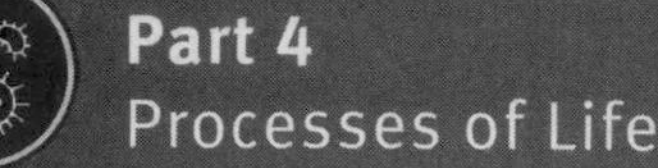

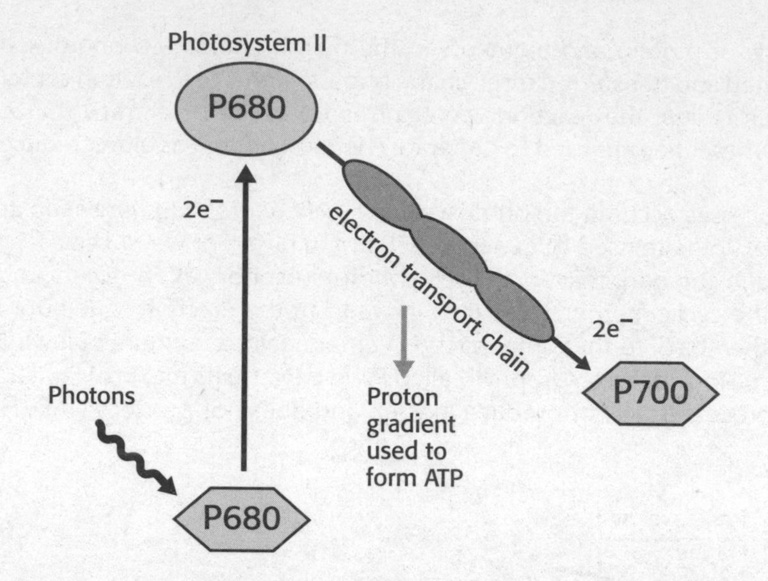

Noncyclic Photophosphorylation

So far, we have not addressed the mechanism used to produce ATP during photosynthesis. As the electrons work their way through the electron transport chains, protons are pumped out of the stroma and into the interior of the thylakoid membranes, creating a proton gradient. This electron transport chain–generated proton gradient is similar to the pH gradient created in mitochondria during aerobic respiration and is used in the same way to produce ATP. Protons flow down this gradient back out into the stroma through an ATP synthase to produce ATP, similar once again to mitochondria. The NADPH and ATP produced during the light reactions are used to complete photosynthesis in the Calvin cycle; NADPH and ATP supply the reducing power and energy necessary to make sugars from carbon dioxide.

The oxygen produced in the light reactions is released from the plant as a byproduct of photosynthesis. Starting about 1.5 billion years ago, photosynthesis helped to create the oxygen-rich atmosphere found on Earth today, which allowed the evolution of animals requiring the efficient energy metabolism provided by aerobic respiration. The oxygen produced through photosynthesis today maintains Earth's oxygen and is a key to the continued functioning of the biosphere.

Calvin-Benson Cycle

In the second phase of photosynthesis, the energy captured in the light reactions as ATP and NADPH is used to drive carbohydrate synthesis. This cycle, often called the **Calvin cycle** but also known as the Calvin-Benson cycle, fixes CO_2 into carbohydrates, reducing the fixed carbon to carbohydrates through the addition of electrons. The NADPH provides the reducing power for the reduction of CO_2 to carbohydrates, and air provides the carbon dioxide. CO_2 first combines with, or is fixed to, ribulose bisphosphate (RuBP), a five-carbon sugar with two phosphate groups. The enzyme that catalyzes this reaction is called rubisco and is the most abundant enzyme on Earth. The resulting six-carbon compound is promptly split, resulting in the formation of two molecules of 3-phosphoglycerate, a three-carbon compound. The 3-phosphoglycerate is then phosphorylated by ATP and reduced by NADPH, which leads to the formation of phosphoglyceraldehyde (PGAL). This can then be utilized as a starting point for the synthesis of glucose, starch, proteins, and fats.

✔ AP Expert Note

The steps of the Calvin cycle and the structure of the molecules involved are not required knowledge for the AP Biology exam.

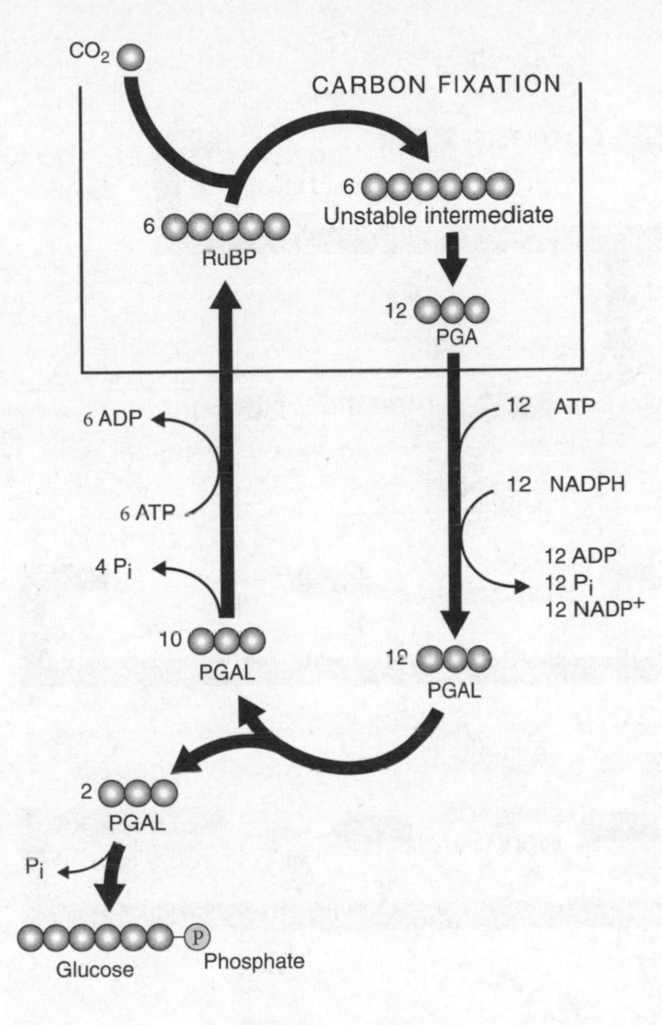

Calvin-Benson Cycle

ANAEROBIC RESPIRATION

13.3 Explain energy pathways for glycolysis and fermentation

Overview of Cellular Respiration

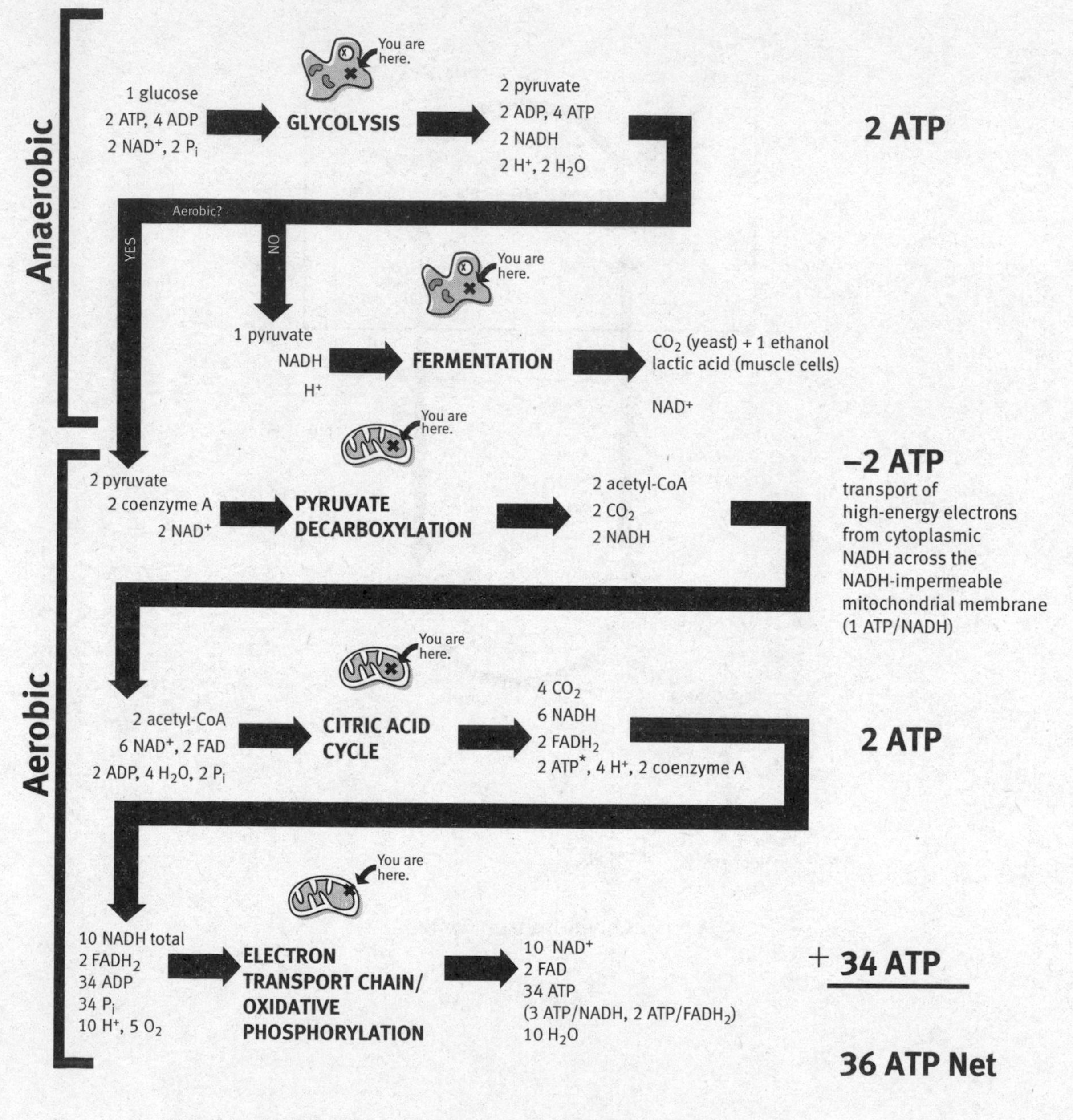

*The citric acid cycle yields a direct product of 2 GTP. The 2 GTP subsequently donate their phosphate to 2 ADP to form 2 ATP and regenerate the original 2 GDP.

General Overview of Cellular Respiration

Cellular **respiration** is the most efficient catabolic pathway used by organisms to harvest the energy stored in glucose. Whereas glycolysis yields only 2 ATP per molecule of glucose, cellular respiration can yield 36–38 ATP. Cellular respiration is an **aerobic** process; oxygen acts as the final acceptor of electrons that are passed from carrier to carrier during the final stage of glucose oxidation. The metabolic reactions of cellular respiration occur in the eukaryotic **mitochondria** and are catalyzed by reaction-specific enzymes.

Cellular respiration can be divided into five stages: glycolysis, fermentation, pyruvate decarboxylation, the citric acid cycle, and the electron transport chain.

Glycolysis

The first stage of glucose catabolism is **glycolysis.** Glycolysis is a series of reactions that lead to the oxidative breakdown of glucose into two molecules of pyruvate (the ionized form of pyruvic acid), the production of ATP, and the reduction of NAD^+ into NADH. All of these reactions occur in the cytoplasm and are mediated by specific enzymes. The glycolytic pathway is as follows:

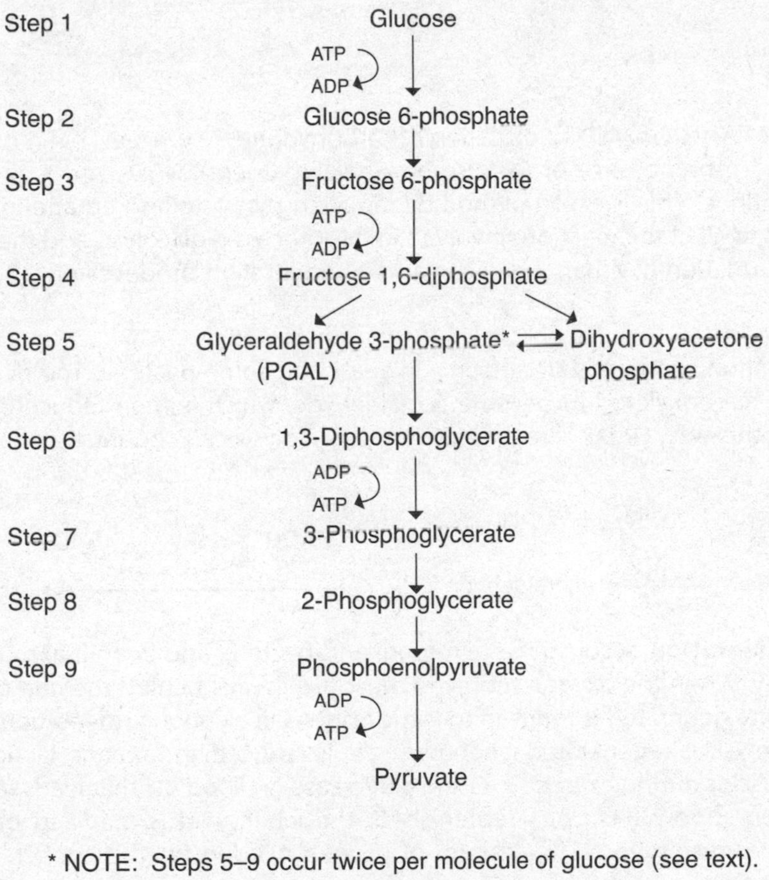

Step 1 Glucose
(ATP → ADP)
Step 2 Glucose 6-phosphate
Step 3 Fructose 6-phosphate
(ATP → ADP)
Step 4 Fructose 1,6-diphosphate
Step 5 Glyceraldehyde 3-phosphate* (PGAL) ⇄ Dihydroxyacetone phosphate
Step 6 1,3-Diphosphoglycerate
(ADP → ATP)
Step 7 3-Phosphoglycerate
Step 8 2-Phosphoglycerate
Step 9 Phosphoenolpyruvate
(ADP → ATP)
Pyruvate

* NOTE: Steps 5–9 occur twice per molecule of glucose (see text).

Glycolysis

✔ AP Expert Note

You do not need to know all these steps or the intermediates.

From one molecule of glucose (a six-carbon molecule), two molecules of pyruvate (a three-carbon molecule) are obtained. During this sequence of reactions, 2 ATP are used (in steps 1 and 3) and 4 ATP are generated (2 in step 6 and 2 in step 9). Thus, there is a net production of 2 ATP per glucose molecule. This type of phosphorylation is called substrate level phosphorylation because ATP synthesis is directly coupled with the degradation of glucose without the participation of an intermediate molecule, such as NAD^+. One NADH is produced per PGAL, for a total of 2 NADH per glucose.

The net reaction for glycolysis is:

$$\text{Glucose} + 2\ \text{ADP} + 2\ \text{P}_i + 2\ \text{NAD}^+ \rightarrow 2\ \text{Pyruvate} + 2\ \text{ATP} + 2\ \text{NADH} + 2\ \text{H}^+ + 2\ \text{H}_2\text{O}$$

This series of reactions occurs in both prokaryotic and eukaryotic cells. However, at this stage, much of the initial energy stored in the glucose molecule has not been released and is still present in the chemical bonds of pyruvate. Depending on the capabilities of the organism, pyruvate degradation can proceed in one of two directions. Under **anaerobic** conditions (in the absence of oxygen), pyruvate is reduced during the process of fermentation. Under aerobic conditions (in the presence of oxygen), pyruvate is further oxidized during cellular respiration in the mitochondria.

Fermentation

The production of ATP involves the oxidation of carbohydrates, a process that requires NAD^+ as an oxidizing agent. In the absence of O_2 (also an oxidizing agent), NAD^+ must be regenerated for glycolysis to continue. This is accomplished by reducing pyruvate into ethanol or lactic acid. Fermentation refers to all of the reactions involved in this process—glycolysis and the additional steps leading to the formation of ethanol or lactic acid. Fermentation produces only 2 ATP per glucose molecule.

Alcohol fermentation commonly occurs only in yeast and some bacteria. The pyruvate produced in glycolysis is decarboxylated to become acetaldehyde, which is then reduced by the NADH to yield ethanol. In this way, NAD^+ is regenerated and glycolysis can continue.

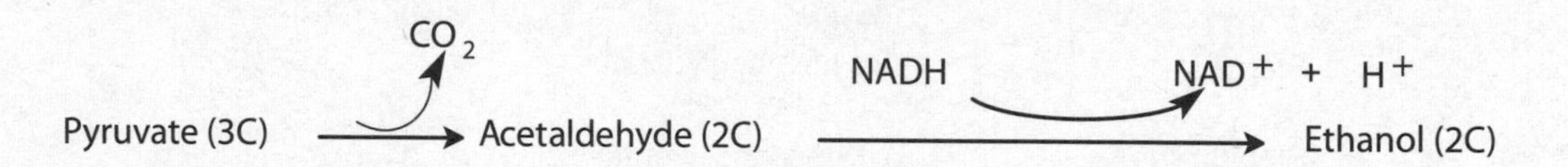

Lactic acid fermentation occurs in certain fungi and bacteria and in animal muscle cells during strenuous activity. When the oxygen supply to muscle cells lags behind the rate of glucose catabolism, the pyruvate generated is reduced to lactic acid. As in alcohol fermentation, the NAD^+ used in step 5 of glycolysis is regenerated when pyruvate is reduced. In humans, lactic acid may accumulate in the muscles during exercise, causing a decrease in blood pH that leads to muscle fatigue. Once the oxygen supply has been replenished, the lactic acid is oxidized back to pyruvate and enters cellular respiration. The amount of oxygen needed for this conversion is known as the oxygen debt.

AEROBIC RESPIRATION

13.4 Explain the function of the electron transport chain

13.5 Compare input and output molecules and energy

Pyruvate Decarboxylation

The pyruvate formed during glycolysis is transported from the cytoplasm into the mitochondrial matrix, where it is decarboxylated; i.e., it loses a CO_2, and the acetyl group that remains is transferred to coenzyme A to form acetyl-CoA. In the process, NAD^+ is reduced to NADH.

$$\text{Pyruvate (3C)} + \text{Coenzyme A} \xrightarrow{NAD^+ \rightarrow NADH + H^+} \text{Acetyl-CoA (2C)}$$

Citric Acid Cycle

The citric acid cycle is also known as the **Krebs cycle** or the tricarboxylic acid cycle (TCA cycle). The cycle begins when the two-carbon acetyl group from acetyl-CoA combines with oxaloacetate, a four-carbon molecule, to form the six-carbon citrate. Through a complicated series of reactions, 2 CO_2 are released, and oxaloacetate is regenerated for use in another turn of the cycle.

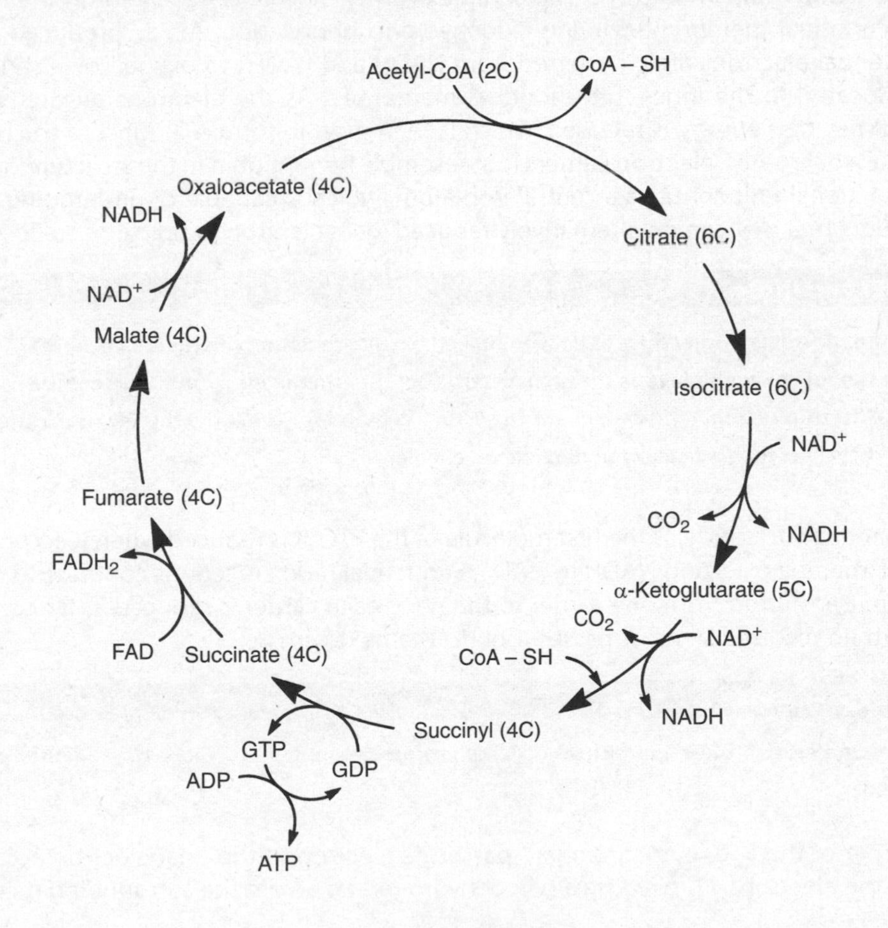

The Citric Acid Cycle

For each turn of the citric acid cycle, 1 ATP is produced by substrate level phosphorylation via a GTP intermediate. In addition, electrons are transferred to NAD^+ and FAD, generating NADH and $FADH_2$, respectively. These coenzymes then transport the electrons to the electron transport chain, where more ATP is produced via oxidative phosphorylation. To be clear, the most important function of the citric acid cycle is the production of a large number of NADH and $FADH_2$, which ultimately contribute to the production of a large number of ATP molecules.

✔ AP Expert Note

The exam will not test you on the details of pyruvate decarboxylation or the steps and intermediates of the citric acid cycle, so just be aware of the role each process plays in the larger process of aerobic respiration.

The net reaction of the citric acid cycle per glucose molecule is:

$$2\text{ Acetyl-CoA} + 6\text{ NAD}^+ + 2\text{ FAD} + 2\text{ GDP} + 2\text{ P}_i + 4\text{ H}_2\text{O} \rightarrow$$
$$4\text{ CO}_2 + 6\text{ NADH} + 2\text{ FADH}_2 + 2\text{ ATP} + 4\text{ H}^+ + 2\text{ CoA}$$

Electron Transport Chain

The **electron transport chain** (ETC) is a complex carrier mechanism located on the inside of the inner mitochondrial membrane. During oxidative phosphorylation, ATP is produced when high-energy-potential electrons are transferred from NADH and $FADH_2$ to oxygen by a series of carrier molecules located in the inner mitochondrial membrane. As the electrons are transferred from carrier to carrier, free energy is released, which is then used to form ATP. Most of the molecules of the ETC are cytochromes, electron carriers that resemble hemoglobin in the structure of their active site. The functional unit contains a central iron atom, which is capable of undergoing a reversible redox reaction; that is, it can be alternatively reduced (gain electrons) and oxidized (lose electrons).

✔ AP Expert Note

Everything the human body does to deliver inhaled oxygen to tissues (discussed in chapter 11) comes down to the role oxygen plays as the final electron acceptor in the electron transport chain. Without oxygen, ATP production is not adequate to sustain human life. Similarly, the CO_2 generated in the citric acid cycle is the same carbon dioxide we exhale.

FMN (flavin mononucleotide) is the first molecule of the ETC. It is reduced when it accepts electrons from NADH, thereby oxidizing NADH to NAD^+. Sequential redox reactions continue to occur as the electrons are transferred from one carrier to the next; each carrier is reduced as it accepts an electron and is then oxidized when it passes it on to the next carrier.

✔ AP Expert Note

Note that the electron transport chain (ETC) is like an assembly line where the majority of ATP is generated.

The last carrier of the ETC, cytochrome a_3, passes its electron to the final electron acceptor, O_2. In addition to the electrons, O_2 picks up a pair of hydrogen ions from the surrounding medium, forming water.

✔ AP Expert Note

The names of the specific electron carriers in the electron transport chain are not required knowledge for the AP Biology exam.

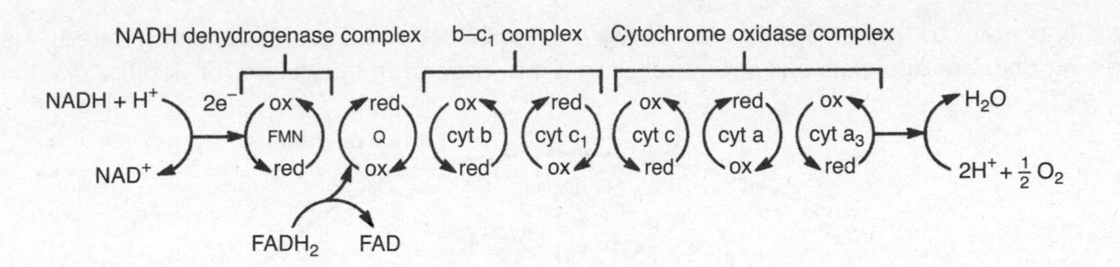

Electron Transport Chain

Without oxygen, the ETC becomes backlogged with electrons. As a result, NAD^+ cannot be regenerated and glycolysis cannot continue unless lactic acid fermentation occurs. Likewise, ATP synthesis comes to a halt if respiratory poisons such as cyanide or dinitrophenol enter the cell. Cyanide blocks the transfer of electrons from cytochrome a_3 to O_2. Dinitrophenol uncouples the electron transport chain from the proton gradient established across the inner mitochondrial membrane.

ATP Generation and the Proton Pump

The electron carriers are categorized into three large protein complexes (NADH dehydrogenase, the cytochrome $b–c_1$ complex, and cytochrome oxidase), as well as the carrier molecule Q. There are energy losses as the electrons are transferred from one complex to the next; this energy is then used to synthesize 1 ATP per complex. Thus, an electron passing through the entire ETC supplies enough energy to generate 3 ATP. NADH delivers its electrons to the NADH dehydrogenase complex so that for each NADH, 3 ATP are produced. However, $FADH_2$ bypasses the NADH dehydrogenase complex and delivers its electrons directly to carrier Q (ubiquinone), which lies between the NADH dehydrogenase and cytochrome $b–c_1$ complexes. Therefore, for each $FADH_2$, there are only two energy drops, and only 2 ATP are produced.

$$3 \times 6 \text{ NADH} \rightarrow 18 \text{ ATP}$$

$$2 \times 2 \text{ FADH}_2 \rightarrow 4 \text{ ATP}$$

$$1 \times 2 \text{ GTP (ATP)} \rightarrow 2 \text{ ATP}$$

The operating mechanism in this type of ATP production involves coupling the oxidation of NADH and $FADH_2$ to the phosphorylation of ADP. The coupling agent for these two processes is a proton gradient across the inner mitochondrial membrane, maintained by the ETC. As NADH and $FADH_2$ pass their electrons to the ETC, hydrogen ions (H^+) are pumped out of the matrix, across the inner mitochondrial membrane, and into the **intermembrane space** at each of the three protein complexes. The continuous translocation of H^+ creates a positively charged, acidic environment in the intermembrane space. This electrochemical gradient generates a proton-motive force, which drives H^+ back across the inner membrane and into the matrix. However, to pass through the membrane (which is impermeable to ions), the H^+ must flow through specialized channels provided by enzyme complexes called ATP synthetases. As the H^+ pass through the ATP synthetases, enough energy is released to allow for the phosphorylation of ADP to ATP. The coupling of the oxidation of NADH and $FADH_2$ with the phosphorylation of ADP is called **oxidative phosphorylation**.

Review of Glucose Catabolism

It is important to understand how all of the previously described events are interrelated. The following diagram depicts a eukaryotic cell with a mitochondrion magnified for detail.

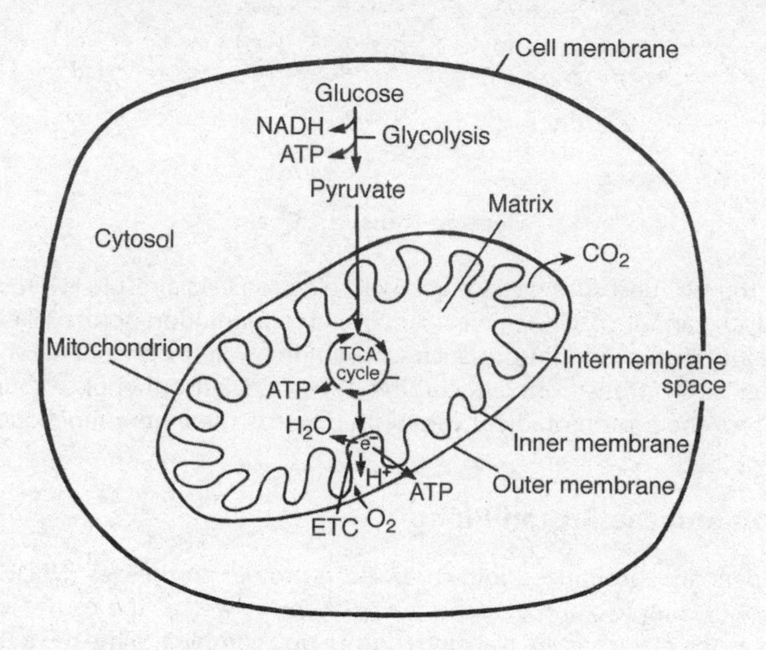

Locations of Glucose Catabolism

To calculate the net amount of ATP produced per molecule of glucose, we need to tally the number of ATP produced by substrate level phosphorylation and the number of ATP produced by oxidative phosphorylation.

Substrate Level Phosphorylation

Degradation of one glucose molecule yields a net of 2 ATP from glycolysis and 1 ATP for each turn of the citric acid cycle with two turns per glucose. Thus, a total of 4 ATP are produced by substrate level phosphorylation.

Oxidative Phosphorylation

Two pyruvate decarboxylations yield 1 NADH each for a total of 2 NADH per glucose. Each turn of the citric acid cycle yields 3 NADH and 1 $FADH_2$, for a total of 6 NADH and 2 $FADH_2$ per glucose molecule. Each $FADH_2$ generates 2 ATP, as previously discussed. Each NADH generates 3 ATP except for the 2 NADH that were reduced during glycolysis; these NADH cannot cross the inner mitochondrial membrane and must transfer their electrons to an intermediate carrier molecule, which delivers the electrons to the second carrier protein complex, Q. Therefore, these NADH generate only 2 ATP per glucose. Thus, the 2 NADH of glycolysis yield 4 ATP, the other 8 NADH yield 24 ATP, and the 2 $FADH_2$ produce 4 ATP, for a total of 32 ATP by oxidative phosphorylation.

The total amount of ATP produced during eukaryotic glucose catabolism is 4 via substrate level phosphorylation plus 32 via oxidative phosphorylation, for a total of 36 ATP. (For prokaryotes the yield is 38 ATP because the 2 NADH of glycolysis don't have any mitochondrial membranes to cross and therefore don't lose energy.) See the following table for a summary of eukaryotic ATP production.

A QUICK REFERENCE TO ENERGY PRODUCTION IN CELLULAR RESPIRATION

Glycolysis	
2 ATP invested (steps 1 and 3)	−2 ATP
4 ATP generated (steps 6 and 9)	+4 ATP (substrate)
2 NADH × 2 ATP/NADH (step 5)	+4 ATP (oxidative)
Pyruvate Decarboxylation	
2 NADH × 3 ATP/NADH	+6 ATP (oxidative)
Citric Acid Cycle	
6 NADH × 3 ATP/NADH	+18 ATP (oxidative)
2 $FADH_2$ × 2 ATP/$FADH_2$	+4 ATP (oxidative)
2 GTP × 1 ATP/GTP	+2 ATP (substrate)
	Total: + 36 ATP

AP BIOLOGY LAB 5: PHOTOSYNTHESIS INVESTIGATION

13.6 Investigate effects on the rate of photosynthesis

In this investigation, you will learn how to measure the rate of photosynthesis by determining oxygen production. To review, photosynthesis is the process by which autotrophs capture free energy (in the form of sunlight) to build carbohydrates. The process is summarized by the following reaction:

$$2\,H_2O + CO_2 + \text{light} \rightarrow \text{carbohydrate } (CH_2O) + O_2 + H_2O$$

To determine the rate of photosynthesis by a plant cell, you can measure the production of O_2 or the consumption of CO_2. The process is not that simple, though, because photosynthesis is coupled with aerobic respiration, in which the oxygen produced is simultaneously consumed. Because measuring the consumption of carbon dioxide typically requires expensive equipment and complex procedures, you will use the floating disk procedure to measure the production of oxygen.

In this procedure, you use a vacuum to remove all air and then add a bicarbonate solution to plant (leaf) disk samples. These leaves sink until placed in sufficient light, when photosynthesis produces enough oxygen bubbles to change the buoyancy of the disk, causing it to float to the surface. Many different factors can affect the rate of photosynthesis in the real world (i.e., intensity of light, color of light, leaf size, type of plant), but the results of this experiment can also be influenced by different procedural factors (i.e., depth of solution, method of cutting disks, size of leaf disks). You will effectively be measuring net photosynthesis. The standard measurement to use after determining how long it takes each disk to float to the surface is ET_{50}, the estimated time for 50 percent of the disks to rise. This measurement will help you to aggregate your data for the second half of the investigation.

After learning how to use the floating disk procedure, you will choose one factor that affects the rate of photosynthesis and then develop and conduct an investigation of that variable. When you compare the ET_{50} for different levels of your chosen variable, you should observe that ET_{50} goes down as rate of photosynthesis goes up. This creates a nontraditional graph that is not the best display of your data (below left). Alternatively, you can use $1/ET_{50}$, which will show increasing rates of photosynthesis and, therefore, a graph with a positive slope (below right).

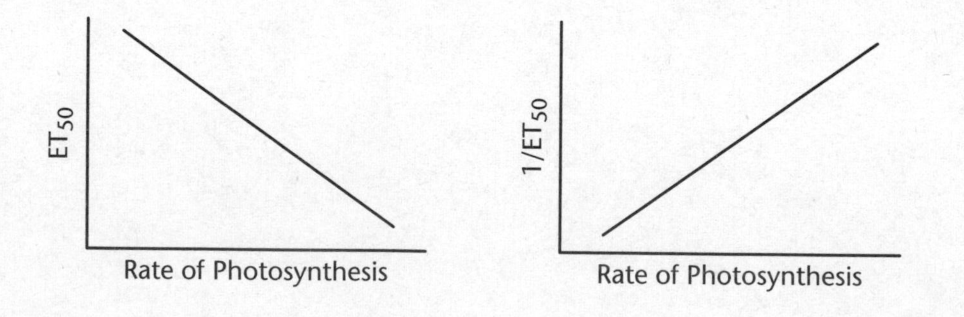

AP BIOLOGY LAB 6: CELLULAR RESPIRATION INVESTIGATION

13.7 Investigate effects on the rate of cellular respiration

This investigation is much more about the experimental method than cellular respiration. In this investigation, you learn how to set up and use a respirometer to measure the change in volume of a gas, which can be assumed to be O_2, from the germinating seeds placed in water. The effect of increasing temperature on the rate of gas volume change (O_2 utilization) is also measured during the lab. One of the problems with this lab is that gas expands when it is heated. Sometimes the volume increases so much it blows the dye out of the end of the respirometer tube. You will also have the opportunity to ask your own questions and conduct your own investigations about cellular respiration.

This experiment is an excellent example of using control specimens to isolate experimental variables. Put quite simply, the dependent variable being measured is the change in gas volume in the respirometer. The hypothesis is that the change in gas volume is being caused by the utilization of O_2 by the germinating seeds. Other factors can contribute to a change in gas volume (i.e., temperature and pressure), so these variables need to be isolated from the variables you are interested in (i.e., the rate of respiration of the seeds). To isolate variables you are interested in from other variables, glass beads should be subjected to the same experimental treatment (change in temperature) that the target specimens, the live seeds, are subjected to. The glass bead sample is the control group. Any measurable change in the dependent variable in the control group must be removed from the dependent variable of the target group.

Let's say that when you heated your seeds to 35°C, the volume of gas in the respirometer of the target group (live seeds) decreased by 0.3 mL, but the volume of gas in the respirometer of the control group (glass beads) increased by 0.1 mL. Something occurred in the respirometer of the control group that caused an increase in gas volume. Perhaps the change in volume in the respirometer was created by the expansion of heated gas, or expansion was caused by CO_2 not being absorbed by the soda lime or other CO_2 absorbent. Either way, the dependent variable of temperature must be taken away from the target group because a change in volume should have occurred in the respirometer of the target group as well as the respirometer of the control group. As a result, it can be assumed that an additional 0.1 mL (for a total of 0.4 mL) of O_2 was likely used by the germinating seeds. In the experiment, 0.4 mL of gas were not measured because 0.1 mL of gas was obscured by the expansion of gas from some unknown factor, likely increased temperature. An accurate measure of the target group can never be obtained if the control group isn't included in the experiment.

Though this investigation is typically conducted during the study of Big Idea Two (energy and cellular processes), it also connects to content and processes described under Big Idea One (evolution) and Big Idea Four (ecology). Think about how cellular respiration is a conserved evolutionary process or how different ecological habitats have modified the capture and use of free energy by different organisms. By thinking outside the box, you will be better able to apply your knowledge across the entire AP Biology exam. This is particularly helpful when addressing free-response questions, which often ask you to make connections between content and process.

RAPID REVIEW

If you take away only 6 things from this chapter:

1. Photosynthesis is the energy foundation for almost all living systems. In addition, photosynthesis provides almost all of the oxygen present in the Earth's atmosphere.
2. All photosynthetic organisms use chloroplasts and mitochondria to perform photosynthesis and cellular respiration. Photosynthesis and cellular respiration are essentially reverse operations of each other.
3. Photosynthesis has two main parts—the light cycle and the dark cycle (the latter is usually called the Calvin or Calvin-Benson cycle). Light reactions produce energy and dark reactions make sugars. The light reactions occur in the interior of the thylakoid, while the Calvin-Benson cycle occurs in the stroma.
4. Cellular respiration is an efficient catabolic pathway and yields ATP. Cellular respiration is an aerobic process. The metabolic reactions of respiration occur in the eukaryotic mitochondria and are catalyzed by reaction-specific enzymes.
5. Cellular respiration can be divided into several stages: glycolysis, pyruvate decarboxylation, the citric acid cycle, and the electron transport chain.
6. Cellular respiration is a complex process that requires many different products and specialized molecules. Focus on the requirements and overall net production for the major steps.

TEST WHAT YOU LEARNED

1. Which of the following figures correctly illustrates the light-dependent and light-independent reactions of photosynthesis?

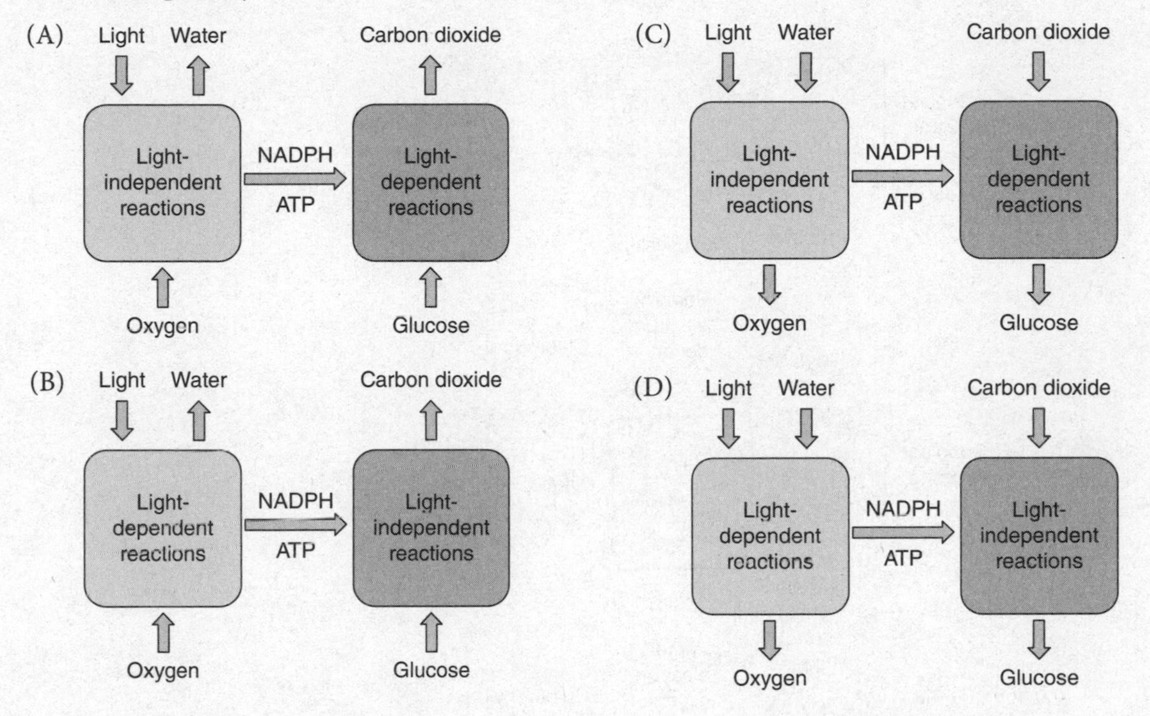

2. Autotrophs capture free energy to produce organic compounds for food while heterotrophs metabolize carbon compounds produced by other organisms as sources of free energy. Which of the following best demonstrates the function of CO_2 in autotrophic and heterotrophic metabolism?

   (A) Autotrophs use CO_2 to make glucose in the Calvin cycle, and heterotrophs produce CO_2 as a waste product in the Krebs cycle.

   (B) Autotrophs use CO_2 to make glucose in the Krebs cycle, and heterotrophs produce CO_2 as a waste product in the Calvin cycle.

   (C) Autotrophs produce CO_2 as a waste product in the Krebs cycle, and heterotrophs use CO_2 to make glucose in the Calvin cycle.

   (D) Autotrophs produce CO_2 as a waste product in the Calvin cycle, and heterotrophs use CO_2 to make glucose in the Krebs cycle.

3. Chlorophyll absorbs light energy and stores it as chemical energy by catalyzing redox reactions. Which of the following figures most accurately illustrates the structures required for photosynthesis?

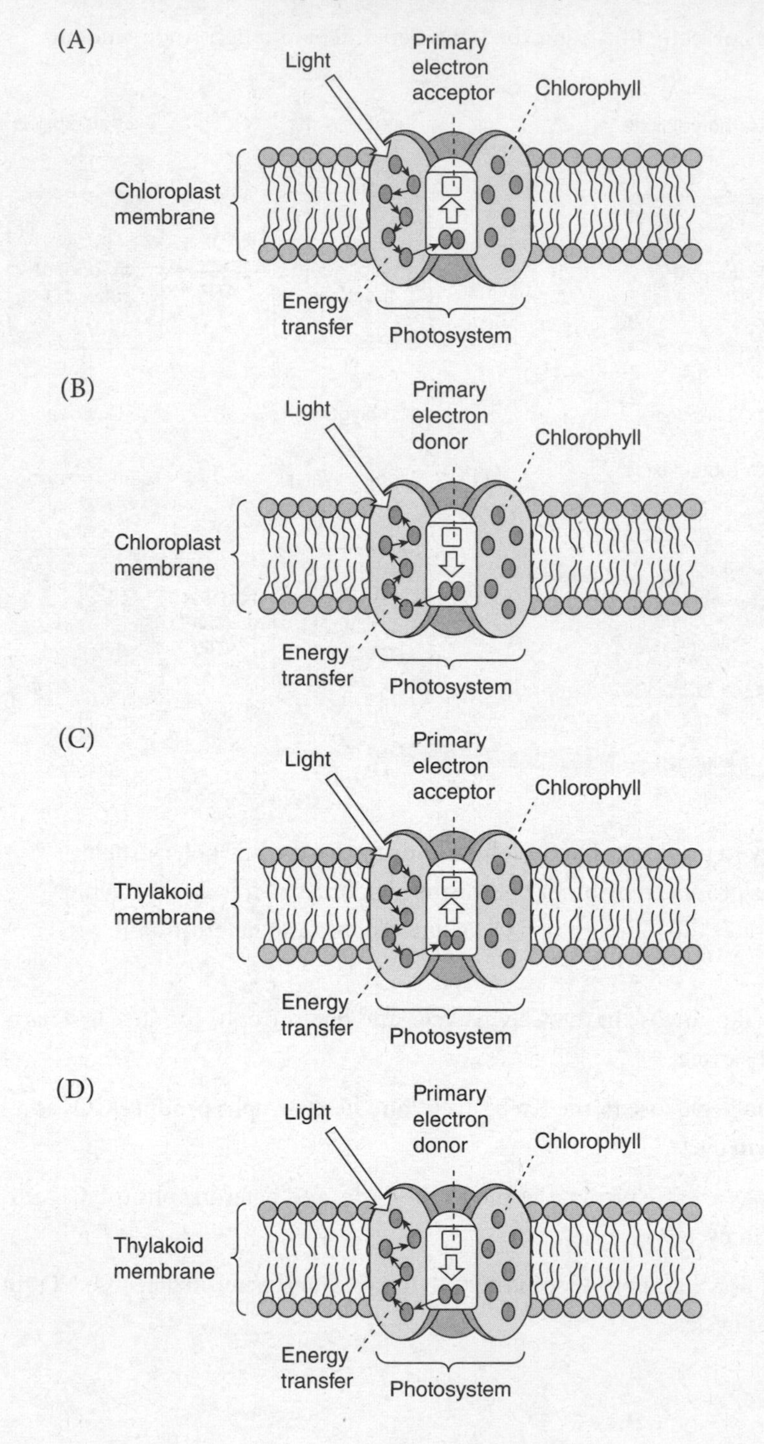

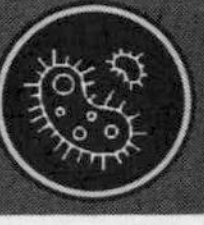

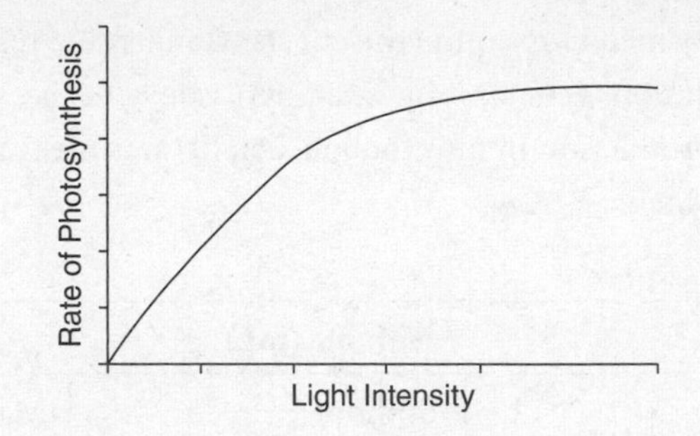

4. The graph above shows the rate of photosynthesis as a function of light intensity. The rate of photosynthesis as a function of carbon dioxide concentration exhibits similar behavior. Which of the following best explains the relationship between carbon dioxide concentration and the rate of photosynthesis?
   (A) At low concentrations, an increase in carbon dioxide has no effect on the rate of photosynthesis.
   (B) At low concentrations, carbon dioxide is a limiting factor of photosynthesis.
   (C) At high concentrations, carbon dioxide concentration is directly related to the rate of photosynthesis.
   (D) At high concentrations, carbon dioxide is a limiting factor of photosynthesis.

5. According to the Warburg effect, cancer cells have been observed to undergo aerobic glycolysis. Unlike normal cells, which carry out aerobic respiration when oxygen is available, cancer cells metabolize glucose to lactic acid via anaerobic fermentation in the presence of oxygen. Cancer cells produce energy at glycolytic rates up to 200 times higher than those of normal cells. Which of the following best explains why cancer cells consume glucose at accelerated rates?
   (A) Aerobic glycolysis generates up to 36 ATP per molecule of glucose.
   (B) Converting glucose to lactate is less efficient than mitochondrial oxidative phosphorylation.
   (C) Lactic acid fermentation consumes more ATP than mitochondrial oxidative phosphorylation.
   (D) Oxygen is readily available in cancer cells to be consumed via glycolysis.

6. An experiment was performed to examine the effects of anaerobic respiration in yeast. Yeast was suspended in various solutions (containing water, pyruvate, glucose, sodium fluoride, and magnesium sulfate), and CO_2 production in mm (bubble height) was measured after incubation. The following data were obtained.

Tube	Yeast	Volume (mL)					CO_2 Produced (mm)
		Water	Glucose	Sodium Fluoride	Pyruvate	Magnesium Sulfate	
1	yes	12.0	0.0	0.0	0.0	0.0	0.52
2	no	8.0	4.0	0.0	0.0	0.0	0.00
3	yes	8.0	4.0	0.0	0.0	0.0	1.49
4	yes	7.4	4.0	0.6	0.0	0.0	1.27
5	yes	2.0	4.0	6.0	0.0	0.0	0.80
6	yes	2.0	4.0	3.0	3.0	0.0	1.00
7	yes	2.0	4.0	0.0	0.0	6.0	1.98

According to the data, yeast in solution with both sodium fluoride and pyruvate produced more CO_2 than yeast in solution with sodium fluoride and no pyruvate. Which of the following statements provides the most reasonable explanation for this difference?

(A) The yeast oxidized the pyruvate, converting it to CO_2 and lactic acid.

(B) Pyruvate is a product of glycolysis, which is reduced to ethanol by yeast, so adding pyruvate would increase CO_2 production.

(C) Magnesium sulfate activates enzymes in fermentation in yeast, increasing CO_2 production.

(D) Sodium fluoride inhibits anaerobic respiration and promotes aerobic respiration, thus decreasing CO_2 production.

7. Since mitochondria are absent, the electron transport chain (ETC) in bacteria is located in the plasma membrane and is generally shorter than the mitochondrial transport chain in eukaryotes. Which of the following best describes the ETC in bacteria under aerobic conditions?

(A) Oxygen donates an electron to initiate the bacterial ETC.

(B) The bacterial ETC generates ATP via alcoholic fermentation.

(C) Electrons are pumped from one side of the membrane to the other along the bacterial ETC.

(D) The bacterial ETC has a lower ATP-to-oxygen ratio than the mitochondrial ETC.

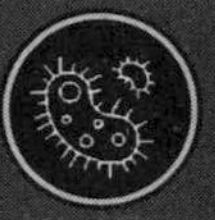

Processes

The answer key to this quiz is located on the next page.

Answer Key

Test What You Already Know

1. **C** **Learning Objective:** 13.1
2. **A** **Learning Objective:** 13.2
3. **D** **Learning Objective:** 13.3
4. **A** **Learning Objective:** 13.4
5. **D** **Learning Objective:** 13.5
6. **C** **Learning Objective:** 13.6
7. **B** **Learning Objective:** 13.7

Test What You Learned

1. **D** **Learning Objective:** 13.2
2. **A** **Learning Objective:** 13.5
3. **C** **Learning Objective:** 13.1
4. **B** **Learning Objective:** 13.6
5. **B** **Learning Objective:** 13.3
6. **B** **Learning Objective:** 13.7
7. **D** **Learning Objective:** 13.4

Detailed solutions can be found in the Answers and Explanations section at the back of this book.

REFLECTION

Test What You Already Know score: __________

Test What You Learned score: __________

Use this section to evaluate your progress. After working through the pre-quiz, check off the boxes in the "Pre" column to indicate which Learning Objectives you feel confident about. Then, after completing the chapter, including the post-quiz, do the same to the boxes in the "Post" column. Keep working on unchecked Objectives until you're confident about them all!

Pre	Post	
❑	❑	**13.1** Describe structures required for photosynthesis
❑	❑	**13.2** Explain the light-dependent reactions of photosynthesis
❑	❑	**13.3** Explain energy pathways for glycolysis and fermentation
❑	❑	**13.4** Explain the function of the electron transport chain
❑	❑	**13.5** Compare input and output molecules and energy
❑	❑	**13.6** Investigate effects on the rate of photosynthesis
❑	❑	**13.7** Investigate effects on the rate of cellular respiration

FOR MORE PRACTICE

Complete more practice online at kaptest.com. Haven't registered your book yet? Go to kaptest.com/booksonline to begin.

CHAPTER 14

The Cell Cycle

LEARNING OBJECTIVES

In this chapter, you will review how to:

14.1 Describe the three phases of interphase

14.2 Explain the process and function of mitosis

14.3 Explain the process and function of meiosis

14.4 Explain how meiosis promotes genetic diversity and evolution

14.5 Investigate genetic probabilities in cell division

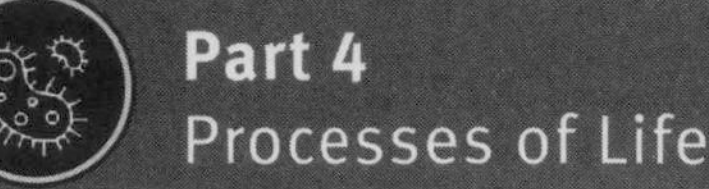

TEST WHAT YOU ALREADY KNOW

1. Below is a diagram of the cell cycle, in which the length of each labeled arc is roughly proportional to the time that a cell spends in that corresponding stage of the cell cycle.

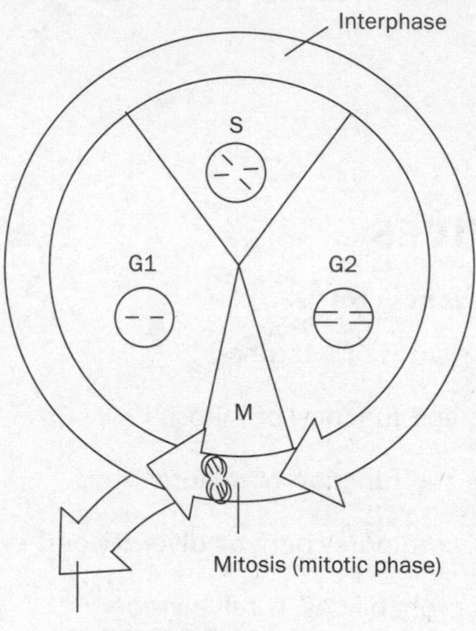

Which of the following most accurately describes a reason for the relative lengths of these stages?

(A) During the S phase, the cell synthesizes all the proteins necessary to carry out the various functions of the cell, so the S phase must be shorter than the G phases so that newly synthesized proteins have time to carry out their functions.

(B) Most of the time of the cell cycle is dedicated to the G phases so that the cell can grow, absorb nutrients, and synthesize the various biomolecules and organelles necessary for the survival and effective functioning of daughter cells.

(C) The M phase is shorter than the S phase because if the M phase took longer, the newly-synthesized chromosomes would become so unstable that one of the two daughter cells would inherit an incomplete set of chromosomes.

(D) The two G phases must be exactly the same length to ensure that the cell cycle repeats at perfectly even intervals, as the cell cannot control the cell cycle except by dividing at very consistent intervals of time.

2. A cell culture is treated with a drug that effectively prevents DNA synthesis, but which affects no other process in the cell. How would the administration of this drug affect mitosis in a culture of otherwise healthy adult epithelial cells?
   (A) It would not affect the cells because each cell already has two copies of the whole genome, which is enough to allow for division.
   (B) It would not affect the cells because cells cease to divide once an organism reaches its full size at adulthood.
   (C) It would affect the cells because they would still be able to divide once but would end up with only half the ordinary amount of DNA.
   (D) It would affect the cells because they would not progress beyond the synthesis phase of the cell cycle and would, therefore, fail to divide.

3. The diagram below illustrates the effects of nondisjunction during meiosis I and II.

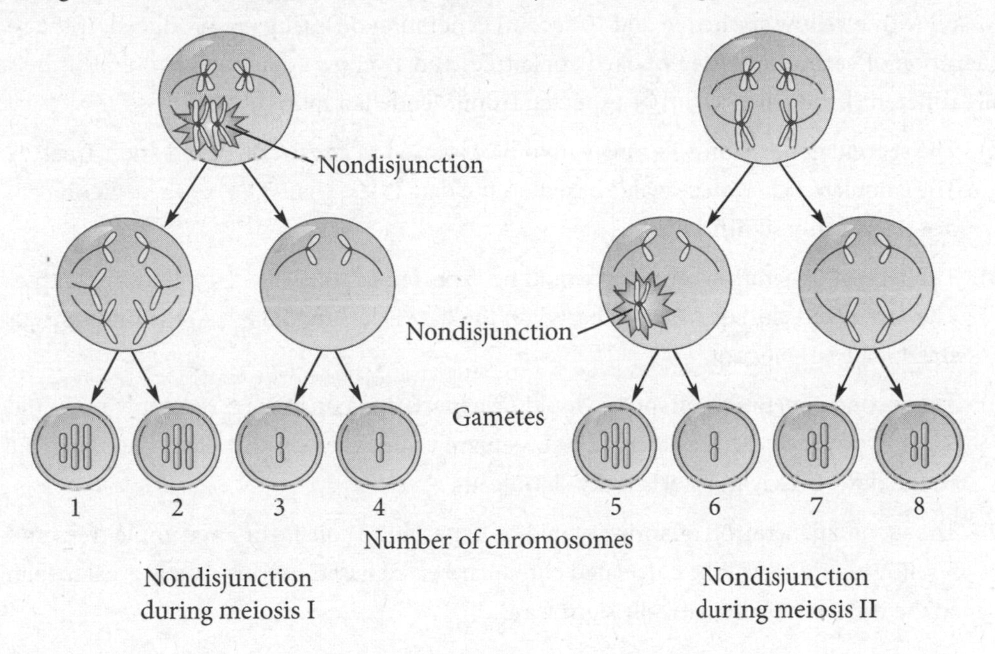

What would happen if gamete 2 undergoes fertilization by combining with a normal gamete from the same species?
   (A) Gamete 2 has an extra chromosome. If it undergoes fertilization, there would be three copies of the longer chromosome in the offspring.
   (B) Gamete 2 has an extra chromosome. If it undergoes fertilization, there would be two copies of the longer chromosome in the offspring.
   (C) Gamete 2 has one fewer chromosome. If it undergoes fertilization, there would be one copy of the longer chromosome in the offspring.
   (D) Gamete 2 has one fewer chromosome. If it undergoes fertilization, there would be two copies of the longer chromosome in the offspring.

4. Which of the following best characterizes the evolutionary purpose of the chromosomal "crossing over" that occurs during meiotic division?

    (A) The structural integrity of chromosomes is optimized by fitting together the most stable pieces of different chromosomes.

    (B) The organism is able to swap out damaged or mutated portions of chromosomes, thereby improving the health and fitness of the resultant gametes.

    (C) Organisms can create superior offspring by combining only the best genes of each chromosome and passing on only those copies with good genes.

    (D) The random shuffling of genetic material allows for greater genetic diversity among offspring, thereby increasing the general fitness of a population.

5. A homozygous red apple (*RR*) tree was crossed with a homozygous yellow apple (*yy*) tree resulting in first generation offspring that produced only red apples. Next, a first generation offspring is crossed with a yellow apple tree and 10 second generation offspring are produced. If the second generation offspring consisted of 6 red apple trees and 4 yellow apple trees, would this be statistically different from what would be expected from Mendelian ratios?

    (A) The second generation offspring would be expected to consist of 10 red apple trees. The calculated chi-square value based on the data is less than 3.84, so the difference is not statistically significant.

    (B) The second generation offspring would be expected to consist of 10 yellow apple trees. The calculated chi-square value based on the data is less than 3.84, so the difference is not statistically significant.

    (C) The second generation offspring would be expected to consist of 5 red apple trees and 5 yellow apple trees. The calculated chi-square value based on this data is less than 3.84, so the difference is not statistically significant.

    (D) The second generation offspring would be expected to consist of 5 red apple trees and 5 yellow apple trees. The calculated chi-square value based on this data is greater than 3.84, so the difference is statistically significant.

Answers to this quiz can be found at the end of this chapter.

INTERPHASE

14.1 Describe the three phases of interphase

The cell cycle is the life cycle of the cell, including reproduction. In **somatic cells**, the cell cycle consists of the three stages of **interphase**, followed by mitosis, as shown in the following figure.

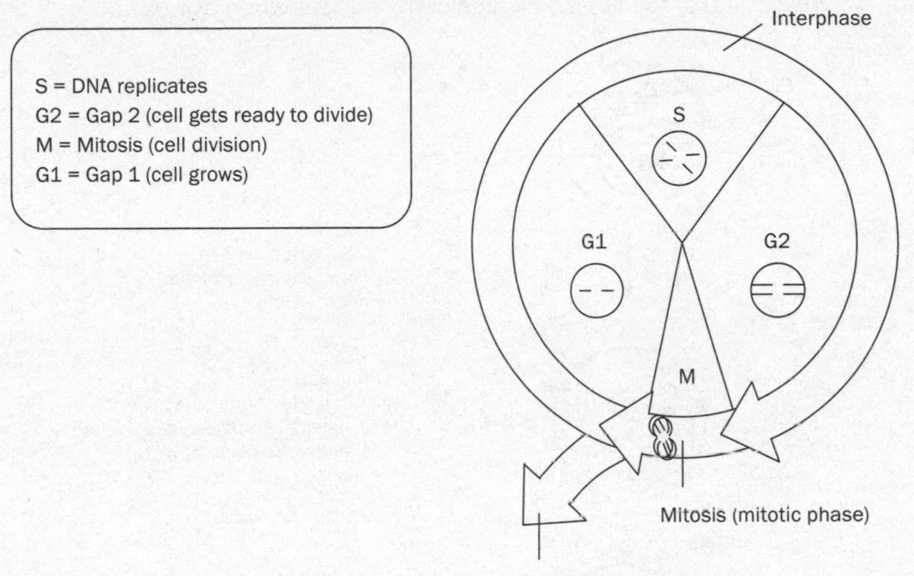

The Cell Cycle

Most somatic cells spend about 90 percent of the cell cycle in interphase. Interphase begins with G_1, the first gap phase, in which the cell grows and synthesizes RNA, proteins, and organelles. After sufficient growth has occurred, the cell passes the G_1 checkpoint and proceeds to the next phase. During the S phase, the synthesis phase, DNA replication occurs. Before replication, each chromosome in the cell consists of a single **chromatid** but, after replication, each consists of two identical sister chromatids, attached at the centromere. Next comes G_2, the second gap phase, in which the cell continues to grow and synthesize more proteins. After the cell grows enough to pass the G_2 checkpoint, it will then proceed on to the phases of mitosis.

✔ **AP Expert Note**

G_0

Some somatic cells cease dividing after they become specialized, entering a resting phase known as G_0. Specific cellular conditions can cause these resting cells to reenter the cell cycle and begin dividing again.

MITOSIS

HIGH YIELD

14.2 Explain the process and function of mitosis

Mitosis (also known as the M phase) is the process through which a cell replicates and divides. The following diagram illustrates the four phases of mitosis (**prophase**, **metaphase**, **anaphase**, and **telophase**) and situates them within the larger cell cycle, which also includes interphase and cytokinesis, the division of cytoplasm in animal cells that follows the division of the nucleus when two distinct cells are produced.

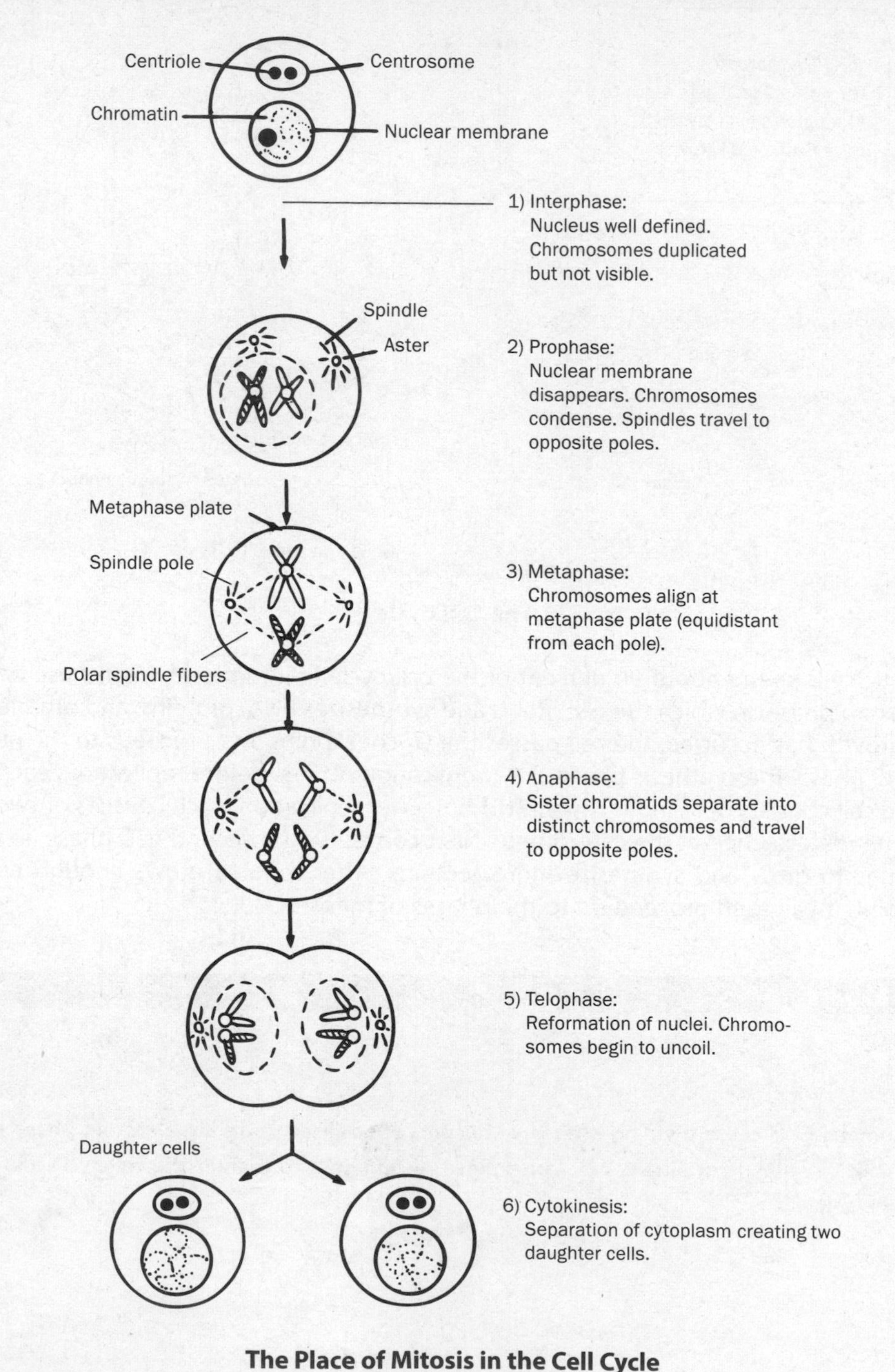

The Place of Mitosis in the Cell Cycle

Processes

✔ AP Expert Note

You are not required to know the names of the phases of mitosis or meiosis for the AP exam. Instead, focus on the overall processes (replication, alignment, and separation) and the theme of heredity.

Cell division is primarily controlled through genetics. The DNA in a cell controls whether a cell divides at all and at what rate. Certain hormones and other chemicals need to be present for a cell to divide. Cell division can be controlled in tissue cultures by inhibiting protein synthesis or affecting nutrient availability. Cell division is reduced by crowding through contact inhibition.

Phases of Mitosis

Among the key aspects of mitosis are the following:

- Chromosomes shorten and thicken in the nucleus, and the nuclear membrane dissolves.
- The mitotic spindle of microtubules is formed.
- The contractile ring of actin develops around the center of the cell.

For mitosis to work, a single pair of centrioles will copy themselves during the S phase, and the two pairs will move to opposite poles of the cell. These pairs of centrioles form the foundation for centrosomes, microtubule organizing centers that will shoot linked tubulin proteins across the cell as mitosis begins. Keep in mind, however, that the centrioles themselves are not necessary for microtubules to form from the centrosome areas of the cell. In fact, plant cells have centrosomes without centrioles. The thing to remember is that the centrosome regions form the two poles (like north and south) on opposite ends of the cells, between which microtubule spindle fibers will form.

Despite the conventional division into distinct stages, mitosis is a continuous process that does not stop between each phase. Four of these stages comprise mitosis, and the final stage—**cytokinesis**—completes the process of cell division as the cell pinches in two.

Prophase

In prophase, **chromatin** shows up under the microscope as well-defined chromosomes. These chromosomes are an X shape, two sister chromatids connected by a centromere, a specific DNA sequence. The mitotic **spindle** begins to form and elongate from the centrosome regions during prophase.

Metaphase

During metaphase, kinetochore microtubules push equally from opposite poles so that chromosomes are aligned in the middle of the cell. This center area where the alignment occurs is called the metaphase plate.

Anaphase

In anaphase, paired sister chromatids separate as kinetochore microtubules shorten rapidly. The polar microtubules lengthen as the kinetochore microtubules shorten, thereby pushing the poles of the cell farther apart.

Telophase

During telophase, separated sister chromatids group at opposite ends of the cell, near the centrosome region, having been pulled there by the receding microtubules. A new nuclear envelope forms around each group of separated chromosomes. At this point, mitosis has ended.

Cytokinesis

In cytokinesis, the contractile ring of actin protein fibers shortens at the center of the cell. A cleavage furrow, or indentation, is created as the ring contracts. As one side of the cell contacts the other, the membrane pinches off and two cells exist where one did before.

MEIOSIS

HIGH YIELD

14.3 Explain the process and function of meiosis

14.4 Explain how meiosis promotes genetic diversity and evolution

Meiosis is the process that sexually reproducing organisms use to generate **gametes**, **haploid** cells that combine (usually with gametes from another member of the same species) to produce offspring with a full **diploid** set of chromosomes. The overarching purpose of meiosis is to facilitate sexual reproduction, which increases the genetic diversity of a population. This allows for greater variation within a species, which can help that species adapt to changing circumstances and evolve.

✔ AP Expert Note

Ploidy

Ploidy refers to the number of sets of chromosomes within the cells of an organism. An organism or cell that has two sets of chromosomes is called diploid (designated as 2*N*). A diploid organism has one set of chromosomes from each parent, for a total of two sets. When a cell or organism has only one set of chromosomes, it is called haploid (1*N* or *N*). Note that diploid organisms will produce haploid gametes, but their somatic cells will be diploid. In some cases, cells have several sets of chromosomes and are called polyploid (3*N*, 4*N*, etc.). There are a few organisms that exist in a natural state with haploid or polyploid cells, but most organisms are diploid.

Meiosis occurs in two steps, *meiosis I* and *meiosis II*. Each step includes a round of division and is divided into four phases. Meiosis II is essentially mitosis with half the number of chromosomes, but meiosis I contains some key differences from both meiosis II and mitosis.

Meiosis I

The four phases of meiosis I are described below.

Prophase I

The chromatin condenses into chromosomes, the spindle apparatus forms, and the **nucleoli** and **nuclear membrane** disappear. Homologous chromosomes (chromosomes that code for the same traits, one inherited from each parent) come together and intertwine in a process

called **synapsis**. Because at this stage each chromosome consists of two sister chromatids, each synaptic pair of homologous chromosomes contains four chromatids and is therefore often called a **tetrad**.

Sometimes chromatids of homologous chromosomes break at corresponding points and exchange equivalent pieces of DNA; this process is called **crossing over**. Note that crossing over occurs between homologous chromosomes and not between sister chromatids of the same chromosome. (The latter are identical, so crossing over would not produce any change.) Those chromatids involved are left with an altered but structurally complete set of genes. The chromosomes remain joined at points, called chiasmata, where the crossing over occurred. Such genetic recombination can unlink linked genes, thereby increasing the variety of genetic combinations that can be produced via gametogenesis. Recombination among chromosomes results in increased genetic diversity within a species. Note that sister chromatids are no longer identical after recombination has occurred.

Metaphase I

Homologous pairs (tetrads) align at the equatorial plane, and each pair attaches to a separate spindle fiber by its kinetochore.

Anaphase I

The homologous pairs separate and are pulled to opposite poles of the cell. This process is called disjunction, and it accounts for a fundamental Mendelian law: independent assortment. During disjunction, each chromosome of paternal origin separates (or disjoins) from its homologue of maternal origin and either chromosome can end up in either daughter cell. Thus, the distribution of homologous chromosomes to the two intermediate daughter cells is random with respect to parental origin. Each daughter cell will have a unique pool of alleles (genes coding for alternative forms of a given trait; e.g., yellow flowers or purple flowers) from a random mixture of maternal and paternal origin. In other words, each daughter cell will almost certainly have some chromosomes of paternal origin and some of maternal origin, rather than all paternal or all maternal.

Telophase I

A nuclear membrane forms around each new nucleus. At this point, each chromosome still consists of sister chromatids joined at the centromere. The cell divides (by cytokinesis) into two daughter cells, each of which receives a nucleus containing the haploid number of chromosomes. Between cell divisions, there may be a short rest period, or interkinesis, during which the chromosomes partially uncoil.

Meiosis II

This second division is very similar to mitosis, except that meiosis II is not preceded by chromosomal replication.

Prophase II

The centrioles migrate to opposite poles, and the spindle apparatus forms.

Metaphase II

The chromosomes line up along the equatorial plane. The centromeres divide, separating the chromosomes into two sister chromatids.

Anaphase II

The sister chromatids are pulled to opposite poles by the spindle fibers.

Telophase II

A nuclear membrane forms around each new (haploid) nucleus. Cytokinesis follows and two daughter cells are formed. Thus, by the completion of meiosis II, four haploid daughter cells are produced per gametocyte. (In human females, only one of these becomes a functional gamete.)

Comparison with Mitosis

In addition to having different functions (mitosis allows for the replication of somatic cells, while meiosis enables sexual reproduction), there are a number of notable differences between mitosis and meiosis. These differences are summarized in the following table and figure.

MITOSIS VS. MEIOSIS						
Process	**Number of Chromosomes in Parent Cell**	**Prophase/ Prophase I**	**Metaphase/ Metaphase I**	**Anaphase/ Anaphase I**	**Number of Daughter Cells**	**Number of Chromosomes in Daughter Cells**
Mitosis	$2N$	Replicated chromosomes come into view as sister chromatids	Individual chromosomes align at the metaphase plate	Centromeres separate and sister chromatids travel to opposite poles	2	$2N$
Meiosis	$2N$	Chromosomes form tetrads by synapsis; crossing over at chiasmata	Pairs of homologous chromosomes align at the metaphase plate	Synapsis ends and homologous chromosomes travel to opposite poles; sister chromatids travel to the same pole	4	N

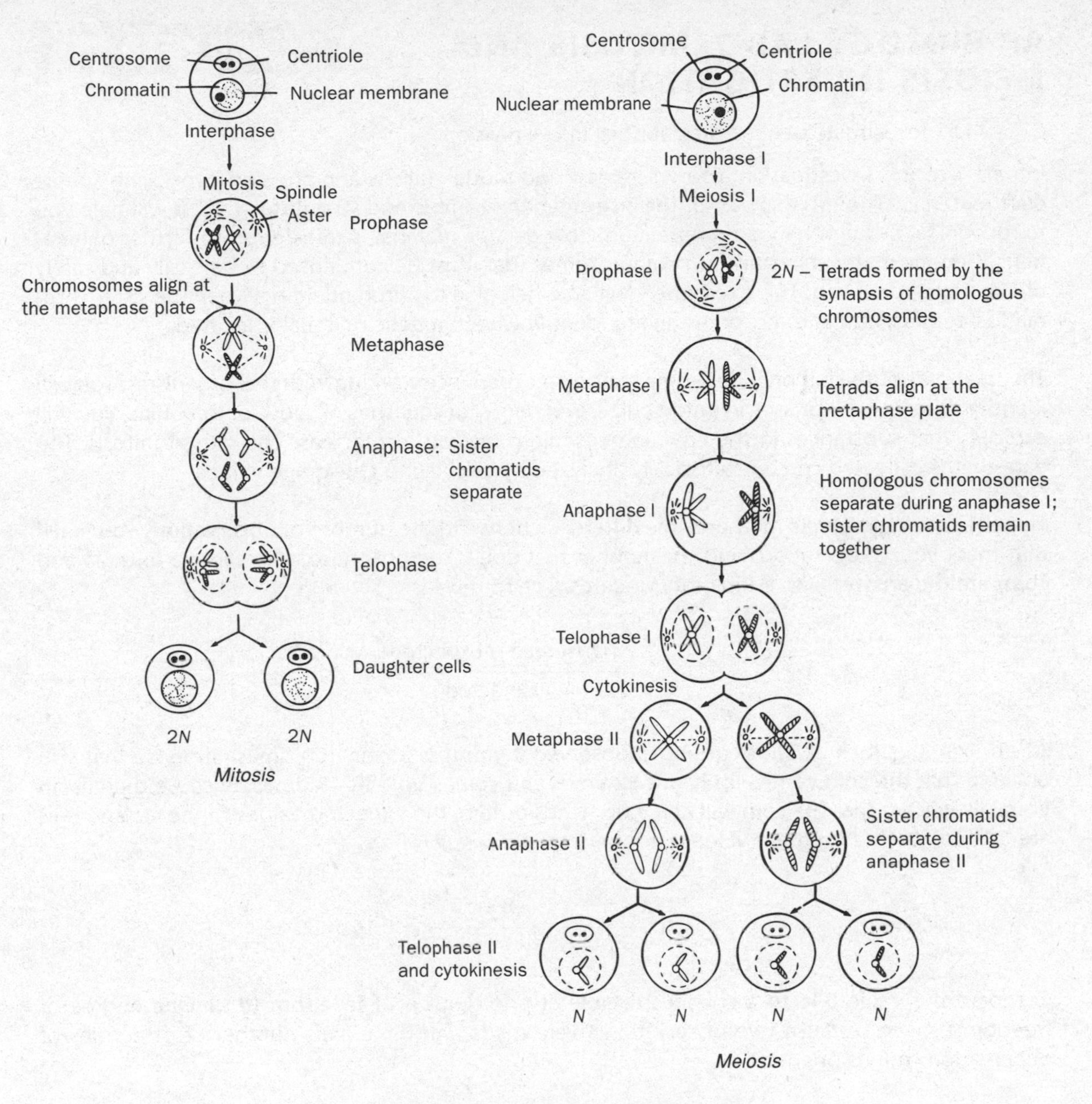

Comparison of Mitosis and Meiosis

AP BIOLOGY LAB 7: MITOSIS AND MEIOSIS INVESTIGATION

HIGH YIELD

14.5 Investigate genetic probabilities in cell division

In part 1 of this investigation, you will review and model mitosis and crossing over using simple craft materials. Though this part of the investigation is simple and straightforward, it will help you to recognize, through hands-on repetition, how genetic material is transferred to further generations. Genetic material in cells is in chromosomes that must be condensed to be easily and safely duplicated and moved. This experiment will also help you to differentiate between the sister chromatids, tetrads, and chromosomes and to identify where genetic material is located.

The next part of the lab provides some hands-on experience working with real organisms. You will identify different cell phases in onion bulbs. Working with squashes of those onion bulbs, you will explore what substances in the environment might increase or decrease the rate of mitosis. The data that is collected can be statistically analyzed by calculating chi-square values.

The **chi-square analysis** measures the difference between the number of observations you make that meet your expectations and the number that don't. You put these values into a formula and compare the answer with a table of standards. The formula is:

$$\chi^2 = \sum \frac{(\text{Observed} - \text{Expected})^2}{\text{Expected}}$$

As an example, look at the frequencies observed if you toss a coin 100 times. Suppose that you observe that the coin comes up heads 52 times and comes up tails 48 times. Because there is an equal likelihood that the coin will come up heads or tails, the expected values for heads and tails are 50 and 50. Putting these values into the formula:

$$\chi^2 = \frac{(52-50)^2}{50} + \frac{(48-50)^2}{50} = 0.16$$

Compare the value 0.16 to a chi-square table at one **degree of freedom (d.f.)**. One degree of freedom is used because two observations were made and d.f. equals number of categories of observations minus one.

Probability (p)	Degrees of Freedom (d.f.)				
	1	2	3	4	5
0.05	3.84	5.99	7.82	9.49	11.1

At $p = 0.05$ in the chi-square table, the value obtained from the previous equation would have to be at least 3.8 for your observations to be statistically different from what is expected. The value calculated was only 0.16, so it can be concluded that although your observations weren't exactly $\frac{50}{50}$, they were not statistically significantly different from the expected outcome. There are many considerations and assumptions to make when performing statistical analyses, which would be covered in greater depth in a class on statistical analysis. For the chi-square analysis, there are three key points to remember:

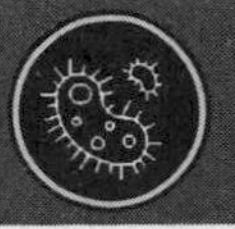

- The formula for the chi-square statistic:

$$\chi^2 = \sum \frac{(\text{Observed} - \text{Expected})^2}{\text{Expected}}$$

- The formula for degrees of freedom: d.f. = # of categories of observations −1
- Your calculated χ^2 must be greater than the corresponding p value in the table for the outcome to be significantly different than what you expected.

✔ AP Expert Note

The formula for chi-square and the degrees of freedom table will be provided to you in the equations sheet on Test Day, so you do not need to memorize them. You are most likely to see chi-square tested in grid-in questions, but it could also appear in other question types.

Part 3 of the lab asks you to explore the differences between normal cells and cancer cells. Remember that cancer typically results from a mutation in a gene or protein that controls the cell cycle. Comparing pictures of chromosomes from normal and HeLa cells, you will be able to count the number of chromosomes found in each type of cell and identify differences in appearance. The main conclusion to draw from this part of the investigation is that in normal cells, division is blocked if there is damage to or mutations in DNA. In contrast, this ability to block division is lacking in cancer cells, and they divide uncontrollably.

Finally, you will review meiosis and nondisjunction events by modeling. Based on this review, you will measure crossover frequencies and genetic outcomes in a fungus. This part of the investigation will also give you solid experience with the microscope by preparing slides and counting fungal asci (singular: ascus) under the microscope. Though you will not need to use a microscope or describe how to use a microscope on the AP Biology exam, these experiences are helpful for answering short and long free-response questions, which often ask you to describe experimental setups and procedures.

RAPID REVIEW

If you take away only 3 things from this chapter:

1. Interphase consists of three stages: G_1 (first gap phase), S (synthesis, in which DNA replication occurs), and G_2 (second gap phase). Interphase is followed in somatic cells by mitosis.
2. Mitosis is the process in which a cell produces two identical daughter cells. Its stages include prophase, metaphase, anaphase, and telophase. Cytokinesis occurs immediately following mitosis and refers to the splitting of the cell into two new cells.
3. Meiosis refers to the process by which sexually reproducing organisms produce sex cells (gametes) with half the chromosomes (haploid) of the rest of the organism's cells (which are diploid). It consists of two rounds of division, meiosis I and meiosis II, and it results in the creation of four gametes. In sexual reproduction, the male and female gametes join to create a new organism with the normal diploid number of chromosomes.

TEST WHAT YOU LEARNED

1. The effects of a particular insecticide on vegetable roots were studied. Two batches of parsnip roots were immersed in water, and one batch was treated with 0.5 M solution of insecticide while the other was untreated. After 5 days, samples were collected and prepared from both batches and the tips were observed under a microscope. The total number of cells and those undergoing mitosis were counted. The data is shown in the table below.

Parsnip Tips with Insecticide			Parsnip Tips without Insecticide		
Phase	**Cell Count**	**% Cells in Phase**	**Phase**	**Cell Count**	**% Cells in Phase**
Interphase	575	88%	Interphase	330	75%
Mitosis	78	12%	Mitosis	110	25%
Total Cell Count	653		**Total Cell Count**	440	

Does the insecticide have a statistically significant effect on the mitotic division of cells in parsnip root tips?

(A) The calculated chi-square value based on this data is greater than 3.84. Therefore, the 0.5 M solution of insecticide does not have a significant negative effect on the mitotic division of cells in parsnip root tips.

(B) The calculated chi-square value based on this data is less than 3.84. Therefore, the 0.5 M solution of insecticide does not have a significant negative effect on the mitotic division of cells in parsnip root tips.

(C) The calculated chi-square value based on this data is greater than 3.84. Therefore, the 0.5 M solution of insecticide has a significant negative effect on the mitotic division of cells in parsnip root tips.

(D) The calculated chi-square value based on this data is less than 3.84. Therefore, the 0.5 M solution of insecticide has a significant negative effect on the mitotic division of cells in parsnip root tips.

2. Which is the most likely cause of an error during meiosis that results in daughter cells with an incorrect number of chromosomes?

(A) Displacement of sister chromatids to opposite poles of the dividing cell

(B) Recombination of genetic material during the earlier stages of meiosis

(C) Activation of the securin proteins that separate chromatids during anaphase II

(D) Nondisjunction of the genetic material at any point during meiosis

3. Certain cell types may become senescent, meaning that they no longer divide. A simple example of this phenomenon can be found in skin cells, which originate from live progenitor cells beneath the surface layers of skin. Progenitor cells will undergo a simple mitotic division in which one daughter cell remains anchored to the basilar membrane, while the other daughter cell travels toward the surface of the skin and ceases to undergo further mitotic divisions. How do the genetic contents of these two daughter cells compare?

   (A) During any normal mitotic division, all the genetic information is copied and passed along to both daughter cells equally.

   (B) Modifications are made to the genetic information passed on to the mobile cell, which renders it unable to divide further.

   (C) The anchored daughter cell retains all the genetic information so that it can continue to create new cells, while the mobile cell retains none and later dies.

   (D) Neither cell retains the full set of genetic information after division, leading to the gradual aging of the skin.

4. Before a cell can undergo cellular division, it must progress through an interphase stage in which the cell matures and produces proteins needed for division. Which of the following best describes the interphase stage?

   (A) The cell grows, replicates its DNA, and prepares to divide during interphase. If a genetic mutation is introduced during interphase and not repaired, the daughter cells will inherit the mutation.

   (B) The cell grows, replicates its DNA, and prepares to divide during interphase. None of the genetic mutations introduced during interphase are inherited by the daughter cells.

   (C) The cell replicates its DNA and prepares to divide during interphase, but does not grow. None of the genetic mutations introduced during interphase are inherited by the daughter cells.

   (D) The cell replicates its DNA and prepares to divide during interphase, but does not grow. If a genetic mutation is introduced during interphase and not repaired, the daughter cells will inherit the mutation.

5. A couple with black hair has three children with black hair and one child with red hair. How did this most likely occur?

   (A) A child with red hair is possible through genetic drift, in which two chromosomes exchange genetic information, increasing the diversity of the individual's genome.

   (B) Each member of the couple was heterozygous, carrying one dominant gene for black hair and one recessive gene for red hair.

   (C) Having a child with red hair is possible if genes on the father's Y chromosome mutated and the mutated Y chromosome was passed on to the red-haired child.

   (D) Giving birth to a child with red hair can occur through gene flow, in which new genes enter a population as a result of migration.

Answer Key

Test What You Already Know

1. **B** **Learning Objective:** 14.1
2. **D** **Learning Objective:** 14.2
3. **A** **Learning Objective:** 14.3
4. **D** **Learning Objective:** 14.4
5. **C** **Learning Objective:** 14.5

Test What You Learned

1. **C** **Learning Objective:** 14.5
2. **D** **Learning Objective:** 14.3
3. **A** **Learning Objective:** 14.2
4. **A** **Learning Objective:** 14.1
5. **B** **Learning Objective:** 14.4

Detailed solutions can be found in the Answers and Explanations section at the back of this book.

REFLECTION

Test What You Already Know score: __________

Test What You Learned score: __________

Use this section to evaluate your progress. After working through the pre-quiz, check off the boxes in the "Pre" column to indicate which Learning Objectives you feel confident about. Then, after completing the chapter, including the post-quiz, do the same to the boxes in the "Post" column. Keep working on unchecked Objectives until you're confident about them all!

Pre	Post	
☐	☐	**14.1** Describe the three phases of interphase
☐	☐	**14.2** Explain the process and function of mitosis
☐	☐	**14.3** Explain the process and function of meiosis
☐	☐	**14.4** Explain how meiosis promotes genetic diversity and evolution
☐	☐	**14.5** Investigate genetic probabilities in cell division

FOR MORE PRACTICE

Complete more practice online at kaptest.com. Haven't registered your book yet? Go to kaptest.com/booksonline to begin.

CHAPTER 15

Homeostasis

LEARNING OBJECTIVES

In this chapter, you will review how to:

15.1 Describe energy flow in biological systems

15.2 Explain how the laws of thermodynamics apply to living systems

15.3 Describe how changes in energy affect living systems

15.4 Describe coupled reactions important for life

15.5 Contrast positive and negative feedback mechanisms

15.6 Describe thermoregulation processes

TEST WHAT YOU ALREADY KNOW

1. The following diagrams represent biotic and abiotic components of the arctic tundra. Which diagram represents the most accurate flow of energy between the components of the ecosystem?

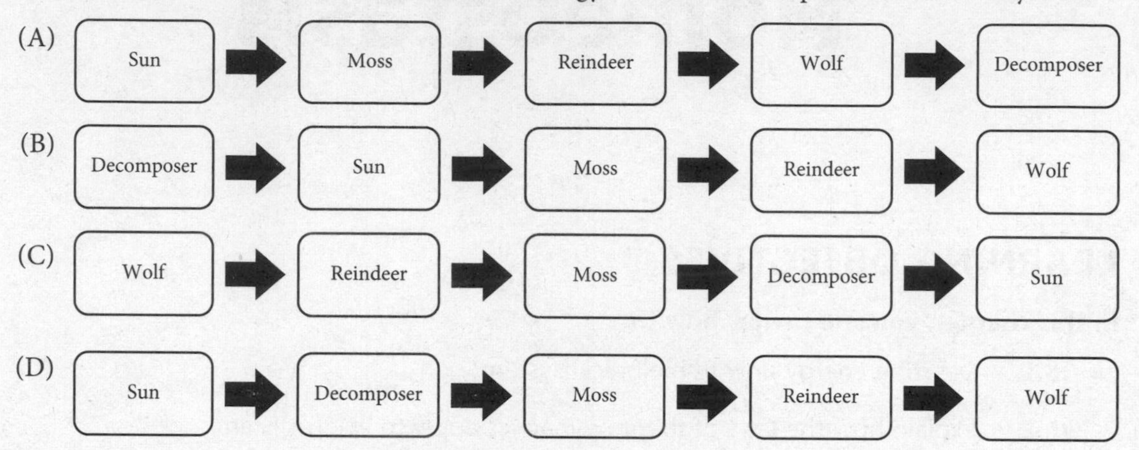

2. A small pond is populated by algae, paramecia, and amoebae. Sunlight is absorbed by the microscopic algae. Paramecia eat the algae. Amoebae feed on the paramecia and algae. Which of the following statements is correct about the pond, based on the laws of thermodynamics?
   (A) Algae transform all of the solar energy into chemical energy in the form of sugars.
   (B) Solar energy is only converted into chemical energy in this ecosystem.
   (C) Photosynthesis is an endergonic process because algae have more energy at the end of the process.
   (D) Photosynthesis is an exergonic process because algae expend the energy from photosynthesis to perform cellular work.

3. Yeast cells obtain energy by metabolizing sugars produced by other organisms. In an experiment, yeast cells are mixed with barley seeds, which supply a source of starch for fermentation. Fermentation is measured by estimating the production of CO_2 in a respirometer, a device that measures changes in gas volume. Which of the following experimental conditions would most likely yield the highest level of fermentation?
   (A) Mixing barley seeds and yeast in water
   (B) Scratching barley seeds to release starch before adding yeast
   (C) Presoaking barley seeds in water to activate enzymes that break down starch
   (D) Germinating barley seeds until seedlings emerge before mixing them with the yeast

4. The following diagram illustrates a classic experiment in chemiosmosis in which isolated thylakoid membranes are incubated in the dark at pH 4 and, after equilibration, transferred to a buffer at pH 8 containing ADP and P_i. ATP was synthesized and accumulated in the medium.

Processes

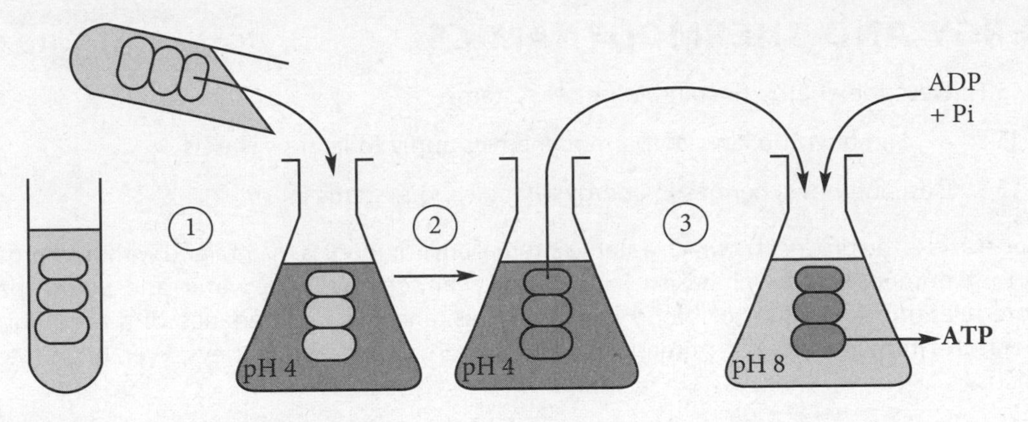

Which of the following hypotheses about the production of ATP in the thylakoid membrane is best supported by the experiment?

(A) ATP formation in the thylakoid membrane is driven by the oxidation of NADPH.

(B) ATP synthesis depends on the light reactions of photosynthesis.

(C) ATP synthesis does not require an intact thylakoid membrane.

(D) ATP synthesis depends on an H^+ gradient across the thylakoid membrane.

5. Platelets circulate in the blood in an inactive state. Damage in a blood vessel exposes and releases molecules not normally found in blood, which activates platelets. The activated platelets adhere to each other and to the blood vessel. This sets off an enzymatic cascade that activates additional platelets. Plugging of the hole in the blood vessel finally shuts off the response of the platelets. Which of the following statements best describes the response of the platelets?

(A) The response of the activated platelets is a negative feedback mechanism because it stops the loss of blood.

(B) Activation of platelets is a positive feedback loop because it acts as an amplifier of the response.

(C) Platelets do not respond to positive feedback because there would be no way to shut off activation once the damage is repaired.

(D) The response of the activated platelets promotes an increase in the total number of platelets.

6. The temperature setpoint of the body is determined by the hypothalamus. When core body temperature rises above the setpoint, responses such as sweating bring about cooling. When the temperature drops below it, warming mechanisms, such as narrowing of blood vessels and shivering, increase body temperature. A child develops a temperature of 40°C (104°F) because an ear infection has reset her hypothalamus to a higher core body temperature. Which of the following symptoms would the child most likely experience while the infection persists?

(A) Profuse sweating

(B) Chills and shivering

(C) Feeling overheated

(D) Thirst and decreased urination

Answers to this quiz can be found at the end of this chapter.

ENERGY AND THERMODYNAMICS

15.1 Describe energy flow in biological systems

15.2 Explain how the laws of thermodynamics apply to living systems

15.3 Describe how changes in energy affect living systems

Homeostasis is the process by which a stable internal environment is maintained within an organism. Important homeostatic mechanisms include the maintenance of a water and solute balance (osmoregulation), regulation of blood glucose levels, and the maintenance of a constant body temperature. In mammals, the primary homeostatic organs are the kidneys, liver, large intestine, and skin.

The maintenance of homeostasis within an organism uses up a considerable amount of energy. Thus, before considering specific homeostatic processes, it is essential to understand a few points on energy flow and thermodynamics.

Energy Flow

The ultimate energy source for living organisms is the Sun. Autotrophic organisms, such as green plants, convert sunlight into energy stored in the bonds of organic compounds (chiefly glucose) during the anabolic process of photosynthesis. Autotrophs do not need an exogenous supply of organic compounds. Heterotrophic organisms obtain their energy catabolically, via the breakdown of organic nutrients that must be ingested. Note in the following energy flow diagram that some energy is dissipated as heat at every stage.

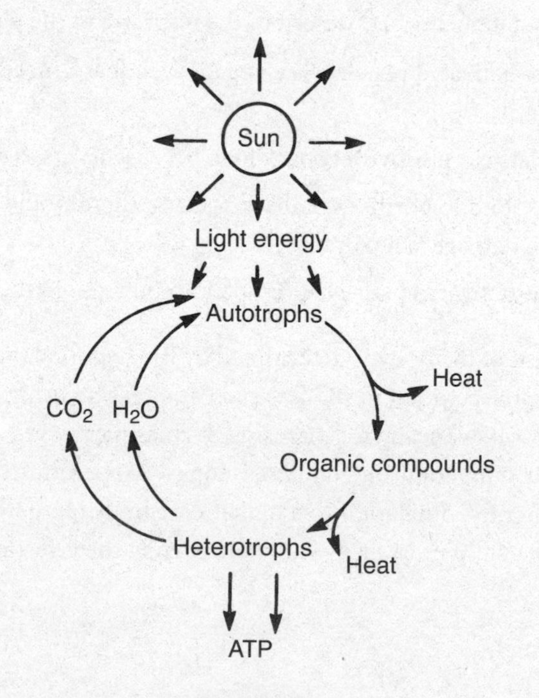

Energy Flow

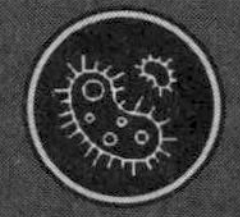

Thermodynamics

Thermodynamics deals with heat, energy, and work and can be discussed in terms of temperature, internal energy, entropy, and pressure. For the AP Biology exam, you should be familiar with the three laws of thermodynamics.

The First Law of Thermodynamics

The first law maintains that the increase in internal energy of a closed system is equal to the difference of the heat supplied to the system and the work done by the system. This is a variation of the law of the conservation of energy that states that energy cannot be created or destroyed, but can only change form. The classic example of this is how kinetic energy can be converted to potential energy or (within the body) chemical energy can be converted to kinetic energy.

The Second Law of Thermodynamics

The second law states that the entropy of a closed system tends to increase. In other words, energy spreads out over time rather than spontaneously consolidating. Eventually, differences in temperature, pressure, and chemical potential tend to even out in a physical system that is isolated from the outside world. Consider, for example, how "normal" each of the following seems: hot coffee cools down, iron rusts, and balloons deflate. The thermal energy in the hot coffee is spreading out to the cooler air that surrounds it. The chemical energy in the bonds of elemental iron and oxygen is released and dispersed as a result of the formation of the more stable, lower-energy bonds of iron oxide (rust). The energy of the pressurized air is released to the surrounding atmosphere as the balloon deflates. These examples have a common denominator: in each of them, energy of some form is going from being localized or concentrated to being spread out or dispersed.

The Third Law of Thermodynamics

According to the third law, the entropy of a system approaches a constant value as the temperature approaches zero. Therefore, at absolute zero (the coldest possible temperature), entropy reaches its minimum value. At absolute zero, nothing can be colder and no heat energy remains in a system.

Bioenergetics

Biological thermodynamics, also known as bioenergetics, is the study of energy transformation in biology. This involves looking at energy transformations and transductions in and between living things, including their major functions down to the cellular level, and understanding the function of the chemical processes underlying these transductions.

Changes in free energy availability affect the ability of organisms to maintain organization, grow, and reproduce. Organisms that acquire excess free energy will either use the energy for growth or store the energy, while organisms that acquire insufficient free energy will lose mass and ultimately die. Since the energetic costs of reproduction are large, depending on energy availability, different organisms use various reproductive strategies. For instance, plants and animals may reproduce seasonally during the most energetically favorable part of the year (determined by ambient temperature and food availability). A biennial plant takes two years to complete its life cycle. In the first year, the plant germinates and grows, and then becomes dormant in the colder months. During the subsequent year, the plant produces flowers and fruits. Animals may undergo reproductive diapause, a period during which growth or development is suspended and metabolism is decreased in response to adverse environmental conditions such as temperature extremes or reduced food and water availability. Changes in energy also cause disruptions at the population level by affecting population size and number.

✔ **AP Expert Note**

The most important energy transformations in organisms are found in the processes of photosynthesis and cellular respiration, discussed in chapter 13 of this book. At a larger scale, energy transformations can also be studied at the ecosystem or biome level—see chapter 22 for more on this ecological perspective.

COUPLED REACTIONS AND CHEMIOSMOSIS

15.4 Describe coupled reactions important for life

A **coupled reaction** is one in which transport across a membrane is coupled with a chemical reaction. Of all the things to remember here, there is one that is particularly important: **chemiosmosis** is used by cells to generate ATP by moving H^+ ions across a membrane, down a concentration gradient. Special membrane proteins called **ATPases** create proton channels to convert ADP to ATP when a proton passes through. Aerobic respiration and photosynthesis utilize chemiosmosis, whereas glycolysis and other forms of ATP creation do not.

The mitochondrion moves H^+ into the **intermembrane space** via the **electron transport chain**, which creates a proton gradient across the inner membrane. The energy to do this comes from the breakdown of food. The chloroplast moves H^+ into the **thylakoid space** in a way very similar to the mitochondrion, but it drives the oxidative phosphorylation process with light energy. This process is called **photophosphorylation**.

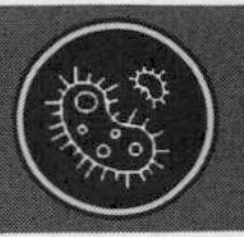

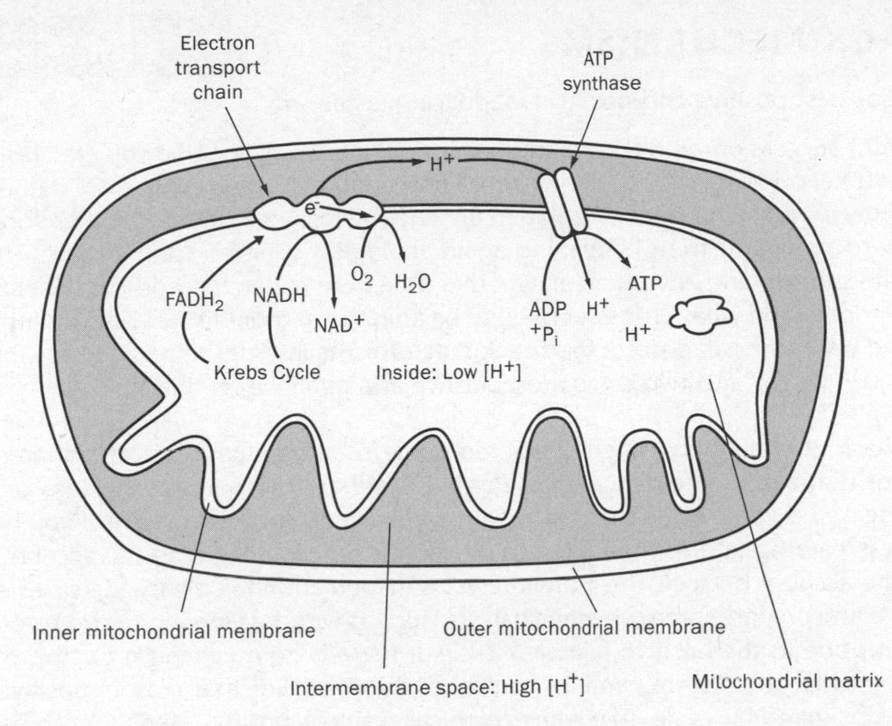

Chemiosmosis in a Mitochondrion

> ✔ **AP Expert Note**
>
> **The processes that drive electron transport and chemiosmosis are similar in photosynthesis and cellular respiration.**

The structures that perform chemiosmosis are examples of form specialized for function. The invaginations of the inner membrane (cristae) inside the mitochondrion provide an increased surface area. Even the proteins in the membrane are aligned in a way that spans the membrane width, and their juxtaposition provides the proper architecture to allow chemical reactions to take place. The most important thing to remember about fine protein structures is that their location and form are what allow chemiosmosis to happen.

Catabolism

Catabolism is the breaking down of complex substances into simple substances, making energy available in the process. In **anaerobic catabolism**, there is no electron transport chain or oxygen (O_2) available to carry out a reaction. Anaerobic reactions require almost as much energy to carry out as they yield. From an evolutionary perspective, it is much more advantageous to have the ability to utilize the oxidative properties of oxygen, as reactions in **aerobic catabolism** do, because more energy can be generated. For more on the aerobic and anaerobic catabolism of glucose, see chapter 13.

Keep in mind that not all energy comes from sugar in the form of glucose. Organisms use food in the form of carbohydrates (starch), protein, and fat. Carbohydrates ultimately break down or are transformed into glucose. The amino acids from proteins ultimately enter the citric acid cycle as pyruvate and acetyl-CoA. Fats are broken down into glycerols that can be converted to pyruvate and fatty acids that are converted to acetyl-CoA. All of the pyruvate and acetyl-CoA from the breakdown of food enter the citric acid cycle and the electron transport chain.

FEEDBACK MECHANISMS

HIGH YIELD

15.5 Contrast positive and negative feedback mechanisms

Input is vital for success on both the organismal and cellular level. Imagine you are playing a video game and you keep losing. All you know is that the video game suddenly says "game over," but you never know why. You have no information on why you lost your game and, therefore, you have no idea how to prevent it from happening again. Biological systems are the same. They require input from the surrounding environment, whether it be from within the body or the external environment, to understand what processes need to be adapted in order to survive. This input is called feedback, and organisms all possess **feedback mechanisms** in order to respond to stimuli. These mechanisms usually fall into two categories: positive and negative feedback.

Positive feedback results when the effects of feedback from a system result in an increase in the original factor that causes the disturbance. Positive feedback mechanisms increase or accelerate the output created by a stimulus that has already been activated. For example, if you have a bank account that is earning interest, the amount of interest grows every time the account increases. The higher the account balance, the more interest is earned and the balance increases even more. This can go on and on until some other mechanism (such as withdrawing money from your account) causes a disruption in the positive feedback cycle. If there is no mechanism to stop the positive feedback loop, then the system can lose control of the cycle. For that reason, positive feedback systems are considered unstable. To use an ecological example, positive feedback can be seen with climate change. An initially small perturbation in the environment can positively feed back onto itself, growing and growing until the problem yields huge effects on the climate.

Negative feedback is the opposite of positive feedback. As more feedback is received, it causes the processes that brought about the initial change to slow down or stop altogether. Self-regulating systems tend to function by using negative feedback. It allows for stability within a system because it reduces the effects of fluctuations. Negative feedback loops allow a system to have the necessary amount of correction at the most important time. One of the simplest examples of negative feedback is one of the human body's methods of thermoregulation. An increase in core body temperature will stimulate the body to produce sweat. As the body's sweat production increases, it causes a drop in body temperature. This decrease in body temperature will turn off the mechanism for sweat production.

THERMOREGULATION

15.6 Describe thermoregulation processes

Thermoregulation is one aspect of homeostasis that is commonly featured in the AP Biology exam. The hypothalamus is the brain region that acts to control the body temperature of an organism that is able to set its internal temperature. Mammals fall into this category. Hormones such as **adrenaline** and the thyroid hormones can increase metabolic rate and, subsequently, heat production. Muscles can generate heat by contracting rapidly (shivering). Heat loss is regulated through the contraction or relaxation of precapillary sphincters.

Alternative mechanisms are used by some mammals to regulate body temperature. Panting is a cooling mechanism that results in the evaporation of water from the respiratory passages. Sweating is also important in increasing evaporative heat loss so that the body cools. Fur is used to trap heat, and hibernation during the winter conserves energy by decreasing heart rate, breathing, and metabolism. Animals able to regulate their internal temperature even in the face of a changing external temperature are called endotherms, or homeotherms. Mammals and birds are capable of this type of regulation, yet reptiles, amphibians, and most other animals are not and are known to be cold-blooded, or ectotherms.

There are two laws you should know that relate heat and body size:

1. Bigger bodies produce less body heat per pound per hour.
2. Bigger bodies lose less body heat per pound per hour.

Metabolic heat production drops in a very specific manner as body size increases. Compared with an elephant, which might weigh 10,000 pounds, a small mammal weighing only 1 pound produces about 10 times more heat per pound than the elephant does. Yet of the two animals, the elephant stays warmer because it has much less overall surface area compared to internal volume than does the small mammal. In other words, the small mammal gives off much more heat to its surroundings.

RAPID REVIEW

If you take away only 5 things from this chapter:

1. Homeostasis is the process by which a stable internal environment is maintained within an organism. Our primary homeostatic organs are the kidneys, liver, large intestine, and skin.
2. There are three laws of thermodynamics. Together they discuss the conservation of energy, entropy, and absolute zero.
3. Bioenergetics, or biological thermodynamics, studies how chemical energy is broken down and converted to usable energy within the biological system. This can be at the ecosystem, organismal, or cellular level.
4. Chemiosmosis produces energy from the movement of H^+ ions across a membrane down a concentration gradient in both photosynthesis and respiration. Catabolism is the breaking down of complex substances into simple substances, making energy available in the process.
5. Thermoregulation refers to the physiological processes that come together to maintain a stable body temperature in warm-blooded animals.

TEST WHAT YOU LEARNED

1. Consider the chemical reactions X and Y. The progress of each reaction is plotted as a function of the free energy of the compounds involved in the reaction.

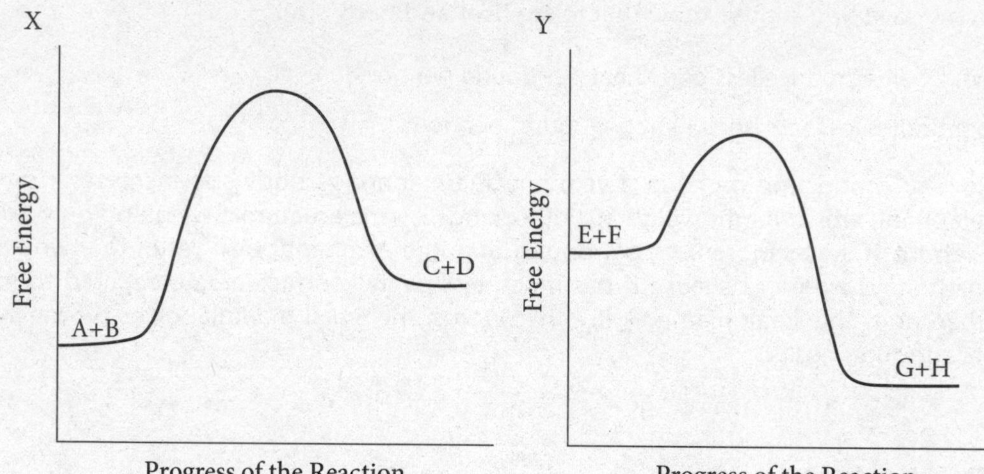

Based on the free energy plots, a biochemist considers the following hypotheses on whether the reactions would take place in a living organism. Which of them is best supported by the graphs?

(A) Reaction X is endergonic and cannot take place in a living cell.

(B) Reaction Y would proceed on its own without the input of energy.

(C) Reaction X would proceed if an enzyme lowered the free energy of C+D.

(D) Reaction X would proceed if it is coupled to reaction Y.

2. The second law of thermodynamics states that the entropy of a closed system must always increase. This means that such systems always move to a more disordered state. However, cells constantly synthesize macromolecules that are complex and highly ordered. Which of the following statements best explains the connection between the second law of thermodynamics and the biosynthesis of macromolecules in a living cell?

(A) Releasing water molecules in the biosynthesis of macromolecules increases entropy.

(B) The entropy of the surroundings increases as the cell synthesizes macromolecules.

(C) The entropy in the cell increases as it synthesizes macromolecules.

(D) The second law of thermodynamics does not apply to biological systems.

3. The Cretaceous-Paleogene boundary coincides with a mass extinction that led to the disappearance of more than three-fourths of all plant and animal species living on Earth. One popular hypothesis proposes that a large meteorite collided with the Earth, sending large clouds of debris into the atmosphere that blocked sunlight. How would such a catastrophic event most likely have changed the levels of energy in the biosphere?
   (A) The clouds blocking the Sun caused a long period of cold temperatures, which stopped energy flow.
   (B) The asteroid impact caused major earthquakes and tsunamis that wiped out ecosystems.
   (C) Darkness made photosynthesis nearly impossible, removing the foundation of the food web.
   (D) Most life was extinguished upon meteorite impact or soon thereafter, stopping the flow of energy.

4. A bicyclist eats cereal for breakfast before a race. The flow of energy for this scenario can be represented by a diagram that shows energy flowing from the Sun to the grain to the bicyclist's muscles. In which of the following shapes should the diagram be made to best represent the flow of energy up these three levels?
   (A) A rectangle, because energy is neither created nor destroyed as it flows up the levels
   (B) A circle, because energy cycles through the levels
   (C) An upright pyramid, because energy is lost at each level as heat
   (D) An inverted pyramid, because the biker's muscles accumulated energy from all of the lower levels

5. Zoologists estimated the variation in the body sizes of woodrats over the last 20,000 years by dating pellets with ^{14}C. They plotted their results as the relative increase or decrease in body size relative to present size. Temperatures are below average during glacial periods and above average during interglacial periods.

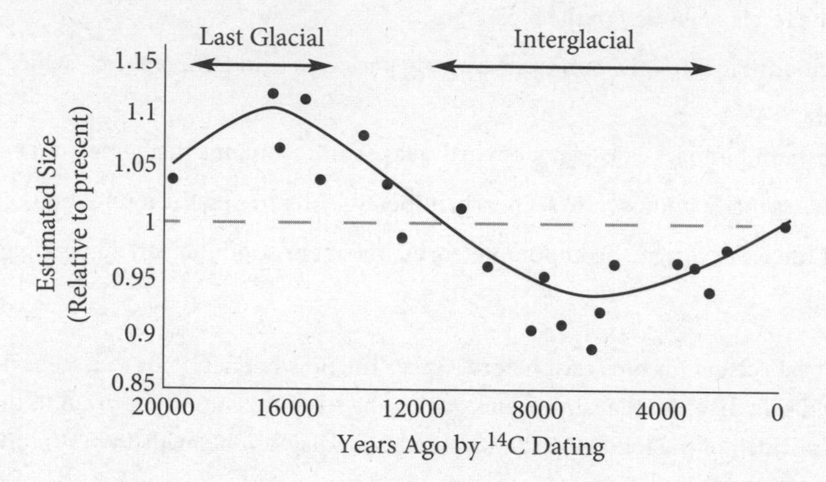

Adapted from Felisa A. Smith, Julio L. Betancourt, and James H. Brown, "Evolution of Body Size in the Woodrat Over the Past 25,000 Years of Climate Change," *Science* 270, no. 5244 (December 1995): 2012–2014.

Which of the following statements is best supported by the data?

(A) Body size is independent of ambient temperature.

(B) Body size decreases when ambient temperature increases.

(C) Body size increases when ambient temperature increases.

(D) Body size increases as a function of evolution.

6. The hypothalamus secretes thyrotropin-releasing hormone (TRH), which causes the anterior pituitary to release thyroid-stimulating hormone (TSH) in the bloodstream. The thyroid gland responds to TSH by secreting the hormone thyroxine, which contains iodine. When iodine is missing in a diet, thyroxine is not secreted by the gland. Individuals whose diet is deficient in iodine may develop a goiter, an enlarged thyroid gland. Which of the following hypotheses best explains the formation of a goiter?

(A) TSH is regulated by a negative feedback loop and keeps stimulating the thyroid when thyroxine production is low.

(B) Without enough iodine present, cells in the gland keep dividing and the organ becomes enlarged.

(C) Thyroxine precursors accumulate in the gland and begin to form new compounds that cause it to become enlarged.

(D) If the dietary content of iodine is too low, the gland enlarges to maximize uptake from the bloodstream.

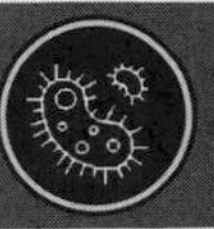

The answer key to this quiz is located on the next page.

Answer Key

Test What You Already Know

1. **A** **Learning Objective:** 15.1
2. **C** **Learning Objective:** 15.2
3. **C** **Learning Objective:** 15.3
4. **D** **Learning Objective:** 15.4
5. **B** **Learning Objective:** 15.5
6. **B** **Learning Objective:** 15.6

Test What You Learned

1. **D** **Learning Objective:** 15.4
2. **B** **Learning Objective:** 15.2
3. **C** **Learning Objective:** 15.3
4. **C** **Learning Objective:** 15.1
5. **B** **Learning Objective:** 15.6
6. **A** **Learning Objective:** 15.5

Detailed solutions can be found in the Answers and Explanations section at the back of this book.

REFLECTION

Test What You Already Know score: __________

Test What You Learned score: __________

Use this section to evaluate your progress. After working through the pre-quiz, check off the boxes in the "Pre" column to indicate which Learning Objectives you feel confident about. Then, after completing the chapter, including the post-quiz, do the same to the boxes in the "Post" column. Keep working on unchecked Objectives until you're confident about them all!

Pre	Post	
❑	❑	**15.1** Describe energy flow in biological systems
❑	❑	**15.2** Explain how the laws of thermodynamics apply to living systems
❑	❑	**15.3** Describe how changes in energy affect living systems
❑	❑	**15.4** Describe coupled reactions important for life
❑	❑	**15.5** Contrast positive and negative feedback mechanisms
❑	❑	**15.6** Describe thermoregulation processes

FOR MORE PRACTICE

Complete more practice online at kaptest.com. Haven't registered your book yet? Go to kaptest.com/booksonline to begin.

PART 5

Transformations of Life

CHAPTER 16

Molecular Genetics

LEARNING OBJECTIVES

In this chapter, you will review how to:

- **16.1** Differentiate between the structures of DNA and RNA
- **16.2** Explain the function of DNA and RNA
- **16.3** Describe the process of DNA replication
- **16.4** Describe the process and product of transcription
- **16.5** Describe the steps of translation
- **16.6** Describe post-translational modifications to polypeptides
- **16.7** Investigate analysis of DNA

TEST WHAT YOU ALREADY KNOW

1. A scientist feeds fluorescently labeled nucleotides to cells and measures the fluorescence emission to quantify the amount of RNA synthesized. Which of the following fluorescently labeled compounds would give the most reliable measure of the amount of newly synthesized RNA?

   (A) Deoxyribose adenine phosphate

   (B) Deoxyribose uracil phosphate

   (C) Ribose thymine phosphate

   (D) Ribose uracil phosphate

2. Lactose fermenting strains of *Streptococcus pneumoniae* are designated as *lac*$^+$, whereas strains deficient in lactose fermentation are designated as *lac*$^-$. Extracts from *lac*$^+$ cells are incubated separately with three different enzymes: extract A with protease K (which digests proteins), extract B with DNases (which digests DNA), and extract C with RNase H (which digests RNA). After treatment with enzymes, each extract is incubated with live *lac*$^-$ cells and grown on a medium containing an indicator. Blue colonies indicate that lactose is being fermented. Colorless colonies indicate that the cells cannot ferment lactose.

   Which of the following would most likely be observed after growing the cultures?

   (A) Only colorless colonies appear from cells incubated with extract A.

   (B) Only colorless colonies appear from cells incubated with extract B.

   (C) Only colorless colonies appear from cells incubated with extract C.

   (D) Blue colonies appear from cells grown with all three extracts.

3. A molecular biologist is studying a mutant strain of *E. coli* that has a defect occurring during DNA replication. In order to understand the cells' mutation, she labels the cells with radioactive thymidine to monitor the synthesis of DNA. After several hours of incubation, she recovers both long and short strands of labeled DNA molecules. The results are not modified by treatment of the sample with mung bean nuclease, an enzyme that digests single-stranded DNA.

   These results best support which of the following hypotheses?

   (A) The *E. coli* are deficient in gyrases, which unwind DNA molecules.

   (B) The *E. coli* are deficient in DNA polymerase I, which fills in gaps in the DNA molecules.

   (C) The *E. coli* are deficient in DNA polymerase III, which synthesizes two new strands simultaneously.

   (D) The *E. coli* are deficient in ligases, which catalyze the formation of phosphodiester bonds.

4. Researchers discovered that two proteins with different primary structures, calcitonin (a parathyroid hormone) and the neurotransmitter peptide CGRP (expressed in neurons), can be mapped to the same gene. Two distinct mRNA molecules are transcribed from the gene in question. The two mRNA sequences differ in length and sequence, with the exception of identical sequences found in the region preceding the protein-encoding regions. Which of the following hypotheses would best explain these observations?

   (A) The sequences of the two mRNA molecules have different mutations.

   (B) The proteins are encoded by the sense and antisense strands of DNA.

   (C) Alternative splicing takes place in different tissues.

   (D) The proteins undergo extensive post-translational modifications.

5. A part of an mRNA molecule with the following sequence is being read by a ribosome.

   5′-UCG-GCA-CAU-UUA-UAU-GUU-3′

CODON TABLE

	SECOND POSITION								
	U		C		A		G		
FIRST POSITION	code	amino acid	code	amino acid	code	amino acid	code	amino acid	THIRD POSITION
U	UUU	phe	UCU	ser	UAU	tyr	UGU	cys	U
	UUC		UCC		UAC		UGC		C
	UUA	leu	UCA		UAA	STOP	UGA	STOP	A
	UUG		UCG		UAG	STOP	UGG	trp	G
C	CUU	leu	CCU	pro	CAU	his	CGU	arg	U
	CUC		CCC		CAC		CGC		C
	CUA		CCA		CAA	gln	CGA		A
	CUG		CCG		CAG		CGG		G
A	AUU	ile	ACU	thr	AAU	asn	AGU	ser	U
	AUC		ACC		AAC		AGC		C
	AUA		ACA		AAA	lys	AGA	arg	A
	AUG	met	ACG		AAG		AGG		G
G	GUU	val	GCU	ala	GAU	asp	GGU	gly	U
	GUC		GCC		GAC		GGC		C
	GUA		GCA		GAA	glu	GGA		A
	GUG		GCG		GAG		GGG		G

Using the codon table above, determine the amino acid sequence encoded.

   (A) ser-ala-his-leu-tyr-val

   (B) ser-ala-gln-leu-(stop signal)

   (C) ser-arg-val-asn-ile-gln

   (D) leu-tyr-ile-tyr-thr-ala

6. Cellular proliferation in cancer can be regulated by signaling receptors belonging to the receptor tyrosine kinase (RTK) family. Binding of a ligand to RTKs triggers phosphorylation of tyrosine residues in the cytoplasmic domain of the protein. One approach to cancer treatment has been the use of small molecule inhibitors that prevent phosphorylation of RTKs.

   Which of the following statements correctly represents how the small molecule inhibitors are regulating RTKs?

   (A) The RTKs are regulated at the transcription level.

   (B) The RTKs are regulated at the RNA processing level.

   (C) The RTKs are regulated at the translation level.

   (D) The RTKs are regulated at the post-translational level.

7. The diagram below shows a DNA plasmid with a total length of 2,000 base pairs. The tick marks indicate cleavage sites for restriction enzymes (enzyme EcoRI and enzyme HaeIII).

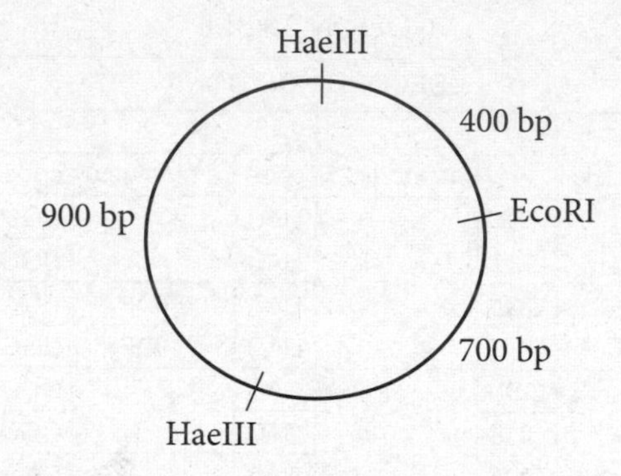

   The plasmid was digested with the enzyme HaeIII to completion and the fragments were separated by gel electrophoresis. Which of the following results would most likely be observed?

   (A) Two fragments measuring 400 bp and 900 bp

   (B) Two fragments measuring 700 bp and 900 bp

   (C) Two fragments measuring 900 bp and 1,100 bp

   (D) Three fragments measuring 400 bp, 700 bp, and 900 bp

Answers to this quiz can be found at the end of this chapter.

NUCLEIC ACIDS

16.1 Differentiate between the structures of DNA and RNA

16.2 Explain the function of DNA and RNA

✔ AP Expert Note

Themes in Molecular Genetics

There are three key themes of molecular genetics to remember:

1. **Be familiar with the structures of DNA and RNA and how these building blocks provide both a simple system for duplication and a wealth of genetic diversity that codes for variation in all existing and extinct organisms.**
2. **Know that ultimately DNA is a code for the translation of proteins, that these proteins impart function to all of the biological processes that make an organism alive, and that control of expression of the genes for these proteins largely controls bioactivity.**
3. **As a complement to your understanding of evolution, understand DNA's role as the root of genetic change in the form of mutations.**

DNA and RNA Structures

You have undoubtedly learned that **DNA** and **RNA** molecules are long strands of nucleotides linked together by their sugar-phosphate backbone between the 5′ and 3′ carbons of deoxyribose or ribose. DNA has a **deoxyribose** sugar backbone while RNA has a **ribose** backbone. RNA also contains the nitrogenous base uracil instead of thymine. Both strands of complementary DNA serve as templates for duplication during mitosis and meiosis, whereas only one strand serves as a template for transcription. The simple sugar-phosphate backbone allows for exact copies of itself to be reproduced during propagation.

There are only four different **nitrogenous bases** that make up the **nucleotides** in DNA: **adenine, cytosine, guanine**, and **thymine**. These bases pair along complementary strands of DNA. Guanine pairs with cytosine, and thymine pairs with adenine. In RNA, thymine is replaced with **uracil**, which pairs with adenine. Therefore, four different nucleotides form a code for protein synthesis in all of the organisms that have ever existed on Earth. It is amazing that the combination of nitrogenous bases in the DNA of all organisms on Earth leads to such genetic variety.

Nucleotides are the basic building blocks of DNA and RNA. Millions of nucleotides make up the DNA in each cell. However, because there is so much DNA, our DNA is packaged into smaller tightly wound structures, **chromosomes**. Because each nucleotide has a nitrogen base (A, C, G, T) and it takes three bases to code for one amino acid, the average gene that codes for 300 amino acids is approximately 900 nucleotides long. Each chromosome may have hundreds to thousands of genes. Chromosomes are made up of millions of nucleotides wound tight and held together by histone proteins.

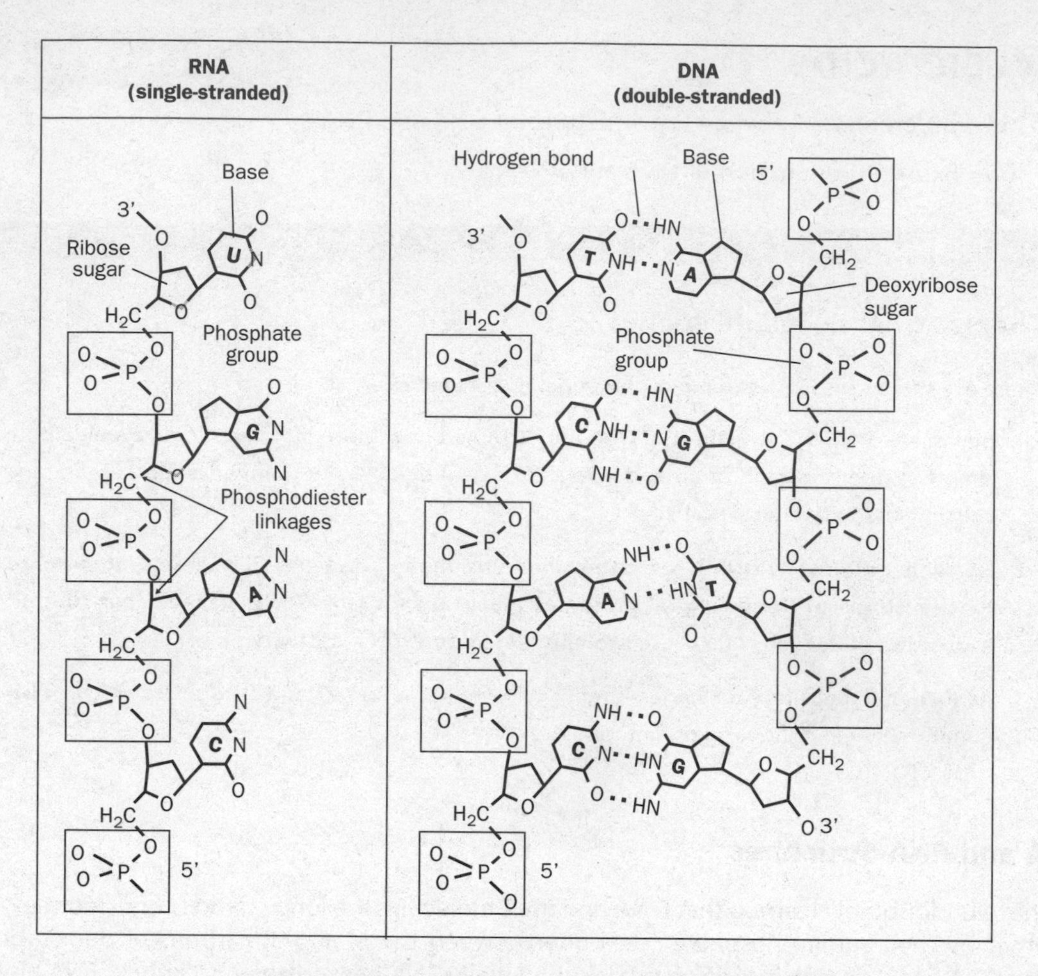

Structural Comparison of DNA and RNA

✔AP Expert Note

The molecular structures of the different nucleotides are outside the scope of the AP Biology exam.

DNA and RNA Functions

The primary function of DNA is to serve as heritable genetic material, containing the recipes for all of the proteins an organism can produce. RNA, in contrast, has a variety of forms with differing functions:

- *mRNA*: messenger RNA; delivers genetic instructions from nucleus to ribosomes
- *tRNA*: transfer RNA; brings amino acids to ribosomes for translation
- *rRNA*: ribosomal RNA; structural component of ribosomes
- *hnRNA*: heterogeneous nuclear RNA; synthesized from a DNA template by transcription—sometimes called pre-mRNA
- *RNAi*: RNA interference; when small segments of RNA bind to specific mRNA to block gene expression

DNA REPLICATION

HIGH YIELD

16.3 Describe the process of DNA replication

It is important to understand that the strands of the DNA double helix are oriented in an antiparallel manner to each other—with one strand 5′ to 3′ and the other side 3′ to 5′. The 3′ to 5′ designations refer to the carbons that make up the sugar in each nucleotide, and the nucleotides of each strand are connected by hydrogen bonds to a nucleotide in the complementary strand. DNA polymerase II adds from the 5′ end of the incoming nucleotide to the 3′ position of the ribose in the last nucleotide on the DNA strand being synthesized. However, DNA polymerase cannot bind until an RNA primer, a short strand of complementary RNA nucleotides, has been attached to a separated strand by an enzyme called primase.

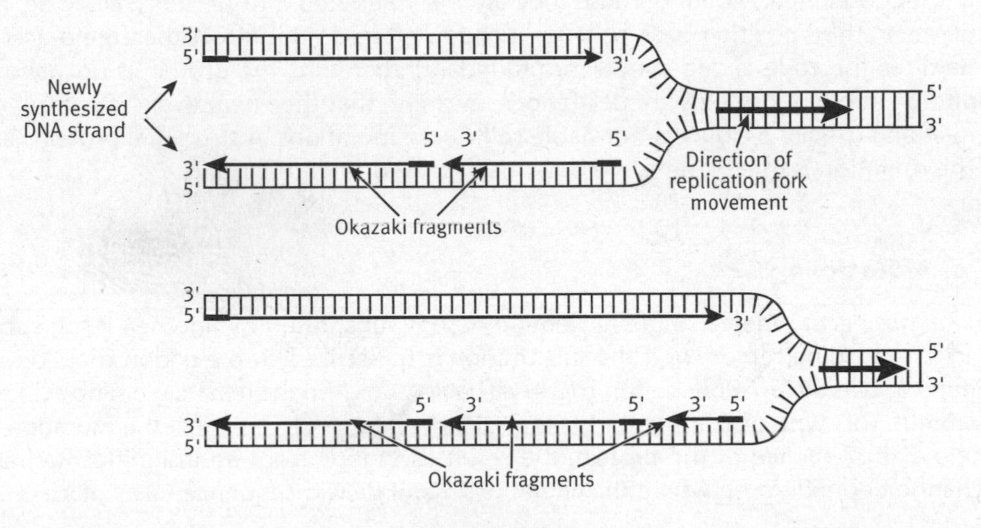

DNA Replication

DNA helicase unzips the DNA to ready it for replication. DNA replication is semiconservative, meaning that half of the original DNA is conserved in each daughter molecule. This means that the new DNA uses the existing DNA strands as a template. When DNA splits and copies, although the two resulting DNA strands that are created are identical to each other, one strand is made up of the "old" DNA strand while the other daughter strand is made of newly synthesized DNA. Okazaki fragments are formed because DNA polymerase can only read the older DNA in a 3′ to 5′ direction and synthesize a new strand in a 5′ to 3′ direction. It works away from the origin of replication on one side and then jumps back to follow the unzipping DNA, but can only continue to work in a direction away from the replication fork, thus leaving fragments of DNA that have to be linked later by DNA ligase.

> ✔ **AP Expert Note**
>
> **Replication always adds new nucleotides in the 5′ to 3′ direction—no exceptions! To help you remember, use the mnemonic device: *You read up on a topic and write down your notes.* Polymerase reads up (3′ to 5′) and writes down (5′ to 3′).**

Mutations

The goal of a cell is usually to maintain the same code in its DNA, base pair by base pair. Sometimes changes occur spontaneously during replication or are due to environmental factors such as irradiation. These changes in the DNA code are called **mutations**. Mutations can involve a change in only one base or several bases. Mutations are also placed in one of two categories, base-pair substitutions or insertions and deletions. **Base-pair substitutions** occur when one base pair is incorrectly reproduced and exchanged with a different base pair. An **insertion** is when any number of extra base pairs are added to the code, and a **deletion** is when any number are removed from the code.

The effect of mutations depends upon where they occur in the code. Mutations of introns often have no effect on an individual because they are not translated into proteins. Likewise, base-pair substitutions in third position base pairs for a codon often do not affect the amino acid it codes for (remember, the code is degenerate, or redundant); therefore, the protein is unchanged. Most mutations in structural proteins are deleterious, meaning that they negatively affect the nature of the protein and usually produce a nonviable cell. Some mutations in structural proteins are viable and, if the mutation is in a gamete, are potentially passed on to offspring.

Types of Mutations

Point mutations occur when a single nucleotide base is substituted by another. If the substitution occurs in a noncoding region, or if the substitution is transcribed into a codon that codes for the same amino acid as the previous codon, there will be no change in the resulting amino acid sequence of the protein. This type of point mutation is a "silent" mutation. However, if the mutation changes the amino acid sequence of the protein, the result can range from an insignificant change to a lethal change depending on where the alteration in amino acid sequence takes place.

Frameshift mutations involve a change in the reading frame of an mRNA. Because ribosomes and tRNAs "read" the mRNA in sections of three bases (codons), if a base is inserted or deleted due to faulty transcription or a mutation in the actual DNA, the reading of the resulting mRNA will shift, and this is called a frameshift mutation. Base insertions and deletions, particularly toward the start of the protein's amino acid sequence, can render the remaining structure nonfunctional as almost every amino acid along the sequence gets changed.

Nonsense mutations produce a premature termination of the polypeptide chain by changing one of the codons to a stop codon. Beta-thalassemia is a hereditary disease in which red blood cells are produced with little or no functional hemoglobin for oxygen carrying. The different forms of this disease can be produced by a variety of mutations, including point mutations, frameshift mutations, and nonsense mutations.

The following diagram is useful for understanding these different types of mutations.

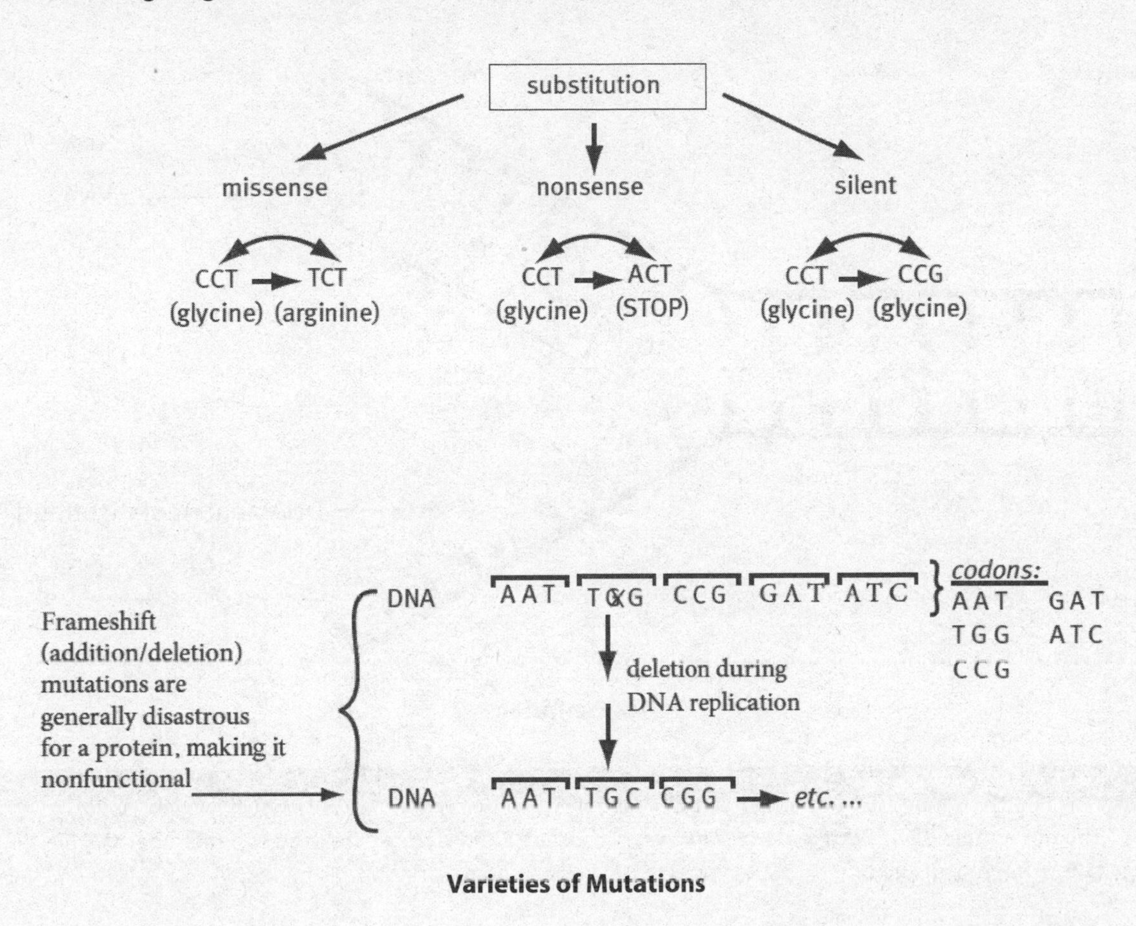

Varieties of Mutations

TRANSCRIPTION

HIGH YIELD

16.4 Describe the process and product of transcription

Transcription is the process of "mirroring" a sequence of nucleotides on an original strand of DNA with a strand of **complementary bases** of mRNA (substituting uracil for thymine). Occurring in the nucleus, transcription starts at a special region called a promoter and ends at another special region called a terminator. The mRNA then undergoes a sequence of post-transcriptional modifications. For reasons not fully understood, the original DNA contains many regions that are not coded for in protein synthesis. These regions are called **introns**, and they are excised before the mRNA enters the cytoplasm for protein translation. **Exons** are the regions that code for protein synthesis in mRNA and remain after the introns have been excised. Transcription factors and regulatory proteins, which play an important role in transcription control, are discussed in chapter 17.

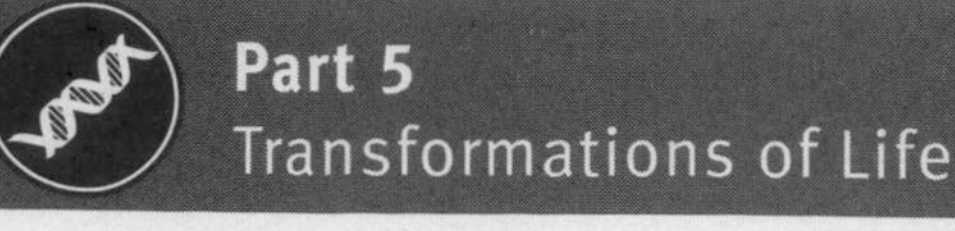

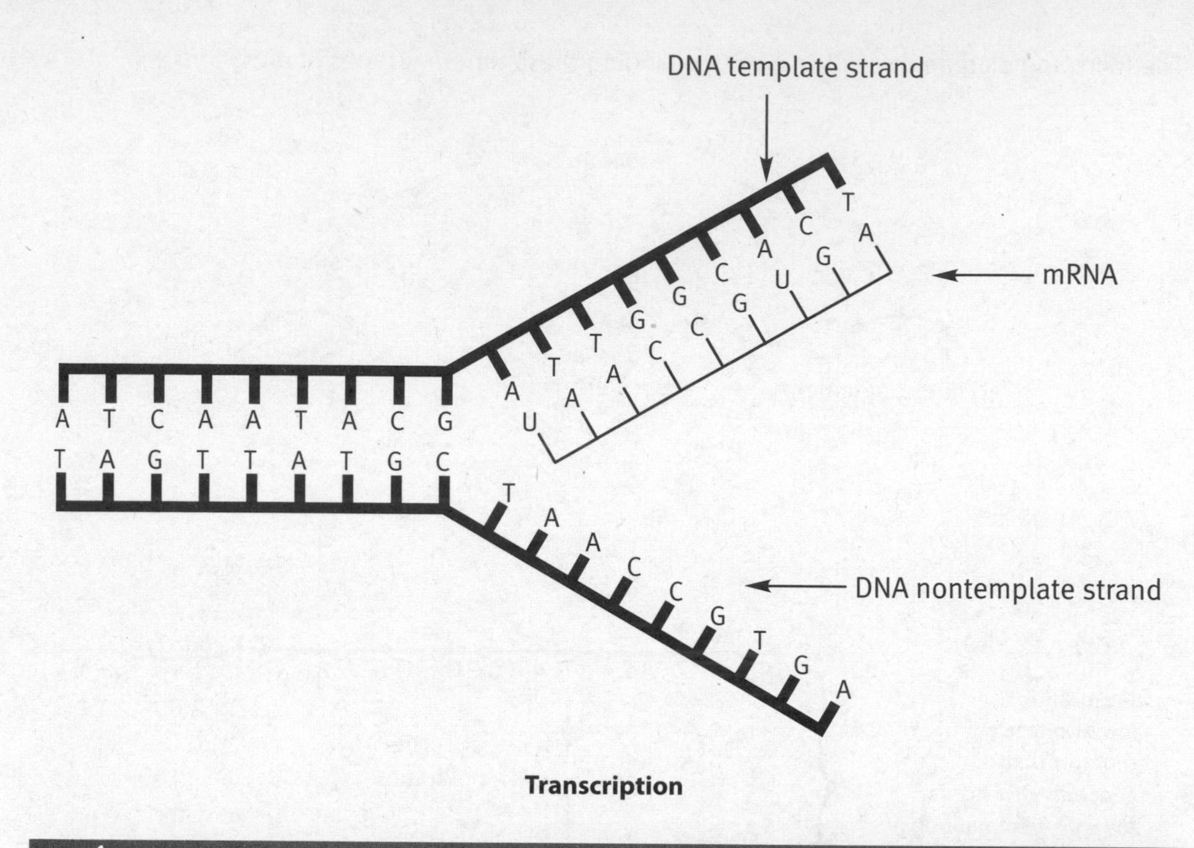

Transcription

✔ **AP Expert Note**

The mnemonic device *INtrons are IN the way, EXons are EXpressed* can be used to remember the different regions in DNA.

Stages of Transcription

Initiation

The first stage of transcription is *initiation*. During this stage, RNA polymerase binds to the **promoter** located upstream from the genes to be "read." The bound RNA polymerase causes DNA to unwind, exposing a single-strand that serves as the template for transcription.

One of the most common elements in eukaryotic promoters is the **TATA box**. Although not present in "housekeeping genes" and other developmental genes, such as the homeotic genes, TATA boxes are A-T-rich regions of DNA that are involved in positioning the start of transcription. The reason for this is that regions of DNA rich in adenine and thymine tend to separate more easily than those rich in Cs and Gs. Adenine and thymine form only two hydrogen bonds across the double helix, while C and G form three. Separation of DNA at the TATA boxes in the promoter regions of genes allows DNA to unzip at those regions for RNA polymerase to access the DNA template. Other types of promoters known as *internal promoters* can be found within the introns of genes. These promoters occur especially in genes that encode rRNA and tRNA molecules.

Elongation

In the *elongation* stage, RNA polymerase transcribes the DNA template strand to produce a complementary antiparallel RNA strand. Reading from the 3′ end of the template strand, RNA polymerase builds an hnRNA (pre-mRNA) chain in the 5′ to 3′ direction.

Termination

The *termination* stage occurs when RNA polymerase encounters a **terminator** sequence located downstream from the promoter. The terminator signals RNA polymerase to detach from the template strand and release the hnRNA chain for post-transcriptional processes. Introns are removed and a 5′ methylguanosine cap and a 3′ poly-A tail are added to complete the mRNA. The mRNA then exits via the nuclear pores and goes to the ribosomes in the cytoplasm for translation.

TRANSLATION

HIGH YIELD

16.5 Describe the steps of translation

16.6 Describe post-translational modifications to polypeptides

In a eukaryotic organism, DNA is found in the **nucleus** of the cell. **Protein synthesis** takes place in the cytoplasm, so information from the coding regions of DNA in the nucleus must be moved to the cytoplasm. Information is moved to the cytoplasm through **mRNA** (**messenger RNA**).

Once in the cytoplasm, mRNA acts as the template for protein **translation**. A short series of three bases, called a triplet or **codon**, codes for a **tRNA (transfer RNA)** that carries a specific amino acid. There are 64 (4^3) possible triplets, but there are only 20 amino acids. Even allowing for the start signal codon AUG as a site to begin protein translation and several codons acting as stop signals, there is considerable redundancy (also known as degeneracy) in the **genetic code**.

The **ribosome** performs protein synthesis. Once an mRNA enters the cytoplasm, several ribosomes attach to it, creating multiple lengths of the protein simultaneously. Elongation continues until a stop codon is reached and the ribosome disengages the mRNA. The newly synthesized polypeptide usually undergoes transformation (i.e., removal of terminal amino acids) before achieving its active state.

✔ **AP Expert Note**

You do not need to memorize the genetic code, codons, or the structure of the 20 amino acids for the AP Biology exam.

Stages of Translation

Initiation

The first stage of translation is *initiation*. During this stage, the mRNA attaches itself to the ribosome. The first codon on the mRNA is always AUG (the start sequence), which codes for the amino acid methionine. tRNA brings the first amino acid and places it in its proper place.

Elongation

Subsequent amino acids are then brought to the ribosome and joined together by peptide bonds in the *elongation* stage.

Termination

The *termination* stage is always triggered by one of three stop codons: UAA, UGA, or UAG. They stop protein synthesis and release the protein from the ribosome.

✔AP Expert Note

To remember the start codon (AUG), think of the month that is the start of the school year for many students: AUGust. A mnemonic to help remember the three stop codons (UAA, UGA, UAG) is: U Are Annoying, U Go Away, U Are Gone.

Post-Translational Modifications

After translation, synthesized polypeptides or proteins may undergo post-translational modifications, which generally include covalent enzymatic alterations to amino acids or the C-terminus or N-terminus. Post-translational modifications influence a protein's structure and specify a protein's function. Examples of post-translational modifications are phosphorylation, ubiquitination, and glycosylation. Phosphorylation is involved in various cellular processes, such as the cell cycle, apoptosis, differentiation, and enzyme activity. By adding a phosphate group (via kinase) and/or removing a phosphate group (via phosphatase), phosphorylation and its reverse, dephosphorylation, can activate or inactivate a protein. Similarly, ubiquitination adds ubiquitin, a small regulatory protein, which acts as a signal that turns transcription levels on, off, up, or down. Glycosylation—the attaching of a carbohydrate molecule—promotes protein folding, improves protein stability, and is involved in immune recognition.

AP BIOLOGY LAB 9: BIOTECHNOLOGY: RESTRICTION ENZYME ANALYSIS OF DNA INVESTIGATION

16.7 Investigate analysis of DNA

This experiment involves splicing DNA using restriction enzymes and visualizing these fragments using gel electrophoresis. A gel is a matrix composed of a polymer that acts like filter paper. Think of the gel as a thick, tangled jungle. DNA fragments are negatively charged, so they will move away from the negative pole of an applied current and move toward a positive pole. As the DNA fragments move away from the negative terminal of the charge applied, they get tangled in the jungle. The smaller the fragments of DNA, the faster they can move through the tangle. The movement of the fragments can be adjusted by changing the density of the gel (the tangles in the jungle) or by changing the quantity of charge (the pushing/pulling power) applied to the gel.

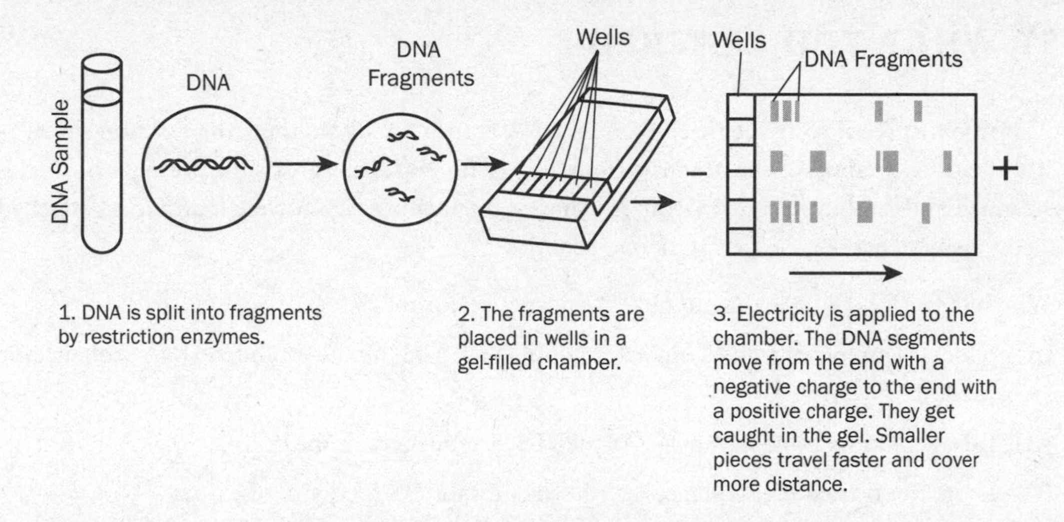

Diagram of Gel Electrophoresis

Restriction enzymes are manufactured by molecular genetics laboratories to have a high specificity for a series of base pairs on a strand of DNA or RNA. When the enzyme finds and binds to these base pairs, it snips the DNA. The more this series of base pairs appears on the DNA or RNA strand, the more cuts the restriction enzyme makes in the strand. After the strand is digested by the restriction enzyme, the fragments with different lengths can be separated using gel electrophoresis. This technique is commonly used to determine genetic differences between organisms, as long as the appropriate region of DNA can be isolated and the right combination of restriction enzymes can be determined. The most important thing to remember from this lab is that smaller fragments move farther on the gel. They move farther *because* they are smaller. All of the fragments move for the same reason: they are charged molecules moving away from an applied charge of the same kind (positive or negative).

RAPID REVIEW

If you take away only 6 things from this chapter:

1. Nitrogenous base pairs make up DNA and RNA: adenine pairs with thymine (DNA only) or uracil (RNA only) and cytosine pairs with guanine.
2. DNA replication is a semiconservative process, in which one of the antiparallel strands of DNA is preserved and the other strand is newly synthesized. New nucleotides are added in a 5′ to 3′ direction.
3. In transcription, the DNA strands separate and mRNA copies one side. The mRNA takes the information to the ribosome, where protein synthesis occurs.
4. In translation, tRNA carries amino acids to the mRNA and assembles them into proteins based on the mRNA code. Proteins are often modified after translation, giving them their final structure.
5. Mutations are the source of genetic change. Types include base-pair substitutions, which affect one amino acid, as well as insertions and deletions, which shift the genetic code and affect many amino acids.
6. Scientists can modify an organism's DNA by adding new genes.

TEST WHAT YOU LEARNED

1. A biochemist is studying the mechanism of DNA replication by stopping the reaction shortly after its initiation. Analysis of the products from the reaction on a gel reveals the presence of single-stranded and double-stranded DNA. The shorter fragments of DNA were found to be attached to a complementary sequence of 10 ribonucleotides.

   Which of the following is the most likely explanation for the data?

   (A) Short fragments of mRNA bind at random to available single-stranded DNA sequences in the cell.

   (B) tRNAs bind to available single-stranded DNA sequences in the cell.

   (C) Short ribonucleotide sequences are the primers for DNA polymerases.

   (D) Short sequences of RNA attached to DNA represent an artifact of the isolation procedure.

2. Hormone response elements are short sequences of DNA to which a hormone-hormone receptor complex can bind. In a mammalian cell, a mutation inserts an estrogen-response element upstream of the transcription initiation site of the gene encoding the protein tubulin. Tubulin is a component of the cytoskeleton. Which of the following results would most likely occur as a result of this mutation?

   (A) The expression of tubulin would become estrogen responsive.

   (B) Baseline expression of tubulin would increase.

   (C) The amino acid sequence of tubulin would be changed.

   (D) Transcription of the tubulin gene could not undergo termination.

3. A researcher wishes to synthesize insulin *in vitro* and decides to use a cytoplasmic wheat germ extract that contains ribosomes, tRNAs, amino acids, enzymes, and translation factors. She isolates several types of macromolecules from pancreatic cells and incubates them with the wheat germ extract.

   Based on this information, the insulin will only be produced when

   (A) DNA from pancreatic cells is added to the mixture

   (B) ribosomes from pancreatic cells are added to the mixture

   (C) mRNA from pancreatic cells is added to the mixture

   (D) both DNA and mRNA from pancreatic cells are added to the mixture

4. Several point mutations in a DNA sequence changed the mRNA transcript encoded by a gene as shown below.

   Original sequence:

   5′-UCG-GCA-CAU-UUA-UAU-GUU-3′

   New sequence:

   5′-AGC-GCA-CAU-UUG-UAA-GUU-3′

   CODON TABLE

FIRST POSITION	SECOND POSITION: U		C		A		G		THIRD POSITION
	code	amino acid	code	amino acid	code	amino acid	code	amino acid	
U	UUU	phe	UCU	ser	UAU	tyr	UGU	cys	U
	UUC		UCC		UAC		UGC		C
	UUA	leu	UCA		UAA	STOP	UGA	STOP	A
	UUG		UCG		UAG	STOP	UGG	trp	G
C	CUU	leu	CCU	pro	CAU	his	CGU	arg	U
	CUC		CCC		CAC		CGC		C
	CUA		CCA		CAA	gln	CGA		A
	CUG		CCG		CAG		CGG		G
A	AUU	ile	ACU	thr	AAU	asn	AGU	ser	U
	AUC		ACC		AAC		AGC		C
	AUA		ACA		AAA	lys	AGA	arg	A
	AUG	met	ACG		AAG		AGG		G
G	GUU	val	GCU	ala	GAU	asp	GGU	gly	U
	GUC		GCC		GAC		GGC		C
	GUA		GCA		GAA	glu	GGA		A
	GUG		GCG		GAG		GGG		G

   Using the codon table provided, which of the following statements is accurate?

   (A) The changes in the sequence caused a substitution.

   (B) The changes in the sequence generated a shorter peptide.

   (C) The changes in the sequence generated a longer peptide.

   (D) The changes in the sequence did not change the peptide.

5. The mRNA encoding the insulin polypeptide is translated on ribosomes attached to the rough endoplasmic reticulum (RER). Studies show that the protein encoded by the mRNA contains almost twice as many amino acids as the active insulin. Insulin is secreted into the bloodstream, along with an inactive peptide, at a 1:1 molar ratio.

   Which of the following best explains the observed results?

   (A) The precursor of insulin contains a signal sequence and an internal region that are excised after translation.

   (B) Insulin contains hydrophobic core residues that allow the molecule to be embedded in the RER membrane.

   (C) Insulin is translated as a dimer and digested into monomers.

   (D) Insulin is degraded in the RER into amino acids that are later reassembled prior to secretion.

6. A study shows that an isolated nucleic acid polymer contains 32% adenine, 15% guanine, and 25% cytosine. Which of the following conclusions is most probable based on this data?

   (A) The percentage of uracil is 32%.

   (B) The percentage of thymine is 32%.

   (C) The nucleic acid is a double-stranded DNA molecule.

   (D) The nucleic acid is a single-stranded RNA molecule.

7. The diagram below shows a DNA plasmid with a total length of 2,000 base pairs. The tick marks indicate cleavage sites for restriction enzymes (enzyme EcoRI and enzyme HaeIII).

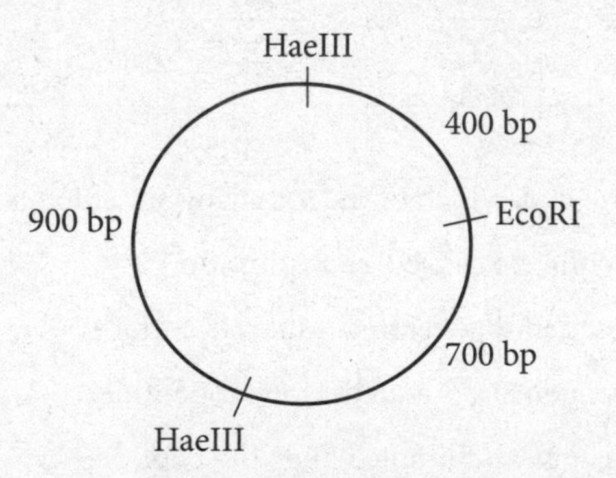

   The plasmid was digested with the enzyme EcoRI to completion and separated by gel electrophoresis. Which of the following results would most likely be observed?

   (A) One 1,600 bp fragment

   (B) One 2,000 bp fragment

   (C) Two fragments of 400 bp and 1,600 bp

   (D) Two fragments of 700 bp and 1,300 bp

Transformations

The answer key to this quiz is located on the next page.

Answer Key

Test What You Already Know

1. **D** **Learning Objective:** 16.1
2. **B** **Learning Objective:** 16.2
3. **D** **Learning Objective:** 16.3
4. **C** **Learning Objective:** 16.4
5. **A** **Learning Objective:** 16.5
6. **D** **Learning Objective:** 16.6
7. **C** **Learning Objective:** 16.7

Test What You Learned

1. **C** **Learning Objective:** 16.3
2. **A** **Learning Objective:** 16.4
3. **C** **Learning Objective:** 16.2
4. **B** **Learning Objective:** 16.5
5. **A** **Learning Objective:** 16.6
6. **D** **Learning Objective:** 16.1
7. **B** **Learning Objective:** 16.7

Detailed solutions can be found in the Answers and Explanations section at the back of this book.

REFLECTION

Test What You Already Know score: __________

Test What You Learned score: __________

Use this section to evaluate your progress. After working through the pre-quiz, check off the boxes in the "Pre" column to indicate which Learning Objectives you feel confident about. Then, after completing the chapter, including the post-quiz, do the same to the boxes in the "Post" column. Keep working on unchecked Objectives until you're confident about them all!

Pre	Post	
☐	☐	**16.1** Differentiate between the structures of DNA and RNA
☐	☐	**16.2** Explain the function of DNA and RNA
☐	☐	**16.3** Describe the process of DNA replication
☐	☐	**16.4** Describe the process and product of transcription
☐	☐	**16.5** Describe the steps of translation
☐	☐	**16.6** Describe post-translational modifications to polypeptides
☐	☐	**16.7** Investigate analysis of DNA

FOR MORE PRACTICE

Complete more practice online at kaptest.com. Haven't registered your book yet? Go to kaptest.com/booksonline to begin.

CHAPTER 17

Development

LEARNING OBJECTIVES

In this chapter, you will review how to:

- **17.1** Describe the process of fertilization and genetic transfer
- **17.2** Describe positive and negative gene control mechanisms
- **17.3** Explain how gene regulation results in gene expression
- **17.4** Explain how gene expression leads to cell specialization

TEST WHAT YOU ALREADY KNOW

1. *Lymnaea stagnalis* snails can have left-coiling (sinistral) or right-coiling (dextral) shells, shown in the figure as A and B, respectively.

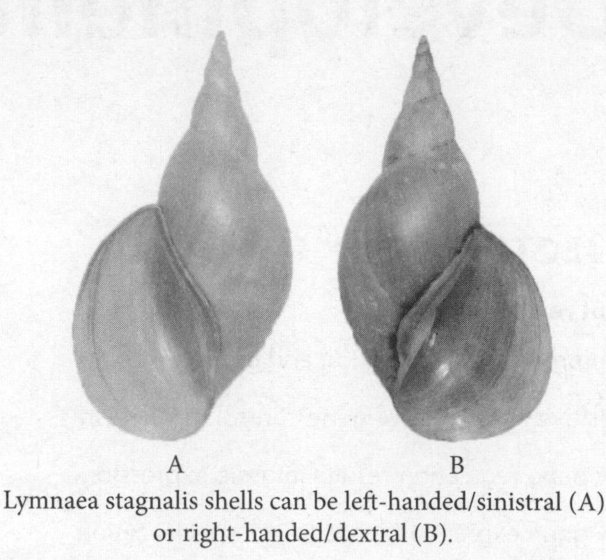

Lymnaea stagnalis shells can be left-handed/sinistral (A) or right-handed/dextral (B).

Adapted from Edmund Gittenberger, Thomas D. Hamann, and Takahiro Asami, "Chiral Speciation in Terrestrial Pulmonate Snails," *PLoS ONE* 7, no. 4 (April 2012): e34005.

Although researchers know that shell development is heavily influenced by genetics, they are interested in the effects of the environment on shell development as well. To study this, researchers could raise snails in different laboratory environments. Which of the following possibilities would be least likely from crosses of true-breeding, male, left-coiling snails and true-breeding, female, right-coiling snails?

(A) After the offspring were randomly divided into groups to be raised in two different laboratory environments, those in one environment all resembled type A while those in the other environment all resembled type B.

(B) After the offspring were randomly divided into groups to be raised in two different laboratory environments, all of the snails in both environments resembled type A.

(C) After the offspring were randomly divided into groups to be raised in two different laboratory environments, all of the snails in both environments resembled type B.

(D) After the offspring were randomly divided into groups to be raised in four different laboratory environments, there were similar amounts of each shell type but slightly more of type A.

2. If a particular gene product is needed during early development, the gene must be turned on at the appropriate time. The default setting is off and then the binding of an activator turns on the gene when it is needed. Which statement gives the best description of this type of gene control?
   (A) It is constitutive expression because the gene is expressed throughout development.
   (B) It is positive control because the gene is turned on by the presence of an activator once it is needed.
   (C) It is negative control because the gene is turned on and is capable of being turned off.
   (D) It is an operon because operons are always controlled by repressors.

3. The table below compares gene expression in larval honey bee workers and queens. The honey bee genes were compared with genes of known function in *Drosophila melanogaster* in order to predict their functions within honey bees. The light bars represent the percent of genes that are differentially expressed in queens. The dark bars represent values for workers.

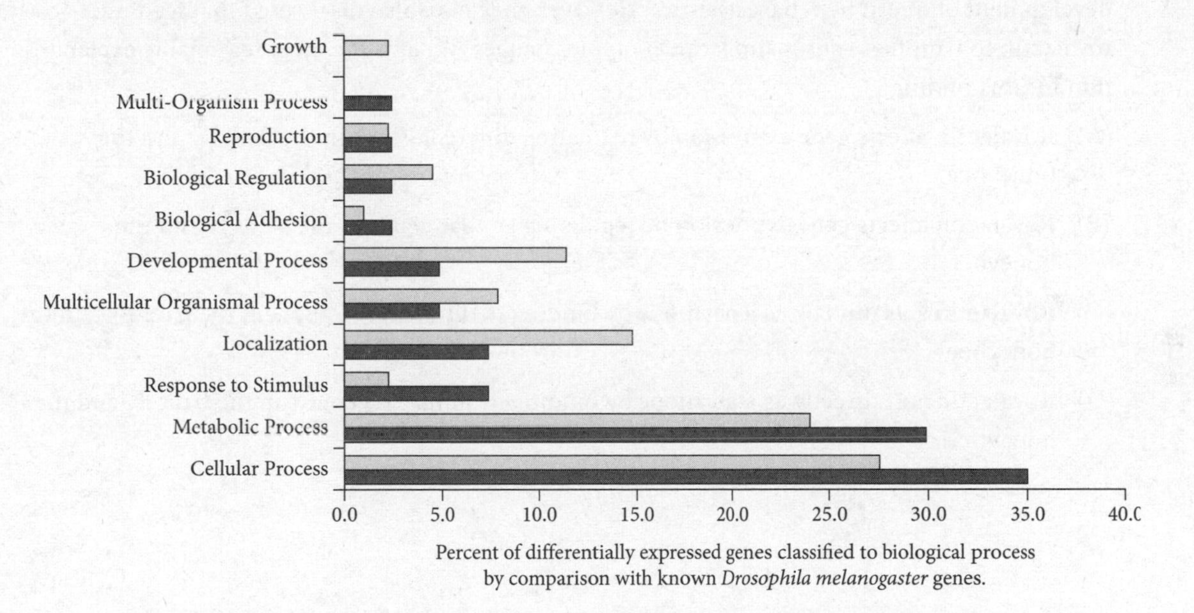

Adapted from Ana Durvalina Bomtorin et al., "Hox Gene Expression Leads to Differential Hind Leg Development between Honeybee Castes," *PLoS ONE* 7, no. 7 (July 2012): e40111.

Based on this data, what can most reasonably be concluded about the difference between queens and workers?

(A) Workers may need to express a greater proportion of regulatory genes than queens as they are exposed to so many complex demands while outside of the hive.

(B) Queens may need a lower proportion of genes expressed during development than workers as they have a relatively simple lifestyle that does not require leaving the hive regularly.

(C) Workers may need to express a higher proportion of genes related to sensory organs to allow them to respond to a variety of stimuli while foraging.

(D) Queens may need a wider range of metabolic functions compared with workers because of their important role in the colony.

4. Honeybees have two female castes, workers and queens, which are phenotypically very different. Researchers have discovered that feeding bee larvae a diet rich in a protein called royalactin increases the amount of juvenile hormone that they produce. Higher hormone levels result in the development of queen bee characteristics. However, they have also discovered that feeding royalactin to fruit flies causes similar phenotypic changes. What is the most reasonable explanation for this finding?

(A) Royalactin affects gene expression by regulating different genes in the fruit fly and the honeybee.

(B) Royalactin affects gene expression by regulating similar genes in the fruit fly and the honeybee.

(C) Royalactin acts directly as a hormone by binding to different receptors in the fruit fly and the honeybee.

(D) Royalactin acts directly as a hormone by binding to similar receptors in the fruit fly and the honeybee.

Answers to this quiz can be found at the end of this chapter.

FERTILIZATION AND EARLY DEVELOPMENT

17.1 Describe the process of fertilization and genetic transfer

Sexual reproduction of diploid (2*N*) organisms involves the recombination and transmission of genetic material from parents to offspring via vertical gene transfer and **fertilization**. (Note: horizontal gene transfer is the transmission of genetic material from a donor to a recipient organism that is not its offspring.) For sexual reproduction to occur, cells undergo **gametogenesis**—the process that forms **egg (ovum)** and **sperm (spermatozoon)** cells or gametes via meiosis. Recall that meiosis, which was discussed in chapter 14, is the process that forms haploid cells (containing one set of chromosomes, 1*N*). During fertilization, a sperm cell fertilizes the egg cell to produce a single cell called a **zygote.** (Sexual reproduction can also occur when a hermaphroditic organism fertilizes its own eggs to form offspring.) The fusion of one sperm with one egg, both of which are haploid, restores the diploid number of chromosomes. Since each zygote is genetically different, fertilization increases genetic variation and contributes to the survival of a species.

Fertilization is the first step in embryogenesis—the process by which an embryo forms and develops. After fertilization, the zygote undergoes **cleavage** (cell division without an increase in mass) and cellular differentiation to form a multicellular organism.

Fertilization can occur within the body or outside of the body. Internal fertilization typically occurs in the female reproductive organs of mammals and birds, while external fertilization usually occurs in aquatic environments and can be seen in frogs, sea urchins, and most fish. In plants, fertilization is preceded by **pollination**, the transferring of pollen to the female reproductive organs of a plant. Additionally, in angiosperms (flowering plants), double fertilization occurs—one sperm cell fertilizes the egg cell, and the other fuses to form the endosperm, which provides nutrients to the developing embryo in the seed.

✔ AP Expert Note

The details of embryonic development are not tested on the AP Biology exam.

GENE REGULATION AND CELL SPECIALIZATION

HIGH YIELD

17.2 Describe positive and negative gene control mechanisms

17.3 Explain how gene regulation results in gene expression

17.4 Explain how gene expression leads to cell specialization

In multicellular organisms, the developmental pathways from zygote to fully formed organism (adult) involve regulatory systems and differential **gene expression** governed by internal and external signals. Though each eukaryotic cell contains the same genome, each cell does not express the same set of genes. Some genes may be continually expressed, but most gene expression is regulated to maximize energy usage and increase metabolic fitness. Otherwise, cells would not be able to obtain enough resources to stay alive.

Gene Regulation and Expression

Regulatory systems in both prokaryotes and eukaryotes consist of genes that encode proteins (**repressors** and **activators**), which interact with DNA sequences to control gene expression via negative and positive control mechanisms. In **negative control**, genes are expressed until they are switched off by a repressor; in **positive control**, genes are expressed when they are turned on by an activator. Thus, repressors decrease expression and activators increase expression. **Inducers**, which bind to repressors and activators, increase gene expression by disabling repressors and enabling activators.

Prokaryotic cells generally control gene expression at the **transcriptional level**, while eukaryotic cells regulate gene expression not only at the transcriptional level but also at the post-transcriptional, translational, and post-translational levels. **Transcription factors** include hundreds or thousands of proteins that exert transcriptional control over the genome. Each protein has a highly conserved DNA-binding domain, which allows it to attach to specific DNA sequences or regulatory proteins. Together with regulatory proteins, transcription factors can act as repressors or activators.

An example of a regulatory DNA sequence is an **enhancer** sequence—a noncoding region of DNA located upstream or downstream of a promoter. As discussed in chapter 16, DNA is transcribed when RNA polymerase binds to a promoter sequence. The binding of transcription factors to enhancers enhances the transcription of an associated gene. Experimental removal of enhancers has been shown to cause drastic decreases in gene transcription.

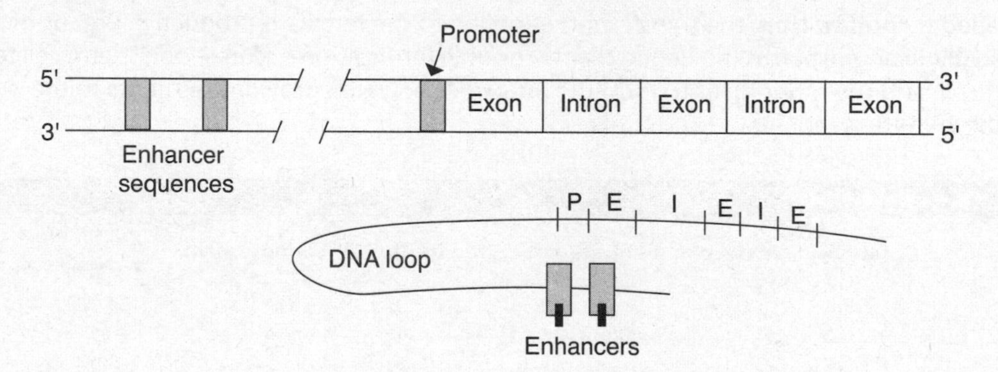

One model of regulated gene expression that illustrates inducible (positive) and repressible (negative) systems is the *lac* (lactose) **operon** in *E. coli*, which is required to transport and metabolize lactose for carbon and energy. An operon is a series of genes (including a promoter, an operator, and a terminator) that synchronize to perform a biological function. The following diagram shows that when lactose is not present in the growth medium, a protein repressor binds to the operator (regulatory site) to inhibit the transcription of the operon. When lactose (the inducer) is present, it binds to the repressor. The release of the repressor from the operator causes the transcription of the operon by RNA polymerase to commence. The transcribed mRNA codes for the proteins that metabolize lactose, which eventually leads to the removal of lactose from the medium. When lactose is gone, the repressor binds to the operator again and the operon is back to square one. This is a classic example of a negative feedback loop. After all, there is no need for the proteins that metabolize lactose unless the substrate is available. This is also an example of positive control because the expression of the genes is induced by the presence of the substrate.

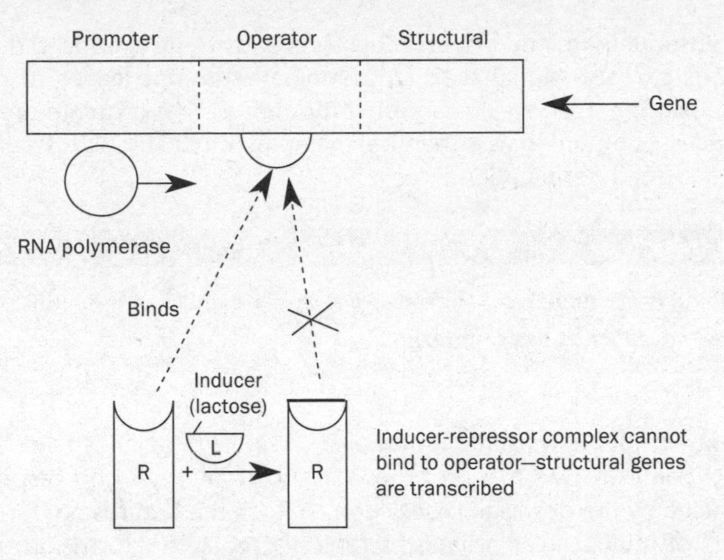

The *Lac* Operon in *E. coli*

Another important concept to appreciate from this model of gene expression is that the promoter and operator of an operon are "unaware" of what happens at another part of the DNA strand, unless the DNA performs a specific activity on them. For example, if the gene to produce human insulin were placed within a bacterium's operon that produces enzymes for metabolizing glucose, the insulin gene would also be transcribed when glucose signals the transcription of the glucose metabolizing genes. Scientists can then harvest the insulin from the bacterium. This type of genetic manipulation has revolutionized the discipline of molecular genetics.

Cell Differentiation and Specialization

Every cell in multicellular organisms carries out very specific and different functions in different structures. How does a cell in the eye of an eagle express the proper genes to make proteins for eye development and function, while a cell in the pectoral muscle of an eagle expresses proteins to build muscle tissue and use energy? The regulation of gene expression in eukaryotes causes cells to differentiate. Since most eukaryotic cells in a given organism contain the same set of genetic material, differential expression determines the cells' structure and function. Each cell type undergoes **differentiation** and **specialization** to produce specific proteins with particular functions according to its needs. For instance, regulation determines whether cells become lens cells that are sensitive to light, muscle cells that contract, or nerve cells that transmit information. Another example is the expression of the SRY (sex-determining region Y) gene on the Y chromosome, which initiates male sexual development. Errors or changes in gene regulation during development may lead to severe and detrimental consequences.

Both **intercellular** (external) and **intracellular** (internal) signal transmissions regulate gene expression. Signals may activate transcription and tell the developing organism which genes to express and when to express them. Additionally, external signals may trigger intracellular signaling pathways or signal transduction cascades. For example, in eukaryotes cAMP is a small molecule that is involved in amplifying the message in intracellular signaling pathways. In bacteria, cAMP regulates metabolic gene expression. When glucose concentration is low, cAMP accumulates and transcription factors are activated to turn on the lac operon. In plant cells, ethylene triggers fruit-ripening. Special receptors bind to ethylene, which triggers a cascade that turns off anti-ripening genes. Ripened fruits invite animals to consume them and disperse their seeds. Ethylene produced during certain developmental conditions can also signal seeds to germinate, prompt leaves to change colors, and trigger flower petals to die. Gibberellins are plant hormones that regulate growth and influence developmental processes by acting as a chemical messenger that stimulates cell elongation, breaking and budding, and seed germination.

> ✔ **AP Expert Note**
>
> **The important thing to remember is that genes are always present in an organism's DNA, but certain environmental variables affect their expression.**

Molecules in the extracellular environment may also be signal molecules released from other cells. Haploid yeast cells come in two mating types—a and α (alpha)—and respond to the mating pheromones produced by the opposite type. Upon contact, the two fuse to form a diploid zygote. Signals in embryos of multicellular organisms typically regulate expression of certain genes in targeted cells via **induction**. In cell-to-cell interactions, cytokines, molecules secreted by cells, signal other cells to begin cell replication and division. During pattern formation, cells of the developing embryo communicate their relative position. The concentration of morphogens tells cells the proximity of the cell releasing the signal and determines the spatial organization of tissues and organs. Cells exposed to higher concentrations of morphogen will develop differently than cells exposed to lower concentrations of morphogen. Hox genes (homeotic genes), which are also involved in pattern formation, control the development of appendages like antennae and legs. When one of the Hox genes is mutated, the wrong body part forms.

RAPID REVIEW

If you take away only 5 things from this chapter:

1. Fertilization, the first step in embryonic development, is the fusion of one sperm with one egg (gametes) to produce a zygote that becomes a fully formed organism.
2. In gene regulation, negative control occurs when a repressor switches genes off by inhibiting transcription, while positive control occurs when an activator switches genes on by promoting transcription. Inducers are molecules that bind to repressors and/or activators to increase gene expression.
3. Operons generally occur in bacterial genomes and are sets of genes that perform a biological function like metabolizing lactose. When the substrate is absent, a repressor binds to the operator to inhibit transcription of the operon, and when present, the substrate binds to the repressor and induces transcription.
4. Though each contains the same set of genetic material, cells in multicellular organisms have specific and different functions. The regulation of gene expression causes cells to undergo differentiation and specialization.
5. Intercellular and intracellular signals regulate gene expression. For example, in embryos of multicellular organisms, the concentration of morphogens determines spatial organization of tissues and organs.

TEST WHAT YOU LEARNED

1. Queen, worker, and drone honeybees have dramatically different morphologies and roles in the hive, even though all these individuals possess a similar genome and live in the same hive. A researcher finds that bee morphology is driven by an external chemical signal. Based on this finding, which of the following hypotheses would be the most appropriate to test next?

   (A) Differences in incubation temperature during development affect which genes are expressed.

   (B) Differences in diet during larval development affect which genes are expressed.

   (C) The number of mutations in the coding DNA regions of queens versus workers affect which genes are expressed.

   (D) The differences in the ways that queens, workers, and drones use their hindlimbs affect which genes are expressed.

2. *Lymnaea stagnalis* snails usually have right-coiling shells, and left-coiling shells are rare. Studies have shown that the allele for a right-coiling shell is dominant to the allele for a left-coiling shell. A cross between a true-breeding left-coiling female snail and a homozygous right-coiling male snail produces entirely left-coiling offspring. However, a cross between a right-coiling female and a homozygous-left coiling male produces entirely right-coiling offspring. Based on these findings, which of the following explanations is most probable?

   (A) The male snail does not contribute any genetic material during fertilization in this species, meaning that the species must be parthenogenetic.

   (B) The direction of shell coiling is controlled solely by environmental factors, so the alleles contributed by the mother and father do not affect the shell morphology of the offspring.

   (C) During the earliest stages of development, only mRNA from the male's sperm is used for gene expression to determine the direction of shell coiling.

   (D) During the earliest stages of development, mRNA from the mother's genes affects the direction of shell coiling.

3. Researchers have found that a gene called Tbx5 is important in vertebrate forelimb development. Activation of this gene can occur along the entire region of the body where limbs develop. However, a repressive complex including Hoxc9 prevents *Tbx5* expression in areas of the body posterior to the forelimb region. The figure below shows the length of a developing embryo with the head (cephalic) region at the top and the tail (caudal) region at the bottom. It shows a neural tube (NT), which will become the spinal cord; somites; and lateral plate mesoderm (LPM), which will give rise to limbs and other structures.

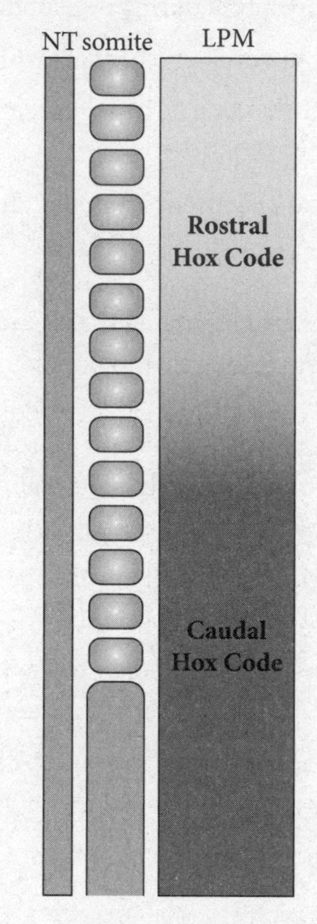

Adapted from Satoko Nishimoto et al., "A Combination of Activation and Repression by a Colinear Hox Code Controls Forelimb-Restricted Expression of *Tbx5* and Reveals Hox Protein Specificity," *PLoS Genet* 10, no. 3 (March 2014): e1004245.

Based on the information provided, which of the following is the most reasonable conclusion?

(A) The Hoxc9 regulatory complex will exert negative control in the caudal region, so the forelimb will develop in the rostral region.

(B) The Hoxc9 regulatory complex will exert negative control in the rostral region, so the forelimb will develop in the rostral region.

(C) The Hoxc9 regulatory complex will exert positive control in the caudal region, so the forelimb will develop in the caudal region.

(D) The Hoxc9 regulatory complex will exert positive control in the rostral region, so the forelimb will develop in the caudal region.

4. The figure shows the expression of four different genes associated with hindlimb development in honeybee workers and queens. The developmental stages shown are those most critical for hindlimb development, which is believed to begin around stage L4.

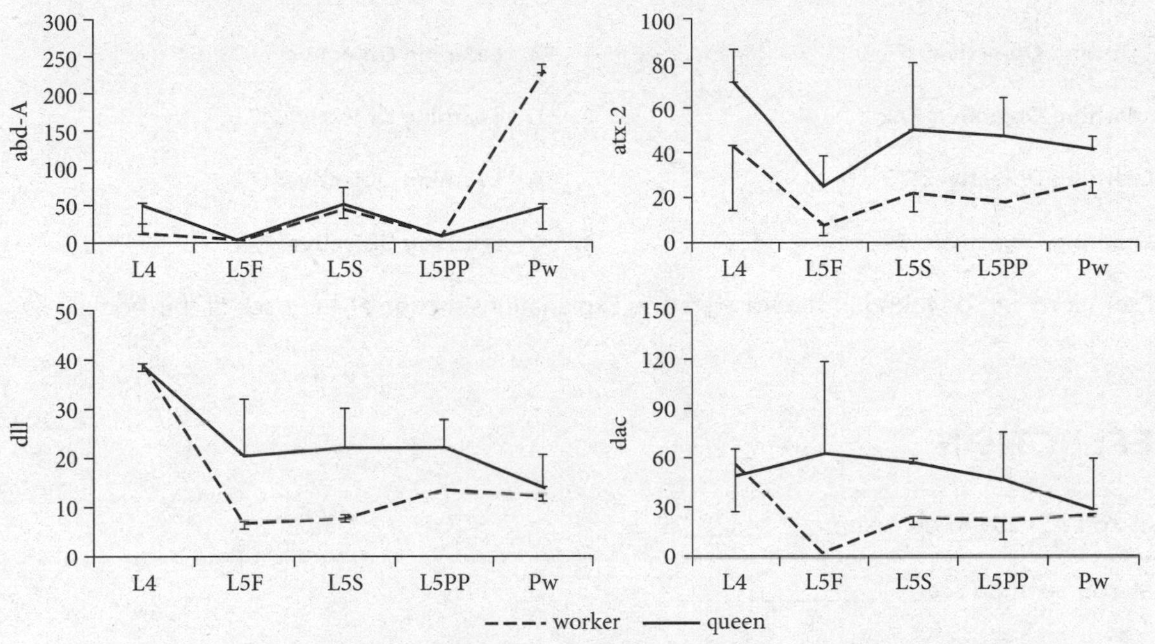

Expression of genes (identified on the *y*-axes) as development progresses from a larval stage (L4) to the white-eyed pupa stage (Pw). The *x*-axis represents stage of development and the *y*-axis represents relative transcript levels (a measure of gene expression).

Adapted from Ana Durvalina Bomtorin et al., "Hox Gene Expression Leads to Differential Hind Leg Development between Honeybee Castes," *PLoS ONE* 7, no. 7 (July 2012): e40111.

Which gene appears to be most important for the difference in hindlimb development in workers compared with queens, and why?

(A) The atx-2 gene is probably the most important for worker hindlimb specialization because it is expressed the most in workers at the L4 stage.

(B) The dac gene is probably the most important for worker hindlimb specialization because it is expressed at a relatively constant level throughout the stages shown.

(C) The abd-A gene is probably the most important for worker hindlimb specialization because it increases more dramatically in workers than in queens at the white-eyed pupa stage.

(D) The dll gene is probably the most important for worker hindlimb specialization because its expression dramatically drops after the L4 stage in workers but drops less in queens.

Answer Key

Test What You Already Know

1. **A** **Learning Objective:** 17.1
2. **B** **Learning Objective:** 17.2
3. **C** **Learning Objective:** 17.3
4. **B** **Learning Objective:** 17.4

Test What You Learned

1. **B** **Learning Objective:** 17.3
2. **D** **Learning Objective:** 17.1
3. **A** **Learning Objective:** 17.2
4. **C** **Learning Objective:** 17.4

Detailed solutions can be found in the Answers and Explanations section at the back of this book.

REFLECTION

Test What You Already Know score: __________

Test What You Learned score: __________

Use this section to evaluate your progress. After working through the pre-quiz, check off the boxes in the "Pre" column to indicate which Learning Objectives you feel confident about. Then, after completing the chapter, including the post-quiz, do the same to the boxes in the "Post" column. Keep working on unchecked Objectives until you're confident about them all!

Pre	Post	
❑	❑	**17.1** Describe the process of fertilization and genetic transfer
❑	❑	**17.2** Describe positive and negative gene control mechanisms
❑	❑	**17.3** Explain how gene regulation results in gene expression
❑	❑	**17.4** Explain how gene expression leads to cell specialization

FOR MORE PRACTICE

Complete more practice online at kaptest.com. Haven't registered your book yet? Go to kaptest.com/booksonline to begin.

CHAPTER 18

Inheritance

LEARNING OBJECTIVES

In this chapter, you will review how to:

18.1 Describe chromosome structure and function

18.2 Compare chromosomal mutations and abnormalities

18.3 Explain and predict results of Mendelian inheritance

18.4 Explain why some traits are non-Mendelian

TEST WHAT YOU ALREADY KNOW

1. To understand more about the function of chromatin compaction, researchers examined the effects of radiation exposure on condensed and decondensed chromatin. The figure below shows the results from chromatin that was exposed to different amounts of radiation in Gray (Gy) units. The left column for each dose shows all DNA, whereas the right column shows only DNA that has experienced double strand breaks (DSBs). DSBs indicate radiation damage. The rows show DNA that was condensed, decondensed, or previously decondensed and then later recondensed at the time of irradiation.

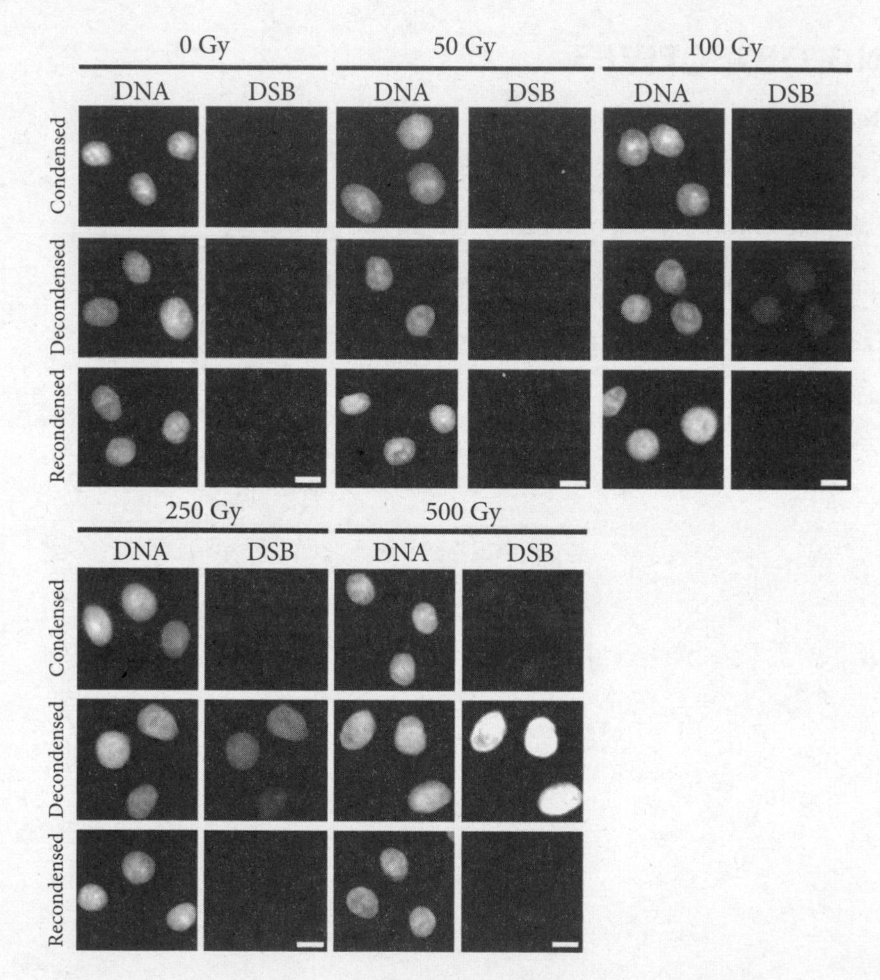

Adapted from Hideaki Takata et al., "Chromatin Compaction Protects Genomic DNA from Radiation Damage," *PLoS ONE* 8, no. 10 (October 2013): e75622.

Which of the following conclusions can most justifiably be drawn from these results?

(A) Condensed DNA increasingly experienced damage due to double strand breaks as radiation intensity increased, but decondensed DNA was protected from this damage.

(B) Decondensed DNA increasingly experienced damage due to double strand breaks as radiation intensity increased, but condensed DNA was protected from this damage.

(C) Both decondensed and recondensed DNA experienced damage due to double strand breaks in a dose-dependent manner.

(D) Decondensed DNA showed no damage in even the highest radiation treatment, suggesting that decondensation offered some type of protection from radiation damage.

2. To understand gene regulation, researchers have carefully examined the structure and function of the *lac* operon in *E. coli*. The *lac* operon contains three genes that code for proteins associated with lactose metabolism. It also contains an operator region that binds to a repressor and turns off the expression of all three genes. If a mutation occurred in the operator region of the *lac* operon, what would most likely happen?

   (A) Because a mutation had occurred, the operon could not work properly and the genes would not be expressed.

   (B) Because a mutation had occurred, the proteins associated with lactose metabolism might not work properly.

   (C) Because the repressor would be affected by the mutation, the repressor would bind too strongly to the operator and the genes could not be expressed at all.

   (D) Because the repressor could not recognize the operator, the repressor would not bind and the genes would be expressed.

3. Purebred Birman cats have distinctive white feet (called "gloves") that develop from having two copies of a Mendelian gloving allele. What could most reasonably be concluded if two cats that did not have gloves were crossed and produced a kitten with gloving?

   (A) The gloving allele is dominant and a mutation must have occurred for the kitten to have gloving.

   (B) The gloving allele is dominant and it was passed by one parent to the kitten.

   (C) The gloving allele is recessive and both of the parents must have been carriers.

   (D) The gloving allele is recessive and a single parent must have passed it to the kitten.

4. Glucose-6-phosphate dehydrogenase (G6PD) is an X-linked gene whose mutation is associated with a reduced risk of malaria. The figure shows G6PD enzyme activity in men and women by genotype.

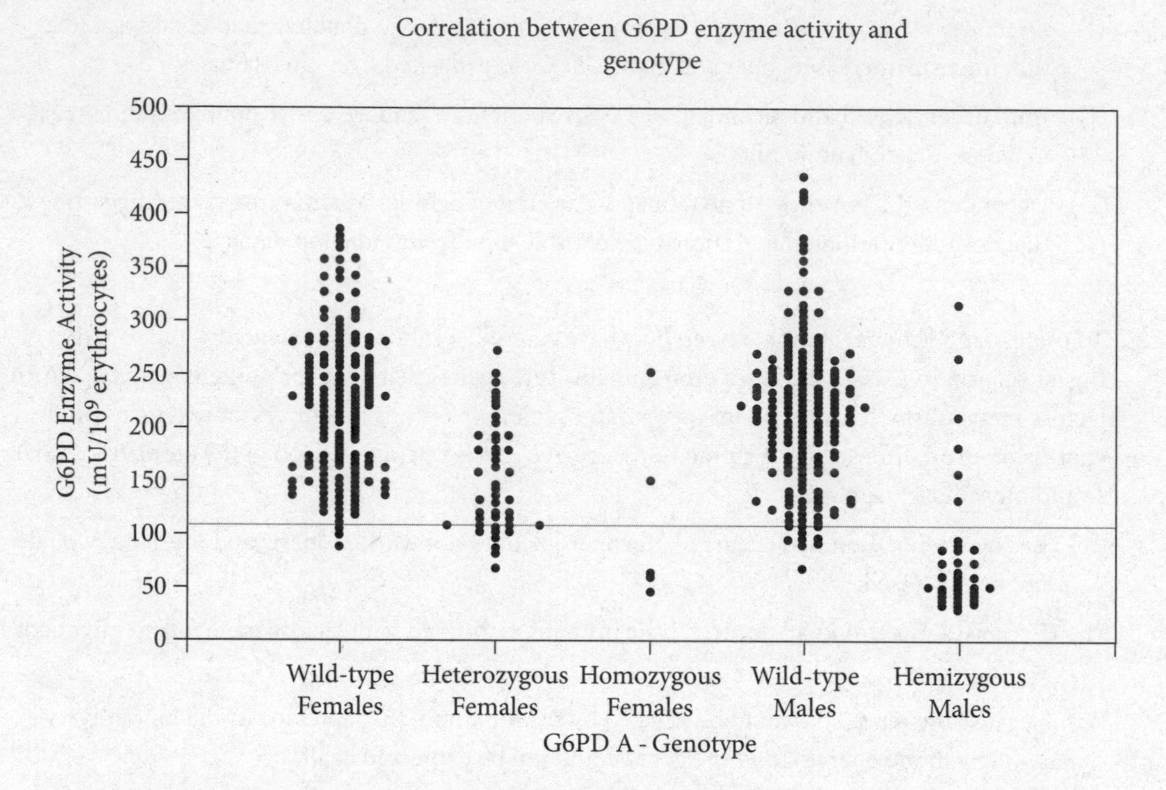

Adapted from M. K. Johnson et al., "Impact of the Method of G6PD Deficiency Assessment on Genetic Association Studies of Malaria Susceptibility," *PLoS ONE* 4, no. 9 (September 2009): e7246.

The researchers found that one copy of the mutant allele was sufficient to reduce the risk of malaria in men but homozygosity of the mutant allele was required to reduce the risk of malaria in women. What is the most likely explanation for these results?

(A) Women possess two copies of the X chromosome, so both copies must contain the recessive mutant allele to yield resistance to malaria.

(B) Gene expression is impacted by whether an individual is likely to be affected by malaria, meaning that individuals in high-risk areas are more likely to have the mutant allele.

(C) Hemizygous individuals have the greatest protection against malaria because they produce more G6PD enzyme.

(D) Women who are homozygous for the mutant allele and men who are hemizygous for the mutant allele are not able to produce any functional G6PD enzyme.

Answers to this quiz can be found at the end of this chapter.

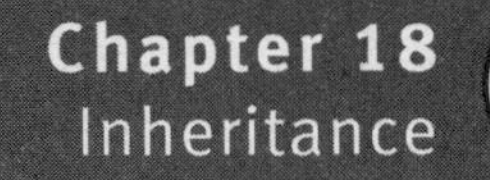

CHROMOSOMES

HIGH YIELD

18.1 Describe chromosome structure and function

18.2 Compare chromosomal mutations and abnormalities

Chromosomes are condensed bodies of DNA molecules that store codes for the translation of several different kinds of proteins. These proteins dictate how an organism is put together and functions. In the most simplified way of looking at the chromosomal theory of inheritance, each section of DNA that translates a different protein is called a **gene**. For example, a region of DNA may code for a protein that controls eye color, so this region of DNA is called the gene for eye color. In other words, a *gene* refers to a specific location on a chromosome.

Alleles specifically code for the traits of an organism. In the case of eye color, there might be an allele for blue eye color and an allele for green eye color. The eye color that an organism ends up with (its **phenotype**) is dictated by which alleles are placed in the gene location for eye color (its **genotype**). Sometimes an organism will have a genotype that includes an allele that is not expressed in its phenotype.

In a typical diploid (2*N*) eukaryotic organism, each cell has two copies of each chromosome. These pairs of chromosomes are a result of the combination of two haploid gametes formed by meiosis, one from the mother and one from the father. The mother and father each provided one allele for each gene, leaving the offspring with either two of the same allele (e.g., two for blue eye color or two for green eye color) or one of each allele (e.g., one allele for blue eye color and one allele for green eye color). In many genes, there is one allele that is **dominant** over the other and hides the expression of the other allele in the phenotype of the offspring. The other allele is called **recessive**. Geneticists generally use uppercase letters for dominant alleles (*B*) and lowercase letters for recessive alleles (*b*). If blue eye color is the dominant allele (indicated as *B*) and green eye color is the recessive allele (indicated as *b*), an offspring could have one of three different genotypes (*BB*, *Bb*, and *bb*) when its parents' gametes combine. The phenotype of an offspring with the genotype *BB* is blue eyes. An offspring with the genotype *Bb* will also have blue eyes because the blue eye color allele is dominant. Only offspring with the genotype *bb* will have a green eye phenotype.

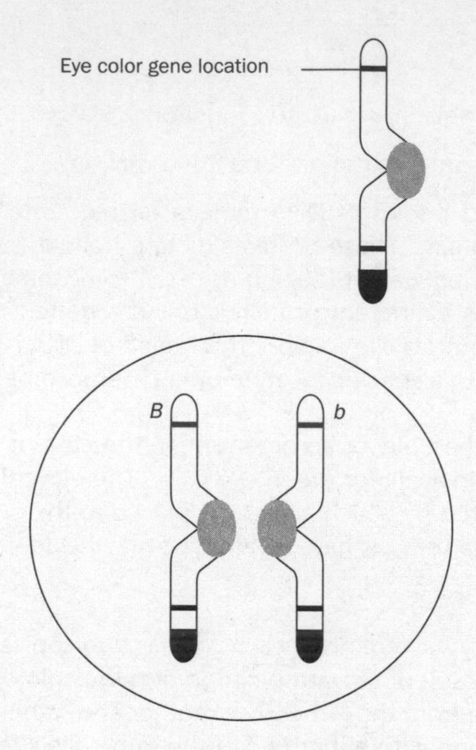

The diagram above shows a cell from an animal with one allele for blue eye color (*B*) and one allele for green eye color (*b*). This animal's genotype for eye color is *Bb,* and its phenotype is blue eyes because the blue eye color allele is dominant. The chromosomes are essentially identical in appearance, but the alleles have a slightly different DNA code.

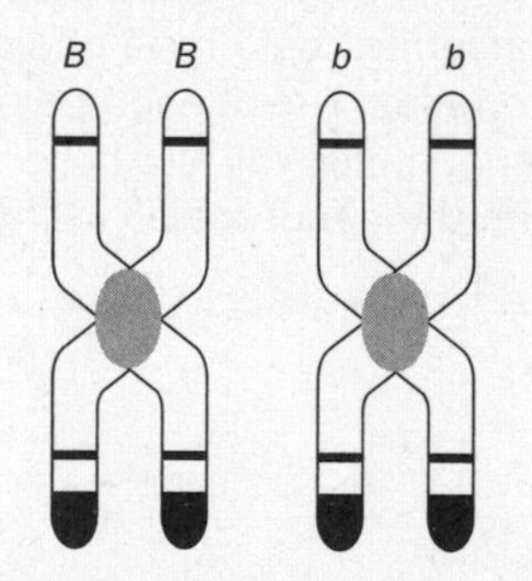

The previous figure shows chromosomes after DNA replication. The DNA strands have been doubled so that there is an exact duplicate of each original chromosome linked by a centromere to sister chromatids. There are now two copies of the *B* allele and two copies of the *b* allele.

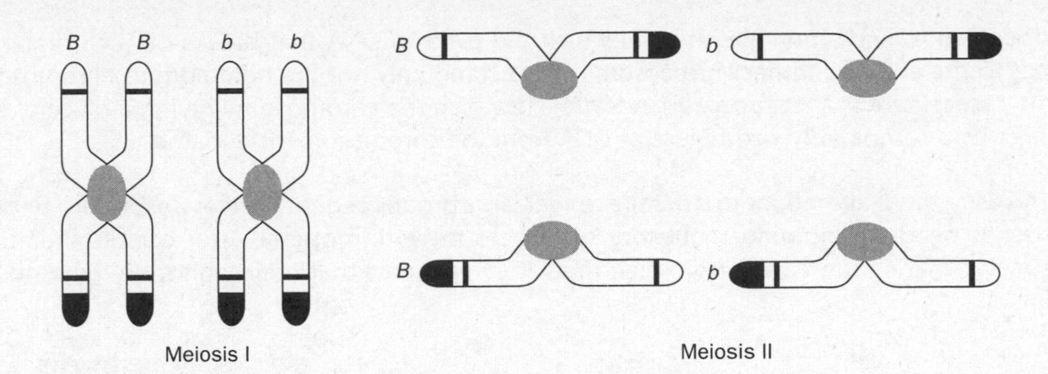

After the cell undergoes complete meiosis, the original cell with 2*N* chromosomes has divided into four daughter cells, each with 1*N* chromosomes. The daughter cells are four gametes, two with the *B* allele and two with the *b* allele. Because the original animal cell had both a *B* allele and a *b* allele, it can produce gametes with alleles for either blue or green eye color. This occurs in all individuals of the same species. An individual with the recessive green eye color (its genotype is *bb*) can only produce gametes with *b* alleles. When an organism has two of the same alleles (e.g., *BB* or *bb*), it is called **homozygous** for that gene. If both alleles are the dominant allele, the genotype is called **homozygous dominant**. If both alleles are recessive, the genotype is called **homozygous recessive**. If the organism has different alleles (e.g., *Bb*), it is called **heterozygous** for the gene.

Because the green eye color allele is recessive, it can be discerned that an individual in the population with green eyes has the genotype *bb*. An individual with blue eyes can have either a *BB* genotype or a *Bb* genotype. The genotype of a blue-eyed individual can be determined by mating the blue-eyed individual with a green-eyed individual. This is called a **test cross**. If all of the offspring have blue eyes, then the blue-eyed individual was homozygous dominant. The genotypes of all of the offspring from a mating between a *BB* genotype and a *bb* genotype can only be *Bb*.

If any of the offspring have green eyes, the blue-eyed adult must have been heterozygous. The genotypes of the offspring from a mating between a *Bb* genotype and a *bb* genotype are either *Bb* or *bb*. Mating between two heterozygous adults can produce offspring with three different genotypes: *BB*, *Bb*, and *bb*. In this case, two blue-eyed adults can produce a green-eyed offspring.

Chromosome Alterations

Several types of chromosomal breakage can occur in the course of DNA replication or at other points in the cell cycle. Some of these alterations can cause drastic changes in the ability of a cell to produce certain proteins. Many birth defects can be traced to defects in chromosome structure passed down through sperm or egg cells, in which one of the following alterations has taken place:

Deletion is when a chromosomal fragment, either from the end of a chromosome or from somewhere in the middle, is lost as the chromosome replicates.

If the fragment that detaches from a chromosome during a deletion event reattaches itself to the homologous chromosome, that chromosome will then have two sets of identical genes in a particular region (a *duplication*), while the original chromosome where the deletion occurred will be shorter than normal.

There is also the possibility that the deleted fragment could attach back into the chromosome from which it came, but in a reverse direction. This is known as *inversion*.

Translocation is a common alteration in which the piece of DNA that breaks off a chromosome attaches to the end of another chromosome, most commonly not the homologous chromosome. In some cases, there is a *reciprocal translocation*, in which the chromosome giving a piece of DNA also receives a comparably-sized piece of DNA from the chromosome it donated to.

In many cases, these alterations in structure render large groups of genes useless, especially because most genes need neighboring regulatory sequences to work properly. After certain alterations, genes may be far enough away from their regulatory elements that they cannot be transcribed.

MENDELIAN INHERITANCE

HIGH YIELD

18.3 Explain and predict results of Mendelian inheritance

After learning the basics of genetics, it is important to review some specific information about inheritance that might appear on the exam. You need to be familiar with mathematical principles of simple probability. Relax; it's not as hard as it sounds. Even if there are more than two alleles for a gene, an individual can only have two alleles on each pair of chromosomes. Ending up with a combination of alleles is about as simple as tossing a coin.

An organism that is homozygous for a particular gene (*AA*) produces four gametes each with an *A* allele. Each of the gametes is one out of a possible four gametes, but because all of the gametes have an *A*, four out of four of the gametes have an *A*. Four out of four is a probability of one because $\frac{4}{4} = 1$. You can look at the question more intuitively by saying that because there are only *A* alleles in the parent cell, all of the daughter cells (100%, or $\frac{1}{1}$) will have the *A* allele. If an organism is heterozygous (*Aa*), it produces two gametes with an *A* allele (2 out of 4) and two gametes with an *a* allele (2 out of 4). The probability of either the *A* or *a* allele in the gametes is $\frac{2}{4} = \frac{1}{2}$. Or again, intuitively, if the parent cell has one *A* and one *a* allele, it will produce gametes that are half *A* (50%, or $\frac{1}{2}$) and half *a* (50%, or $\frac{1}{2}$).

Now consider two genes and an individual that is homozygous for both (*AABB*). Although there are two genes to consider, the organism still produces four gametes from a single parent cell. Each gamete gets an *A* allele and a *B* allele. What is the probability that a gamete will have both an *A* and *B* allele? Four out of four will have an *A* and four out of four will have a *B* so all (four out of four) will have *AB* ($\frac{4}{4} \times \frac{4}{4} = \frac{16}{16} = 1$).

If the individual is heterozygous for one of the genes (*AABb*), the probability of the *A* allele stays the same, but now only half of the gametes get a *B* allele ($\frac{2}{4} = \frac{1}{2}$) and the other two get a *b* allele. What is the probability of producing a gamete with both an *A* and *B* allele from this individual? Four out of four will have an *A* and two out of four will have a *B*, so only half (two out of four) will have *AB* ($\frac{4}{4} \times \frac{2}{4} = \frac{8}{16} = \frac{1}{2}$).

If the individual is heterozygous for both genes (*AaBb*), the probability of each allele in the gametes (*A*, *a*, *B*, or *b*) is $\frac{1}{2}$, so the probability of *AB* is $\frac{1}{2} \times \frac{1}{2} = \frac{1}{4}$. The probability of *Ab* would also be $\frac{1}{4}$, as would be the probability of *aB* and the probability of *ab*.

AP Expert Note

The Two Laws of Mendelian Inheritance

Law of Segregation—This describes the separation of alleles in the parent genotype during the process of gametogenesis. There can be a maximum of only two different alleles in a single diploid parent; half the gametes get one allele and the other half get the other allele.

Law of Independent Assortment—This suggests that different genes sort into different gametes, independently of each other. For example, the sorting of alleles for eye color is not affected by the sorting of alleles for hair color. These laws explain the 3 : 1 and 9 : 3 : 3 : 1 ratios of phenotypes observed in monohybrid and dihybrid crosses.

Once you can figure out what gametes will be produced from which individuals, you can figure out the genotypes of offspring from a **Punnett square**. The following is a simple cross between two heterozygous individuals for one gene. A mating of this kind is called a **monohybrid cross.**

	A	*a*
A	*AA*	*Aa*
a	*aA*	*aa*

Monohybrid Cross

Typically, the gametes of the sperm are recorded along the left side of the Punnett square, and those of the egg are recorded across the top. The allele from each sperm is paired with the allele from each egg where the columns and rows meet, showing the genotypes of the offspring. The different indications of *Aa* and *aA* are used only to demonstrate the source of the alleles; they are the same genotype.

This cross results in three different genotypes (*AA*, *Aa*, and *aa*), but only two phenotypes because the dominant trait is expressed in both the *AA* and *Aa* individuals. The dominant trait shows up in the offspring in a ratio of 3 to 1.

You can put together a **dihybrid cross** (tracking two genes—*A* and *B*—rather than one) just as easily as a monohybrid cross. Just put the gametes of the male in the left column and the gametes of the female across the top.

	AB	*Ab*	*aB*	*ab*
AB	*AABB*	*AABb*	*AaBB*	*AaBb*
Ab	*AAbB*	*AAbb*	*AabB*	*Aabb*
aB	*aABB*	*aABb*	*aaBB*	*aaBb*
ab	*aABb*	*aAbb*	*aabB*	*aabb*

Dihybrid Cross

The dihybrid cross produces the famous phenotypic ratio of 9 : 3 : 3 : 1. If you consider a large number of offspring, there will be on average 9 out of every 16 that express the dominant phenotypes for both genes, 3 out of every 16 that express the dominant phenotype of one gene, 3 out of every 16 that express the dominant phenotype of the other gene, and only 1 out of every 16 that expresses the recessive phenotypes for both genes.

✔ **AP Expert Note**

Notice patterns in the ratios that occur with specific types of crosses. For example, crossing a heterozygous dominant (*Rr*) with a homozygous recessive (*rr*) always gives a 1 : 1 ratio of dominant to recessive phenotypes (half dominant phenotype; half recessive phenotype). This can speed up the process of figuring out phenotypic ratios and probabilities.

These are the basics of inheritance. However, genes do not operate in isolation from one another; this makes genetics more complex than Mendel's experiments and the Punnett square might suggest.

Pedigrees

Pedigrees are family trees that enable us to study the inheritance of a particular trait across many related generations. Males are typically designated as squares on the pedigree, while females are designated as circles. Those who phenotypically show a trait are shaded, while those who carry a trait but do not show it are either half-shaded or given a dot inside their circle or square.

In humans, genetic traits can be classified by whether they are transmitted on autosomal chromosomes (numbers 1–22) or on **sex chromosomes** (almost always the X, because the Y chromosome carries a limited number of genes). In addition, traits can be dominant or recessive. For the AP Biology exam, you should be familiar with the characteristics of different inheritance patterns so you can easily spot patterns. (Note that sex-linked traits, including both X-linked and Y-linked, count as non-Mendelian inheritance patterns.)

Autosomal Dominant

- Males and females are equally likely to have the trait.
- Traits do not skip generations.
- The trait is present if the corresponding gene is present.
- There is male-to-male and female-to-female transmission.

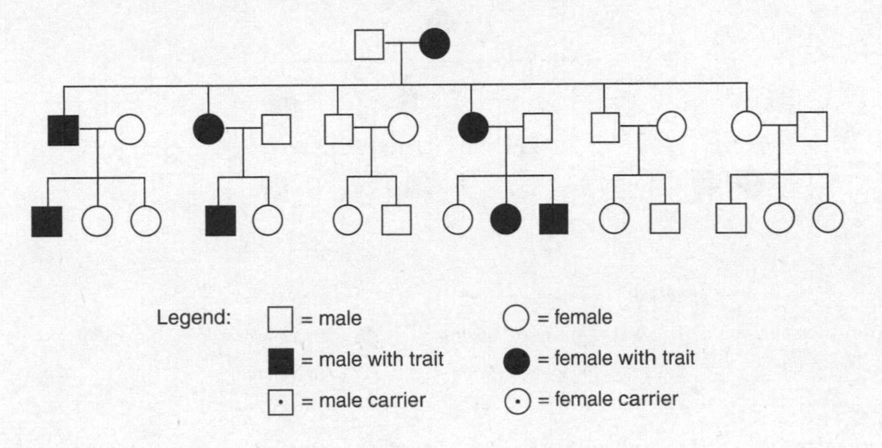

Autosomal Recessive

- Males and females are equally likely to have the trait.
- Traits often skip generations.
- Only homozygous individuals have the trait.
- Traits can appear in siblings without appearing in parents.
- If a parent has the trait, those offspring who do not have it are heterozygous carriers of the trait.

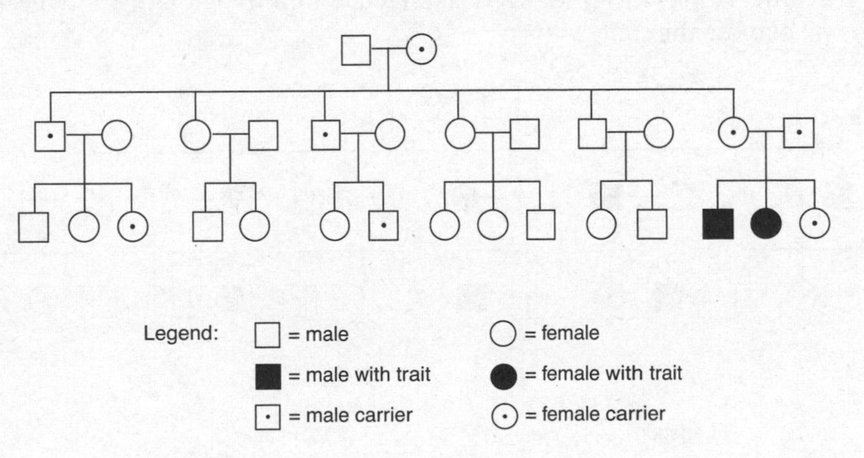

X-Linked Dominant

- All daughters of a male who has the trait will also have the trait.
- There is no male-to-male transmission.
- A female who has the trait may or may not pass on the affected X to her son or daughter (unless she has two affected Xs).

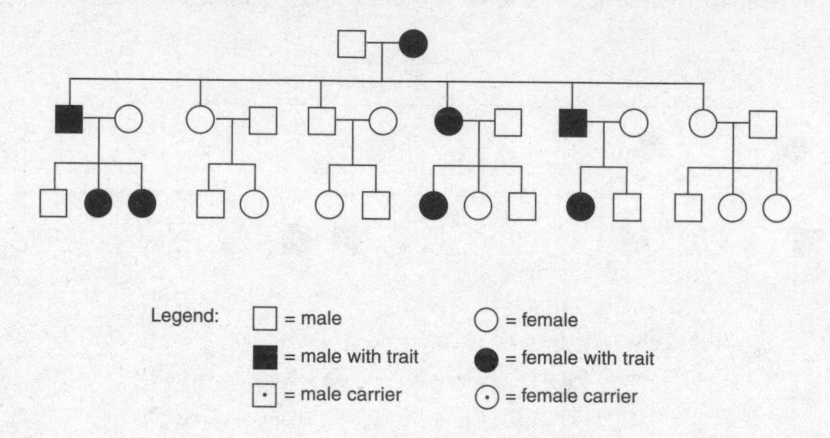

X-Linked Recessive

- Males are more likely to express the trait than females because males only need one copy for expression while females need two.
- Males cannot be carriers.
- All daughters of a male who has the trait will be carriers for the trait (unless they also receive an affected X from their mother as well, in which case they will express the trait).
- There is no male-to-male transmission.
- All sons of an affected female will have the trait, while all daughters of an affected female will be carriers (unless they received an affected X from their father as well, in which case they will express the trait).

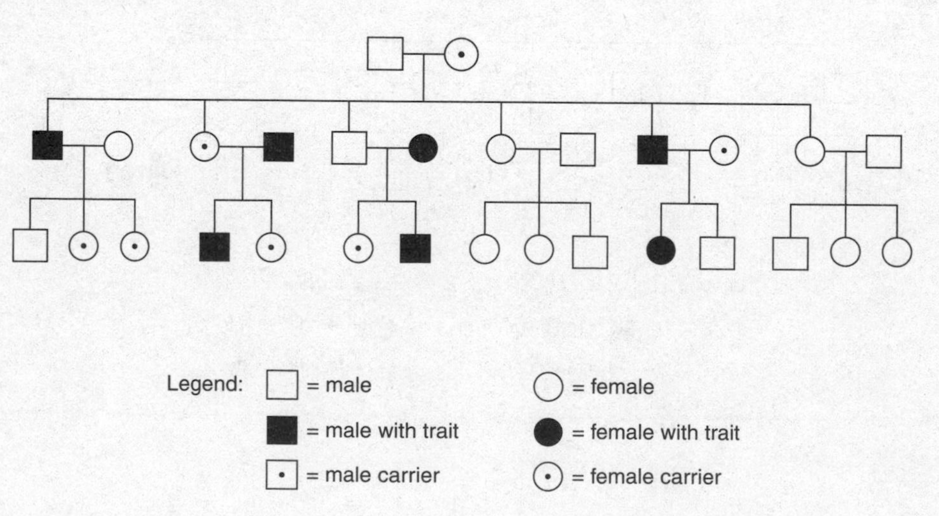

NON-MENDELIAN INHERITANCE

18.4 Explain why some traits are non-Mendelian

Mendelian genetics, however, does not apply to all genes. Inheritance patterns that do not follow **Mendelian laws** fall under non-Mendelian inheritance. For instance, genes may be linked to other genes on the same chromosome. Thus, the linked genes may not sort independently and may be inherited together. If linked genes are far apart on the chromosome, **crossing over** (exchange of genes between homologous chromosomes) during meiosis may separate them. The farther away genes are from each other, the more likely they are to be separated as a result of crossing over. The phenotypic ratios for linked genes, accordingly, will differ from those based on Mendelian inheritance.

Sex-linked traits also do not follow ratios predicted by Mendel's laws. Determined by genes on sex chromosomes (X and Y), sex-linked traits often appear more frequently in one sex (females are XX and males are XY). Since the Y chromosome in mammals and flies is small and carries few genes, X-linked recessive traits, such as hemophilia, are always expressed in males. The genotype of males for X-linked genes is called **hemizygous**, because they only have one copy of genes exclusive to the X chromosome. Sex-limited genes—though present in both sexes—are only expressed in one sex, so the same genotype can result in the expression of different phenotypes. Milk production, for example, is expressed only in female mammals, whereas pattern baldness is expressed exclusively in human males.

Extranuclear inheritance is another example of non-Mendelian inheritance since it describes the inheritance of genes outside the nucleus. Extranuclear inheritance occurs typically in mitochondria and chloroplasts, which contain genes that replicate independently of cell division. In vegetative segregation, the parent cell randomly distributes genes from mitochondria and chloroplasts to daughter cells, whereas in single-parent inheritance, offspring only receive genes from one parent. In humans, mitochondrial DNA is inherited entirely from the mother (maternal inheritance).

✔ AP Expert Note

Phenotype and Genotype: Other Factors to Consider

Many different factors beyond those encompassed in Mendel's laws can affect phenotype and genotype in offspring. You should be familiar with the following:

- **Incomplete dominance** is a form of inheritance in which *both* heterozygous alleles are expressed. This means that the offspring will display a combined phenotype that is distinct from both parent organisms. For example, a plant with red flowers and a plant with white flowers might produce offspring that have pink flowers.
- **Codominance** is an inheritance pattern that occurs when neither of the alleles is completely recessive or dominant. In the heterozygote, both alleles are expressed equally in the phenotype. For instance, the offspring of a plant with red flowers and a plant with white flowers might express both red and white petals. Another example of codominance is the ABO blood groups of humans. The *A* and *B* alleles are codominant to each other (*O* is recessive), so a person with AB blood expresses both A and B antigens on red blood cells.
- **Multiple allele inheritance** involves more than two alleles coding for a certain trait, which results in more than two phenotypes. Human blood type (ABO) is an example of multiple allele inheritance.
- *Polygenic inheritance* is a type of inheritance in which several interacting genes control a single trait. Many traits result from the additive influences of multiple genes; skin color is one common example of a polygenic trait.
- *Genetic recombination* is a molecular process by which an organism's genes are rearranged in its offspring. Through this process, two alleles can be separated and replaced by different alleles, thereby changing the genetic makeup but preserving the structure of the gene. Chromosomal crossing over is an example of a mechanism by which this process takes place.
- *Gene transfer* can be vertical or horizontal. Vertical gene transfer occurs when an organism receives genetic material (i.e., DNA) from a parent organism or from a predecessor species. Horizontal gene transfer occurs when an organism transfers genetic material to cells that are not its offspring. Examples of horizontal gene transfer in bacteria were discussed in chapter 6.

RAPID REVIEW

If you take away only 5 things from this chapter:

1. Chromosomes are the basic units of inheritance. How they reorganize and combine directly influences the genetic material present in an offspring.
2. Different versions of a gene that code for the same trait are called alleles. In classical (Mendelian) genetics, an individual receives one allele from each parent. Individuals with matching alleles are homozygous for that trait while those with different alleles are heterozygous. Usually, one version of the allele is dominant (e.g., brown eye color) and the other is recessive (e.g., blue eye color). Heterozygotes are "ruled" by the dominant allele.
3. Geneticists perform test crosses to determine the genetic makeup (genotype) of organisms displaying the dominant phenotype. A Punnett square is used to illustrate a test cross. Mendel's Law of Segregation states that an individual's alleles separate during meiosis, and either may be passed on to the offspring. Mendel's Law of Independent Assortment states that inheritance of a particular allele for one trait does not affect inheritance of other traits.
4. Sex-linked genes in humans occur on the X or Y chromosome. Males who inherit recessive X-linked genes from their mothers always express the trait in question, since men have only one X chromosome.
5. Pedigrees are family trees that enable us to study the inheritance of a particular trait across many related generations. They can also help us determine the type of transmission through generations: autosomal recessive/dominant, X-linked recessive/dominant, and so on.

TEST WHAT YOU LEARNED

1. In order to develop better melons, researchers want to use a mutagen, ethyl methanesulfonate (EMS), to induce mutations in a melon crop. The numbers and rates of mutations caused by EMS are shown in Table 1 and the rates of different mutation types are shown in Table 2.

Table 1

EMS Dose	M1 Plants	Induced Mutations	Mutation Frequency
1%	617	19	1/588 kb
2%	1473	67	1/356 kb
3%	40	3	1/146 kb

Table 2

Missense	Nonsense	Splicing	Silent
65.1%	2.4%	1.2%	31.3%

Both tables adapted from Fatima Dahmani-Mardas et al., "Engineering Melon Plants with Improved Fruit Shelf Life Using the TILLING Approach," *PLoS ONE* 5, no. 12 (December 2010): e15776.

Based on the information in the tables, how could the researchers best design their mutation protocol?

(A) Researchers should use the highest possible EMS dose to reduce the mutation rate to a manageable amount and then begin by examining splicing mutations to determine whether any produce new enzymes that improve melon quality.

(B) Researchers should use the highest possible EMS dose and then begin by examining the silent mutations to determine whether any are associated with improved melon quality.

(C) Researchers should use a relatively low EMS dose and then begin by examining the missense mutations to determine whether any are associated with improved melon quality.

(D) Researchers should use a relatively low EMS dose and then begin by examining the nonsense mutations to determine whether any produce new enzymes that improve melon quality.

2. Human ABO blood types are inherited through a codominant pattern of inheritance. The possible blood types include A, B, AB, and O, representing different antigens on the red blood cell membrane. Individuals with type A blood have an A antigen and individuals with type B blood have a B antigen. Individuals with type O blood do not have either of these antigens on the surface of their red blood cells. What most likely happens in individuals with type AB blood?

   (A) Individuals with AB blood have a unique antigen that combines features of the A and B antigens within a single molecule, resulting in AB antigens that are present on the surface of their blood cells.

   (B) Individuals with AB blood have some cells that have A antigens on them and other cells that have B antigens on them.

   (C) Individuals with the AB blood type have a gene that codes for the A antigen and another gene that codes for the B antigen, but it is unpredictable which antigens will appear on the surface of their blood cells.

   (D) Individuals with AB blood have one gene that codes for the A antigen and another gene that codes for the B antigen, so both antigens are present on their blood cells.

3. Eukaryotic chromosomes can be tightly packed as heterochromatin or can be more loosely packed as euchromatin. What is the best prediction regarding the expression of genes in heterochromatin and euchromatin?

   (A) RNA polymerase can access the promoters in euchromatin more easily than it can access the promoters in heterochromatin, so euchromatin is transcribed more than heterochromatin.

   (B) RNA polymerase can access the promoters in heterochromatin more easily than it can access the promoters in euchromatin, so heterochromatin is transcribed more than euchromatin.

   (C) DNA polymerase can access promoters in euchromatin more easily than it can access the promoters in heterochromatin, so euchromatin is transcribed more than heterochromatin.

   (D) DNA polymerase can access promoters in heterochromatin more easily than it can access the promoters in euchromatin, so heterochromatin is transcribed more than euchromatin.

4. Coat color in cats is controlled by multiple genes. For example, cats can have a dominant autosomal allele called "agouti" for a particular coat pattern or can have a recessive allele for a solid coat color. Additionally, they can have a standard, full-color coat or a dilute, pale coat depending on whether they have the recessive dilute allele. A test cross was performed between an agouti, full-color male and a female that was homozygous recessive for both traits. Which of the following would be true regarding the results of the cross?

   (A) If all of the offspring had agouti, full-color coats, then the male is probably homozygous dominant for both traits.

   (B) If all of the offspring had agouti, full-color coats, then the male is probably heterozygous for both traits.

   (C) If all of the offspring had solid, dilute coats, then the male is probably homozygous dominant for both traits.

   (D) If half of the offspring had agouti, full-color coats and the other half had solid, dilute coats, then the male is probably homozygous dominant for both traits.

The answer key to this quiz is located on the next page.

Transformations

Answer Key

Test What You Already Know

1. **B** **Learning Objective:** 18.1
2. **D** **Learning Objective:** 18.2
3. **C** **Learning Objective:** 18.3
4. **A** **Learning Objective:** 18.4

Test What You Learned

1. **C** **Learning Objective:** 18.2
2. **D** **Learning Objective:** 18.4
3. **A** **Learning Objective:** 18.1
4. **A** **Learning Objective:** 18.3

Detailed solutions can be found in the Answers and Explanations section at the back of this book.

REFLECTION

Test What You Already Know score: __________

Test What You Learned score: __________

Use this section to evaluate your progress. After working through the pre-quiz, check off the boxes in the "Pre" column to indicate which Learning Objectives you feel confident about. Then, after completing the chapter, including the post-quiz, do the same to the boxes in the "Post" column. Keep working on unchecked Objectives until you're confident about them all!

Pre	Post	
❑	❑	**18.1** Describe chromosome structure and function
❑	❑	**18.2** Compare chromosomal mutations and abnormalities
❑	❑	**18.3** Explain and predict results of Mendelian inheritance
❑	❑	**18.4** Explain why some traits are non-Mendelian

FOR MORE PRACTICE

Complete more practice online at kaptest.com. Haven't registered your book yet? Go to kaptest.com/booksonline to begin.

CHAPTER 19

Evolution

LEARNING OBJECTIVES

In this chapter, you will review how to:

- **19.1** Differentiate between types of selection affecting evolution
- **19.2** Describe the support for the common ancestry theory
- **19.3** Describe genetic drift and gene flow
- **19.4** Explain speciation and extinction
- **19.5** Explain how molecular biology and biochemistry support evolution
- **19.6** Explain how common structures and features support evolution
- **19.7** Describe how populations continue to evolve
- **19.8** Investigate the effects of artificial selection on evolution
- **19.9** Investigate allele distribution mathematical modeling

TEST WHAT YOU ALREADY KNOW

1. A population of rabbits settled in a desert environment where the soil is covered with small boulders and shrubs. Desert foxes are the major predators that the rabbits encounter. Small rabbits hide under rocks and shrubs to avoid detection. Desert foxes usually avoid large rabbits that can inflict serious wounds by kicking and biting. Ecologists survey the population size of the rabbits. Which of the following graphs is most likely to represent the distribution of body sizes in the rabbit population after several generations?

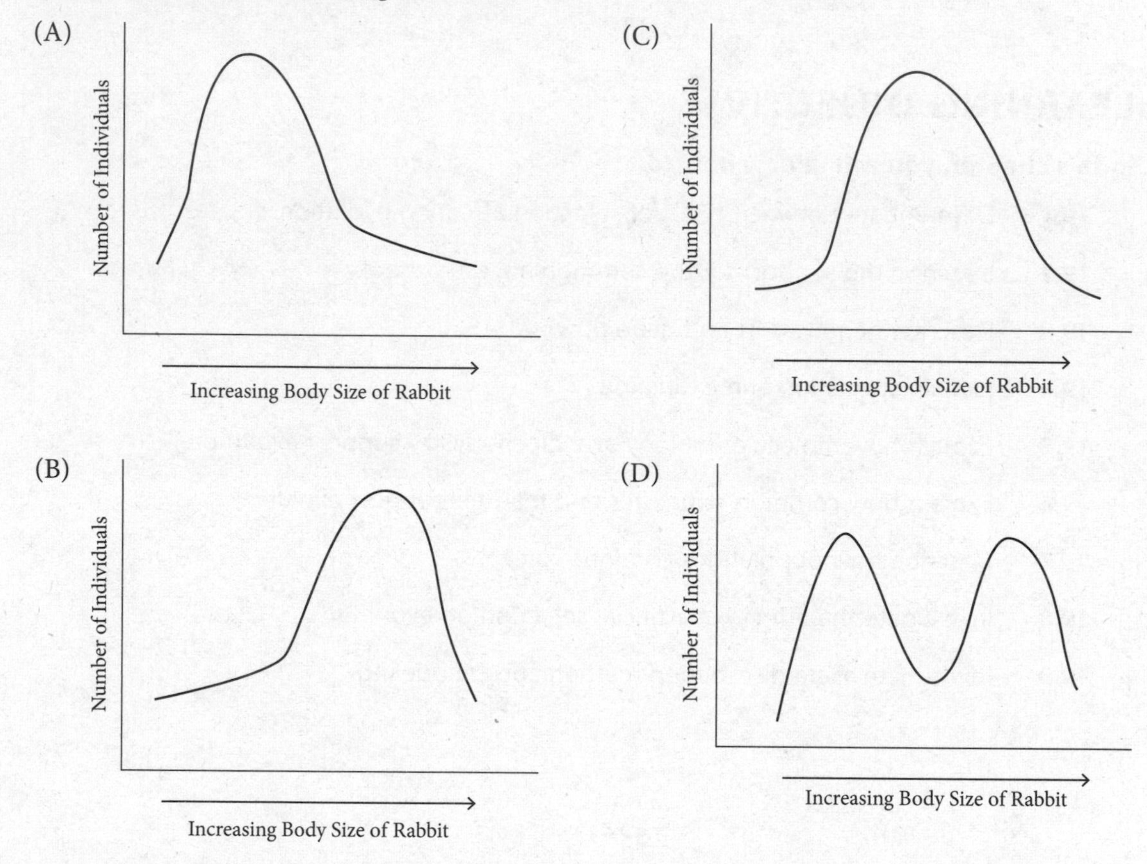

2. Research has shown that the glycolysis pathway exists in almost all organisms. Glycolysis exists among Bacteria, Archaea, and Eukarya and is catalyzed by homologous enzymes in all three domains. Which of the following hypotheses is supported by this finding?
   (A) Glycolysis is an inefficient process to produce ATP in all extant organisms.
   (B) Glycolysis is a primitive process that appeared in a common ancestor of all present-day organisms.
   (C) Glycolysis is present in many different organisms because it has appeared several times during evolution.
   (D) Glycolysis appeared only after organisms developed mitochondria to support metabolism.

3. A small flock of finches was blown over from the mainland onto a neighboring island. On the mainland, a small beak is the most abundant trait, but a large proportion of birds that landed on the island happened to have large beaks. On the island, many types of seeds are available, giving birds of all beak sizes the same opportunity for feeding. Many years later, ecologists survey the distribution of the beak sizes in the island finches. Which of the following statements represents the most likely finding of the survey?
   (A) The size of beaks is equally distributed between large and small because the population experiences stabilizing selection.
   (B) Most birds have a large beak as a result of genetic drift because the trait was overrepresented in the small founding population.
   (C) There are more birds with small beaks because it reflects the makeup of the mainland bird population.
   (D) Most birds have a large beak because it gives a competitive advantage over other birds.

4. The soil in areas surrounding mines is often polluted with toxic heavy metals. These toxins prevent most grasses from growing. Buffalo grass is an unusual type of grass that can tolerate heavy metals in the soil. Populations in polluted areas have evolved a distinct genotype for tolerating the metal toxins. This resistant buffalo grass grows in close proximity to nonresistant buffalo grass that can only propagate in unpolluted areas, but resistant buffalo grass does not breed with nonresistant buffalo grasses.

   Based on these findings, which best describes the resistant buffalo grass populations?
   (A) Resistant buffalo grass has not undergone speciation because it is not a distinct geographical population.
   (B) Resistant buffalo grass is not a new species of grass because it has the morphological appearance of neighboring buffalo grasses.
   (C) Resistant buffalo grass has undergone speciation because it is not interbreeding freely with neighboring buffalo grasses.
   (D) Resistant buffalo grass is a new species of grass because it is planted next to mines to detoxify the ground.

5. Fungi, the kingdom that includes mold, yeast, and mushrooms, was formerly considered a part of the kingdom Plantae, which includes land plants. Published studies drew a comparison of base substitutions in actin, a major cytoskeleton protein. A summary of the data is shown in the table below.

Percent of Nucleotide Substitutions in Actin Genes

	Yeast	**Soybean**	**Maize**	**Fruit Fly**	**Sea Urchin**	**Rat Skeletal Muscle**	**Human Cardiac Muscle**
Yeast		17.7	18.7	10.9	11.6	11.6	11.8
Soybean	17.7		10.3	14.5	13.3	13.6	13.7
Maize	18.7	10.3		14.8	12.2	13.9	14.1
Fruit Fly	10.9	14.5	14.8		5.0	6.0	5.8
Sea Urchin	11.6	13.3	12.2	5.0		5.5	5.5
Rat Skeletal Muscle	11.6	13.6	13.9	6.0	5.5		0.57
Human Cardiac Muscle	11.8	13.7	14.1	5.8	5.5	0.57	

Adapted from Robin C. Hightower and Richard B. Meagher, "The Molecular Evolution of Actin," *Genetics* 114, no. 1 (September 1986): 315–332.

Based on the data, was the reclassification of Fungi as a distinct kingdom appropriate?

(A) The classification should have been revised because yeast are more closely related to maize than to humans.

(B) The classification should not have been revised because yeast are more closely related to soybeans than to animals.

(C) The classification should have been revised because fungi are more closely related to animals than to plants.

(D) The classification should not have been revised because fungi occupy a distinct evolutionary branch.

6. Snakes have no visible limbs and move by slithering on their ventral scales. The skeletons of boas and pythons show the remnants of small bones which are characterized as embryonic hind limb buds. Some scientists call them vestigial structures that indicate that snakes descended from ancient lizards that lost their limbs during the course of evolution. Which of the following findings would best support the hypothesis that snakes are descendants of lizard ancestors rather than a new class of animals?

(A) Snakes with fully developed hind limbs were discovered in the fossil record.

(B) Snakes can develop hind limbs if they exercise by climbing trees.

(C) A snake is found that is born with hind legs that they lose soon after birth.

(D) Lizards and snakes have similar scales that demonstrate their shared lineage.

7. The common cuckoo lays its eggs in the nests of other birds so that other species will take on the costs of raising the baby cuckoos. The ashy-throated parrotbill is one species of bird that is parasitized by the cuckoo in this way. Based on this information, what selection pressures are most likely faced by cuckoo populations that parasitize parrotbills?

   (A) There is strong selection pressure on the cuckoos to produce distinctive eggs that can be quickly recognized as unique.

   (B) There is strong selection pressure on the cuckoos to produce large eggs so that their offspring can outcompete parrotbill offspring.

   (C) There is strong selection pressure on the cuckoos to produce eggs with variable coloration to increase the likelihood that some will resemble parrotbill eggs.

   (D) There is strong selection pressure on the cuckoos to produce eggs that are as similar as possible to those of the parrotbills.

8. Researchers have developed two lines of rats, named low intrinsic aerobic running capacity (LCR) and high intrinsic aerobic running capacity (HCR). After 19 generations of artificial selection, these rats differ greatly in their athletic ability. HCR rats can run much farther than LCR rats before reaching exhaustion. Although the researchers selected for running capacity, the rats differ in other ways. The table below shows some characteristics of the two lines. The figure below shows the time, in minutes, to heart failure in hypoxic conditions.

Trait	HCR	LCR
Heart Weight (g)	1.27 ± 0.0	1.46 ± 0.1
Body Weight (g)	416.8 ± 14.5	570.3 ± 23.7
HW/BW ratio ($\times 10^{-3}$)	3.1 ± 0.1	2.6 ± 0.2

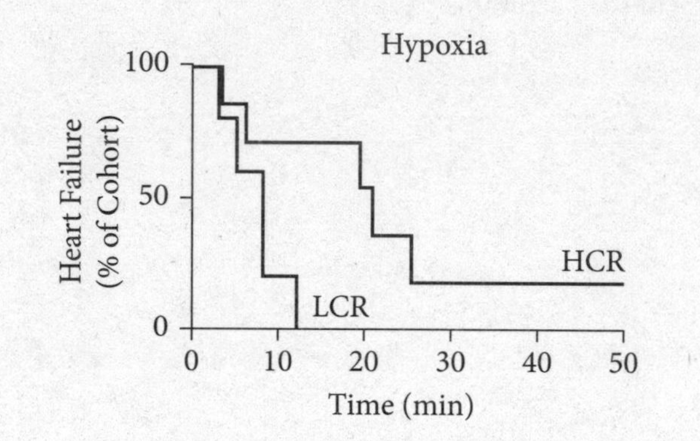

Table and figure adapted from Nathan J. Palpant et al., "Artificial Selection for Whole Animal Low Intrinsic Aerobic Capacity Co-Segregates with Hypoxia-Induced Cardiac Pump Failure," *PLoS ONE* 4, no. 7 (July 2009): e6117.

If the researchers selected for running capacity, why do the rats differ in these other ways?

(A) Artificial selection for a particular trait also affects related traits, such as cardiac function and muscle in these rats.

(B) The researchers deliberately selected multiple traits simultaneously when the lines were developed.

(C) All of these characteristics are associated with the same phenotype and therefore developed in association with limited running ability.

(D) The greater heart weight of LCR rats suggests that the researchers were selecting for overall reduced fitness.

9. Researchers are studying a population that has historically appeared to be in Hardy-Weinberg equilibrium based on the calculation of its allelic and genotypic frequencies. The population is unusual in that it is very large, does not experience immigration or emigration, does not show sexual selection, and is not exposed to mutagens. After one generation, the percentage of individuals with a particular dominant trait increases. After two generations, the percentage of individuals with the dominant trait slightly increases again. This pattern continues in the third generation.

   What is the most likely explanation of these findings?

   (A) The small changes are simply a result of random variation.

   (B) The population is experiencing a new selective pressure for the dominant trait.

   (C) The population is undergoing an increase in its rate of mutation.

   (D) The small changes are merely a result of genetic drift.

Answers to this quiz can be found at the end of this chapter.

TYPES OF EVOLUTIONARY CHANGE

19.1 Differentiate between types of selection affecting evolution

19.2 Describe the support for the common ancestry theory

19.3 Describe genetic drift and gene flow

19.4 Explain speciation and extinction

There is no biological concept more controversial and more misunderstood than biological evolution. The concept can be simply defined with the phrase "descent with modification." This means that over time, populations of organisms exhibit changes in characteristics that are passed on through inheritable (i.e., genetic) means. The important distinction between this simple definition and what is commonly accepted as biological evolution is mechanism. Mechanisms will be covered on the exam. You should also clarify the semantic distinction between evolution and biological evolution. Personalities evolve. Societies evolve. But only populations of organisms undergo biological evolution by means of a modification of inheritable characteristics. In this chapter, the term *evolution* will always refer to biological evolution.

✔ AP Expert Note

For more on evolution, see the discussions of life origin hypotheses and Darwin's theory of natural selection in chapter 4.

Evolutionary mechanisms

In the simplest sense of the term, evolution occurs whenever there is a change in gene frequencies within a population. Imagine a species of fish in which 50 percent of the population have green fins and 50 percent have blue fins. If, five years later, 51 percent of the population have green fins and 49 percent have blue fins, then that species has evolved. Evolution in this sense has two basic kinds of mechanisms: sources of variation (which increase genetic diversity) and selective pressures (which decrease genetic diversity).

Sources of variation have already been discussed in multiple places throughout this book. At the most foundational level, all heritable variations ultimately arise from mutations (discussed in chapter 16), direct changes to the genetic code, which impact the proteins produced by an organism's cells and, consequently, the phenotypes of that organism. Genetic variation can also arise from other sources, such as the shuffling of genes that occurs when sexual reproduction leads to a new organism with a novel combination of parental genes. For instance, immigration of new organisms into a population, followed by interbreeding, can lead to an influx of new genes in a population.

If sources of variation acted alone, then species would just become more and more diverse over time, but selective pressures serve to weed out some phenotypes (and their associated genotypes), acting as a force that decreases variation within a population. Selection effectively acts as a filter on the gene pool: organisms with phenotypes that allow them to survive and produce fertile offspring within their specific environmental niche pass through the filter successfully, while unfit organisms that are not well adapted to their environment are filtered out. This process of filtering is generally known as **natural selection**, a theory established by Darwin and discussed in chapter 4.

In addition to natural selection, Darwin also put forth a theory of **sexual selection** (in his 1871 book, *The Descent of Man, and Selection in Relation to Sex*). Sexual selection has specifically to do with selective pressures that impact the capacity for successful reproduction. Many male birds have bright mating plumage that makes them conspicuous in a forest and more prone to attack from predators. Male ungulates often have huge antlers that grow every year and make it more difficult to travel through dense woods. The bright feathers on the male bird and the large antlers on the male deer are characteristics that are **selective disadvantages** in the animals' natural environment but are **selective advantages** when it comes to courtship and mating. This is because females of those species have a strong preference for excessive adornment, refusing to mate with males who lack such augmentation. Because these female preferences are also often heritable, sexual selection can lead to rapid evolution of a population.

✔ **AP Expert Note**

The balance between survival and the ability to mate dictates how species evolve.

One other type of selection is *artificial selection*, a process that plays an active role in the domestication of plants and animals. Similar to sexual selection (in which mate preference shapes evolution), choice also plays a role in artificial selection but, in this case, it is the decisions of members of another species (usually human beings), which are able to exert control over which pairs of organisms mate. The wide variety of breeds of dogs, which share a common ancestor with wolves, is in large part due to extensive artificial selection.

These joint processes of variation and selection have spawned the tremendous amounts of diversity found in the natural world today. Indeed, biologists generally believe that life originated only once on this planet and that all living organisms ultimately share a single common ancestor. This theory is supported by a number of common biochemical processes that are shared by all living things, such as genetic information encoded in DNA or RNA, the transcription of RNA from DNA, and the translation of mRNA codons into specific amino acids. What's more, all extant (surviving) organisms use ATP as an energy carrier. Accordingly, ATP is known as the energy currency of life.

Two homeostatic processes that illustrate common ancestry are excretion and osmoregulation. Earthworms, arthropods, and vertebrates all use the same principles of filtration and active transport to excrete waste products from the body, and fish, protists, and bacteria use analogous osmoregulatory mechanisms to control water and solute concentrations in order to maintain internal water balance. Similar morphological features (for example, skeletal components of vertebrates) can also be seen across living organisms. Homologous structures, however, may or may not be used for the same function. Vestigial structures like the wings of a flightless bird have lost their original function. Phylogenetic trees, which represent evolutionary relationships, also provide evidence for common descent and are discussed in chapter 20.

Genetic Drift and Gene Flow

HIGH YIELD

Biodiversity arises from **genetic drift**, a random sampling process that occurs in small populations. Two examples of genetic drift are the bottleneck effect and the founder effect. The bottleneck effect generally occurs when a catastrophe like a natural disaster only leaves behind a small group of individuals. Similarly, the founder effect occurs when a small group of individuals breaks off from the larger population and becomes isolated. In both cases, the allele frequencies in the small group differ from those of the original population. Over time, the changes in allele frequencies become prominent in the small population (but would have little effect in a large population). Genetic drift may lead to the formation of a new organism or species. However, since the new species is a result of random changes and not natural selection, it may not be well adapted to survive in its environment.

Unlike genetic drift, which takes place between two generations of one population and in only one species, **gene flow**, also known as gene migration, takes place between two populations of one species or between two species. Gene flow is the process of moving genes between populations via individuals entering or leaving the populations. Movement, for example from migration and pollination, may eliminate or introduce new alleles to the gene pool, causing significant changes in the offspring. On the other hand, geographical barriers such as oceans, mountain ranges, and deserts, restrict gene flow. Both genetic drift and gene flow can be analyzed using Hardy-Weinberg equilibrium, which is discussed later in this chapter.

Speciation and Extinction

HIGH YIELD

Over a dozen different concepts of speciation have been promoted in scientific literature, but the most prevalent by far is the **biological species concept (BSC)**. This concept states that a species is defined by a naturally interbreeding population of organisms that produces viable, fertile offspring. In other words, two species are distinct if they can't breed with each other or don't naturally breed with each other due to certain barriers. There are two kinds of barriers to interbreeding: prezygotic and postzygotic.

Prezygotic barriers to interbreeding include **isolation** of species due to ecological, temporal, behavioral, or mechanical factors, or physiological incompatibility of gametes. **Postzygotic barriers** include ultimate inviability or sterility of **hybrid** organisms from the interbreeding of two species. Hybrid organisms may not die off in one generation, but ultimately the offspring of the mating of two species will die without producing offspring of their own.

Geographic isolation is not the only factor that can cause speciation, so additional terms have been defined to describe how species evolve due to isolating mechanisms. **Allopatric speciation** is when one population is separated into two distinct populations by some **geographic barrier** such as the movement of a tectonic plate or the elevation of a mountain range. After the original population is no longer able to share its alleles, it evolves into distinct populations that have a high probability of acquiring distinctive traits. In contrast, **sympatric speciation** occurs when individuals within a population acquire distinctively different traits while in the same geographic area. Sympatric speciation requires some other form of reproductive isolation, such as those previously mentioned. **Parapatric speciation** is less definitive. This occurs when two populations are able to interbreed along a border, but the exchange of alleles is negligible compared to the amount of genetic exchange occurring within each population. A narrow zone of hybridization exists at the meeting of the two populations, but the two populations never coalesce into one.

A variety of factors such as species interactions, environmental changes, and population size can lead to **extinction**—the termination of an organism or species. For two species whose niches overlap, typically the species that is better adapted will drive the other species to extinction. Human activities (for example, hunting, habitat destruction, introduction of disease, and changing the climate) have become a predominant cause of extinction.

Species that are unable to adapt to a changing environment may die out and become extinct. In particular, species with specialized diets or habitat requirements (like the giant panda that feeds mainly on bamboo) are more vulnerable to environmental changes. Generalist species like raccoons, on the other hand, are more able to survive because they feed on a wide variety of food and live in various habitats. With respect to population size, species with long generation times that produce few offspring (like rhinoceroses) are more subject to extinction than species with short generation times that produce many offspring (like rodents). The ability to increase in population quickly allows species to recover from low populations caused by disturbances and diseases.

The following are examples of species that became extinct in the wild but are being bred in captivity and then released back into the wild. Their low genetic diversity, however, keeps them at risk for extinction and makes them endangered species. California condors, prior to humans hunting and poisoning them to near extinction, were a genetically diverse population. Today, the genetic diversity is extremely low because all existing California condors descended from just a handful of individuals. Likewise, black-footed ferrets, whose population once dropped to alarmingly low levels, have been bred from only a few individuals, making them especially vulnerable to threats. Tasmanian devils (alike on the brink of extinction due to low genetic diversity) are being bred on an island with disease-free animals to protect them from a contagious cancer. Due to habitat loss, prairie chickens live in small groups isolated from one another. The isolation of prairie chicken populations has in turn contributed to a loss in their genetic variance. On a related note, the Irish potato blight and southern corn rust (diseases caused by fungi) were the consequence of low genetic diversity.

Mass extinctions involve deadly events such as volcanic eruptions, asteroid collisions, or global warming/cooling that cause at least half of the species to die out in a relatively short amount of time. Following mass extinctions, new life-forms may emerge and evolve as new habitats open via **adaptive radiation**. Five major mass extinction events occurred on Earth in prehistoric times: Ordovician-Silurian Extinction (small marine organisms died out), Devonian Extinction (tropical marine species died out), Permian-Triassic Extinction (many vertebrates died out), Triassic-Jurassic Extinction (dinosaurs flourished after other species died out), and Cretaceous-Tertiary Extinction (dinosaurs died out). The extinction of the dinosaurs made room for other animals to diversify and evolve rapidly.

EVIDENCE FOR EVOLUTION

19.5 Explain how molecular biology and biochemistry support evolution

19.6 Explain how common structures and features support evolution

19.7 Describe how populations continue to evolve

Types of Evidence

Even though evolution is sometimes described as a "theory," it is more than a mere hypothesis. A wealth of evidence from a variety of fields indicates that evolution is a well-established scientific fact.

The Fossil Record

The geological layers in the Earth's crust stack on top of each other, with the oldest layers deeper and the youngest layers closer to the surface. Older geological layers hold more "primitive" fossils. Below is a table displaying different major biological events that have occurred through geologic time, the period in which these events occurred, and approximately how many millions of years ago they happened. You are not required to know the dates of these events, but be prepared to analyze given data and develop scientific inquiries to investigate certain claims.

The Fossil Record over Geologic Time

Period	Millions of Years Ago	Events
Precambrian	$> 3{,}500$	First prokaryotes
Precambrian	$> 1{,}000$	Earliest eukaryotes
Cambrian	540–490	Origin of all extant and some extinct animal phyla, including chordates
Ordovician	489–446	Continued evolution of ocean life
Silurian	445–415	First terrestrial organisms
Devonian	415–360	Diversification of bony fishes; first insects; first seed plants
Carboniferous	360–300	First gymnosperms
Permian	300–250	Diversification of reptiles
Triassic	250–200	First mammals and dinosaurs
Jurassic	200–145	Diversification of dinosaurs; first birds
Cretaceous	145–65	Origin and diversification of angiosperms; extinction of dinosaurs at end of period
Paleogene and Neogene	65–1.8	Diversification of all major living groups of birds and mammals, including hominids
Quaternary	1.8–present	Extinction of large land mammals; rise of humans

Biogeography

Organisms are more like other organisms in their geographic vicinity. Organisms in adjacent dissimilar environments are more similar than organisms in similar environments on opposite sides of the Earth. This suggests that organisms in adjacent dissimilar environments are descended from recent common ancestors, rather than evolving randomly and independently.

Comparative Anatomy

Organisms have very different structures that are composed of the same basic components. For example, the human arm has the same bones as the wing of a bat. These structures are called **homologous** structures because they are considered to have arisen from a common ancestor. **Analogous** structures are structures that may perform a similar function but have not arisen from the same ancestral condition.

✔ **AP Expert Note**

The wings of a bat and the wings of a butterfly are analogous structures.

Embryology or Ontogeny

Organisms that share a more recent common ancestor have similar modes of development. A classic example is that all vertebrate embryos have a stage of development in which they possess gills, whether they are aquatic or terrestrial. The presence of these more "primitive" characteristics in the embryos of "advanced" organisms suggests that these organisms share genetically controlled developmental physiologies that have been passed on from their common ancestors. The process through which an organism develops from an embryo to an adult is called **ontogeny.**

Taxonomy

Organisms are classified into smaller and smaller subgroups based on similar and dissimilar characteristics. This hierarchy is an implicit illustration of the tree of life, leading to common ancestry by linkage to a superseding group. For example, a plant in the family Euphorbiaceae is more closely related to other plants in Euphorbiaceae than it is to plants in the family Cactaceae. The housefly, *Musca domestica*, is more closely related to other flies in the genus *Musca* than it is to flies in the genus *Stomoxys*.

Phylogenetic trees and cladograms are illustrations that can be used to represent the relationships between similar and dissimilar organisms. They are constructed using morphological, molecular, or DNA evidence. For example, in the simplified cladogram that follows, you can see that Mollusca and Arthropoda are the most closely related phyla, as they shared the most recent common ancestor. However, the phyla Porifera, Cnidaria, Echinodermata, Arthropoda, and Mollusca all share a single common ancestor.

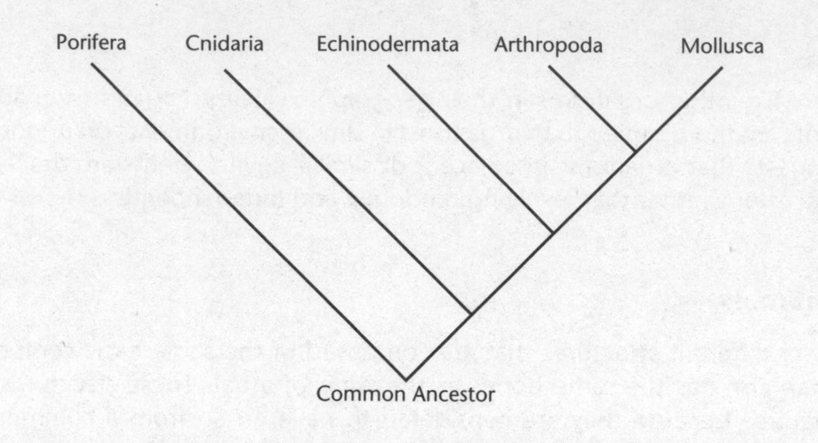

✔ AP Expert Note

For more on phylogenetic trees and cladograms, see chapter 20.

Molecular Biology

Siblings share more similar DNA with each other than they do with other members of the same species. Similar species have more similar DNA than do more distantly related species (in different genera, for example). Related genera share more similar DNA with one another than they do with genera in another family.

Rates of Evolution

Prior to Darwin, and even for scientists today, there is an underlying assumption that evolution takes a long time. The reasoning behind this theory is that because mutation is the ultimate source of variation and mutations that allow viable offspring are extremely rare, the probability of accumulating enough mutations to cause considerable change in organismal form requires a lot of time. Due to this assumption, prior to chemical dating techniques, geological layers were considered to be very old. Chemical dating has allowed modern scientists to assess the age of geological layers accurately.

Scientists have proposed two more hypotheses related to the rate of evolution. The first hypothesis is **punctuated equilibrium**. The hypothesis of punctuated equilibrium suggests that changes in organismal form did not take millions of years. Instead, very large changes in form happened relatively quickly (i.e., over thousands or tens of thousands of years) and were maintained thereafter over long periods of time. There is some evidence to suggest this hypothesis may be true, such as the Cambrian explosion in the fossil record, the discovery of cascading developmental genes, and the observation of large changes in phenotypic expression caused by single base-pair mutations. The second, the **molecular clock hypothesis**, is the notion that genetic mutations occur in a genome at a linear rate. Assuming the molecular clock hypothesis is true, one could extrapolate the age of divergence of two organisms by counting the number of genetic differences in their genomes. This latter hypothesis assumes that mutation rates are constant over time and between species; a likely incorrect pair of assumptions.

Populations and species continue to evolve today. Modern-day evolution can be seen in recent emergent diseases (those that appear for the first time or increase in incidence). Examples include severe acute respiratory syndrome (SARS), H1N1 influenza, and Ebola. Scientific evidence of chemical resistance also supports the claim that evolution continues to occur. Bedbugs today have evolved resistance to pesticides by developing thicker, waxier exoskeletons to protect them from toxins. Additionally, arising from a hybrid between the European house mouse and the Algerian mouse, which was introduced via human travel, the common house mouse can now survive rodenticides. Evolution in action can also be observed in artificial selection by humans, which has resulted in today's domestic animals, as well as crops with desired traits. Human populations who live in extremely high-altitude areas (for example, Tibet, the Andes, and Ethiopia) have evolved enhanced respiratory mechanisms that allow them to survive in low-oxygen environments.

AP BIOLOGY LAB 1: ARTIFICIAL SELECTION INVESTIGATION

19.8 Investigate the effects of artificial selection on evolution

In this inquiry investigation, you will explore real-time natural selection using Wisconsin Fast Plants (*Brassica*), just as Mendel did with pea plants. This is a long-term laboratory investigation in which you will select, plant, grow, and tend to a population of plants to observe how natural selection acts on phenotypic variations. This is a pretty simple investigation in terms of content and equipment. However, throughout the course of this seven-week investigation, you must keep detailed records in which you quantify variation, record images of that variation, and evaluate and explain your results. These skills will be helpful for you to cultivate for your responses on the free-response questions on the AP Biology exam.

Natural selection is the mechanism that describes the reproductive success or fitness of certain traits. Scientists and farmers use artificial selection to grow preferred agricultural crops. We also see natural selection occur inadvertently when diseases and pests grow resistance to the use of antibiotics or pesticides. In this lab, as a class and independently, you will select for a trait and investigate its reproductive success in two generations. Remember that directional selection tends to increase or decrease the trait in the next population. After growing your plants and recording your observations of the two different generations, you have many options for how to analyze your data. Deciding what to do with your data is the tricky part of this investigation, but these types of analyses will help you with the grid-in and free-response questions on your exam. Let's look at two ways we can analyze sets of data.

You could use different descriptive statistics such as mean, median, range, and standard deviation to describe your populations of study. Let's take a look at a couple of these descriptive statistics. The equation for solving for standard deviation S is:

$$S = \sqrt{\frac{\sum (x_i - \bar{x})^2}{n-1}}$$

where x_i is each data point, $\bar{x}$ is the mean of the data points, and n is equal to the size of the sample. The standard deviation calculates the difference between each of your data points and the mean value of your data, allowing you to see how much variability there is in the population.

For each set of data, you will construct a histogram. A histogram is a graphical representation of the distribution of the data. Using your histograms and other statistical analyses, such as a *t*-test or chi-square test, you can determine the validity of the differences in the populations. The formula for a chi-square test is:

$$\chi^2 = \sum \frac{(o-e)^2}{e}$$

where *o* is the number of individuals observed to have a specific phenotype and *e* is the number of individuals expected to have that particular phenotype.

All of this mathematical problem solving might seem intimidating, but it is essential for real-world biology practice and for the AP Biology exam. You can take a deep breath, though; all of these formulas will be provided to you on your exam. You will not be required to memorize these equations and formulas.

AP BIOLOGY LAB 2: MATHEMATICAL MODELING: HARDY-WEINBERG INVESTIGATION

HIGH YIELD

19.9 Investigate allele distribution mathematical modeling

This lab is easy to perform, but it causes confusion for some students because it involves quantitative and analytical skills. In this investigation, you will develop a mathematical model to investigate the relationship between allele frequencies in populations of organisms. You will use a spreadsheet to build a model based on the Hardy-Weinberg relationship to determine how allele frequencies change from one generation to the next. This model will help you to see how selection, mutation, and migration can affect these inheritance patterns.

Now comes the hard part: putting information into equations. The way to determine the frequencies of the alleles in your breeding population is to assume the population is in **Hardy-Weinberg equilibrium**. If your population is in Hardy-Weinberg equilibrium, two things will be true. First, the addition of the frequencies of the alleles will equal one. The simple formula is:

$$p + q = 1$$

where the frequency of the dominant allele is indicated by p and the frequency of the recessive allele is indicated by q. For this experiment, $p = T$ and $q = t$. The frequencies of the phenotypes in a Hardy-Weinberg population follow the equation:

$$p^2 + 2pq + q^2 = 1$$

The genotypes in a Hardy-Weinberg population are indicated by each term on the left side of the previous equation. The frequency of the homozygous dominant phenotype in the population above is:

$$T \times T = T^2 = p^2$$

The frequency of the heterozygote phenotype is:

$$2 \times T \times t = 2Tt = 2pq$$

Transformations

and the frequency of the homozygous recessive phenotype is:

$$t \times t = t^2 = q^2$$

Normally, success in mastering Hardy-Weinberg problems lies in the reading of the question. You usually have to figure out p and q in order to plug in the terms. If the question states "frequency of alleles in a population," then you should start with $p + q = 1$ and solve for p or q. If, on the other hand, the problem says "frequencies of organisms that express the trait (dominant or recessive)," then you start with $p^2 + 2pq + q^2 = 1$ and calculate your frequencies of p and q from this data. Once you have calculated p and q, you can plug those values into the equation where appropriate to figure out frequencies of homozygotes, heterozygotes, or carriers for a trait.

✔ **AP Expert Note**

Assumptions of Hardy-Weinberg Equilibrium

There are five assumptions that must be met for a population to be under Hardy-Weinberg equilibrium:

1. **The population is very large and not subject to small perturbations in the frequencies of alleles. There are no bottleneck effects.**
2. **The population is isolated from both immigration and emigration. There is no gene flow.**
3. **There is no mutation.**
4. **There is no selective breeding, and mating is random between individuals.**
5. **There is no genetic drift. All genotypes code for phenotypes that have an equal chance of viability and reproduction. There is no selection of phenotypes.**

Keep in mind that the theoretical Hardy-Weinberg population is the benchmark to which all naturally occurring populations are compared. There are probably few if any populations in complete equilibrium. Comparing the expected frequencies under the previously given assumptions to what actually occurs in a population gives insight into which of these assumptions is being violated. In this way, a scientist can determine which evolutionary or environmental forces prevent a population from maintaining equilibrium.

✔ **AP Expert Note**

There is a good likelihood that you will see some sort of Hardy-Weinberg question in either the grid-in or free-response section of the exam.

For this experiment, we can make assumptions about the theoretical alleles of our model population. Let's say that 5 percent of the population is homozygous recessive. This means that the frequency of the recessive allele can be determined by letting $q^2 = 0.05$, which means the frequency of the dominant allele can be determined with the equation:

$$p = 1 - q = 1 - 0.22 = 0.78$$

The frequencies of the homozygous dominant genotype and the heterozygous genotype, respectively, are:

$$p^2 = 0.78 \times 0.78 = 0.61$$
$$2pq = 2 \times 0.78 \times 0.22 = 0.34$$

You can check your work by adding all the genotypic frequencies and making sure that they total 1:

$$0.05 + 0.61 + 0.34 = 1$$

After developing your Hardy-Weinberg model and exploring how random events can affect allele frequencies over generations, you will identify and then test different factors that affect the evolution of allele frequencies. In addition to measuring allele frequencies, your model should track changes in population size, the number of generations, selection (fitness), mutation, migration, and genetic drift.

RAPID REVIEW

If you take away only 5 things from this chapter:

1. Evolution is shaped by a number of mechanisms, including selective pressures, sources of variation, and random effects, such as genetic drift.
2. Biological species concept: a species is a reproductively isolated population able to interbreed and produce fertile offspring.
3. In allopatric speciation, geographically separated populations develop into different species. Sympatric speciation occurs when populations in the same environment adapt to fill different niches. Parapatric speciation occurs with limited interbreeding between two groups.
4. Evidence for evolution comes from comparative anatomy (homologous and analogous structures), biogeography, embryology, the fossil record, biological classification, and molecular biology (relatives share DNA).
5. Hardy-Weinberg equilibrium occurs when genetic distribution remains constant in large, isolated, randomly mating populations with no mutation and no natural selection. These conditions rarely (if ever) occur together.

TEST WHAT YOU LEARNED

1. Although many metabolic reactions are different in distinct organisms, the pathways associated with energy release from nutrients are nearly identical in all organisms and depend on molecules such as ATP, NADH, and $FADH_2$. What is the most likely origin of these small molecules?

   (A) They probably arose very early in a common ancestor of all living organisms.

   (B) They probably came about through convergent evolution.

   (C) They probably exhibit small differences and cannot be considered identical.

   (D) They probably are derived from the breakdown of common chemical compounds.

2. The unique ecology of an island distant from the mainland has made it a tourist attraction, bringing visitors on boats. On the island, finches have large beaks that allow them to crack seeds. Finches on the mainland have a broad range of beak sizes, from very small to very large. Over time, ecologists notice the appearance of several new traits in the island finch population. One of these traits is a small beak, which is effective at obtaining food from sources that large-beaked finches overlook. Which of the following is the most reasonable hypothesis regarding the appearance of this new trait?

   (A) The visitors feed the birds, allowing the population of finches to expand and diversify.

   (B) The boats pollute the island, causing mutations in the finch population.

   (C) The boats also carry mainland finches, adding new genes to the gene pool.

   (D) Extensive tourism destroys the finches' habitat, exerting a selective pressure to adapt.

3. Researchers have been interested in understanding differences between Tibetan chickens, which live at very high altitudes, and lowland chickens, which do not. In one study, researchers examined the frequencies of several single nucleotide polymorphisms (SNPs) among these two populations. The data for one of these SNPs is shown in the table below. (Note that TT indicates an allele that features a genetic sequence with two thymines, while TC features a thymine and a cytosine.)

SNP2 rs14330062	**TT**	**TC**
Tibetan Chicken	152 (96.8%)	5 (3.2%)
Lowland Chicken	139 (100%)	0 (0%)

Adapted from Sichen Li et al., "A Non-Synonymous SNP with the Allele Frequency Correlated with the Altitude May Contribute to the Hypoxia Adaptation of Tibetan Chicken," *PLoS ONE* 12, no. 2 (February 2017): e0172211.

Based on this information, which of the following statements is accurate?

(A) The table shows the allelic frequencies of TT and TC, which can be represented as p^2 and q^2 in Hardy-Weinberg calculations.

(B) The table shows the genotypic frequencies of TT and TC, which can be represented as p^2 and q^2 in Hardy-Weinberg calculations.

(C) The table shows the allelic frequencies of TT and TC, which can be represented as p and q in Hardy-Weinberg calculations.

(D) The table shows the genotypic frequencies of TT and TC, which can be represented as p and q in Hardy-Weinberg calculations.

4. Many physiological activities in plants and animals oscillate according to a circadian rhythm. The circadian rhythm is driven by blue light receptors called cryptochromes. Cryptochromes consist of a protein attached to a pigment. The DNA sequences of the cryptochrome protein show extensive similarities among plants and animals. A search of genome databanks reveals that the cryptochromes share considerable sequence similarities with photolyases, which repair enzymes that repair UV radiation damage in DNA.

Which of the following conclusions is most justified based on these findings?

(A) The light receptor proteins are not homologous; they appeared independently in different lineages because of the selective pressure for all organisms to respond to blue light.

(B) The photolyases and cryptochromes are both receptors of blue light; therefore, their genes should share similar nucleotide sequences because of convergence.

(C) Photolyases appeared early during evolution in a common ancestor; photolyase genes mutated and were repurposed in descendants for circadian rhythm control and other blue light responses.

(D) Photolyases and cryptochromes have similar sequences; all light receptor proteins have the same structure.

5. Throughout human history, there have been considerable efforts to use artificial selection to develop breeds of animals that are useful in agriculture. For example, researchers have worked over the last 50 years to develop dairy cows that produce as much milk as possible. Which of the following best describes the methodology behind these efforts?

   (A) Researchers select for a single gene that enhances milk production, resulting in homozygous herds of dairy cattle.

   (B) Researchers only breed cows with greater than average milk production to increase production in future generations of cows.

   (C) Researchers cross cows with dominant alleles for high milk production, resulting in gradual improvements in milk production across each generation.

   (D) Researchers breed unrelated animals to increase genetic diversity and reduce the risk of inbreeding, which increases milk production.

6. A geological survey of seabed sediments revealed large and diverse populations of animal fossils dating from the Permian through the Triassic periods. Paleontologists compiled their data in the following diagram.

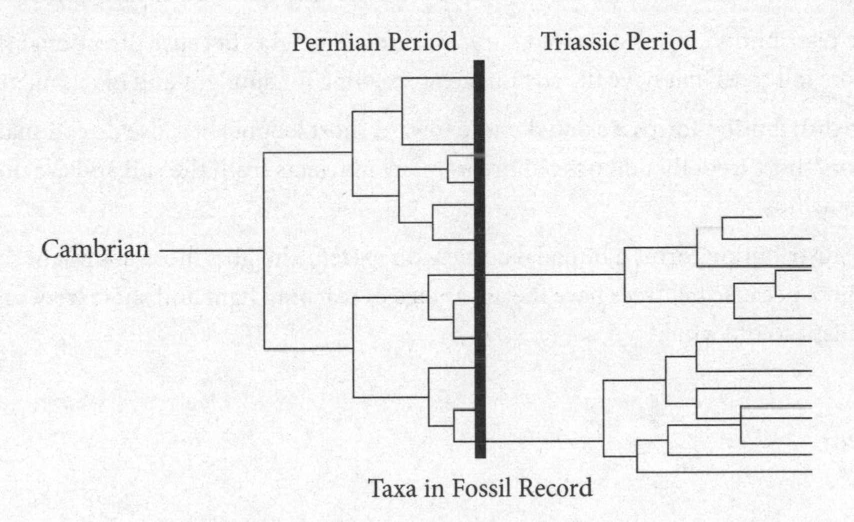

Adapted from P. Hull, "Life in the Aftermath of Mass Extinctions," *Current Biology* 25, no. 19 (October 2015): R941–R952.

The boundary between the Permian and the Triassic periods corresponds to the "Great Dying," when about 90–95% of marine life disappeared from the geological record. What is the most likely explanation for the fossil findings of the late Triassic period?

(A) Brand-new animal life appeared after the mass extinction through spontaneous generation.

(B) The surviving animals colonized newly available niches and underwent adaptive radiation.

(C) Animal life never recovered after the Permian-Triassic extinction event.

(D) A small percentage of each species survived and repopulated the environment.

7. Feathers and hair are associated with birds and mammals, respectively. The bodies of reptiles, on the other hand, are covered by scales. If birds and mammals are both descended from reptiles, scales would be an ancestral trait. Therefore, scales could have undergone modification during the evolution of birds and mammals. Which of the following best supports the hypothesis that scales represent an ancestral trait in birds and mammals?

   (A) Hair and feathers are modified skin structures and fulfill the same function as scales.

   (B) Hair and feathers are not found in present-day reptiles.

   (C) Scales present on bird legs are analogous to the scales on reptiles.

   (D) The same genes control the expression of hair, feathers, and scales.

8. Environmental scientists surveyed a population of spruce trees growing in a dense alpine forest. The population was in the path of high winds. They plotted the distribution of tree heights to investigate the selection pressures at work. Which of the following distributions of tree heights would be expected?

   (A) The distribution forms a bell curve because the selection stabilizes the population to a median height tall enough to reach for sunlight but not tall enough to be toppled by the winds.

   (B) The distribution forms a skewed curve toward tall heights because directional selection favors tall trees that have the advantage of reaching for sunlight and blocking gusts of wind.

   (C) The distribution forms a skewed curve toward short heights because directional selection favors short trees that do not require as many nutrients from the soil and are not uprooted by high winds.

   (D) The distribution forms a bimodal curve with extreme heights more abundant than average heights because tall trees have the advantage of reaching light and short trees are more resilient to the wind.

9. Avian brood parasites lay their eggs in the nests of other birds so that those other birds will incur the costs of raising their offspring. The figure below shows the results of a study of cuckoo parasitism on ashy-throated parrotbills. The *x*-axis shows a contrast score measuring the difference between the egg coloration of the host and parasite, with higher numbers representing greater color difference. The *y*-axis shows the percentage of parasite eggs rejected by the hosts.

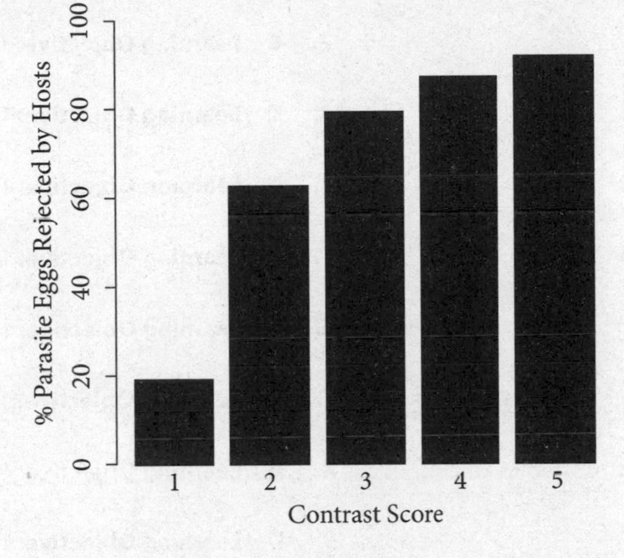

Adapted from Canchao Yang et al., "Coevolution in Action: Disruptive Selection on Egg Colour in an Avian Brood Parasite and Its Host," *PLoS ONE* 5, no. 5 (May 2010): e10816.

Which of the following conclusions is best supported by the data?

(A) There is evolution occurring in the population, but it is impossible to make specific predictions about the direction that evolution is taking.

(B) There is a stable situation in which closely-matching parasitic eggs are most likely to be raised by the host, without selection on coloration patterns.

(C) There is an increase in the amount of contrast with each generation, as evidenced by the increasing height of the bars on the right side of the graph.

(D) There is coevolution between the two species because only closely-matching parasitic eggs are likely to be raised by the host parent.

Answer Key

Test What You Already Know

1. **D** **Learning Objective:** 19.1
2. **B** **Learning Objective:** 19.2
3. **B** **Learning Objective:** 19.3
4. **C** **Learning Objective:** 19.4
5. **C** **Learning Objective:** 19.5
6. **A** **Learning Objective:** 19.6
7. **D** **Learning Objective:** 19.7
8. **A** **Learning Objective:** 19.8
9. **B** **Learning Objective:** 19.9

Test What You Learned

1. **A** **Learning Objective:** 19.5
2. **C** **Learning Objective:** 19.3
3. **C** **Learning Objective:** 19.9
4. **C** **Learning Objective:** 19.2
5. **B** **Learning Objective:** 19.8
6. **B** **Learning Objective:** 19.4
7. **D** **Learning Objective:** 19.6
8. **A** **Learning Objective:** 19.1
9. **D** **Learning Objective:** 19.7

Detailed solutions can be found in the Answers and Explanations section at the back of this book.

REFLECTION

Test What You Already Know score: __________

Test What You Learned score: __________

Use this section to evaluate your progress. After working through the pre-quiz, check off the boxes in the "Pre" column to indicate which Learning Objectives you feel confident about. Then, after completing the chapter, including the post-quiz, do the same to the boxes in the "Post" column. Keep working on unchecked Objectives until you're confident about them all!

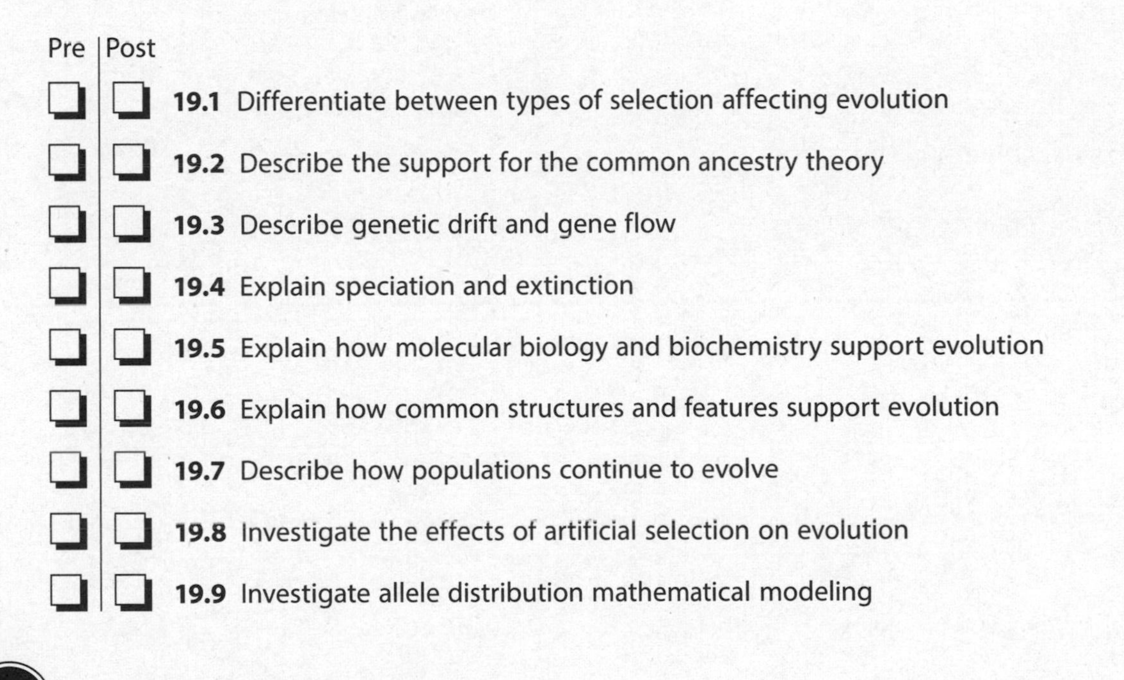

Pre	Post	
☐	☐	**19.1** Differentiate between types of selection affecting evolution
☐	☐	**19.2** Describe the support for the common ancestry theory
☐	☐	**19.3** Describe genetic drift and gene flow
☐	☐	**19.4** Explain speciation and extinction
☐	☐	**19.5** Explain how molecular biology and biochemistry support evolution
☐	☐	**19.6** Explain how common structures and features support evolution
☐	☐	**19.7** Describe how populations continue to evolve
☐	☐	**19.8** Investigate the effects of artificial selection on evolution
☐	☐	**19.9** Investigate allele distribution mathematical modeling

FOR MORE PRACTICE

Complete more practice online at kaptest.com. Haven't registered your book yet? Go to kaptest.com/booksonline to begin.

PART 6

Interactions of Life

CHAPTER 20

Biodiversity

LEARNING OBJECTIVES

In this chapter, you will review how to:

- **20.1** Explain how to read and test a phylogenetic tree
- **20.2** Explain how to read and test a cladogram
- **20.3** Investigate evolutionary changes with cladograms

TEST WHAT YOU ALREADY KNOW

1. The figure below shows a phylogenetic tree for birds.

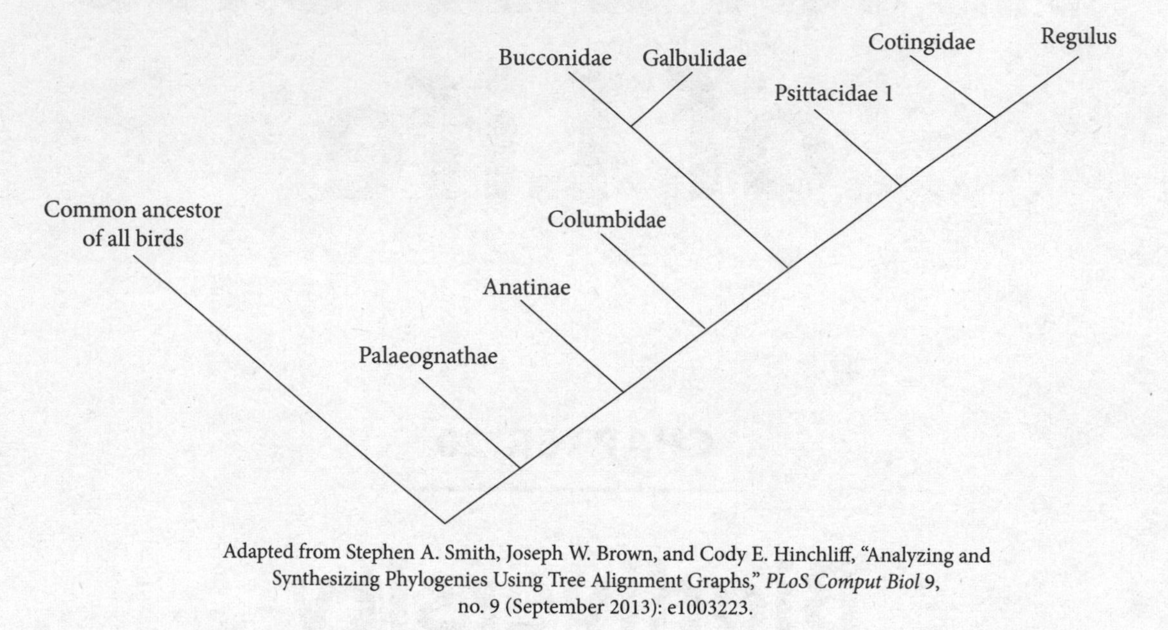

Adapted from Stephen A. Smith, Joseph W. Brown, and Cody E. Hinchliff, "Analyzing and Synthesizing Phylogenies Using Tree Alignment Graphs," *PLoS Comput Biol* 9, no. 9 (September 2013): e1003223.

Based on this tree, which of the following groups are most closely related?

(A) *Psittacidae* 1 and *Cotingidae*

(B) *Bucconidae* and *Galbulidae*

(C) *Palaeognathae* and *Regulus*

(D) *Anatinae* and *Columbidae*

2. The figure below shows a cladogram of fishes across time. The abbreviations at the top represent time periods and the bar at the bottom represents time.

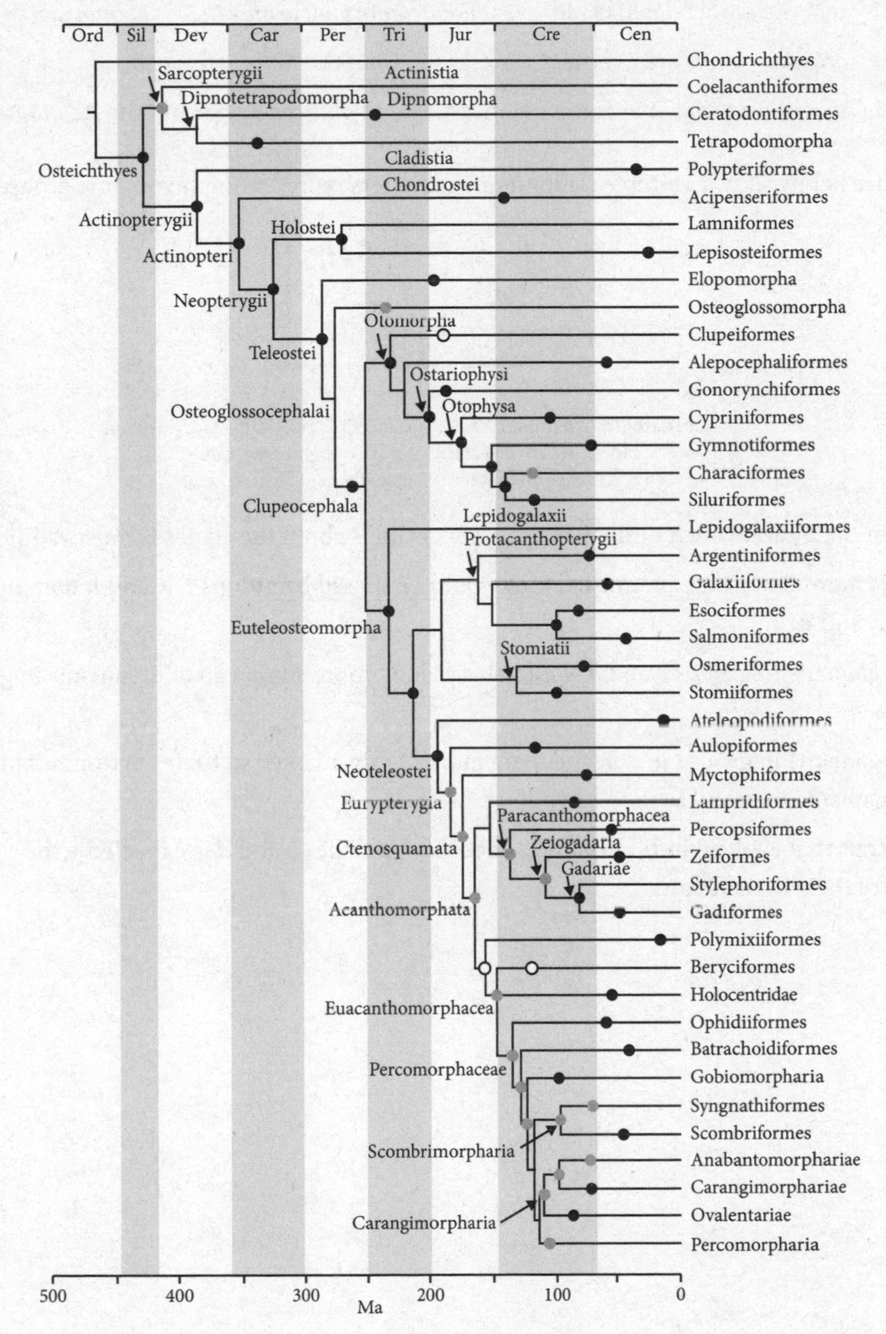

Adapted from Ricardo Betancur-R. et al., "The Tree of Life and a New Classification of Bony Fishes," *PLOS Currents* (April 2013).

Which of the following conclusions is most justified based on the cladogram?

(A) All of the *Sarcopterygians* are extinct.

(B) The most distantly related clades are the *Percomorpharia* and the *Ovalentariae.*

(C) The *Sarcopterygians* are a more diverse group than the *Actinopterygians.*

(D) An *Osteichthyean* was the common ancestor of the *Sarcopterygii* and the *Actinopterygii.*

3. The figure below shows a simple cladogram with letters representing taxonomic groups.

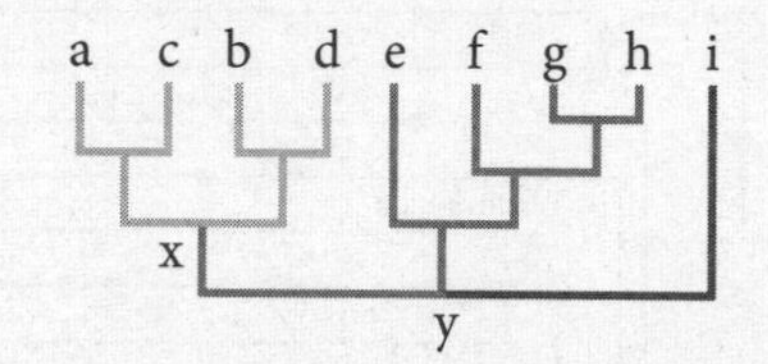

Adapted from S. A. Smith, J. W. Brown, and C. E. Hinchliff, "Analyzing and Synthesizing Phylogenies Using Tree Alignment Graphs," *PLoS Comput Biol* 9, no. 9 (September 2013): e1003223.

Based on the figure, which of the following conclusions about the clades is most valid?

(A) It is more likely that an unusual trait would be shared by groups *f*, *g*, and *h* than by groups *a*, *c*, and *e*.

(B) Organisms in groups *g* and *h* would always look more similar than organisms in groups *f* and *h*.

(C) Organisms in groups *a*, *c*, *b*, and *d* are more highly adapted to their environment than organisms in group *x*.

(D) A trait that evolved in the common ancestor for clades *a* and *d* is expected to be present in all of their descendants.

Answers to this quiz can be found at the end of this chapter.

PHYLOGENETIC TREES

HIGH YIELD

20.1 Explain how to read and test a phylogenetic tree

Biologists maintain that all living organisms are descended from a single common ancestor, that all the great diversity of species existing today share a common lineage that can be traced back to the origin of life. Not only are all eukaryotic organisms descended from a single ancestor, but that first eukaryote is also a descendant of the first prokaryotes. Because of this fact, the emergence of biological diversity can be likened to the growth of a tree, in which the multitude of the tree's branches are ultimately derived from its trunk. This structural analogy explains why biologists sometimes use diagrams known as *phylogenetic trees* to model biodiversity.

✔ **AP Expert Note**

Until recently, the first level of classification of organisms was the kingdom. There were five recognized kingdoms (Monera, Protoctista, Fungi, Plantae, and Animalia). Now all of life is first divided into three domains: Archaea, Bacteria, and Eukarya.

A taxon is a grouping of related organisms, such as a **species**, **genus**, **phylum**, or **kingdom**. Phylogenetic trees show hypotheses of evolutionary relationships among various taxa, indicating common lines of descent from shared ancestors. The following example shows evolutionary relationships among the phyla of kingdom Animalia, the animals.

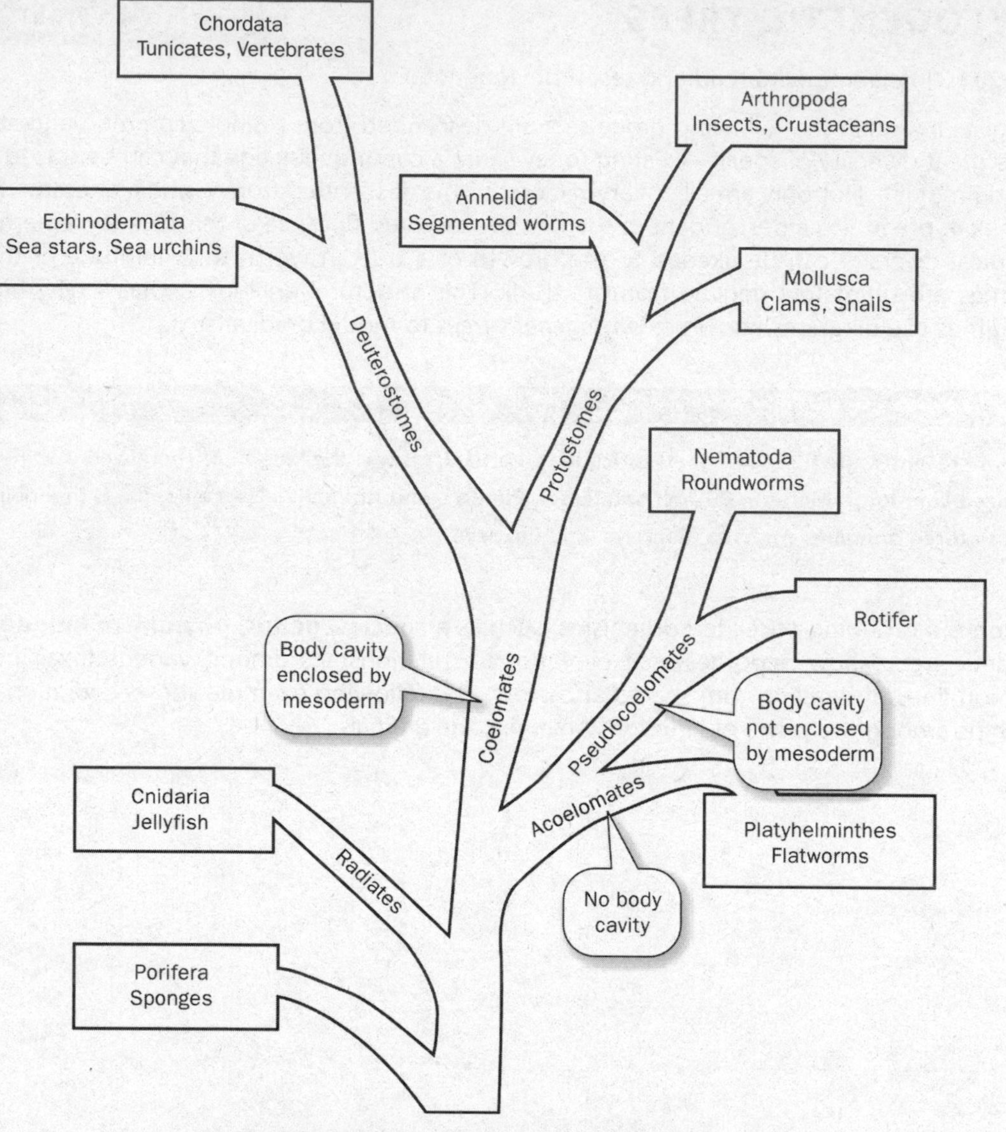

Phylogenetic Tree of Major Animal Groups

> ✔ **AP Expert Note**
>
> **The names and characteristics of specific biological taxa, as well as the relationships between these taxa, are not tested on the AP Biology exam.**

Some phylogenetic trees also give an indication of evolutionary timescale, by keeping the distances between branchings in the tree proportional to the amount of time between species divergences. On a phylogenetic tree, as you move from the root (ancestor) to the tips of the branches (descendants), you move forward in time. Each branch represents speciation. The following example shows the lineage of primates.

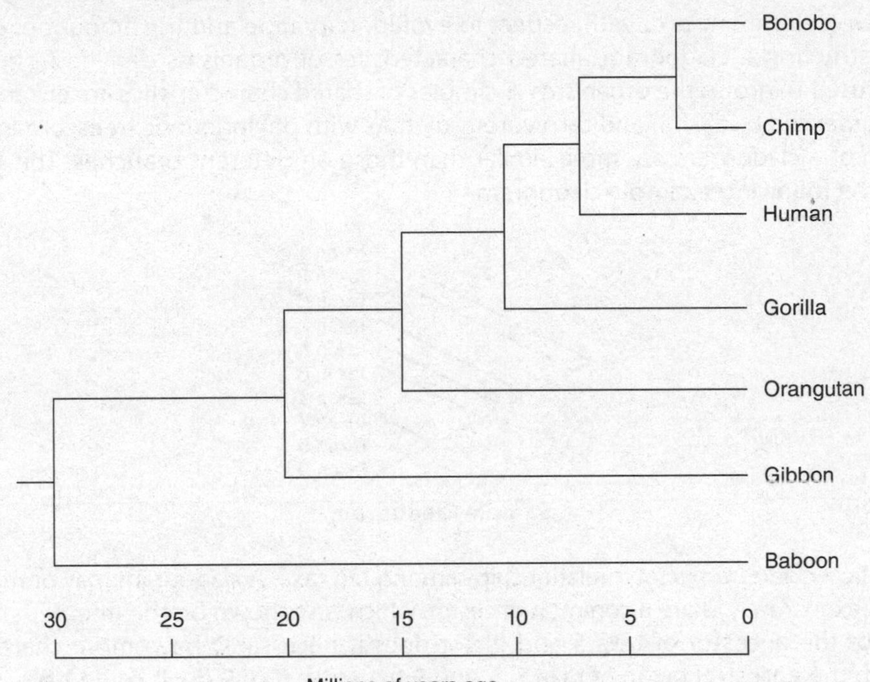

Phylogenetic Tree of Primates Over Time

CLADOGRAMS

HIGH YIELD

20.2 Explain how to read and test a cladogram

Also used to represent hypotheses on evolutionary relationships in a group of organisms, a *cladogram* is similar in appearance to a phylogenetic tree. Cladograms demonstrate similarities between types of organisms with respect to a common ancestor, but they do not show relationships of descent between different taxa with respect to evolutionary time and the amount of change with time. In constructing a cladogram, shared characteristics of organisms derived from a common ancestor are used to group the organisms. Examples of shared characteristics are eukaryotic, warm-blooded, segmented body, fur, and carnivorous diet. As with phylogenetic trees, organisms on the same branch of a cladogram are more similar than those on different branches. This can be illustrated with the following example cladogram.

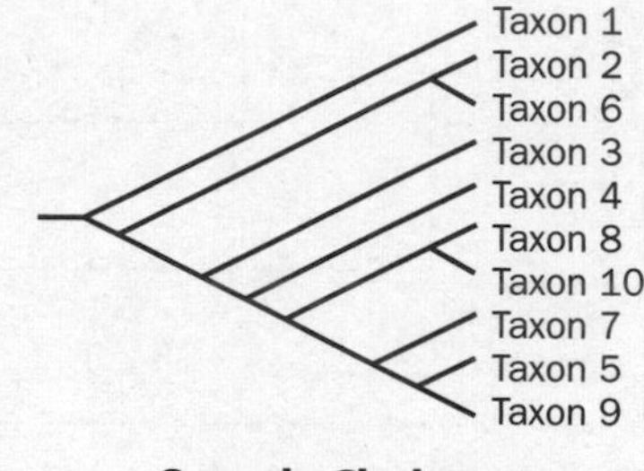

Sample Cladogram

This hypothetical cladogram shows relationships among ten taxa. A cladogram may or may not show a time scale. Taxon 7 may share a common ancestor, which isn't shown on the tree, with taxa 5 and 9. Taxon 7 is not the ancestor of taxa 5 and 9, nor does it necessarily have more characteristics in common with the ancestral taxon of taxa 7, 5, and 9 than taxa 5 or 9 do. It could have just as many *anagenetic* changes (evolutionary events without speciation) along its branch as *cladogenetic* changes (evolutionary events that lead to speciation). Taxon 4 is more closely related to taxa 8, 10, 7, 5, and 9 than it is to taxon 3. Even though taxon 3 is close to taxon 4 on the tree, it is not on the same branch that includes taxa 4, 8, 10, 7, 5, and 9. Likewise, taxon 2 is more closely related to taxon 6 than it is to taxon 1. Taxa 2 and 6 belong to the same branch that does not include taxon 1.

The following is an example of a cladogram showing the relationships among primates.

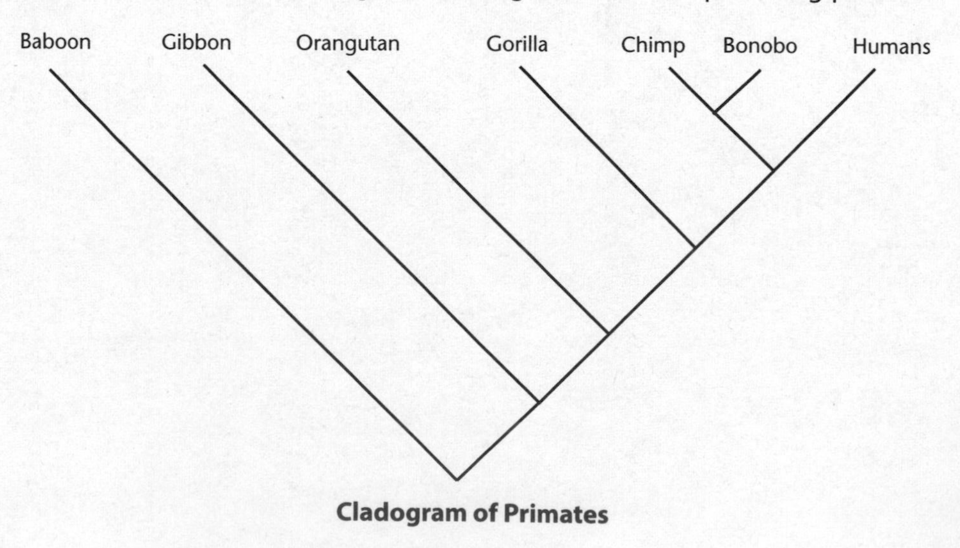

Cladogram of Primates

AP BIOLOGY LAB 3: COMPARING DNA SEQUENCES TO UNDERSTAND EVOLUTIONARY RELATIONSHIPS WITH BLAST INVESTIGATION

20.3 Investigate evolutionary changes with cladograms

Since 1990, scientists have been working to develop a comprehensive library of genes from several species, including humans, mice, fruit flies, and *E. coli.* BLAST (Basic Local Alignment Search Tool) is a powerful bioinformatics program that helps scientists to compare genes from different organisms cataloged in this library. Information derived from BLAST can, therefore, show scientists the evolutionary relationships between different organisms.

In this investigation, you will use BLAST to compare several genes from different organisms and then construct a cladogram to represent the evolutionary relationships among these different species. While cladograms can be constructed using many different factors, including the presence of different morphological traits (wings, gills, etc.), in this investigation you will use DNA evidence to establish the similarity among different organisms.

Locating and sequencing genes in different organisms not only helps us to better understand evolutionary relationships among organisms, but it can also provide important insights into genetic diseases. Species with smaller genomes, such as the fruit fly and mouse, are easier for scientists to study than humans. When scientists locate a disease-causing gene in a fruit fly or mouse, they can then use BLAST to see if there is a similar sequence in the human genome.

Using BLAST requires several specific steps and an abundance of information, none of which you need to memorize for the AP Biology exam. Rather, you should be able to apply similarities and differences in morphology and genetics to determine the evolutionary relationships between different species. Drawing and analyzing cladograms (as well as phylogenetic trees) is a skill you should practice and be able to perform successfully.

Let's use some data from an example in this investigation to construct a cladogram based on genetic data. In humans, the GAPDH gene produces a protein that catalyzes a step in glycolysis. The following table shows the percentage similarity between the GAPDH gene and protein in humans compared with four different species.

Species	Gene Similarity Percentage	Protein Similarity Percentage
Chimpanzee (*Pan troglodytes*)	99.6%	100%
Dog (*Canis lupus familiaris*)	91.3%	95.2%
Fruit fly (*Drosophila melanogaster*)	72.4%	76.7%
Roundworm (*Caenorhabditis elegans*)	68.2%	74.3%

Just as with the cladogram shown earlier in this chapter, we can use this data to construct a cladogram that tells us how these species are related to each other based on GADPH. There are five species being compared here, so our cladogram will have five branches.

Reading the table, we see that humans and chimpanzees have the most similarities in their GADPH genes and proteins. Therefore, humans and chimpanzees would be placed closest together on our cladogram as shown.

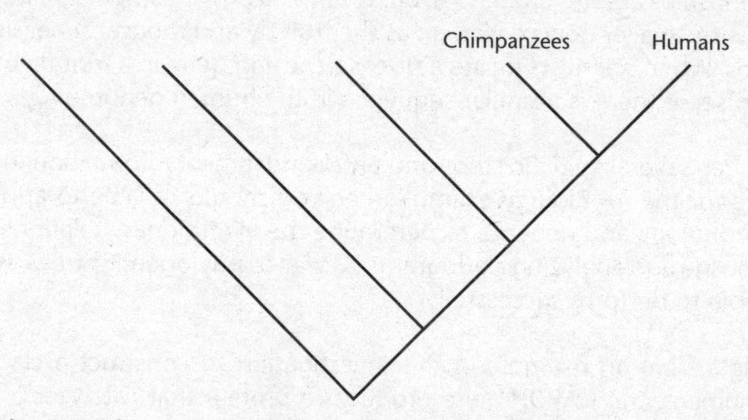

We can make deductions for the remaining three species based on their similarity to humans. Roundworms' GADPH genes have the least in common with humans, so roundworms should be placed farthest from humans on the cladogram. Dogs are closer to humans than fruit flies but farther away than chimpanzees.

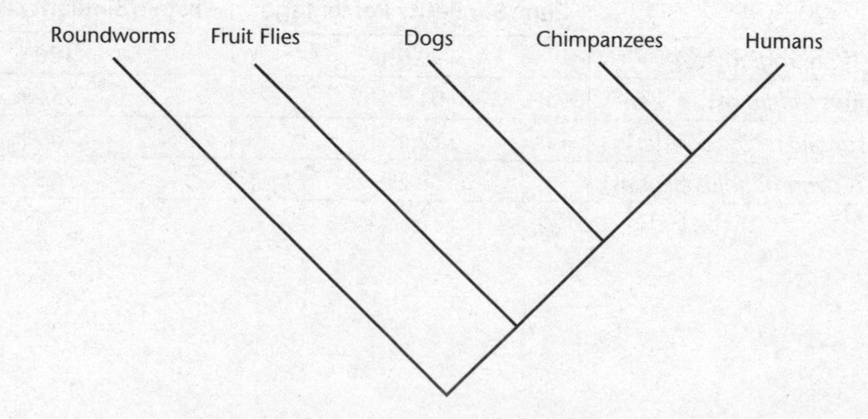

RAPID REVIEW

If you take away only 3 things from this chapter:

1. A phylogenetic tree is a diagram indicating evolutionary relationships between taxa of organisms. It depicts an ancestral species with other species branching off from it. More closely related taxa are nearer to one another in the diagram. Phylogenetic trees are intended to show ancestral connections between species and some include information about the timescale of species divergences.
2. A cladogram is typically a simplified version of a phylogenetic tree, which shows degrees of relatedness without necessarily offering information about the ancestral relationships between species or the timescale of divergences.
3. Phylogenetic trees and cladograms can show anagenetic evolutionary changes (which do not lead to speciation), as well as cladogenetic evolutionary changes (which do).

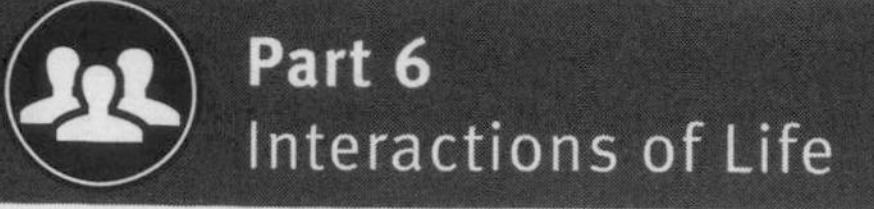

TEST WHAT YOU LEARNED

1. The figure below shows a recent reptile cladogram developed using molecular, morphological, and paleontological data. The *Gekkota, Anguimorpha, Scincidae, Teiidae,* and *Iguania* are lizard groups. The *Mosasauria* is a group of extinct marine reptiles. The *Serpentes* are the snakes.

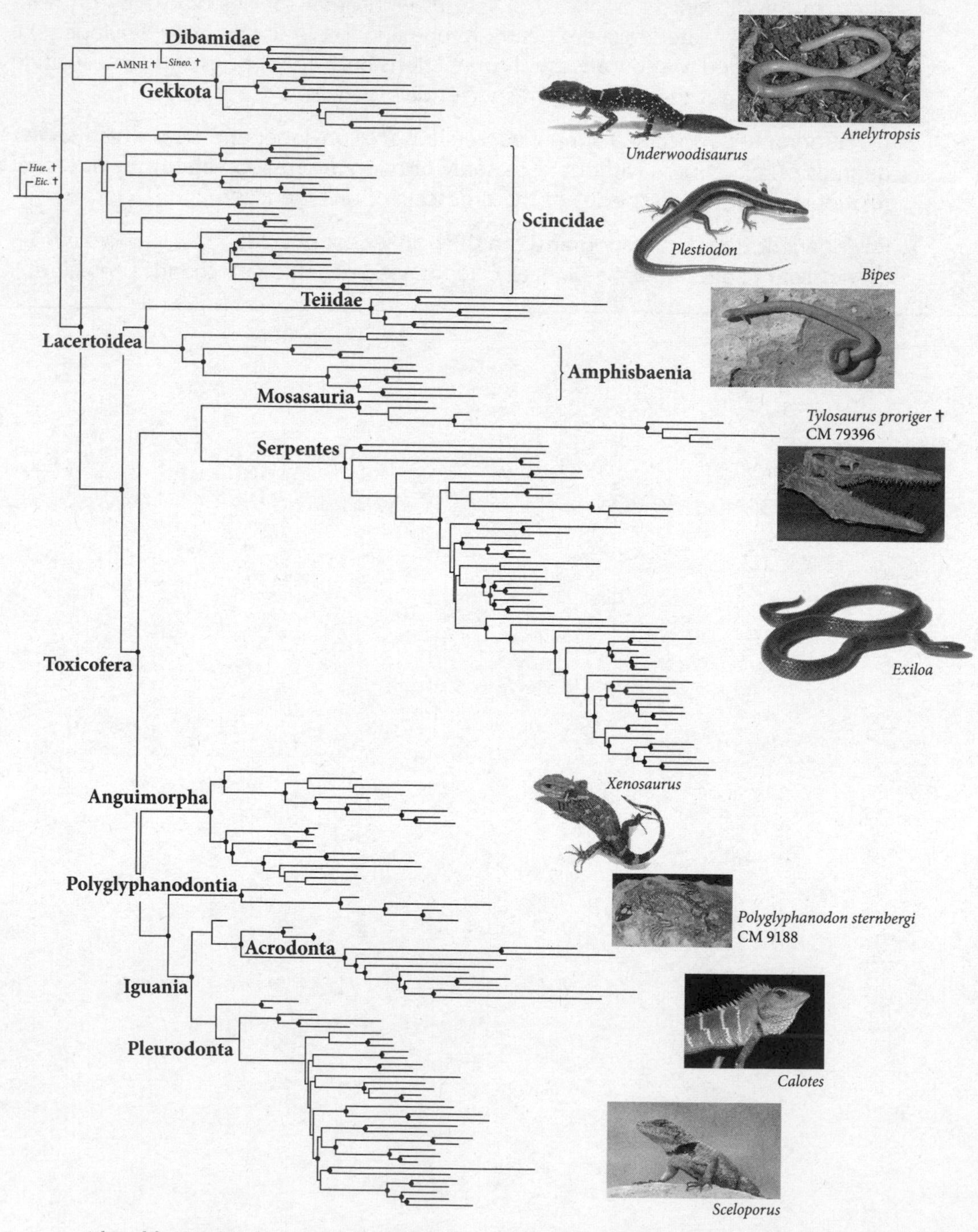

Adapted from Tod W. Reeder et al., "Integrated Analyses Resolve Conflicts over Squamate Reptile Phylogeny and Reveal Unexpected Placements for Fossil Taxa," *PLoS ONE* 10, no. 3 (March 2015): e0118199.

Which of the following statements is best supported by the cladogram?

(A) The snakes are more closely related to the *Iguania* than to the *Gekkota*.

(B) The *Scincidae* is polyphyletic, containing descendants from multiple different ancestors.

(C) All of the lizards are more closely related to each other than to the snakes and mosasaurs.

(D) Extinct groups branch toward the left whereas modern groups are clustered toward the right of the cladogram.

2. The figure below shows a phylogeny for amphipods, a type of crustacean. There is information on two traits, temperature tolerance and tube building. The hour values represent survival times at increased temperature, with larger circles representing greater temperature tolerance.

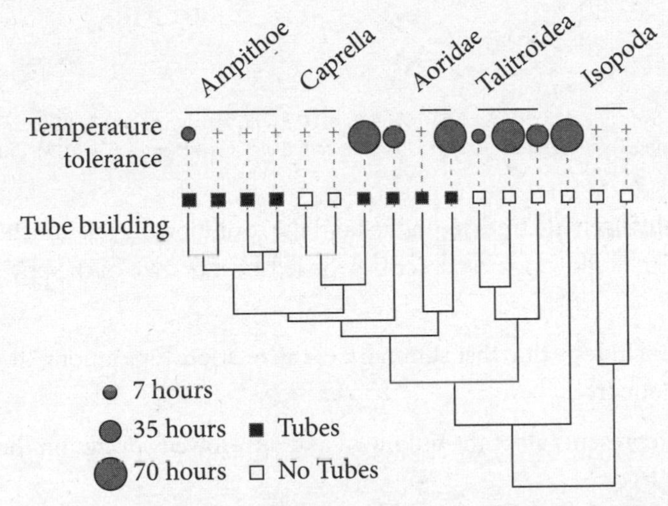

Adapted from Rebecca J. Best and John J. Stachowicz, "Phylogeny as a Proxy for Ecology in Seagrass Amphipods: Which Traits Are Most Conserved?" *PLoS ONE* 8, no. 3 (March 2013): e57550.

Based on this data, which of the following conclusions is most likely to be accurate?

(A) Tube building was a trait possessed by the common ancestor of all of these groups but was secondarily lost in many groups, including *Isopoda*.

(B) Tube building evolved in a common ancestor of the *Ampithoe*, *Caprella*, and *Aoridae*, but was secondarily lost in the *Caprella*.

(C) Temperature tolerance was a trait possessed by the common ancestor of all of these groups but was secondarily lost in some groups, including *Isopoda*.

(D) Temperature tolerance evolved in a common ancestor of the *Caprella*.

Interactions

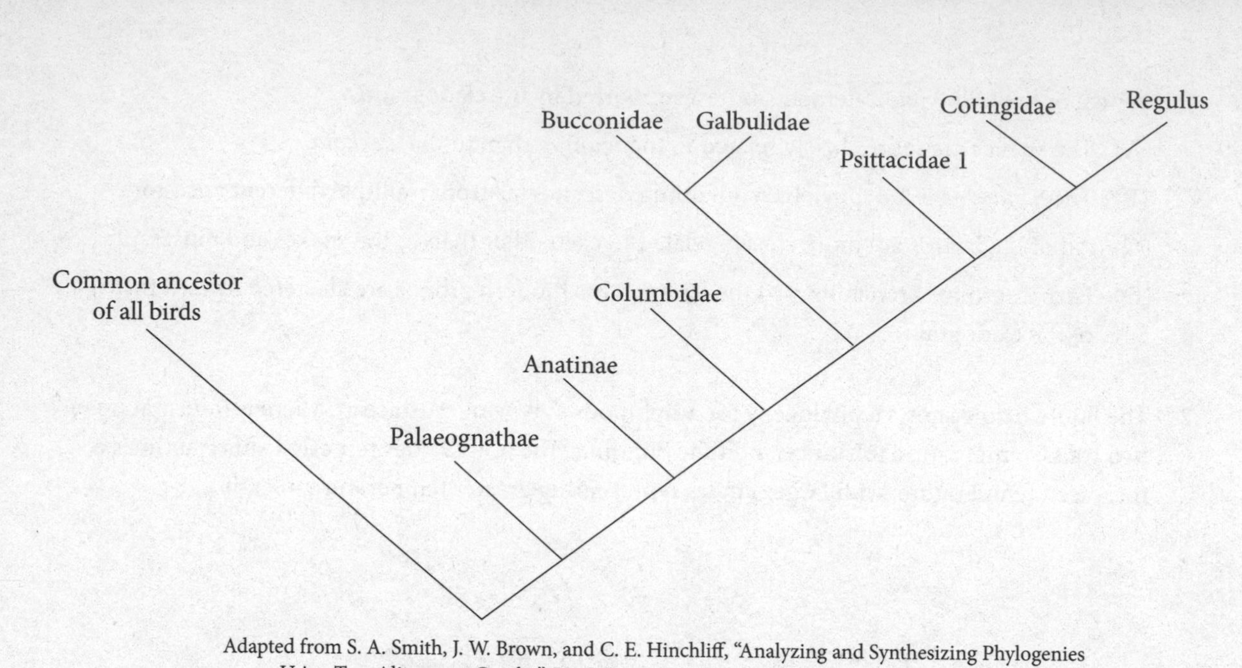

Adapted from S. A. Smith, J. W. Brown, and C. E. Hinchliff, "Analyzing and Synthesizing Phylogenies Using Tree Alignment Graphs," *PLoS Comput Biol* 9, no. 9 (September 2013): e1003223.

3. Researchers use phylogenetic trees to understand the evolutionary relationships between different species. Which of the following best describes the leftmost branch of the phylogenetic tree shown above?

   (A) This branch includes a title that summarizes the relationships among the other branches of the phylogenetic tree.

   (B) This branch represents all of the unknown and unresolved species on the bird phylogenetic tree.

   (C) This branch indicates the most ancestral group on the tree, which evolved from non-bird groups such as reptiles.

   (D) This branch represents extinct groups of birds that cannot be included in the main phylogenetic tree.

The answer key to this quiz is located on the next page.

Interactions

Answer Key

Test What You Already Know

1. **B** **Learning Objective:** 20.1
2. **D** **Learning Objective:** 20.2
3. **A** **Learning Objective:** 20.3

Test What You Learned

1. **A** **Learning Objective:** 20.2
2. **B** **Learning Objective:** 20.3
3. **C** **Learning Objective:** 20.1

Detailed solutions can be found in the Answers and Explanations section at the back of this book.

Test What You Already Know score: ________

Test What You Learned score: ________

Use this section to evaluate your progress. After working through the pre-quiz, check off the boxes in the "Pre" column to indicate which Learning Objectives you feel confident about. Then, after completing the chapter, including the post-quiz, do the same to the boxes in the "Post" column. Keep working on unchecked Objectives until you're confident about them all!

Pre	Post	
❑	❑	**20.1** Explain how to read and test a phylogenetic tree
❑	❑	**20.2** Explain how to read and test a cladogram
❑	❑	**20.3** Investigate evolutionary changes with cladograms

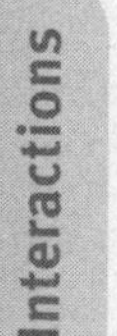

FOR MORE PRACTICE

Complete more practice online at kaptest.com. Haven't registered your book yet? Go to kaptest.com/booksonline to begin.

CHAPTER 21

Behavior

LEARNING OBJECTIVES

In this chapter, you will review how to:

- **21.1** Differentiate between innate and learned behaviors
- **21.2** Contrast how competition and cooperation affect survival
- **21.3** Explain environmental and invasive species effects
- **21.4** Explain how communication improves survival
- **21.5** Investigate how behavior affects survival

TEST WHAT YOU ALREADY KNOW

1. A plant biologist is investigating the response of a Venus flytrap to stimulation using a probing needle. When she lightly touches an open insect-catching leaf, she observes no response. She increases the pressure of the stimulus, but the leaf still does not move. When she touches the leaf quickly in several spots, the leaf slams shut on her probing needle.

   Which of the following conclusions is best supported by her results?

   (A) Chemical compounds accumulate in the tissue causing the leaf to snap shut.

   (B) The stimulus must reach a threshold of pressure before the leaf can respond by shutting.

   (C) The leaf shuts when a rapid stimulus in several locations mimics the presence of an animal.

   (D) The plant becomes sensitized to her presence and responds to her irritating behavior.

2. A vineyard represents a complex ecosystem containing not just the plants, but also many microorganisms associated with the plants and soil. These microorganisms can be both beneficial and detrimental to the vines. For example, beneficial mycorrhizae in the soil are symbiotic fungi that grow in or on plant roots. Conversely, molds in the air will settle on fruit and rot the grapes. Viticulturists can treat vineyards with fungicides to destroy the spores of damaging mold. However, the yield from plots treated with fungicides is usually low, and viticulturists must heavily fertilize to increase the harvest.

   Which is the best hypothesis to test in determining why fertilizers are needed after fungicide use?

   (A) The fungicides deplete the soil of nutrients, which must then be replaced by fertilizers.

   (B) The fungicides damage the roots, so they extract nutrients from the soil less efficiently.

   (C) The fungicides destroy the mycorrhizae, which aid nutrient uptake by the roots.

   (D) The fungicides kill the mold on the fruit, slowing nutrient uptake by the plant.

3. Himalayan blackberries, a small and sour fruit, were introduced in the Pacific Northwest as a potential cash crop that requires low maintenance. The plants spread easily through rhizomes and seeds in the wild. Their brambles cover the ground with thick, thorny tangles that are impenetrable to wildlife and prevent native plants from germinating and growing. Which of the following statements most accurately describes the probable long-term impact of the blackberries?

   (A) The blackberries represent a steady source of revenue because the berries are harvested as food.

   (B) The blackberries decrease biodiversity because they outcompete native plants and are impenetrable.

   (C) The blackberries increase biodiversity because the brambles provide food and shelter for many species.

   (D) The blackberries do not readily spread and cannot be considered an invasive species.

4. In Belding's ground squirrel populations, females rear most of the young, while males mate with several females and wander from territory to territory. They use alarm calls to alert other members of their species to approaching land predators. A researcher investigated the frequency of alarm calls in a population of Belding's ground squirrels and plotted the results as a function of age and sex. The diagram compares the expected frequency of calls (which assumes the calls were random) to the observed frequency.

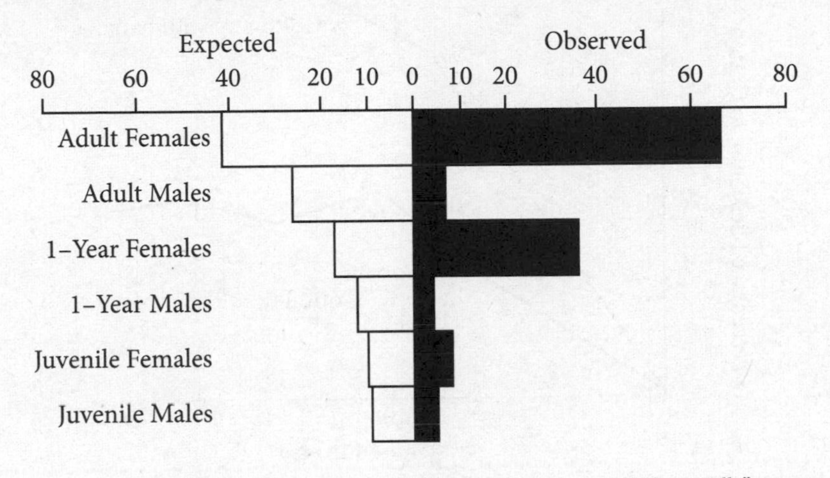

Adapted from Paul W. Sherman, "Nepotism and the Evolution of Alarm Calls," *Science* 197, no. 4310 (September 1977): 1246.

According to the data, which of the following best explains the alarm calls of Belding's ground squirrels?

(A) Alarm calls are altruistic behaviors adapted to protect next of kin.

(B) Alarm calls are intended to alert all members of the group to take cover.

(C) Alarm calls distract the predator's attention and redirect it to other squirrels nearby.

(D) Alarm calls warn the predator that the caller is going to defend itself.

5. In order to study fruit fly behavior, students built a choice chamber made of two empty plastic bottles taped together. At one end, they added a cotton tip coated with a chemical stimulus. At the opposite end, they introduced the fruit flies. They record the distribution of fruit flies as a function of distance to the cotton tip. Their results are summarized in the following graph.

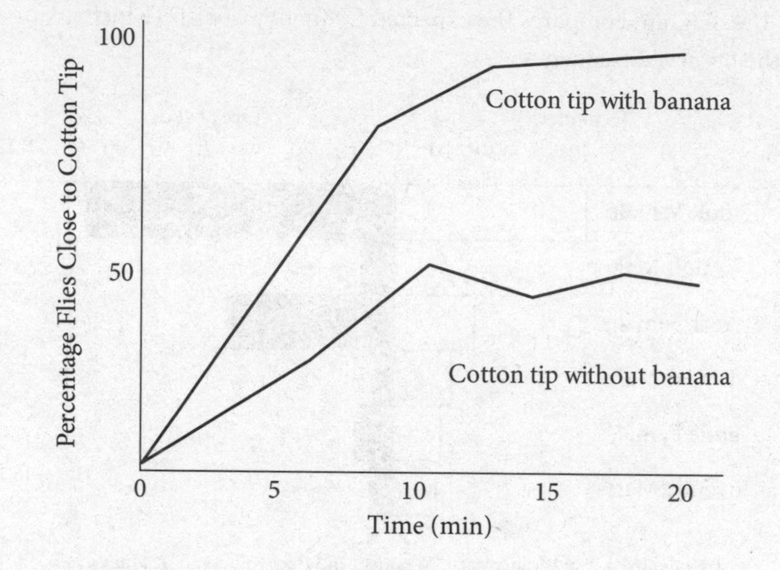

Adapted from Lar L. Vang, Alexei V. Medvedev, and Julius Adler, "Simple Ways to Measure Behavioral Responses of *Drosophila* to Stimuli and Use of These Methods to Characterize a Novel Mutant," *PLoS ONE* 7, no. 5 (May 2012): e37495.

What is the best interpretation of the results of the experiment?

(A) Fruit flies distribute randomly in the bottles because flies have a poor sense of smell.

(B) Flies gather away from the cotton tip because the smell acts as a negative chemotactic stimulus.

(C) Although there are more flies in the chamber with the banana stimulus, a chi-square analysis would show that the movement is random.

(D) Fruit flies gather in the chamber with the banana cotton tip because the smell acts as a positive chemotactic stimulus.

Answers to this quiz can be found at the end of this chapter.

INNATE VERSUS LEARNED BEHAVIOR

21.1 Differentiate between innate and learned behaviors

Ethology is the study of behavior—the way organisms act. Behaviors in organisms are triggered by internal and external stimuli, such as hunger and danger. Natural selection favors behaviors that promote reproduction and survival. Two types of behavior are innate and learned.

Innate behaviors are responses encoded in the genes of organisms, meaning organisms are genetically programmed how to respond to particular stimuli. Birds instinctively know how to build nests, and nest building occurs when both internal stimuli (hormonal signals) and external stimuli (proper nest building materials) are present. Innate behaviors that increase the organism's ability to survive and reproduce will persist in a population and be passed from one generation to the next. A change in innate behavior occurs only if a genetic mutation arises, and the altered behavior would develop over many generations. Other examples of innate behavioral responses include taxes (singular: **taxis**) (movements by the entire body) and reflexes (movements by individual muscles). For instance, chemotaxis involves the response to chemical gradients, and a reflex to the sensation of pain is avoidance.

Learned behaviors, the opposite of innate behaviors, are acquired or lost through interacting with the world or through teaching. Young geese and other birds learn via imprinting, in which they follow an object (usually a parent) that they have become attached to during a receptive period after birth or hatching. Compared to innate behavior, learned behavior is largely independent of inheritance and dependent on the environmental context, so if the context disappears, the behavior will also disappear. As a result, learned behavior spreads through a population, passes within a single generation, and develops/degrades more rapidly than innate behavior. By obtaining knowledge and/or skills, organisms can modify learned behaviors to survive in unpredictable environments. For example, a mouse can learn how to run through a maze to reach a piece of cheese.

Innate and learned behaviors are not mutually exclusive and may influence behavior together. In honeybees, genes associated with foraging behavior determine whether bees are foragers or workers and whether they forage for pollen or for nectar. However, foraging performance is also correlated with learning. The ability to adjust foraging behavior in response to the constantly changing environment allows honeybees to minimize energetic cost and maximize food acquisition.

COMPETITION AND COOPERATION

HIGH YIELD

21.2 Contrast how competition and cooperation affect survival

21.3 Explain environmental and invasive species effects

Innate and learned behaviors may be cooperative and/or competitive. **Cooperation** is the process of acting together for common benefits, whereas **competition** is the process of striving for limited resources. Both cooperation and competition affect survival and occur at various levels of organization: cells, tissues, organs, organ systems, organisms, populations, communities, and ecosystems.

On the cellular level, cooperation and competition increase the efficiency of using matter and energy. Chemical reactions catalyzed by enzymes are affected by the cooperation between enzymes and their substrates, as well as by the competition between inhibitors and substrates. Examples of cooperation between organs and organ systems include the coordination of roots and shoots of plants and of the digestive and excretory systems of animals. In plants, to replace water lost via transpiration from shoots, roots absorb water. In animals, organs in the digestive system work together to process food, and the digestive system works in parallel with the excretory system to eliminate undigested waste.

Contributing to the survival of individuals and the population, cooperation and competition occur between individuals of the same species (intraspecific) and/or between individuals of different species (interspecific). Cooperative behavior within a population can provide protection from predators, acquisition of prey and resources, recognition of offspring, and transmission of learned responses. Cooperation can be seen in schools of fish, in which living together decreases the chance of a predator attack, and among wolves in a pack, in which each wolf has a specific role in the pack's hunting strategy.

Competition arises when resources such as food, territories, and mates that two or more individuals depend upon become limited. Direct competition occurs when individuals compete for the same resource (for example, two males competing to mate with a single female), and indirect competition occurs when organisms use the same resource (for example, a waterhole) but may not interact with each other. The use of resources by one individual decreases the availability for use by others, so the less competitive individual will be forced to go elsewhere, resulting in competitive exclusion. When sunlight is limited, plants unable to outgrow surrounding plants may use energy to produce many seeds in order to spread their offspring to areas with more sunlight. Species may coexist if intraspecific competition is stronger than interspecific competition because then each species limits their own growth and population size.

In addition to cooperation and competition, ecological relationships that affect population dynamics include predation and symbiosis (**mutualism, commensalism,** and *parasitism*). **Predation** involves one organism feeding on another and typically occurs interspecies. For instance, some protozoans prey on bacteria and some plants can trap and digest insects. **Symbiosis** describes the relationship between two different organisms. In mutualism, both organisms in the relationship benefit. Examples include bacteria in digestive tracts of animals and pollination. Bacteria in cows help cows digest cellulose and in turn cows provide bacteria with nutrients and a hospitable place to live. During pollination, plants provide insects with food and insects help plants spread their pollen from one plant to another. In commensalism, one organism benefits and the other is unharmed, whereas in parasitism one benefits while the other is harmed. Remoras attached to sharks obtain protection and leftover food, while the shark is unaffected. Fleas and ticks, in contrast, are external parasites that feed on the blood of their hosts and cause itching and transmit disease.

Environmental Changes and Invasive Species

Changes in the environment, such as environmental catastrophes, geological events, and the sudden influx/depletion of abiotic resources, also affect the behavioral responses of organisms. An environmental catastrophe or disaster due to human activity (like deforestation or overhunting) puts extraordinary pressures on organisms and may result in the loss of species or the selection for altered behavioral traits. Geological events such as volcanic eruptions and earthquakes may also result in the extinction of species. Geologically separating a population (for instance, by the formation of a mountain range) can lead to changes in diversity. Changing the abiotic environment can alter the nutrient availability of a community, which determines whether a species can invade and/or persist in a community. A flood may bring an influx of nutrients, while a drought will diminish the availability of resources and may thus kill the species.

Changes in an ecosystem, even the slightest, can elicit an adaptive response and lead to changes in the behavior of native populations. When a nonnative species is introduced into an ecosystem, it can disrupt the balance of the ecosystem and have a negative impact. Nonnative species that cause harm to the environment, the health of native species, or the local economy are termed **invasive species**. An invasive species can be any kind of living organism—a bird, fish, insect, fungus, bacterium, plant, seed, or egg.

Invasive species can be a direct or indirect threat to the native population in the ecosystem. Direct threats include preying on native species, out-competing for food and resources, out-competing for habitats or breeding sites, and causing or carrying toxins or disease that native species are not adapted to deal with. In response to invasive predatory species and competition, native organisms may change their anti-predator, foraging, and feeding behaviors. For instance, crickets suppress their calling behavior to avoid acoustically-hunting parasites. Invasive species may dominate in a new ecosystem because they grow and reproduce rapidly and/or lack natural enemies or pests. As a result, the invasive species will displace the native species, as illustrated by both kudzu and Dutch elm disease. Kudzu is an invasive plant that has replaced the diverse ecosystem in the southeastern United States with a monoculture. Dutch elm disease, caused by the fungus *Ophiostoma ulmi* and transmitted to trees by elm bark beetles, has killed native populations of elms that are not resistant to the disease. Indirect threats include changing the food web by destroying or replacing native food sources, decreasing the abundance and diversity of native species, and changing the environment. For example, invasive plants can modify nutrient availability (making an environment less inhabitable) by affecting soil pH or causing erosion (because they may not hold soil as well as native species).

COMMUNICATION

21.4 Explain how communication improves survival

In response to stimuli, signals are transmitted and received to exchange information. Individuals can then act on the information and communicate it to others (individuals of the same or different species). Communication among organisms produces changes in behavior that are vital to reproductive success, natural selection, and evolution.

Signals to improve survival may stem from environmental cues like temperature and oxygen levels. Hibernation is regulated by temperature, level of food supply, and/or photoperiod (length of day). When it gets cold or when the food supply gets low, animals get ready to hibernate to save energy. Likewise, animals estivate, or slow their activity in response to high temperatures and arid conditions, to save energy.

Organisms communicate using visual, audible, tactile, electrical, and chemical (pheromone) signals. These signals may be used to locate resources, show dominance, defend territory, coordinate group behavior, and care for young. The following examples illustrate communication in organisms. Male lizards use visual signals to show their territorial dominance by standing up high off the ground and swallowing air to increase their size. Frogs and toads produce auditory signals to attract mates, and the bright colors of toxic species warn predators not to eat them. On the other hand, the bright coloration of flowers and fruits signal animals to pollinate their flowers or disperse their fruits. Honeybees use tactile signals to communicate the location of nectar sources via a waggle dance, and minnows release an alarm pheromone when their skin is damaged to warn other minnows of a predatory fish.

AP BIOLOGY LAB 12: FRUIT FLY BEHAVIOR INVESTIGATION

HIGH YIELD

21.5 Investigate how behavior affects survival

Taxis, mentioned earlier in this chapter, will be explored in this investigation of the relationship between a model organism, *Drosophila* (fruit fly), and its response to different environmental conditions. You can examine the behavior of fruit flies by placing them in an apparatus called a choice chamber, which presents two (or more) choices to the fruit flies. Experiments performed in the choice chamber will investigate chemotaxis (movement in response to a chemical stimulus), geotaxis (movement in response to gravity), and/or phototaxis (movement in response to light). Fruit flies that move toward a stimulus exhibit positive taxis (attract), whereas fruit flies that move away from a stimulus exhibit negative taxis (repel). Undirected, random movement in response to an external stimulus is **kinesis**.

You will design a controlled experiment to explore environmental factors that trigger fruit fly responses. A simple choice chamber may be constructed by joining together two clear plastic bottles with the bottoms cut off. After exposing fruit flies in the chamber to a stimulus (one end) or stimuli (both ends), give them time to respond and then determine the number of flies at each end of the chamber.

From observations, generate your own hypotheses to explore other factors (sex, age, different colors of light, ripeness of fruit, mutations) that affect fruit fly behavior. To verify the results, you need to conduct several trials and change the positions of the stimuli in the chamber. Quantify and express your results graphically. Then complete a chi-square analysis and construct a preference table to identify and compare the preferences of fruit flies. Your goal is to identify a pattern in the behavior of fruit flies and which of the responses (geotaxis, chemotaxis, or phototaxis) is the strongest. Potential challenges may include difficulty in counting fruit flies accurately and fruit flies escaping the choice chamber.

RAPID REVIEW

If you take away only 5 things from this chapter:

1. Behaviors in organisms are responses that are triggered by stimuli (internal and external), and the behaviors that promote reproduction and survival are favored by natural selection.
2. Innate behaviors are inherited and instinctive, whereas learned behaviors are acquired through interactions with the environment. The behaviors are not mutually exclusive.
3. Cooperation involves organisms working together for mutual benefits, while competition involves organisms contending with each other for limited resources.
4. Both environmental changes (such as deforestation and earthquakes) and invasive species (nonnative species introduced into an ecosystem) disrupt the balance of an ecosystem and affect the behavior of native populations.
5. Communication among organisms involves the transmission of signals to produce changes in behavior that are vital to reproductive success, natural selection, and evolution.

TEST WHAT YOU LEARNED

1. The gut microbiome is a crowded ecosystem in which large amounts of bacteria compete for resources from the host. Microbiologists have found that infection by a parasite that is resistant to the host's nonspecific immune system actually elicits a strong nonspecific immune response from the host.

   Which of the following best describes how a resistant parasite would most likely benefit from stimulating its host's immune system in this way?

   (A) Stimulation of the host immune system depletes the energy of the host and makes it more vulnerable to infection.

   (B) Stimulation of the immune system reduces competition by decreasing the population of other parasites susceptible to the immune system.

   (C) Stimulation of the immune system is automatic and does not present an advantage to the parasite.

   (D) Stimulation of the host's nonspecific defenses at the site of infection results in a flow of nutrients which supports parasite growth.

2. Students built a choice chamber to study fruit fly behavior. The chamber consisted of two empty plastic bottles taped together. At one end, they added a cotton tip coated with a chemical stimulus. They introduced the fruit flies at the opposite end. They record the distribution of fruit flies as a function of distance to the cotton tip.

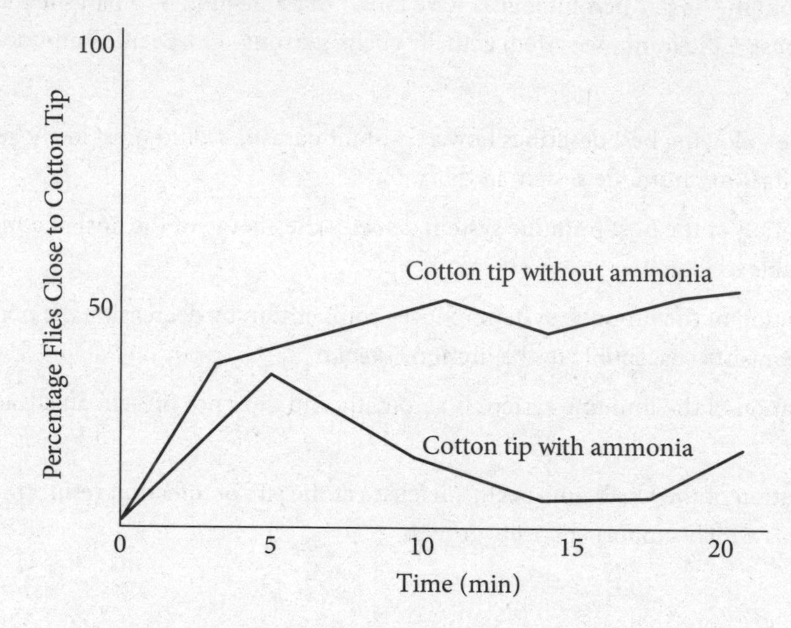

Adapted from Lar L. Vang, Alexei V. Medvedev, and Julius Adler, "Simple Ways to Measure Behavioral Responses of *Drosophila* to Stimuli and Use of These Methods to Characterize a Novel Mutant," *PLoS ONE* 7, no. 5 (May 2012): e37495.

According to the graph, what can be concluded about the reaction of fruit flies to ammonia, a strong base?

(A) Ammonia does not elicit a strong response from fruit flies.

(B) Ammonia paralyzes fruit flies as it diffuses through the chamber.

(C) Ammonia appears to attract and stimulate the fruit flies.

(D) Ammonia is a volatile compound that repels fruit flies.

3. Plants are exposed to many threats from the environment, including predators and diseases caused by infectious agents. Warding off injury presents a distinct survival advantage. While investigating the response of marigolds to predation by golden beetles, researchers detected a noxious chemical compound that signals to the beetle that the plant is not a good source of food. The researchers hypothesized that plants synthesize this compound in response to communication with neighboring plants.

   Which of the following compounds is the most likely means by which the marigolds communicate?

   (A) Jasmonate, a volatile organic compound that is produced when the plant is stressed

   (B) Oligosaccharides, large sugars that act as signaling molecules in response to infection

   (C) Abscisic acid, a water soluble plant hormone that is produced as a response to stress

   (D) Salicylic acid, a water soluble acid that mediates defense against pathogens

4. One of the ecosystem changes that followed the disappearance of wolves from Yellowstone National Park was the substantial erosion of riverbanks along streams, which followed a decrease in the number of trees and other plants. After wolves were reintroduced in the park, the effects of erosion were reversed. Which of the following best explains how the disappearance of wolves would most likely affect the erosion of riverbanks?

   (A) Wolves destroy vegetation along the riverbanks while hunting. The loss of vegetation can cause erosion.

   (B) Wolves hunt elk, which graze on vegetation that protects riverbanks from erosion. Without wolves, elk populations increase, reducing vegetation.

   (C) Wolves competitively exclude other predators. These predators eat plants and thus contribute to the degradation of the riverbanks.

   (D) Wolves eat beavers, which cut down trees to build dams. The dams slow down the flow of rivers, which decreases the amount of erosion.

5. An octopus in an aquarium would occasionally splash water on several summer volunteers. One of the volunteers, named Michael, claims that the octopus does not like him because he always gets drenched when he walks by. The other volunteers insist that the drenching is purely random. Which of the following experimental treatments would help to test the hypothesis that the octopus learned to recognize faces?

   (A) Have each of the volunteers except Michael walk past the tank once to see if any get drenched.

   (B) Arm Michael with a water gun to drench the octopus back if the octopus drenches him.

   (C) Cover the faces of each of the volunteers with masks before they walk past the tank and compare those results to walks without masks.

   (D) Randomize the time at which different volunteers walk by the octopus tank and determine whether the octopus is more likely to splash at specific times.

Answer Key

Test What You Already Know

1. C Learning Objective: 21.1

2. C Learning Objective: 21.2

3. B Learning Objective: 21.3

4. A Learning Objective: 21.4

5. D Learning Objective: 21.5

Test What You Learned

1. B Learning Objective: 21.2

2. D Learning Objective: 21.5

3. A Learning Objective: 21.4

4. B Learning Objective: 21.3

5. C Learning Objective: 21.1

Detailed solutions can be found in the Answers and Explanations section at the back of this book.

REFLECTION

Test What You Already Know score: __________

Test What You Learned score: __________

Use this section to evaluate your progress. After working through the pre-quiz, check off the boxes in the "Pre" column to indicate which Learning Objectives you feel confident about. Then, after completing the chapter, including the post-quiz, do the same to the boxes in the "Post" column. Keep working on unchecked Objectives until you're confident about them all!

Pre	Post	
☐	☐	**21.1** Differentiate between innate and learned behaviors
☐	☐	**21.2** Contrast how competition and cooperation affect survival
☐	☐	**21.3** Explain environmental and invasive species effects
☐	☐	**21.4** Explain how communication improves survival
☐	☐	**21.5** Investigate how behavior affects survival

FOR MORE PRACTICE

Complete more practice online at kaptest.com. Haven't registered your book yet? Go to kaptest.com/booksonline to begin.

CHAPTER 22

Ecology

LEARNING OBJECTIVES

In this chapter, you will review how to:

22.1 Contrast exponential and logistic growth models

22.2 Explain how biotic and abiotic factors affect population

22.3 Explain how environmental factors affect organisms

22.4 Contrast how matter is recycled and energy flows

22.5 Explain how diversity affects stability

22.6 Describe the human impact on biodiversity and climate

22.7 Investigate energy flow through a system

TEST WHAT YOU ALREADY KNOW

1. When species are introduced into a new area, they can sometimes find the environment suitable and establish new populations. If a new species is accidentally released in an area in which there are appropriate climate conditions, abundant food, and no predators, which of the following best describes the most likely growth of its population?

   (A) The population will follow an exponential growth curve as there are no limitations on food and other resources in this new location.

   (B) The population growth curve will rapidly level off and then decline because this species does not belong in this area.

   (C) The population will remain at a steady level since the species has become established but cannot become abundant in a new location that differs from its native habitat.

   (D) The population growth curve may initially appear to be exponential, but will then level off over time to form a logistic growth curve as resources become more limited.

2. A study examined the response of bird communities to Hurricane Iris. Hurricanes can affect communities in a variety of ways, particularly by knocking down trees and causing structural damage. The figure below shows the mean number of birds captured for every 100 hours of effort using a net. Values are given for 1) captures over 58 days before Hurricane Iris ("pre-Iris"), 2) captures over about a month beginning 11 days after Iris ("Post-I"), and 3) captures over 69 days one year after Iris ("Post-II").

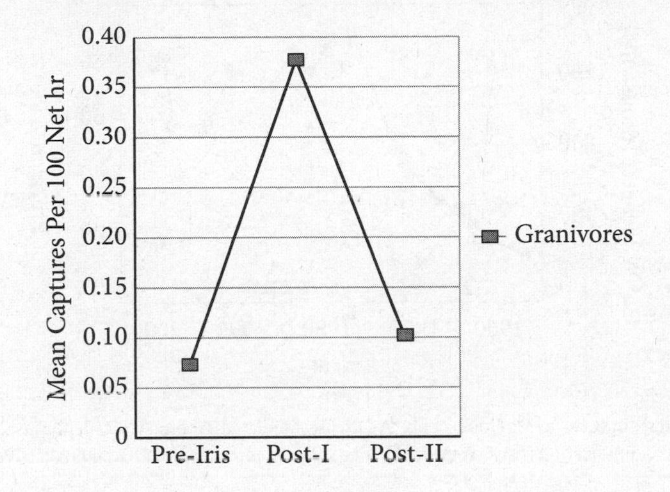

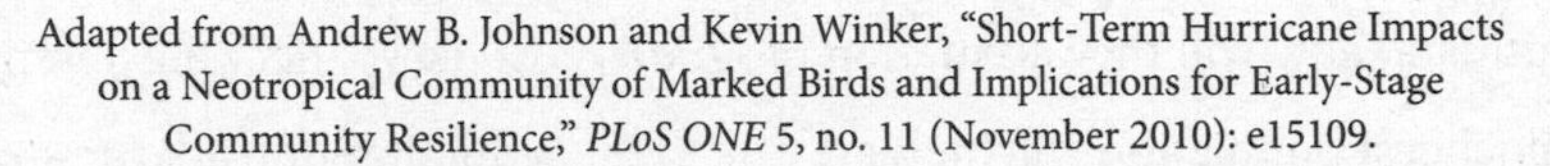
Adapted from Andrew B. Johnson and Kevin Winker, "Short-Term Hurricane Impacts on a Neotropical Community of Marked Birds and Implications for Early-Stage Community Resilience," *PLoS ONE* 5, no. 11 (November 2010): e15109.

Which of the following hypotheses is NOT consistent with these findings?

(A) Birds living in the area experienced high mortality due to the hurricane.

(B) Reduced canopy cover after the hurricane made it easier to capture birds.

(C) Birds had to change their foraging habits due to changes in the forest.

(D) Because the habitat was damaged, birds needed larger home ranges and territories.

3. The figure below shows data from a study in a Serengeti ecosystem. Rinderpest is a virus that causes disease in wildebeest, herbivores that are the dominant grazing animal in this ecosystem. A vaccination program was used to reduce the prevalence of rinderpest in the wildebeest population. The prevalence of the virus is shown with open squares in the figure below. The wildebeest population is represented by closed circles.

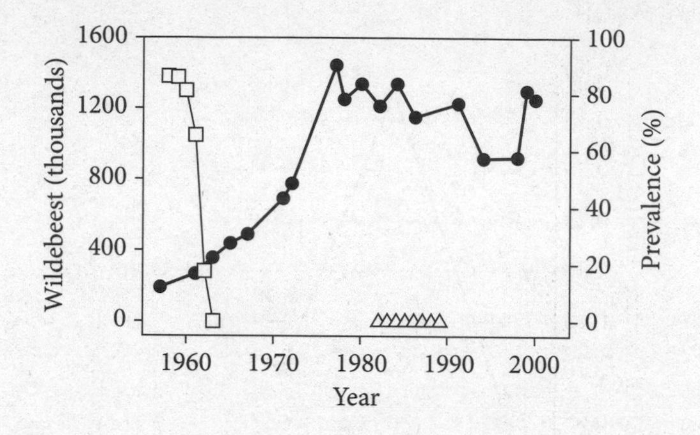

Adapted from Ricardo M. Holdo et al., "A Disease-Mediated Trophic Cascade in the Serengeti and its Implications for Ecosystem C," *PLoS Biol* 7, no. 9 (September 2009): e1000210.

Based on this data, how will the vaccination program most likely affect the Serengeti ecosystem?

(A) The wildebeest population will increase due to the reduction in rinderpest prevalence, which will lead to an increase in grass cover and tree cover.

(B) The wildebeest population will decrease due to the reduction in rinderpest prevalence, which will cause the rinderpest to develop immunity to the vaccination protocol.

(C) The wildebeest population will increase due to the reduction in rinderpest prevalence, which will result in less grass available to generate and spread fires.

(D) The wildebeest population will decrease due to the reduction in rinderpest prevalence, which will result in the emergence of new parasites to lower the population size further.

4. Which of the following best describes the effects, in terms of energy and matter flow and cycling, when plants are fertilized with nitrogen-rich fertilizer?
    (A) Plants take up energy from the fertilizer, which can be recycled. Nitrogen is consumed during plant metabolism.
    (B) Plants obtain energy from the Sun, which flows to primary consumers. Nitrogen in fertilizer is used by plants to make organic molecules that can be recycled through the nitrogen cycle.
    (C) Plants take up energy from fertilizer, which is eventually lost as heat. Nitrogen in fertilizer is used to make organic molecules that can be recycled through the nitrogen cycle.
    (D) Plants obtain energy from the Sun and lose it again as heat. Nitrogen fertilizer slows this process, allowing the plants to conserve more energy.

5. In 1995, there were fewer than 30 Florida panthers remaining in their native ecosystem. Researchers introduced eight panthers from Texas, members of a different subspecies, to southern Florida that year. Why would it be beneficial to introduce panthers from another population to interbreed with the Florida panthers?
    (A) The existing small population was very inbred, increasing the risk that they could go extinct due to a lack of genetic diversity in the population.
    (B) Increasing the size of the population is very important for survival, and subspecies are sufficiently similar that the genetic differences did not matter.
    (C) Texas populations were likely better adapted to the southern Florida habitat, and introducing their genes would increase the survival of the population.
    (D) The population was so small that it was impossible to save, making it a good location to experiment with interbreeding panther subspecies.

6. Wetlands are of considerable environmental importance and researchers actively study wetland restoration methods. Figure A shows the ratio of vertebrate density, vertebrate richness, and macroinvertebrate density in restored wetlands versus natural wetlands. Figure B shows plant density and richness in restored wetlands. The dotted line represents the values for reference wetlands. Negative values represent numbers below those of the reference wetlands.

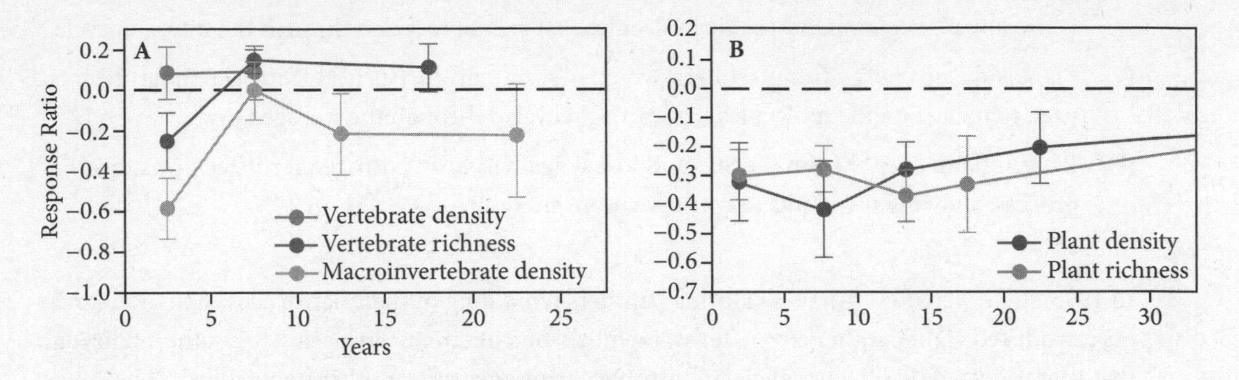

Adapted from David Moreno-Mateos et al., "Structural and Functional Loss in Restored Wetland Ecosystems," PLoS Biol 10, no. 1 (January 2012): e1001247.

Which of the following conclusions is best supported by the data provided?

(A) Although plants have the lowest mean response ratios initially, they stabilize at a similar level to macroinvertebrates after extended periods of time.

(B) Although plants all reach relatively stable levels equivalent to those found in reference wetlands, animals in the studied locations do not.

(C) Macroinvertebrate density in manipulated wetlands never reaches the levels found in reference wetlands.

(D) Although vertebrates may be able to colonize wetlands relatively easily, macroinvertebrates and plants do not successfully stabilize at the levels found in natural wetlands.

Interactions

7. Scientists are interested in understanding how climate change can affect forest biomass. The figures below show the amount of above-ground biomass at three forest locations, with dates on the x-axis, growth rate on the y-axis, and a dotted line representing no net change in biomass.

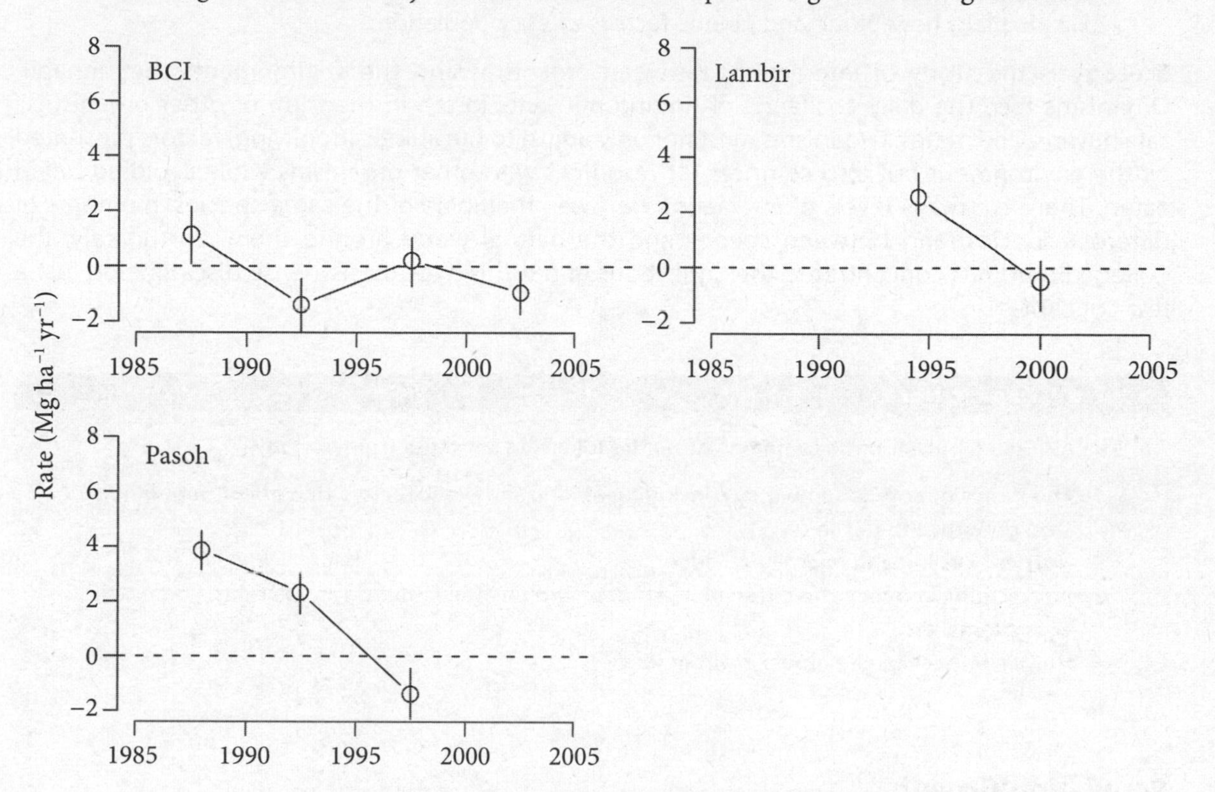

Adapted from Jérôme Chave et al., "Assessing Evidence for a Pervasive Alteration in Tropical Tree Communities," *PLoS Biol* 6, no. 3 (March 2008): e45.

Which of the following conclusions can most reasonably be drawn from the results?

(A) There is an increase in biomass growth rate in these locations. This data may be specific for the study locations and may not be generalizable.

(B) There is no change or a slight decrease in biomass growth rate in these locations. This data may be specific for the study locations and may not be generalizable.

(C) There was a net increase in biomass growth rate in some plots but not in others. More research is needed to determine whether there is a consistent pattern.

(D) There is not a change in biomass growth rate in these locations. More research is needed to determine whether there is a change in underground biomass growth rate.

Answers to this quiz can be found at the end of this chapter.

POPULATION DYNAMICS

HIGH YIELD

22.1 Contrast exponential and logistic growth models

22.2 Explain how biotic and abiotic factors affect population

Ecology is the study of interactions between organisms and the environments they inhabit. Organisms face the daily challenge of finding nutrients (often in the form of other organisms), safe havens, and mates. Organisms must not only adjust to the abiotic (nonliving) factors presented by the environment but also compete for resources with other organisms while avoiding being eaten. There are many levels of interaction between members of the same species, members of different species, and between species and the natural world around them. Fortunately, the College Board has concentrated their questions about this hypothesis-dense discipline on just a few concepts.

✔ AP Expert Note

The AP Biology exam most commonly tests the following concepts from ecology:

1. **The basics of population biology, including abiotic and biotic factors that affect population size, growth, and decline**
2. **Nutrient cycling and energy exchange**
3. **Interactions between the different levels of environmental systems (populations, communities, ecosystems, etc.)**
4. **Human impact on the global environment**

Population Growth

What do you get if you put one bacterium in a lot of space and give it an unlimited amount of resources and time to reproduce? You get a whole lot of bacteria, that's what! As a matter of fact, you get an exponential rise in the number of bacteria according to the equation $N_t = N_0e^{rt}$, where

N_t = the number of bacteria at time t,

N_0 = the number of bacteria at the beginning,

r = the rate of population growth (a difference of the reproduction and death rates),

t = the number of chronological steps of reproduction (seconds, minutes, years, etc.), and

e = Euler's number (approximately 2.71828).

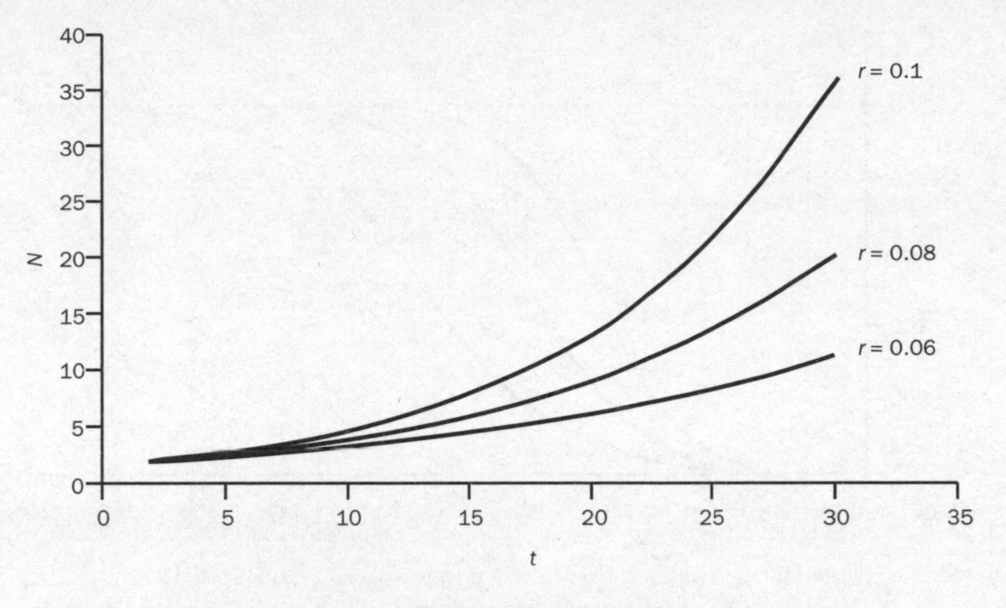

Exponential Population Increase

The three lines on the above graph show the exponential increase in population size with no limits according to the equation $N_t = N_0e^{rt}$. Each line represents a population increasing at the rate indicated, having started with two individuals.

This graph shows how rapidly populations can grow when the increase is exponential. Remember that *t* is a measure of the population's rate of producing another generation, which is very fast for bacteria. Some bacteria can reproduce every 10 to 30 minutes, so 30 time steps would only take between 5 and 15 hours. The upshot is that bacteria can grow to numbers in the millions in only a matter of hours!

In reality, populations do not grow exponentially without limits. Most reach a size at which they plateau, called the **carrying capacity**. This type of limited increase is known as *logistic growth*. The rate of change in size of natural populations can be estimated by the equation $\frac{\Delta N}{\Delta t} = rN(\frac{K - N}{K})$, where

$\frac{\Delta N}{\Delta t}$ = the change in the population size over the given time,

N = the size of the population at the beginning of time *t*,

r = the rate of population growth (again, a difference of the reproduction and death rates), and

K = the carrying capacity.

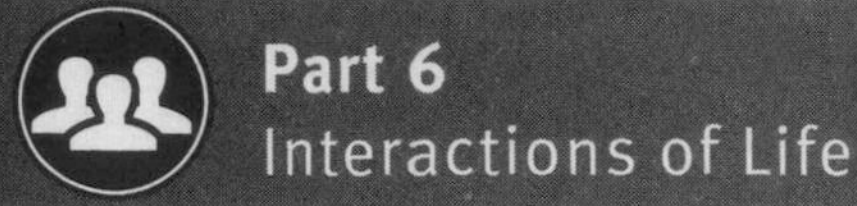

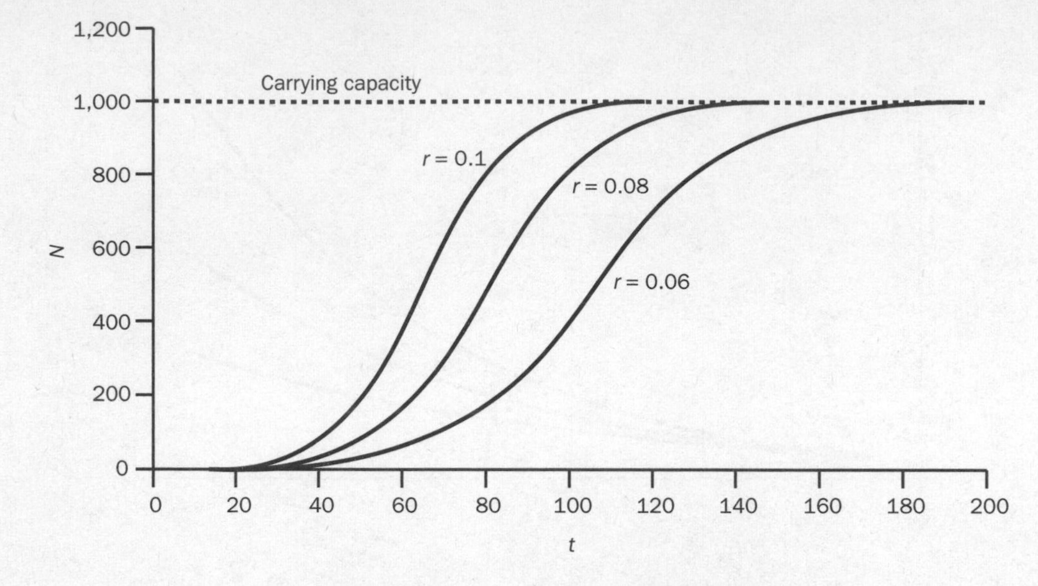

Population Increase to Carrying Capacity

These three lines show the increases in population size with the same rates of population growth as in the previous graph, but with a limited carrying capacity of 1,000 individuals. This graph shows that carrying capacity is the ultimate limit to population size, while the rate of population growth affects how quickly the population increases. A change in carrying capacity will change where the plateau occurs.

✔ AP Expert Note

There is a good possibility that you will see a graph illustrating exponential growth and carrying capacity on the AP Biology exam. Be sure that you are able to interpret major trends such as exponential growth, logistic growth, *K* and *r* selection, and factors that can alter these trends.

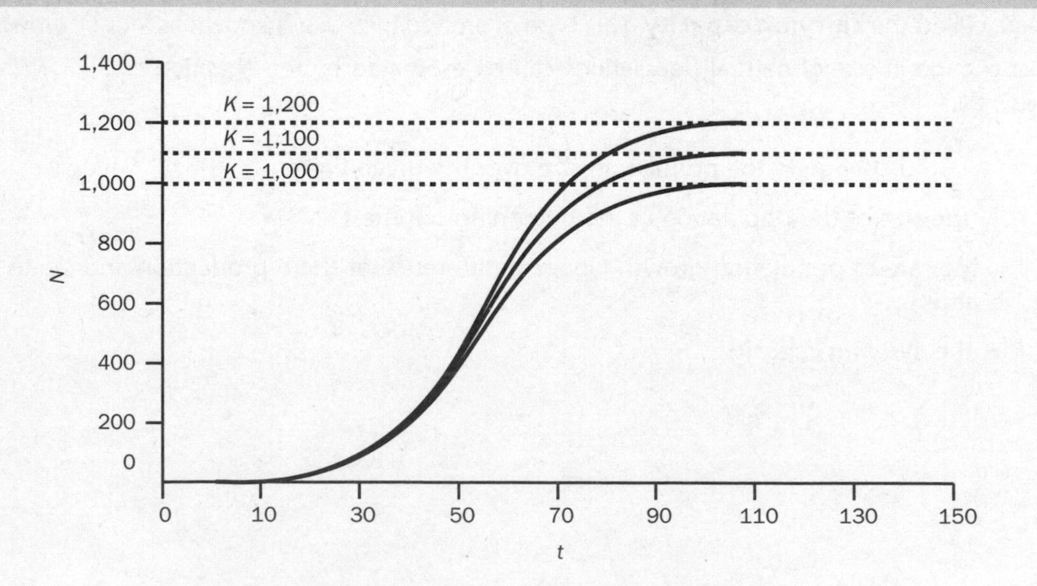

Populations with Different Carrying Capacities

The three lines in the previous graph show the increases in population size with a constant rate of population growth ($r = 0.10$) for three populations whose carrying capacities are not the same. Carrying capacities are indicated on the graph. When the carrying capacity increases for a population with a constant growth rate, the shape of the **growth curve** hardly changes. It is the ultimate population size that is really affected.

✔ **AP Expert Note**

Population Growth Factors

Factors that affect population growth fall into two categories, density-independent and density-dependent.

- **Density-independent factors** affect populations in the same way regardless of how many individuals are in the population at that given time. These factors tend to be more catastrophic, abiotic factors, such as flood, drought, hurricane, or fire.
- The impact of **density-dependent factors** on populations increases as the number of individuals in the population, and therefore the density in a given area, increases. These factors include variables such as competition among or between species, predation, or emigration.

Different **biotic** and **abiotic** factors in the environment have contributed to adaptive strategies that allow organisms to maximize their reproduction. Species that monopolize rapidly changing environments, such as disturbed habitats, produce many offspring quickly. These species are called ***r*-selected**, or *r*-strategists, because their strategy is to increase their *r*-value. Species that are *r*-strategists generally have short life spans, begin breeding early in life, and produce large numbers of offspring.

Species in more stable environments tend to produce fewer offspring and invest more resources into the success of each offspring. These species are called ***K*-selected**, or *K*-strategists. *K*-strategists have longer life spans, begin breeding later in life, have longer generation times, and produce fewer offspring. They take better care of their young than *r*-strategists; they have also evolved to be better at exploiting limited parts of their environments. There is a continuum between *r*-selected and *K*-selected species, as few communities are made up of strictly *r*-strategists or *K*-strategists.

Every species has a set of conditions that are optimal for its reproductive strategies; in any ecosystem, the most abundant species will be the one for which the environment most closely approximates this set of optimal conditions. If these conditions remain constant, all other things being equal, the distribution of organisms in an ecosystem should remain roughly the same.

✔ **AP Expert Note**

Without important changes to the environment or available resources, there is little reason to expect a population to boom or crash. However, these kinds of significant changes are inevitable over time, and they determine which species can thrive or dominate in an ecosystem. This change in the species makeup of an ecosystem is called **ecological succession**.

ENVIRONMENTAL DYNAMICS

HIGH YIELD

22.3 Explain how environmental factors affect organisms

22.4 Contrast how matter is recycled and energy flows

Environmental dynamics deals with the flow of energy and other nutrients through the environment. **Energy flow** is traditionally shown in the form of a **food web**. Arrows are drawn between different species of organisms, with the direction of the arrow indicating the direction of energy flow.

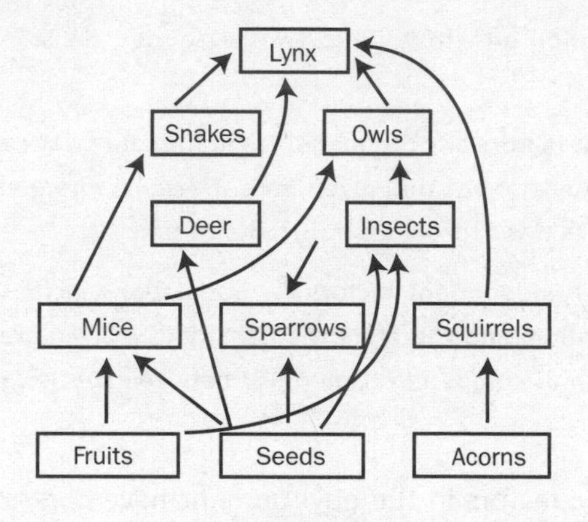

Food Web

In the illustration of a terrestrial food web, the arrows indicate the direction of energy flow between the organisms. The fruits, seeds, and acorns only have arrows pointing away from them, so they are primary producers. The lynx is the top predator (tertiary **consumer**) and only has arrows pointing toward it. There are no decomposers shown in this food web.

> ✔ **AP Expert Note**
>
> **Energy is lost as it moves through the various food webs. Only about 10 percent of the energy received is passed from one trophic level to the next.**

Food webs can quickly become quite complicated because there are so many different organisms interacting in any given ecosystem. Food webs are rarely entirely comprehensive for the same reason. As one kind of organism feeds another, nutrition moves through a food web from one trophic level to the next. Organisms included in a food web may be **primary producers** (**autotrophs**), which produce energy from sunlight, **primary consumers** and **secondary consumers**, and **decomposers**. Because they make their own food, primary producers only have arrows pointing away from them in a food web. Organisms that get energy by consuming other living things (**heterotrophs**) have arrows pointing toward them, indicating the flow of energy from another trophic level. If decomposers are part of the food web, there will also be arrows leading from the various levels of heterotrophs to the decomposers. Primary consumers eat plants, while secondary consumers, tertiary consumers, and quaternary consumers feed on the levels below them, such as the primary consumers. **Herbivores** eat only plants (thus, they are always primary consumers).

However, some secondary and tertiary consumers are omnivores, which eat both animals and plants. Omnivores in a food web are indicated by arrows pointing toward them from both primary producers and primary consumers.

> ✔ **AP Expert Note**
>
> **It is important to understand productivity as it relates to energy flow in food webs—many AP Biology exam questions will deal with differences between gross and net primary productivity and factors that influence productivity within a food chain or food web.**

Other simplified models illustrating how nutrients are cycled include those for water, nitrogen, carbon, and phosphorous. These systems are known as biogeochemical cycles.

In the *water cycle*, the Sun's energy drives the movement of water through the environment. The Sun causes surface water to evaporate and plants to transpire, releasing water vapor into the atmosphere. In the atmosphere, the water vapor cools and condenses into clouds and precipitation. Precipitation returns water to the Earth's surface, where it runs off into lakes and rivers and percolates into groundwater.

There is more nitrogen in the air than oxygen: nitrogen gas makes up over two-thirds of Earth's air. However, most living things cannot get the nitrogen they need from nitrogen gas. Certain groups of bacteria in the soil, called *nitrogen-fixing bacteria*, convert nitrogen gas into nitrates and nitrites; plants then take up the nitrites and nitrates from the soil and from fertilizers. Animals, including humans, get nitrogen by eating plants and other living or dead organic matter. When animals and plants die, groups of bacteria and fungi break down these organic remains and release nitrogen back into the soil. Animal waste contains nitrogen as well. *Denitrifying bacteria* in the soil then change the nitrates and nitrites back into nitrogen gas, which returns to the air. Together, all of these steps make up the *nitrogen cycle*.

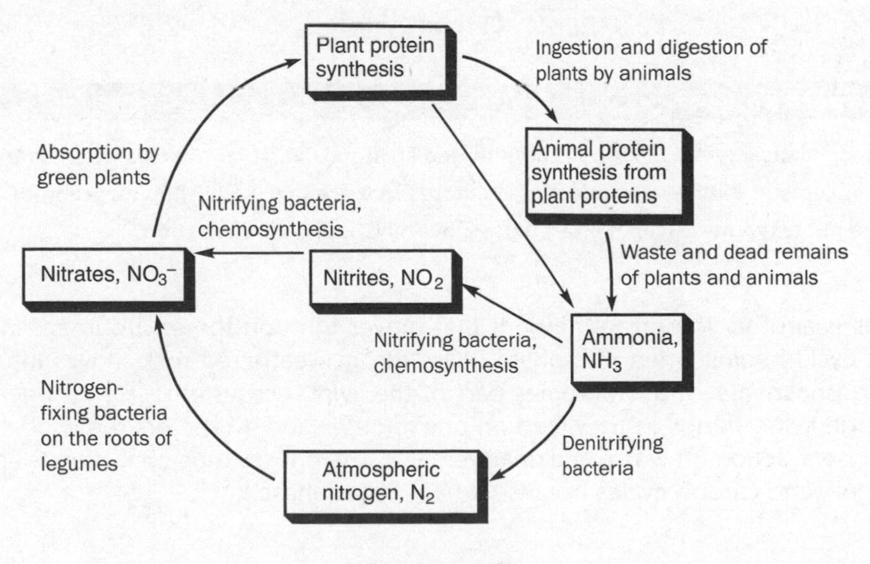

Nitrogen Cycle

Photosynthesis and respiration are the complementary reactions of the **carbon cycle**. In photosynthesis, carbon dioxide and water combine to form sugars and oxygen; the Sun's energy is stored in the bonds among carbon atoms in the sugars. Respiration uses carbohydrates and oxygen to produce carbon dioxide, water, and energy; it releases the energy that is stored by photosynthesis. As animals eat the plants and one another and then breathe, carbon moves through the food web; the carbon that plants and animals contain is returned to the soil when they decompose. The carbon in the soil enters the cycle again, perhaps as part of a microorganism. A great deal of carbon is stored in the oceans and in minerals. Combustion is another way that carbon enters the atmosphere, in the form of carbon dioxide. The widespread and heavy use of fossil fuels has created concerns about how all of this combustion is affecting the environment.

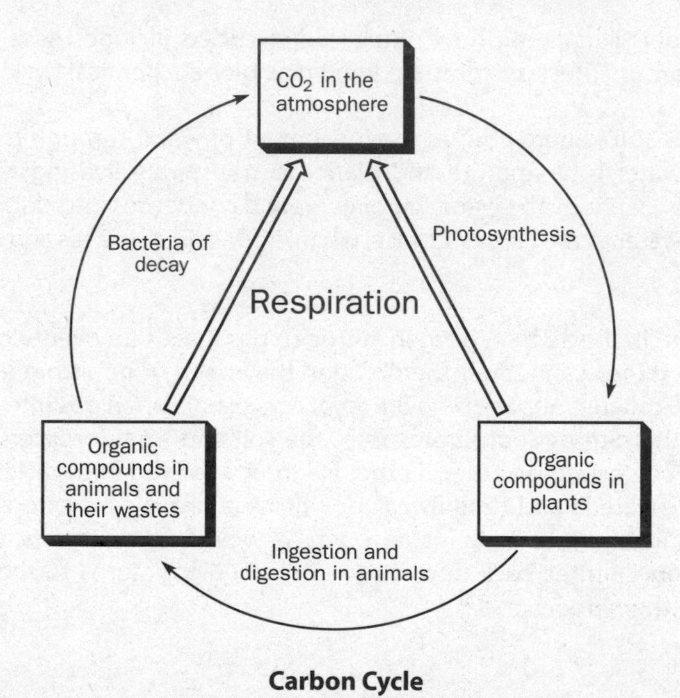

Carbon Cycle

> ✔ **AP Expert Note**
>
> **The biogeochemical cycles of carbon and nitrogen are good examples of areas where you can relate several concepts in biology. Examples of processes that are involved in the recycling of these elements are photosynthesis, cellular respiration, hydrolysis, and condensation.**

Phosphorous is another important nutrient that moves through the environment in a cycle. The phosphorus cycle begins when phosphate (PO_4^{3-}) from weathered rock moves into soils. Plants then take up phosphate, and it becomes part of the living ecosystem. Like nitrogen, phosphate moves through living things as they feed on one another, and it reenters the ecosystem through the decomposers' action on waste and dead remains. The phosphorus cycle stands apart from the water, nitrogen, and carbon cycles because it has no gas phase.

ECOSYSTEMS

22.5 Explain how diversity affects stability

In ecology, the levels of organization from lowest complexity to highest are organism, population, community, ecosystem, biome, and biosphere. **Populations** are all the organisms that occur in a specific **habitat**, and **communities** are all the populations that live in that habitat. **Ecosystems** encompass communities and their abiotic environment, including their interactions through which nutrients and energy flow. The role and position a species has in an ecosystem and its interactions with abiotic and biotic factors of the environment is called an ecological **niche**. Ecosystems within a specific geographic region form **biomes**, and together all ecosystems form a complex web that make up the biosphere.

An ecosystem can be a tropical rain forest, an entire mountain range or a coral reef, but can also be an ephemeral pool of water in the middle of the desert. Due to limiting factors, such as food, nutrients, land, water, and sunlight, ecosystems are unable to support an unlimited number of organisms. The health and balance of ecosystems are sustained by the biodiversity of the ecosystem. Biodiversity includes the genetic diversity of the species, as well as the various ways species interact with each other and their environment. All species in an ecosystem depend on one another. As seen in chapter 19, the level of variation in a population affects population dynamics (size and age). Similarly, the diversity within an ecosystem may influence the stability of the ecosystem. Providing robustness to the ecosystem, diversity increases the ability of ecosystems to tolerate and respond to changes in the environment, as well as preventing widespread diseases. Keystone species, producers, and abiotic and biotic factors all contribute to maintaining the diversity of an ecosystem. Human activities, however, have strained the diversity of ecosystems and could eventually lead to another mass extinction.

Keystone species are species that help to increase the diversity of an ecosystem. Many species are dependent upon keystone species. Thus, the disappearance of keystone species from an ecosystem would drastically affect the balance of the ecosystem and may lead to the disappearance of other species. For example, sea otters, a keystone species, feed on sea urchins, which feed on kelp. The regulation of sea urchin populations, in turn, maintains sufficient kelp for food and habitats of other species like fish, which use kelp to hide from predators. The removal of sea otters would drastically affect the marine ecosystem since, without a predator, sea urchins would consume all the kelp.

Other predators, like jaguars, which have a diet of many different species, keep species in balance by controlling distribution and population of prey species. Without predators, prey would increase and lead to competition with each other for resources and possibly the decline of a species. Beavers alter the environment by building dams and creating wetlands upon which the prevalence and activities of many species depend. Bees and hummingbirds are also keystone species because they contribute to the survival of several plant species through pollination. Plants provide shelter for insects, which are food for other species.

Plants not only provide habitats for a variety of species, they also provide food. Plants, or producers, convert energy from the environment into organic compounds, which begins the energy flow through organisms living in ecosystems. Energy is required for a stable ecosystem and, without producers, ecosystems would have no basis of energy flow.

Abiotic factors including climate (temperature, humidity, day length, rainfall) and physical conditions (pH, water, ions) affect the amount of environmental stress on an ecosystem and, in turn, the stability of the ecosystem. Abiotic factors that increase environmental stress decrease ecosystem stability, whereas abiotic factors that decrease environmental stress increase ecosystem stability.

GLOBAL ISSUES

22.6 Describe the human impact on biodiversity and climate

Other than the major extinction events recorded every 100 million years or so in Earth's geological history, the only variables that seem to have a global impact are related to human activity. You should be familiar with two of these variables.

The Earth is surrounded by a layer of ozone (O_3). The layer lies in the stratosphere, 10–17 kilometers above the Earth's surface, and protects the surface from harmful UV rays emitted by the sun. Over the past few decades, scientists have determined that the ozone layer is being depleted by compounds containing chlorine, fluorine, bromine, carbon, and hydrogen (halocarbons) that seem to be produced by human activity. One particular group of chemicals, CFCs (chlorofluorocarbons), has received a considerable amount of the blame for depletion of the ozone layer.

✔ **AP Expert Note**

The ozone layer may be depleted by as much as 60 percent over the Antarctic in the spring, causing the notorious hole in the ozone layer. Scientists worry that without changes in human activity, the ozone layer will be depleted beyond recovery, preventing certain forms of life from surviving on Earth.

The other major global impact of human activity is the release of greenhouse gases (water vapor, carbon dioxide, methane, and nitrous oxide) into the atmosphere from the inefficient burning of fossil fuels (there are sources from manmade aerosols as well). As these gases become denser in the Earth's atmosphere, they prevent heat from escaping the Earth's surface, causing global warming. Scientists believe that the Earth's atmospheric temperature is steadily rising and that this will cause major shifts in seawater levels, local climates, and world weather patterns.

Habitat destruction occurs as people clear natural areas for natural resources, housing, and recreational areas. This has resulted in the extinction of many species of plants and animals and has an adverse impact on our worldwide water resources, global temperatures, and food availability.

AP BIOLOGY LAB 10: ENERGY DYNAMICS INVESTIGATION

22.7 Investigate energy flow through a system

Using a model ecosystem, this investigation explores how matter and energy flow, the roles of producers and consumers, and the complex interactions between organisms. First, estimate the net primary productivity of Wisconsin Fast Plants growing under lights, and then determine the flow of energy from the plants to cabbage white butterflies as the larvae consume cabbage-family plants.

Remember that the source of almost all energy on Earth is the Sun. Free energy from sunlight is captured by producers that convert the energy to oxygen and carbohydrates through photosynthesis. The net amount of energy captured and stored by the producers in a system is the system's net productivity. Net productivity is calculated by assuming the change in biomass of a plant is due to uptake and use of energy.

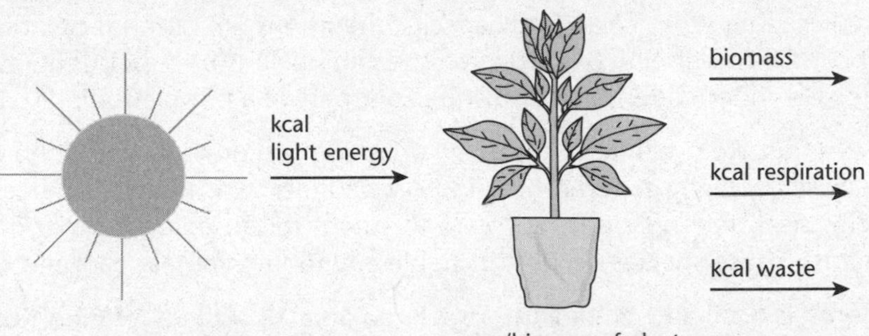

To determine the net primary productivity of your plants, convert the difference in biomass over the growing time to energy according to the following equation:

$$\text{Energy} = \text{Biomass in grams} \times 4.35 \text{ kcal/gram}$$

Let's look at a sample set of data for 10 plants grown over seven days. In seven days, the 10 plants gain 4.2 grams of dry mass. Using the equation, you can determine that 4.2 grams of dry mass is equivalent to 18.27 kcal of energy captured from the Sun. To determine the net primary productivity per plant per day, you must divide that total amount of energy by 10 plants and by seven days. In this example, the net primary productivity is 0.26 kcal per plant per day.

Age of Plants	Dry Mass Gained	Energy (g biomass × 4.35 kcal/g)	Net Primary Productivity per Plant per Day
7 days	4.2 grams	18.27 kcal	0.26 kcal/day

The efficiency of energy transfer from producer to primary consumers varies with the type of organism and with the characteristics of the ecosystem. In the second part of the investigation, determine the biomass change in butterfly larvae that eat your plants to evaluate the energy use of primary consumers and ultimately the energy efficiency of the relationships.

RAPID REVIEW

If you take away only 6 things from this chapter:

1. Populations with infinite resources increase exponentially in a J-shaped curve. Usually, the environment has a carrying capacity, the maximum number of individuals it will support. Factors affecting population size can be density-dependent (overcrowding) or density-independent (a storm wipes out part of a population).
2. Species that are *r*-selected have many offspring and provide almost no parental care (fish, insects). They succeed best in new habitats with many resources. *K*-selected species have few offspring and care for them for an extended period (elephants, humans). Those species do best in stable environments near carrying capacity.
3. A biome is a climatic zone with associated animals and vegetation (tundra, desert, etc.). An ecosystem comprises a community of living organisms and its habitat. Different populations of organisms make up the community. Each organism is adapted to a specific niche or role.
4. Food webs trace energy flow in a community. Different trophic levels consist of primary producers (plants), primary consumers (herbivores), secondary consumers (carnivores), and decomposers (bacteria).
5. Materials such as water, nitrogen, carbon, and phosphorus travel through the environment and are recycled in biogeochemical cycles involving different life forms.
6. Human activity has affected the environment significantly (e.g., ozone depletion, habitat destruction, and global warming).

TEST WHAT YOU LEARNED

1. Oligotrophic lakes are nutrient poor and have very clear water. However, large quantities of fertilizer runoff can cause more biological growth in these lakes. Continuing input of fertilizer can transform oligotrophic lakes into eutrophic lakes that are rich in organic material. How does the addition of fertilizer ingredients, such as ammonia, lead to changes in the lakes?

   (A) Ammonia in fertilizer can be broken down to release energy through cellular respiration, making it important for organisms to be able to carry out metabolism.

   (B) Ammonia can be broken down to release heat, making it important in thermoregulation within the lake.

   (C) Ammonia plays a role in the nitrogen cycle, making nitrogen available for organisms, such as plants and animals.

   (D) Ammonia is a major source of energy for producers and a component of the nitrogen cycle, so its presence is sufficient to promote the growth of photoautotrophs.

2. A study investigated how various disturbances affect coral reef recovery. Data from the study is shown in the graphs below. The asterisks represent disturbances that may affect the amount of coral.

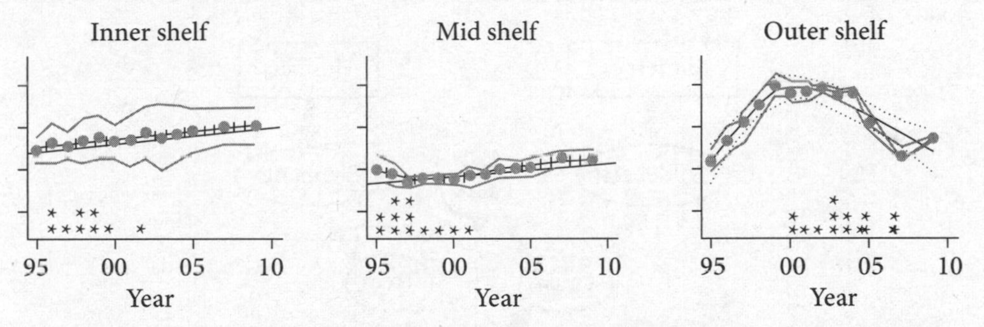

Percent coral cover over the years shown. The average linear trend is depicted with a straight line, whereas the dashed lines represent subregions. Each asterisk shows the timing of a disturbance.

Adapted from Kate Osborne et al., "Disturbance and the Dynamics of Coral Cover on the Great Barrier Reef (1995–2009)," *PLoS ONE* 6, no. 3 (March 2010): e17516.

Which of the following is most justified based on this data?

   (A) The coral populations on the inner shelf and mid shelf increased slightly when the disturbances ceased. The outer shelf population had a period approximating exponential growth followed by a period of disturbance and decrease in growth.

   (B) The coral populations on the inner shelf and mid shelf were relatively stable during the study due to lack of disturbance. The outer shelf population was affected by more disturbances and exhibited a classic logistic growth curve.

   (C) All three figures show classic exponential growth curves.

   (D) All three figures show classic logistic growth curves.

3. A researcher decided to create a trophic pyramid by modeling an ecosystem with plants, caterpillars that eat the plants, and birds that eat the caterpillars. Each part of the pyramid represents the biomass of that level, with primary producers on the bottom. The plants were first grown in low light, and the light intensity was raised over time to the maximum intensity that doesn't harm the plants. How would the trophic pyramid most likely change over the course of this experiment?
   (A) The pyramid would become inverted due to the rapid movement of energy and materials from one level to another.
   (B) The level representing the primary producers would increase, but the upper levels would stay the same size as before because of the limits on energy transfer.
   (C) The pyramid would stay the same because the amount of energy that could be transferred from one level to another would be unchanged.
   (D) There would be an increase in the size of each piece of the pyramid due to gradually increasing biomass at each level.

4. The figure below shows a model generated from a study of interactions in a Serengeti ecosystem. This figure describes hypothesized relationships between a parasite that causes disease in wildebeest, rinderpest, and other important components of the ecosystem. Thick arrows represent dominant effects, and "grass" is in a dotted circle because that variable was not directly measured in the study.

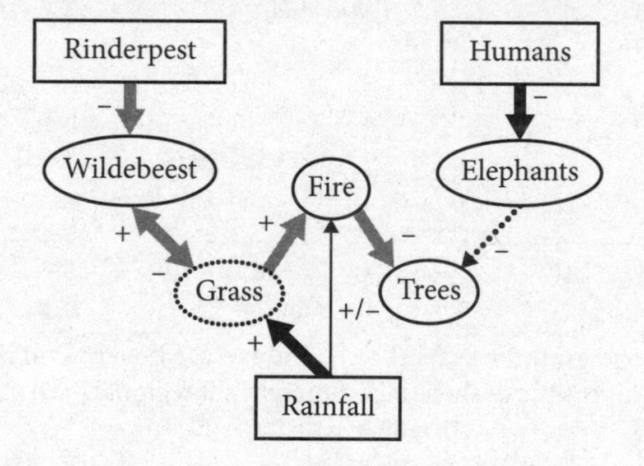

Adapted from Ricardo M. Holdo et al., "A Disease-Mediated Trophic Cascade in the Serengeti and its Implications for Ecosystem C," *PLoS Biol* 7, no. 9 (September 2009): e1000210.

Which of the following can most reasonably be predicted based on this figure?
   (A) Increasing the amount of rinderpest would decrease the amount of grass.
   (B) Increased fire would lead to higher tree and elephant populations.
   (C) Increasing the amount of rinderpest could indirectly lower tree numbers.
   (D) Increased human populations have a net positive effect on the elephant population.

5. A researcher is interested in studying how biotic and abiotic factors influence the population of a particular species. Which of the following experimental designs would most effectively test whether one abiotic and one biotic factor influenced a population?

   (A) The researcher conducts three experiments. In one, she manipulates prey density. In a second, she manipulates population density of the study species. In the third, she manipulates both prey and study species population density to examine the effects of interactions between prey and study species density.

   (B) The researcher conducts three experiments. In one, she manipulates light levels. In a second, she manipulates prey availability. In a third, she manipulates light levels and prey availability to examine the effects of interactions between light levels and prey availability.

   (C) The researcher conducts two experiments. In one, she manipulates the number of cover objects available for the animals to use for shelter. In another, she manipulates the amount of light available in the study areas.

   (D) The researcher conducts a single experiment in which she compares animals with high light levels and high prey density to animals with low light levels and low prey density.

6. As it has become easier for humans to travel around the world, there have been some unintended consequences. Which of the following is NOT an unintended consequence of increased human movement and trade around the world?

   (A) Species have been introduced into new areas with food shipments, allowing them to compete with native species and disrupt ecosystems.

   (B) Pathogens can rapidly reach new locations via airplanes, potentially harming human and animal populations.

   (C) Ship ballast water can contain living organisms, and these can be transported to new destinations between ballast water discharges.

   (D) Bacteria can develop resistance after being exposed to antibiotics, making the illnesses they cause harder to treat.

7. Figure A below shows the relative fitness of a prey species when one or two predators are present. Figure B shows how the prey phenotype is predicted to change depending on which predators are present.

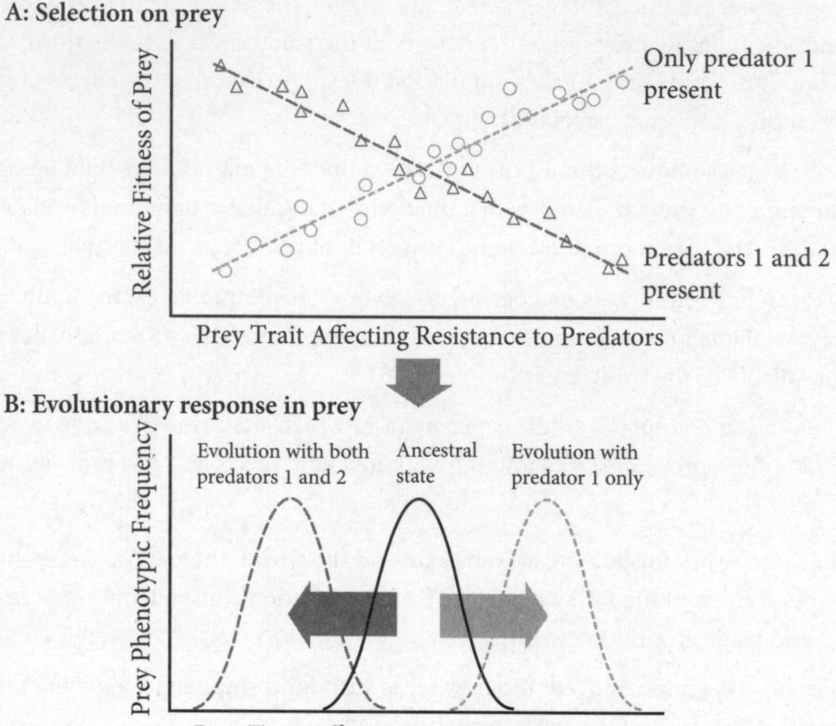

Adapted from Martin M. Turcotte, Michael S. C. Corrin, and Marc T. J. Johnson, " Adaptive Evolution in Ecological Communities," *PLoS Biol* 10, no. 5 (May 2012): e1001332.

Which of the following conclusions can most reasonably be drawn regarding this system?

(A) The change in phenotypic frequency shown can only occur if the trait is ubiquitous in the population.

(B) The change in phenotypic frequency shown can only occur if there is existing genetic variation for the trait in the population.

(C) The trait under selection must be inherited in a Mendelian manner in order for changes in phenotypic frequency to occur.

(D) The plot lines of figure A must have a different shape in order for the changes in phenotypic frequency to occur.

The answer key to this quiz is located on the next page.

Answer Key

Test What You Already Know

1. **D** **Learning Objective:** 22.1
2. **A** **Learning Objective:** 22.2
3. **C** **Learning Objective:** 22.3
4. **B** **Learning Objective:** 22.4
5. **A** **Learning Objective:** 22.5
6. **D** **Learning Objective:** 22.6
7. **B** **Learning Objective:** 22.7

Test What You Learned

1. **C** **Learning Objective:** 22.4
2. **A** **Learning Objective:** 22.1
3. **D** **Learning Objective:** 22.7
4. **C** **Learning Objective:** 22.3
5. **B** **Learning Objective:** 22.2
6. **D** **Learning Objective:** 22.6
7. **B** **Learning Objective:** 22.5

Detailed solutions can be found in the Answers and Explanations section at the back of this book.

REFLECTION

Test What You Already Know score: __________

Test What You Learned score: __________

Use this section to evaluate your progress. After working through the pre-quiz, check off the boxes in the "Pre" column to indicate which Learning Objectives you feel confident about. Then, after completing the chapter, including the post-quiz, do the same to the boxes in the "Post" column. Keep working on unchecked Objectives until you're confident about them all!

Pre	Post		
❑	❑	**22.1**	Contrast exponential and logistic growth models
❑	❑	**22.2**	Explain how biotic and abiotic factors affect population
❑	❑	**22.3**	Explain how environmental factors affect organisms
❑	❑	**22.4**	Contrast how matter is recycled and energy flows
❑	❑	**22.5**	Explain how diversity affects stability
❑	❑	**22.6**	Describe the human impact on biodiversity and climate
❑	❑	**22.7**	Investigate energy flow through a system

FOR MORE PRACTICE

Complete more practice online at kaptest.com. Haven't registered your book yet? Go to kaptest.com/booksonline to begin.

PART 7

Practice

CHAPTER 23

Free Response

LEARNING OBJECTIVES

In this chapter, you will learn how to:

23.1 Use the reading period to outline your responses

23.2 Craft effective responses for long free-response questions

23.3 Craft effective responses for short free-response questions

23.4 Use a scoring rubric to self-score your responses

FREE RESPONSE STRATEGY

23.1 Use the reading period to outline your responses

23.2 Craft effective responses for long free-response questions

23.3 Craft effective responses for short free-response questions

Approaching the 10-Minute Reading Period

There is a 10-minute reading period sandwiched between the multiple-choice and grid-in section and the free-response portion of the AP Biology exam. Note that this is a "reading" period, not a "nap" period. Ten minutes isn't much time, but it does give you an opportunity to read the essay questions. You can write notes in the question booklet, but these will not be seen by graders.

This gives you just over one minute per question to plan your response. Take at least 30 seconds to read and reread each question. Make sure you understand what is being asked. Next, you should jot down any thoughts you have about the answer on a piece of scratch paper or in the test booklet. Write down keywords you want to mention. At some point or other, you have probably brainstormed ideas when writing an essay for your English class. This is exactly what you want to do here as well: Brainstorm ideas about the best way to answer each free-response question.

With the remaining time, make a quick outline of how you would answer each question. You don't need to write complete sentences; just jot down notes that you can understand. If drawings or diagrams are requested, make a brief, crude version of what they will look like.

The following is an example of a free-response question and some notes that could be taken. The actual answer should not be in this outline form, but in a coherent essay, with more detailed descriptions of each concept where appropriate.

> Transcription is the process of generating RNA from a DNA blueprint. It occurs in both prokaryotes and eukaryotes, but the mechanism and results are different.
>
> (a) Transcription in both prokaryotes and eukaryotes involves enzymes and other molecules and sequences. For each (prokaryote and eukaryote), **describe** the function of TWO such components needed for transcription.
>
> (b) Messenger RNA produced by transcription in prokaryotes differs from that in eukaryotes produced by transcription and post-transcriptional processing. **Describe** TWO post-transcriptional modifications in eukaryotes that lead to these differences.
>
> (c) **Explain** how differences in the location of transcription between prokaryotes and eukaryotes affect the translation of a transcript.

Things to Cover:

(a) Prok — same RNA pol for all RNA (sigma factor), –10 and –35 region of promoter

Euk — RNA pol I, II, and III (rRNA, mRNA, tRNA), transcription factors

(b) Euk — Introns removed to form single gene, cap/poly A tail added

Prok — many genes on one mRNA (operon/polycistronic), no introns, cap, or tail

(c) Prok — cytoplasm, translation begins before transcription done → Christmas-tree structure

Euk — nucleus (rRNA in nucleolus), no translation until exits

By the time you've written something like this, your one-to-two minutes on that question should be up. Move on to the next question and repeat the process. When the time comes to start writing your answers, you'll have a good set of notes on which to base your answers to each question.

Approaching Free-Response Questions

For Section II, you'll have 80 minutes (after the reading period) to answer eight questions. That's an average of 10 minutes per question, but you'll want to spend about 20 minutes each on the two long questions and about 6–7 minutes each on the six short ones. Take the time to make your answers as precise and detailed as possible while managing the allotted time.

Important Distinctions

Each free-response question will, of course, be about a distinct topic. However, this is not the only way in which these questions differ from one another. Each question will also need a certain kind of answer, depending on the type of question it is. Part of answering each question correctly is understanding what general type of answer is required. There are several important signal words that indicate the rough shape of the answer you should provide. Here are five of the most common:

- Describe
- Discuss
- Explain
- Compare
- Contrast

Each of these words indicates that a specific sort of response is required; none of them mean the same thing. Questions that ask you to *describe*, *discuss*, or *explain* are testing your comprehension of a topic. A description is a detailed verbal picture of something; a description question is generally asking for "just the facts." This is not the place for opinions or speculation. Instead, you want to create a precise picture of something's features and qualities. A description question might, for example, ask you to describe the results you would expect from an experiment. A good answer here will provide a rich, detailed account of the results you anticipate.

A question that asks you to discuss a topic is asking you for something broader than a mere description. A discussion is more like a conversation about ideas, and—depending on the topic—this may be an appropriate place to talk about the tension between competing theories and views. For example, a discussion question might ask you to discuss which of several theories offers the best explanation for a set of results. A good answer here would go into detail about why one theory does a better job of explaining the results, and it would talk about why the other theories cannot cope with the results as thoroughly.

A question that asks you to explain something is asking you to take something complicated or unclear and present it in simpler terms. For example, an explanation question might ask you to explain why an experiment is likely to produce a certain set of results, or how one might measure a certain sort of experimental result. A simple description of an experimental setup would not be an adequate answer to the latter question. Instead, you would need to describe that setup *and* talk about why it would be an effective method of measuring the result.

✔ **AP Expert Note**

Compare vs. Contrast

Questions that ask you to *compare* or *contrast* are asking you to analyze a topic in relation to something else. A question about comparison needs an answer that is focused on similarities between the two things. A question that focuses on contrast needs an answer emphasizing differences and distinctions.

Three Points to Remember about Free-Response Questions

1. *Most questions are stuffed with smaller questions.* You usually won't get one broad question like, "Are penguins really happy?" Instead, you'll get an initial setup followed by questions labeled (a), (b), (c), and so on. Expect to spend a paragraph writing about each lettered question.
2. *Writing smart things earns you points.* For each subquestion on a free-response question, points are given for saying the right thing. The more points you score, the better off you are on that question. The AP Biology people have a rubric, which acts as a blueprint for what a good answer should look like—we feature similar rubrics in the explanations to our free-response questions, so you have an idea of what these might look like. Every subsection of a question has one to five key ideas attached to it. If you write about one of those ideas, you earn yourself a point. There's a limit to how many points you can earn on a single subquestion, but it boils down to this: Writing smart things about each question will earn you points toward that question.

✔ **AP Expert Note**

Don't forget—you *only* receive points for relevant correct information; you receive *no points* for incorrect information or for restating the question, which also eats up valuable time!

3. *Mimic the Data questions*. Data questions often describe an experiment and provide a graph or table to present the information in visual form. On at least one free-response question, you will be asked about an experiment in some form or another. To score points on this question, you must describe the experiment well and perhaps present the information in visual form.

Beyond these points, there's a bit of a risk in the free-response section because there are only eight questions. If you get a question on a subject you're weak in, things might look grim. Still, take heart. Quite often, you'll earn some points on every question because there will be some subquestions or segments that you are familiar with.

Remember, the goal is not perfection. If you can ace four of the questions and slog your way to partial credit on the other four, you will put yourself in a position to get a good score on the entire test. That's the Big Picture—don't lose sight of it just because you don't know the answer to any particular subquestion!

✔ **AP Expert Note**

Maximize Your FRQ Score

1. **Only answer the number of subsections the long free-response questions call for. For example, if the question has four sections (a, b, c, and d) and says to choose three parts, then choose *only* three parts.**
2. **There are almost always easy points that you can earn. State the obvious and provide a brief but accurate explanation for it.**
3. **In many instances, you can earn points by defining relevant terms. (Example: Writing *osmosis* would not get you a point, but mentioning "movement of water down a gradient across a semi-permeable membrane" would likely get the point).**
4. **While grammar and spelling are not assessed on the free-response portion, correct spellings of words and legible sentences will increase your chances of earning points.**
5. **You do not have to answer free-response questions in the order in which they appear on the exam. It's a good strategy to answer the questions you are most comfortable with first and then answer the more difficult ones.**
6. **The length of your response does not determine your score—a one-paragraph written response containing accurate, succinct, yet detailed information can score the maximum amount of points, while other essays spanning three to four paragraphs of vague, inaccurate materials may not earn any.**
7. **Be careful that you do not over-explain a concept. While the initial explanation gets you points, contradictions can cause points to be taken away.**
8. **Keep personal opinions out of free responses. Base your response on factual researched knowledge.**
9. **Relax and do your best. You know more than you think!**

SAMPLE QUESTIONS

23.4 Use a scoring rubric to self-score your responses

Attempt the following questions on scratch paper. At the end of this chapter, you will find rubrics and sample responses you can use to self-score your work.

1. Students performed an experiment to examine how hypertonic, hypotonic, and isotonic conditions affect osmosis. Eggs soaked in vinegar overnight were placed in solutions containing mixtures of distilled water and corn syrup. Initial and final mass measurements were recorded, and the results are shown in Table 1.

Table 1. Changes in the mass of eggs immersed in varying percent corn syrup solutions

Percent Corn Syrup (%)	0	10	20	30	40	50	60	70	80	90	100
Initial Mass (g)	71.1	67.4	71.7	72.3	69.4	70.3	72.9	71.6	77.0	77.5	76.1
Final Mass (g)	75.9	67.7	65.6	59.4	51.3	47.8	46.4	44.1	46.2	45.8	44.3
Change in Mass (g)	4.8	0.3	−6.1	−12.9	−18.1	−22.5	−26.5	−27.5	−30.8	−31.7	−31.8
Percent Change in Mass (%)	6.8	0.4	−8.5	−17.8	−26.1	−32.0	−36.4	−38.4	−40.0	−40.9	−41.8

(a) The reaction of egg shells (calcium carbonate) with vinegar (4% acetic acid) produces a water-soluble compound (calcium acetate) and carbon dioxide gas. **Explain** why, prior to the experiment, students placed eggs in vinegar overnight.

(b) Based on the data, **determine** whether the 0% and 100% solutions are hypertonic or hypotonic relative to the eggs. **Provide** reasoning for your response.

(c) **Identify** the relationship between the percent change in egg mass and the percentage of corn syrup in solution.

(d) According to the data, **determine** the approximate percentage of corn syrup at which the solution would be isotonic to the egg. **Justify** your response.

(e) Suppose an egg from the 100% corn syrup solution was then immersed into pure distilled water. **Predict** what would happen to the egg, and **justify** your response.

2. To investigate the relatedness of four different species, researchers determined the nucleotide sequences for four homologous genes. The table below shows a 20-nucleotide section of the gene for each species.

Table 1. Nucleotide sequences

	1	2	3	4	5	6	7	8	9	10	11	12	13	14	15	16	17	18	19	20
Species A	C	T	C	A	T	G	A	A	A	A	T	T	C	A	T	A	G	A	T	T
Species B	C	T	C	G	T	G	A	A	A	A	T	T	C	T	T	A	G	A	T	A
Species C	C	T	C	A	T	G	A	C	A	A	T	T	C	T	T	A	G	T	T	A
Species D	C	T	C	A	T	G	A	C	A	A	T	G	C	T	T	A	G	A	T	A

(a) On the template provided, use the information in Table 1 to **construct** a table scoring the number of nucleotide differences in the sequences between each pair of species.

(b) Using the data in the table you constructed in part (a), **create** a phylogenetic tree on the template provided to reflect the evolutionary relationships of the four species. **Provide** reasoning for the placement on the tree of the two species that are most closely related and for the species that is most distantly related to the others.

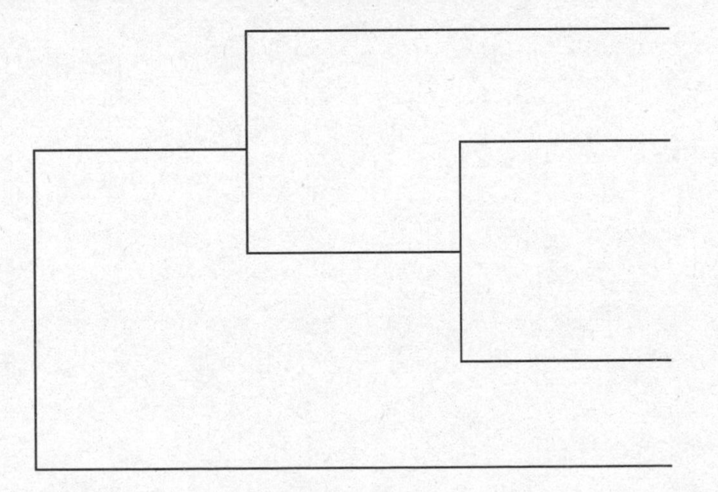

(c) Using the data in the phylogenetic tree you created in part (b) and the information in Table 1, **identify** the nucleotide position that corresponds with the branch point leading to Species C. **Justify** your answer.

3. Students examined the effects of pH on the reactivity of three enzymes (amylase, arginase, and pepsin). The results are shown in Figure 1.

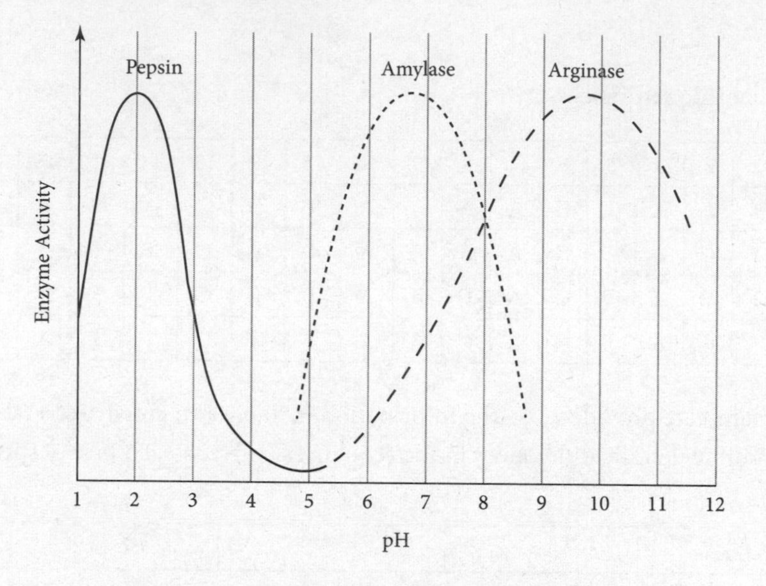

Figure 1. Enzyme activity as a function of pH

(a) Using the data, **determine** the optimum pH of pepsin.

(b) **Explain** why the optimum pH is not the same for each enzyme.

(c) Based on the data, **predict** the optimum pH of chymotrypsin, which has a working pH range from 5.2 to 9.8. **Provide** reasoning for your prediction.

4. The host range of a bacteriophage specifies the spectrum of bacterial species that the virus can infect. The figure shows the evolutionary change of three bacteriophages over time.

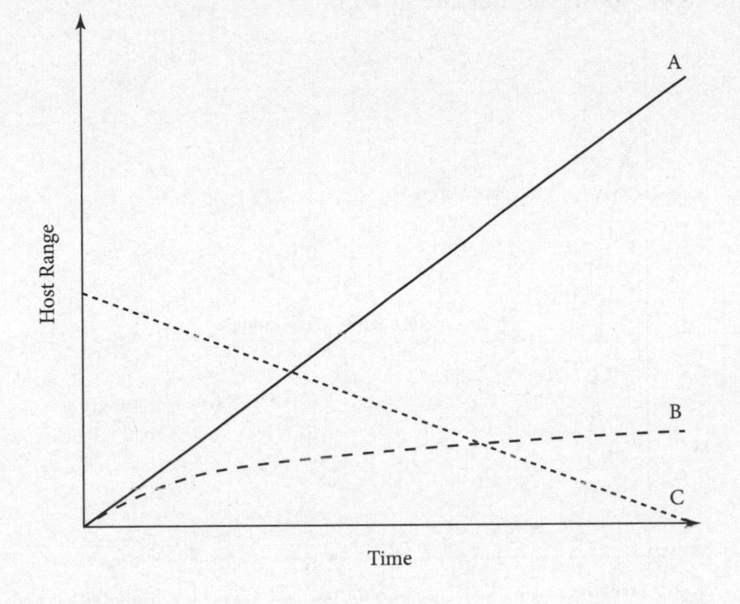

Figure 1. Host range (determined by the number of species infected) over time

(a) Using Figure 1, **describe** the evolution of bacteriophage A and of bacteriophage C.

(b) Based on the curve for bacteriophage B, **propose** ONE adaptation in a host that was produced by natural selection.

(c) **Explain** what determines a bacteriophage's host range.

5. Scientists discovered that during aerobic exercise, muscle fibers obtain ATP (required for muscle contraction) via four sources: free ATP, phosphocreatine, anaerobic respiration, and aerobic respiration. The figure shows the four energy pathways.

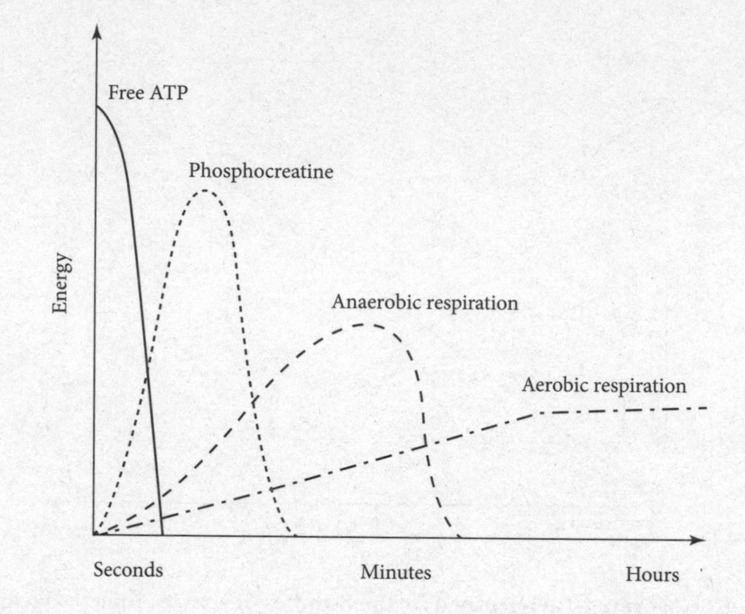

Figure 1. Energy production in muscle fibers during aerobic exercise

(a) **Explain** why cellular respiration is required for long periods of exercise.

(b) Using the template, **graph** the predicted shape of the aerobic respiration energy curve as oxygen becomes limited after a few hours of exercise. **Justify** your prediction.

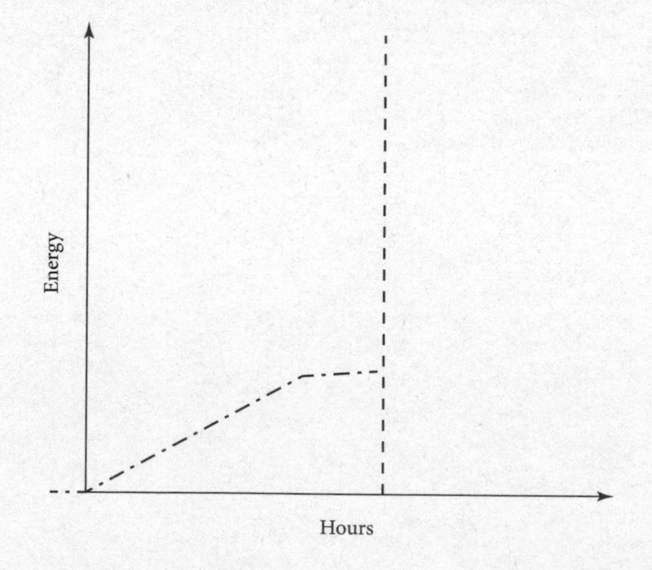

(c) The reaction of phosphocreatine + ADP to ATP + creatine is reversible, and during periods of rest the store of phosphocreatine is regenerated. **Explain** how cellular respiration returns muscles to normal after exercise.

6. Scientists observed that flower color in a plant was controlled by two alleles. The plant produced three flower colors: red, reddish-orange, and yellow. Table 1 shows the results of crosses of some of these plants (reddish-orange is simply listed as "orange").

Table 1. Parental and offspring phenotypes of crosses

Parental phenotype	Offspring phenotype (%)		
	red	**orange**	**yellow**
red × red	100	0	0
red × orange	50	50	0
yellow × orange	0	50	50
yellow × yellow	0	0	100

(a) Based on Table 1, **determine** the mode of inheritance for the observed flower colors. **Provide** reasoning for your answer.

(b) **Determine** the phenotype ratio of offspring produced from the self-fertilization of a reddish-orange plant.

7. The *trp* operon encodes biosynthetic enzymes to produce the amino acid tryptophan in *E. coli*. The expression of the structural genes (*trpE* to *trpA*) is regulated by a repressor protein (encoded by *trpR*) that binds the operator. When tryptophan is present, the operon is not transcribed.

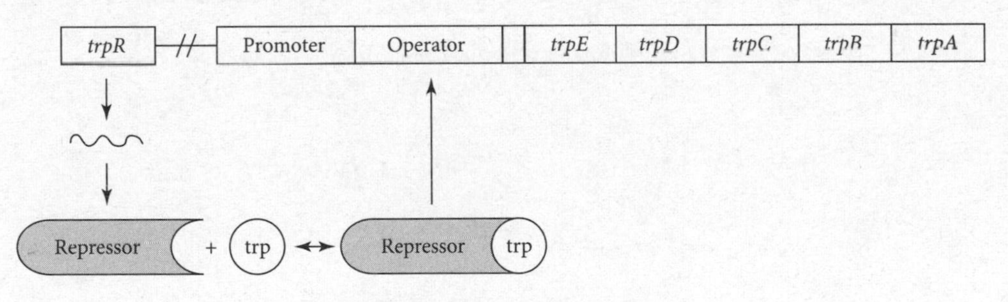

Figure 1. *trp* operon in *E. coli*

(a) **Explain** how the *trp* operon is turned off when tryptophan levels are high.

(b) **Explain** how the *trp* operon is turned on when tryptophan levels are low.

(c) **Predict** ONE effect a mutation in the *trpR* gene might have on gene regulation.

8. To control the population of cane beetles, which were destroying sugar cane crops, the toxic cane toad was introduced in Australia in 1935. The cane toad, however, was unsuccessful in controlling cane beetles. Today, with an estimated population of 200 million, the cane toad, an invasive species, threatens native species in the ecosystem.

   (a) **Propose** ONE reason that the introduction of the cane toad failed to control the cane beetle population.

   (b) **Propose** ONE reason that the population of cane toads has skyrocketed to threatening levels.

   (c) **Describe** ONE contributing factor that maintains the balance of an ecosystem.

ANSWERS AND EXPLANATIONS

1. **Scoring Guidelines for Free-Response Question 1**

   (a) The reaction of egg shells (calcium carbonate) with vinegar (4% acetic acid) produces a water-soluble compound (calcium acetate) and carbon dioxide gas. **Explain** why, prior to the experiment, students placed eggs in vinegar overnight. **(1 point maximum)**

Explanation (1 point)
• The reaction of calcium carbonate with vinegar dissolved the eggshells (exposing the semi-permeable membrane of the eggs) so osmosis through the egg could be examined

Here is a possible response that would receive full credit:

Vinegar dissolved the eggshells and exposed the semi-permeable membrane of the eggs. The semi-permeable eggs could then be used to examine how hypertonic, hypotonic, and isotonic conditions affect osmosis.

   (b) Based on the data, **determine** whether the 0% and 100% solutions are hypertonic or hypotonic relative to the eggs. **Provide** reasoning for your response. **(4 points maximum)**

Determination (1 point each, 2 points maximum)	Reasoning (1 point each, 2 points maximum)
• The 0% corn syrup (100% distilled water) solution is hypotonic relative to the egg • The 100% corn syrup (0% distilled water) solution is hypertonic relative to the egg	• Egg gained mass in 0% corn syrup • Egg lost mass in 100% corn syrup

Here is a possible response that would receive full credit:

The egg gained mass in the 0% corn syrup solution (100% distilled water), indicating water moved into the egg because the water concentration was greater outside the egg than inside. On the other hand, in the 100% corn syrup solution (0% distilled water), the egg lost mass, indicating water moved out of the egg because the water concentration was greater inside the egg than outside. Therefore, the 0% corn syrup solution is hypotonic, and the 100% corn syrup solution is hypertonic relative to the egg.

   (c) **Identify** the relationship between the percent change in egg mass and the percentage of corn syrup in solution. **(1 point maximum)**

Identification (1 point)
• Increasing the percentage of corn syrup decreases the percent change in mass (inverse relationship)

Here is a possible response that would receive full credit:

As the percentage of corn syrup increases from 0% to 100%, the percent change in egg mass decreases from 6.8% to −41.8%. Therefore, the relationship is inverse.

(d) According to the data, **determine** the approximate percentage of corn syrup at which the solution would be isotonic to the egg. **Justify** your response. **(2 points maximum)**

Determination (1 point)	Justification (1 point)
• The solution will be isotonic between 10% and 20% corn syrup	• The solution is isotonic when there is no change in the mass of the egg (0 g falls between 0.3 g and −6.1 g, which corresponds to 10% and 20% corn syrup)

Here is a possible response that would receive full credit:

A solution is isotonic when movement of water in and out of a cell is equal, meaning in this case that the mass of the egg would not change. According to the data, a zero change in mass would occur between the concentrations that caused changes of 0.3 g and −6.1 g, which correspond to the 10% and 20% corn syrup solutions, respectively. Thus, the percentage of corn syrup for the isotonic solution would fall somewhere between 10% and 20%.

(e) Suppose an egg from the 100% corn syrup solution was then immersed into pure distilled water. **Predict** what would happen to the egg, and **justify** your response. **(2 points maximum)**

Prediction (1 point)	Justification (1 point)
• The egg from the 100% corn syrup solution would increase in mass when placed in pure distilled water	• The concentration of water is greater outside the egg than inside, so water will move into the egg

Here is a possible response that would receive full credit:

When the egg from the 100% corn syrup solution is immersed into pure distilled water, the concentration of water is greater outside the egg than inside. As a result, water will move into the egg and cause it to swell.

2. **Scoring Guidelines for Free-Response Question 2**

(a) On the template provided, use the information in Table 1 to **construct** a table scoring the number of nucleotide differences in the sequences between each pair of species. **(2 points maximum)**

Table components (1 point each, 2 points maximum)
• Columns and rows correctly labeled
• Data correctly entered in table

Here is a sample table that would earn full credit:

Species	A	B	C	D
A	0	3	4	4
B		0	3	3
C			0	2
D				0

(b) Using the data in the table you constructed in part (a), **create** a phylogenetic tree on the template provided to reflect the evolutionary relationships of the four species. **Provide** reasoning for the placement on the tree of the two species that are most closely related and for the species that is most distantly related to the others. **(6 points maximum)**

Phylogenetic tree components (1 point for each species, 4 points maximum)
• Species A is the most distantly related species
• Species B is equally related to the other species
• Species C and D are the most closely related species (Note: C and D can be interchanged on the tree due rotation about the branch point)

Here is a sample tree that would earn full credit:

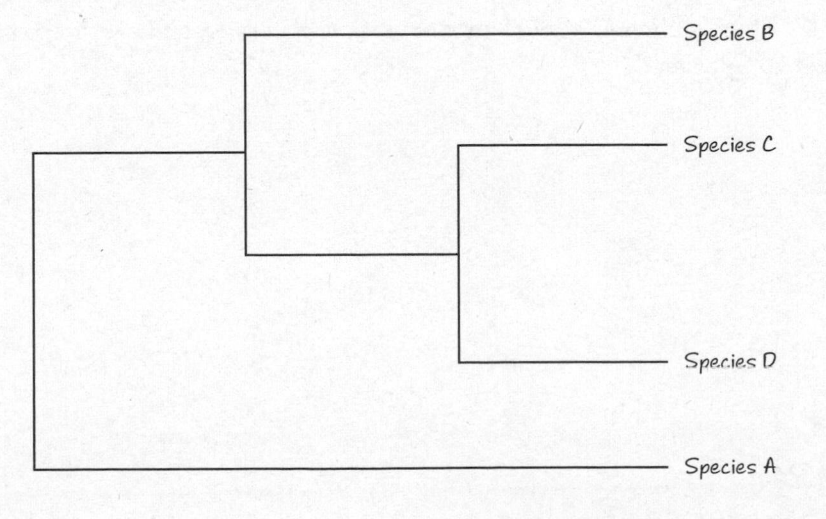

Practice

Reasoning (1 point each, 2 points maximum)
• The number of differences in nucleotides between Species C and D is the smallest (2), so Species C and D are the most closely related species (Note: C and D can be interchanged on the tree due to rotation about the branch point)
• The number of differences in nucleotides between Species A and C and between Species A and D are the largest (4), so Species A is most distantly related to Species C and D

Here is a possible response that would receive full credit:

Since the number of differences in nucleotides between Species C and D is 2, which is the smallest difference on the table, Species C and D are the most closely related. Therefore, the middle two branches on the phylogenetic tree should be labeled Species C and D. Species A is the most distantly related species because the number of differences in nucleotides between Species A and C, as well as between Species A and D, is the largest on the table. Thus, the bottom branch should be labeled A. Species B is equally different from Species A, C, and D, so it belongs on the top branch of the tree.

(c) Using the data in the phylogenetic tree you created in part (b) and the information in Table 1, **identify** the nucleotide position that corresponds with the branch point leading to Species C. **Justify** your answer. **(2 points maximum)**

Identification (1 point)	Justification (1 point)
• A change at nucleotide position 18 in the branch leads to Species C	• Species C and D differ at positions 18 and 12; Species C differs from Species A, B, and D at position 18, while Species D differs from Species A, B, and C at position 12

Here is a possible response that would receive full credit:

According to Table 1, Species C and D differ at positions 18 and 12. Species C differs from Species A, B, and D at position 18, and Species D differs from Species A, B, and C at position 12. Thus, a change at nucleotide position 18 corresponds with the branch point that leads to Species C.

3. **Scoring Guidelines for Free-Response Question 3**

(a) Using the data, **determine** the optimum pH of pepsin. **(1 point maximum)**

Determination (1 point)
• The optimum pH of pepsin, which occurs at maximum activity, is approximately 2

Here is a possible response that would receive full credit:

The optimum pH occurs when an enzyme is at maximum activity. Thus, the optimum pH of pepsin is 2.

(b) **Explain** why the optimum pH is not the same for each enzyme. **(1 point maximum)**

Explanation (1 point)
• The structure (and in turn function) of enzymes depends on the charges of the amino acids, which are influenced by pH

Here is a possible response that would receive full credit:

The optimum pH is not the same for each enzyme because the structure and function of enzymes depend on the amino acids in the chain, and these amino acids have functional groups that are influenced by pH. When pH changes, the charges on the amino acids may change due to the available H^+ and OH^- ions. Thus, the 3-D shape and function of the protein may be altered.

(c) Based on the data, **predict** the optimum pH of chymotrypsin, which has a working pH range of 5.2 to 9.8. **Provide** reasoning for your prediction. **(2 points maximum)**

Prediction (1 point)	**Reasoning (1 point)**
• The optimum pH of chymotrypsin is about 7.5	• Optimum pH occurs at maximum enzyme activity or the midpoint (average) of the working pH range: $(5.2 + 9.8) \div 2$

Here is a possible response that would receive full credit:

Optimum pH occurs at the maximum enzyme activity, which is typically the midpoint of the working pH range. Since the working pH range of chymotrypsin is from 5.2 to 9.8, the optimum pH of chymotrypsin would be the average, or 7.5.

4. **Scoring Guidelines for Free-Response Question 4**

(a) Using Figure 1, **describe** the evolution of bacteriophage A and of bacteriophage C. **(2 points maximum)**

Description (1 point each, 2 points maximum)
• Bacteriophage A exhibits directional selection toward an increased host range over time (more general)
• Bacteriophage C exhibits directional selection toward a decreased host range over time (more specific)

Here is a possible response that would receive full credit:

Bacteriophage A experiences directional selection toward infecting a wider range of host species. Bacteriophage C experiences directional selection toward infecting specific host cells, narrowing its host range.

(b) Based on the curve for bacteriophage B, **propose** ONE adaptation in a host that was produced by natural selection. **(1 point maximum)**

Proposition (1 point)
• Potential hosts of bacteriophage B that developed resistance to bacteriophage B were naturally selected for and limited the bacteriophage's range of infection

Here is a possible response that would receive full credit:

Some of the bacteria that could be infected by bacteriophage B developed resistance to the phage and were naturally selected for. This lessened the phage's ability to infect and limited the host range.

(c) **Explain** what determines a bacteriophage's host range. **(1 point maximum)**

Explanation (1 point)
• A bacteriophage can only infect bacteria that bear surface receptors specific to that bacteriophage

Here is a possible response that would receive full credit:

To infect bacteria, bacteriophages bind to specific receptors on the surface of bacteria. This specificity determines the phage's host range.

5. **Scoring Guidelines for Free-Response Question 5**

(a) **Explain** why cellular respiration is required for long periods of exercise. **(1 point maximum)**

Explanation (1 point)
• Cells consume more energy during long periods of exercise, and aerobic respiration produces more ATP than anaerobic respiration to meet that energy need

Here is a possible response that would receive full credit:

During a long period of exercise, cells consume more energy. To meet that energy need, cells undergo aerobic respiration, which produces a higher ATP yield than anaerobic respiration, approximately 36 molecules of ATP per molecule of glucose, compared to only 2 ATP per glucose for anaerobic.

(b) Using the template, **graph** the predicted shape of the aerobic respiration energy curve as oxygen becomes limited after a few hours of exercise. **Justify** your prediction. **(2 points maximum)**

Graph characteristics (1 point)
• The aerobic respiration energy curve will decrease as oxygen becomes limited

Here is a sample graph that would earn full credit:

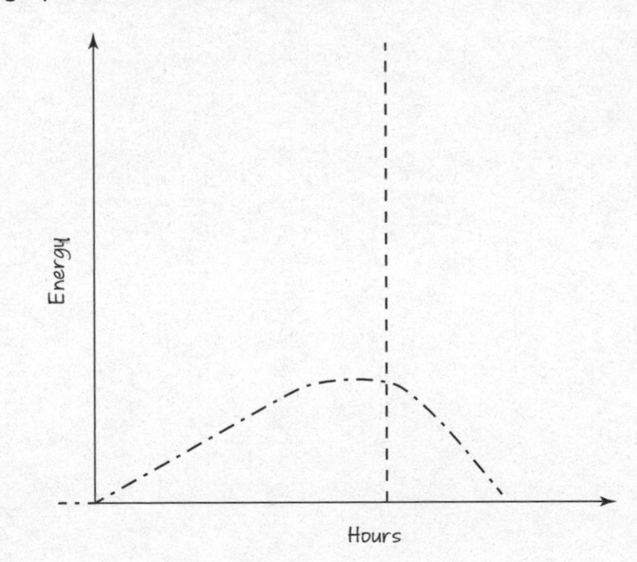

Justification (1 point)
• When oxygen levels decrease, pyruvate in the Krebs cycle will become insufficient for aerobic respiration, so energy produced by aerobic respiration will decrease and muscle fibers will undergo anaerobic respiration, which produces lactic acid

Here is a possible response that would receive full credit:

When oxygen levels decrease, pyruvate in the Krebs cycle becomes insufficient for cellular respiration. For energy, muscle fibers will switch from aerobic respiration to anaerobic respiration, which produces lactic acid. Buildup of lactic acid triggers a cramp during exercise.

(c) The reaction of phosphocreatine + ADP to ATP + creatine is reversible, and during periods of rest the store of phosphocreatine is regenerated. **Explain** how cellular respiration returns muscles to normal after exercise. **(1 point maximum)**

Explanation (1 point)
• During periods of rest, cellular respiration uses excess ATP to replenish stores of ATP and phosphocreatine in muscle fibers

Here is a possible response that would receive full credit:

Since during periods of rest after exercise cells do not require as much ATP, excess ATP produced via cellular respiration is used to replenish stores of ATP and phosphocreatine in muscle fibers.

6. **Scoring Guidelines for Free-Response Question 6**

(a) Based on Table 1, **determine** the mode of inheritance for the observed flower colors. **Provide** reasoning for your answer. **(2 points maximum)**

Determination (1 point)	Reasoning (1 point)
• The mode of inheritance is incomplete dominance	• The offspring phenotype ratios in Table 1 match with those from Punnett squares demonstrating incomplete dominance—the alleles are inherited according to Mendel's laws but the dominant allele is only partially expressed in the heterozygous organism, which exhibits an intermediate phenotype

Here is a possible response that would receive full credit:

The mode of inheritance is incomplete dominance because in incomplete dominance, the dominant allele is only partially expressed in a heterozygous organism. If the two alleles that control flower color are R and y, RR would produce red flowers, yy would produce yellow flowers, and Ry would produce reddish-orange flowers, an intermediate phenotype. The offspring phenotype ratios from Punnett squares demonstrating incomplete dominance match those in Table 1.

(b) **Determine** the phenotype ratio of offspring produced from the self-fertilization of a reddish-orange plant. **(1 point maximum)**

Determination (1 point)
• Self-fertilization of a reddish-orange plant (Ry) would produce a phenotype ratio of 1 : 2 : 1, red : reddish-orange : yellow • Punnett square:

	R	*y*
R	*RR*	*Ry*
y	*Ry*	*yy*

Here is a possible response that would receive full credit:

Self-fertilization of the reddish-orange plant would be a cross between two heterozygotes. A Punnett square shows one homozygous dominant, one homozygous recessive, and two heterozygous offspring would be produced. Since the mode is incomplete dominance, the phenotype ratio would be 1 : 2 : 1, red : reddish-orange : yellow.

7. **Scoring Guidelines for Free-Response Question 7**

(a) **Explain** how the *trp* operon is turned off when tryptophan levels are high. **(1 point maximum)**

Explanation (1 point)
• At high tryptophan levels, tryptophan activates the repressor protein, which binds to the operator and blocks RNA polymerase from binding to the promoter, inhibiting gene expression

Here is a possible response that would receive full credit:

When tryptophan levels are high, *E. coli* does not need to synthesize tryptophan. Tryptophan activates the repressor, which binds to the operator and blocks RNA polymerase from binding to the promoter and thereby blocks gene expression.

(b) **Explain** how the *trp* operon is turned on when tryptophan levels are low. **(1 point maximum)**

Explanation (1 point)
• At low tryptophan levels, the repressor protein is inactive, so RNA polymerase is able to bind to the promoter, promoting gene expression

Here is a possible response that would receive full credit:

When tryptophan levels are low, *E. coli* needs to synthesize tryptophan. Thus, the repressor is inactive and RNA polymerase binds to the promoter, which promotes gene expression.

(c) **Predict** ONE effect a mutation in the *trpR* gene might have on gene regulation. **(1 point maximum)**

Prediction (1 point)
• A mutation in the *trpR* gene could lead to a nonfunctional repressor protein (*trp* operon always on), or • A mutation in the trpR gene could lead to an always active repressor protein (*trp* operon always off), or • A mutation may have no effect

Here is a possible response that would receive full credit:

A mutation in the *trpR* gene could lead to a nonfunctional repressor protein, so the *trp* operon would always be expressed and tryptophan would build up in the cell.

8. **Scoring Guidelines for Free-Response Question 8**

(a) **Propose** ONE reason that the introduction of the cane toad failed to control the cane beetle population. **(1 point maximum)**

Proposition (1 point)
• Cane toads do not strictly feed on cane beetles but feed on a variety of foods, and if given the choice, cane toads prefer feeding on other insects than cane beetles

Here is a possible response that would receive full credit:

The introduction of the cane toad failed to control the cane beetle population because the cane toad did not feed solely on cane beetles. The cane toads found other food sources they preferred.

(b) **Propose** ONE reason that the population of cane toads has skyrocketed to threatening levels. **(1 point maximum)**

Proposition (1 point)
• Cane toads have no natural predators to limit the toads' population growth, or
• Cane toads are able to feed on a variety of foods and outcompete native species, or
• Cane toads produce toxins that kill native animals, or
• Cane toads have evolved longer legs, which allows them to migrate greater distances and spread

Here is a possible response that would receive full credit:

The population of cane toads has skyrocketed to threatening levels because cane toads have no natural predators in Australia. Without predators limiting their growth, the toad population continues to increase.

(c) **Describe** ONE contributing factor that maintains the balance of an ecosystem. **(1 point maximum)**

Description (1 point)
• The balance of an ecosystem is maintained via keystone species, producers, or other abiotic and biotic factors

Here is a possible response that would receive full credit:

One contributing factor that maintains the balance of an ecosystem is keystone species, which help to preserve the diversity in an ecosystem. In turn, the ecosystem is able to tolerate and respond to changes in the environment and thus maintain balance.

Practice Test 1

Section I, Part A

1 Ⓐ Ⓑ Ⓒ Ⓓ through 63 Ⓐ Ⓑ Ⓒ Ⓓ

1 (A)(B)(C)(D)	12 (A)(B)(C)(D)	23 (A)(B)(C)(D)	34 (A)(B)(C)(D)	45 (A)(B)(C)(D)	56 (A)(B)(C)(D)
2 (A)(B)(C)(D)	13 (A)(B)(C)(D)	24 (A)(B)(C)(D)	35 (A)(B)(C)(D)	46 (A)(B)(C)(D)	57 (A)(B)(C)(D)
3 (A)(B)(C)(D)	14 (A)(B)(C)(D)	25 (A)(B)(C)(D)	36 (A)(B)(C)(D)	47 (A)(B)(C)(D)	58 (A)(B)(C)(D)
4 (A)(B)(C)(D)	15 (A)(B)(C)(D)	26 (A)(B)(C)(D)	37 (A)(B)(C)(D)	48 (A)(B)(C)(D)	59 (A)(B)(C)(D)
5 (A)(B)(C)(D)	16 (A)(B)(C)(D)	27 (A)(B)(C)(D)	38 (A)(B)(C)(D)	49 (A)(B)(C)(D)	60 (A)(B)(C)(D)
6 (A)(B)(C)(D)	17 (A)(B)(C)(D)	28 (A)(B)(C)(D)	39 (A)(B)(C)(D)	50 (A)(B)(C)(D)	61 (A)(B)(C)(D)
7 (A)(B)(C)(D)	18 (A)(B)(C)(D)	29 (A)(B)(C)(D)	40 (A)(B)(C)(D)	51 (A)(B)(C)(D)	62 (A)(B)(C)(D)
8 (A)(B)(C)(D)	19 (A)(B)(C)(D)	30 (A)(B)(C)(D)	41 (A)(B)(C)(D)	52 (A)(B)(C)(D)	63 (A)(B)(C)(D)
9 (A)(B)(C)(D)	20 (A)(B)(C)(D)	31 (A)(B)(C)(D)	42 (A)(B)(C)(D)	53 (A)(B)(C)(D)	
10 (A)(B)(C)(D)	21 (A)(B)(C)(D)	32 (A)(B)(C)(D)	43 (A)(B)(C)(D)	54 (A)(B)(C)(D)	
11 (A)(B)(C)(D)	22 (A)(B)(C)(D)	33 (A)(B)(C)(D)	44 (A)(B)(C)(D)	55 (A)(B)(C)(D)	

Section I, Part B

121.

122.

123.

124.

125.

126.

SECTION I
90 Minutes—69 Questions
Part A

Directions: Each of the questions or incomplete statements below is followed by four suggested answers or completions. Select the answer that is best in each case and enter the appropriate letter in the corresponding space on the answer sheet. When you have completed part A, you should continue on to part B.

1. Natural selection ensures that the teeth of particular animal species are specialized to fit the diet of that species. Members of the species that lack the appropriate dentition will be selected against, because it will be more difficult for them to satisfy their nutritional needs. If an animal's jaw contains teeth with broad, rigid surfaces, it is reasonable to conclude that the animal is most likely a(n)

   (A) herbivore

   (B) producer

   (C) top consumer

   (D) carnivore

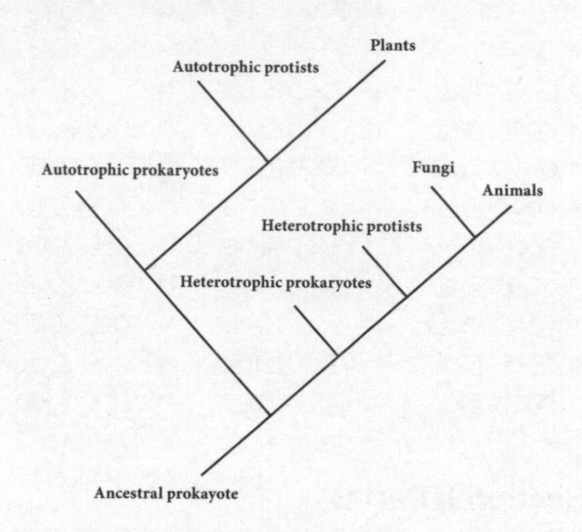

2. Based on the cladogram, it is most reasonable to conclude that

   (A) animals are descended from fungi

   (B) fungi are more similar to animals than to plants

   (C) all protists are closely related

   (D) the first organisms were eukaryotes

3. In humans, short-term energy supplies are stored as glycogen and long-term energy supplies are stored as fat. Which of the following statements best explains this phenomenon?

   (A) Lipids are more readily soluble in water than carbohydrates.

   (B) Lipids are more easily broken down by mitochondria than carbohydrates.

   (C) Lipids are more oxidized and produce less energy per gram.

   (D) Glycogen is more readily converted to glucose than fat.

GO ON TO THE NEXT PAGE.

Site	Plants		Amphibians		Reptiles		Mammals		Total	
	Species	Individuals	Species	Individuals	Species	Individuals	Species	Individuals	Species	Individuals
Bluewater Swamp	15	113	2	8	3	8	5	7	25	136
Papago Buttes	5	27	0	0	2	4	2	3	9	34
Beaver's Bend	8	121	2	2	0	0	3	18	13	141
Sherwood Forest	4	159	1	1	0	0	6	24	11	184
Tortilla Flats	4	63	0	0	3	24	1	5	8	92

4. Based on the table above, which site has the greatest species diversity?
   (A) Bluewater Swamp
   (B) Papago Buttes
   (C) Beaver's Bend
   (D) Sherwood Forest

5. Plants form close associations with mycorrhizae, fungi that colonize plant roots. The plant benefits because the fungus makes soil phosphorus available to the plant. The fungus benefits because the plant provides it with sugars. This is an example of
   (A) commensalism
   (B) competition
   (C) parasitism
   (D) mutualism

6. A man who has a sex-linked recessive disorder carries the gene for the condition on his X chromosome. If he marries a woman who does not have the gene on either of her X chromosomes, what are the chances that their first son will have the disease?
   (A) 0%
   (B) 25%
   (C) 50%
   (D) 100%

7. Transpiration is a process that transports water from roots to leaves in trees and other plants. Transpiration against the force of gravity is possible in trees 100 meters tall because
   (A) water is actively transported from roots to leaves
   (B) evaporation from stomata pulls water up through the tree
   (C) high pressure in the soil pushes the water up
   (D) gravity creates pressure in the xylem, squeezing water out of the stomata

8. During complete aerobic cellular respiration, each molecule of glucose broken down in mitochondria can yield 36 molecules of ATP. Which of the following conditions would most likely lead to a decrease in the amount of ATP produced in a given system?
   (A) An increase in the amount of glucose added to the system
   (B) A decrease in the amount of light the system is exposed to
   (C) A decrease in the amount of oxygen available in the system
   (D) A decrease in the amount of carbon dioxide available in the system

Practice

GO ON TO THE NEXT PAGE.

9. Archaea are unique prokaryotes that are thought to be more closely related to eukaryotes than they are to bacteria. Which of the following characteristics best supports this idea?

   (A) Many archaea are adapted to extreme environments, such as deep-sea thermal vents.

   (B) The cell walls of archaea lack peptidoglycans.

   (C) Archaea have introns in some genes.

   (D) Archaea lack a membrane-bound nucleus and organelles.

Questions 10–12

Recent geological changes in the landscape of a coastal region in southern California allowed multiple species of animals to gain entry to a section of the shoreline region that they previously did not have access to. One of these species is the *Himantopus mexicanus*, more commonly known as the black-necked stilt.

The black-necked stilt consumes a diet consisting of a range of aquatic invertebrates (such as crustaceans and other arthropods), mollusks, small fish, and tadpoles. They have also been known to consume plant seeds, though they do so rarely.

As part of an investigation to observe the effects of the access to the new territory on the black-necked stilt community, researchers tracked the number of individuals in the population over the span of eight years. The results from their data collection are summarized in the graph below.

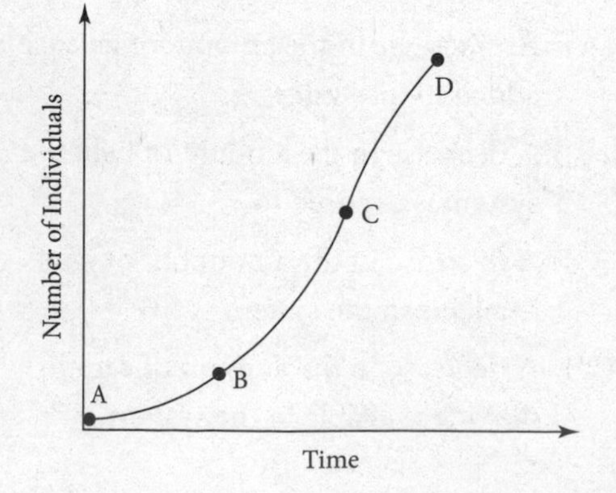

10. The graph indicates that the black-necked stilt population is most likely

    (A) going to stabilize and plateau within the next year as limits from density-dependent and density-independent factors are imposed

    (B) growing at a rapid rate due to the lack of density-dependent limiting factors and environmental constraints

    (C) progressing toward extinction because the data shows their reproduction rate is not reflective of their environmental constraints

    (D) growing in excess of its carrying capacity, since there are no notable fluctuations in population size

11. In addition to tracking the population growth rate, researchers also mapped out the food pyramid (shown below) during the first year of immigration into the new region:

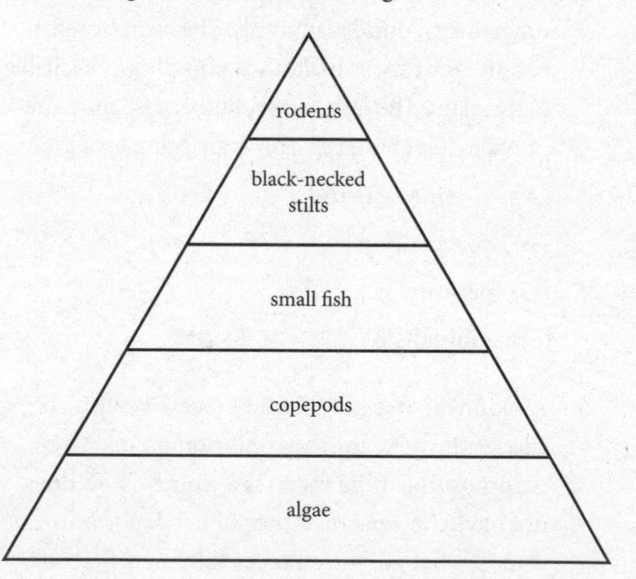

GO ON TO THE NEXT PAGE.

If researchers later realize they need to amend the food pyramid to include large birds of prey, which of the following would most likely be true of the species in the habitat?

(A) The rodent population is larger than the black-necked stilt population.

(B) The large birds of prey have the highest concentration of fat-soluble toxins.

(C) The total biomass of the copepods is greater than that of the algae.

(D) The black-necked stilts can provide more food for the birds of prey than the small fish can.

12. A few years after the initial migration of the black-necked stilt population into the newly accessible coastal region, another species of coastal bird called *Charadrius nivosus*, also known as the snowy plover, began to enter and nest in the new territory as well. After monitoring both populations for five years, researchers were surprised to find that the population size and growth rate of both populations appear unchanged. What is the most likely explanation for this occurrence?

(A) The snowy plover and black-necked stilt are closely related species occupying different niches

(B) The snowy plover and black-necked stilt are related species occupying different habitats

(C) The snowy plover and black-necked stilt are unrelated species occupying different niches.

(D) The snowy plover and black-necked stilt are unrelated species occupying the same niche.

13. Messages delivered by the nervous system are conveyed by a series of neurons. A message is passed from one neuron to the next by

(A) the release of Ca^{2+} ions by the presynaptic neuron that travel across the synaptic gap and initiate the uptake of Na^{+} ions in the postsynaptic neuron

(B) the release of neurotransmitters by the presynaptic neuron that travel across the synaptic gap and initiate an action potential in the postsynaptic neuron

(C) the release of Na^{+} ions by the presynaptic neuron that travel across the synaptic gap and initiate an action potential in the postsynaptic neuron

(D) the mechanical deformation of the postsynaptic neuron by the presynaptic neuron

14. In the cell cycle, G_1 and G_2 are names for the phases in which cell growth occurs, M refers to mitosis, and S is the name for the phase in which DNA replication occurs. Assuming that the cycle proceeds continuously without interruption (returning to the first phase after completing the last), which of the following sequences depicts a correct ordering of the events in the cell cycle?

(A) $G_1 \rightarrow$ cytokinesis $\rightarrow G_2 \rightarrow M \rightarrow S$

(B) $G_1 \rightarrow G_2 \rightarrow S \rightarrow M \rightarrow$ cytokinesis

(C) $S \rightarrow G_2 \rightarrow$ cytokinesis $\rightarrow M \rightarrow G_1$

(D) $S \rightarrow G_2 \rightarrow M \rightarrow$ cytokinesis $\rightarrow G_1$

GO ON TO THE NEXT PAGE.

15. For *E. coli* to utilize lactose as a carbon and energy source, the protein β-galactosidase must be translated. In the presence of both lactose and glucose, *E. coli* will preferentially utilize glucose, conserving the resources necessary to produce β-galactosidase. However, when glucose is absent, lactose will functionally induce the expression of β-galactosidase. This most strongly suggests that
    (A) lactose represses expression of the *lac* operon
    (B) lactose is sufficient to cause an increase in β-galactosidase
    (C) glucose decreases expression of the *lac* operon in the presence of lactose
    (D) the presence of glucose is necessary for β-galactosidase production

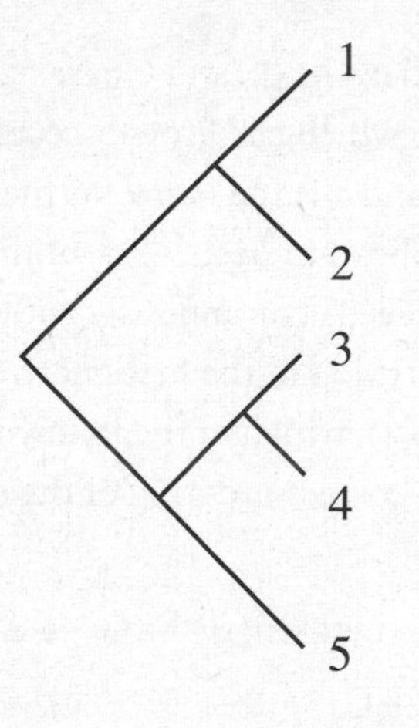

16. The phylogenetic tree above depicts five newly-discovered bacterial species. The branch point separating species 1 and 2 from 3, 4, and 5 is the presence of a gene for enzyme X, required for the metabolism of glucose. The branch point separating species 1 from species 2, and species 3 and 4 from species 5 is the presence of protein Y, a pump that removes a certain antibiotic from cells.

    Which of the following findings, if true, would require the most extensive redrawing of the tree?
    (A) Molecular studies indicate that protein Y evolved 250,000 years before enzyme X.
    (B) Molecular studies indicate that enzyme X evolved 1,000,000 years before protein Y.
    (C) Only species 3 expresses protein Z, a pump that regulates entry of sodium ions into the cell.
    (D) A newly discovered species 6 expresses protein Y but not enzyme X.

17. Nondisjunction, the failure of homologous chromosomes or sister chromatids to separate, results in aneuploidy (an irregular number of chromosomes in a cell). The figure shows the result of one form of nondisjunction.

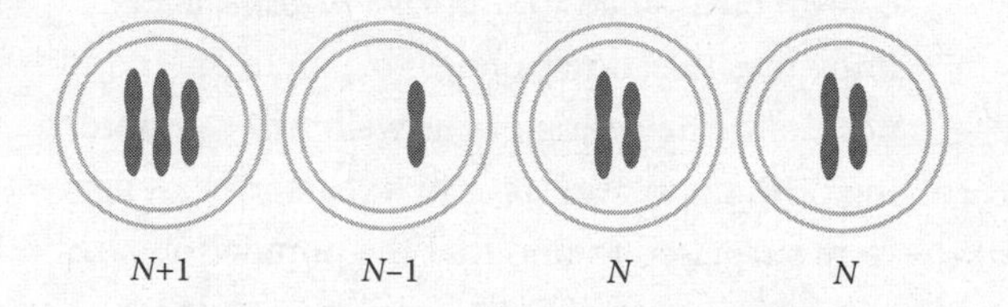

    Based on the figure, which of the following conclusions would be most justified?
    (A) Two homologous chromosomes failed to separate in interphase.
    (B) Sister chromatids failed to separate in mitosis during anaphase.
    (C) Two homologous chromosomes moved to the same pole in meiosis I.
    (D) Sister chromatids moved to the same pole in meiosis II.

GO ON TO THE NEXT PAGE.

18. A retrovirus, such as HIV (human immunodeficiency virus), encodes its genome in RNA, rather than DNA. Which of the following sequences best describes what happens when a retrovirus infects a host cell?
    (A) RNA-DNA hybrid → reverse transcriptase → viral proteins → provirus
    (B) provirus → viral proteins → RNA-DNA hybrid → reverse transcriptase
    (C) viral proteins → RNA-DNA hybrid → reverse transcriptase → provirus
    (D) reverse transcriptase → RNA-DNA hybrid → provirus → viral proteins

19. In a certain terrestrial ecosystem, over a given period of time, producers had 1.1 MJ (megajoules) of energy available in the form of sunlight, whereas primary consumers had approximately 12 kJ (kilojoules) and secondary consumers, 1.1 kJ. Which of the following is the most reasonable conclusion from this data?
    (A) The majority of the energy in a trophic level is transferred to the next higher level.
    (B) The transfer of energy from producers to primary consumers is more efficient than the transfer between consumers.
    (C) Energy within an ecosystem is constantly cycled through the various levels.
    (D) Approximately 90% of the energy available to primary consumers is not transferred to the next highest level.

Questions 20–23

The following diagram shows energy transformations within a cell. Each form of energy is represented by the symbols E I, E II, E III, and E IV. Two cellular organelles are represented by the letters A and B.

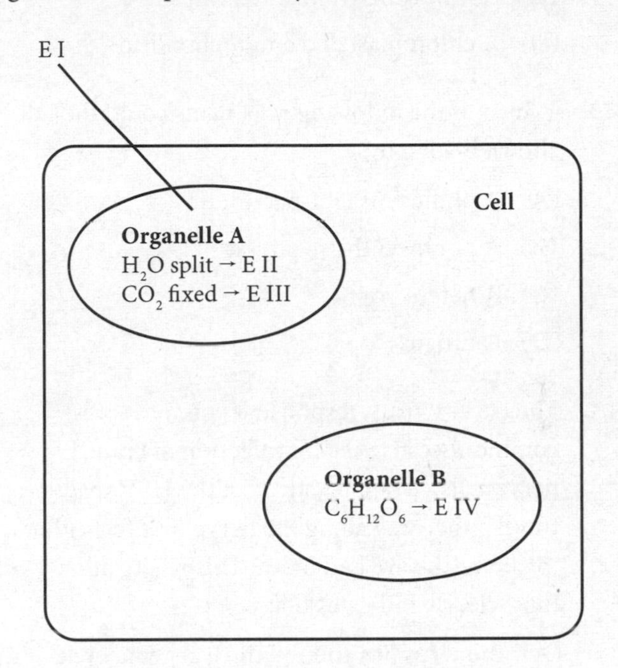

20. What form of energy is represented by E II ?
    (A) Radiant energy in the form of photons
    (B) Chemical energy being stored in the bonds of glucose
    (C) Chemical energy in the form of ATP
    (D) Chemical energy released by glycolysis

21. If the transformation depicted in organelle B requires oxygen, what form of energy is represented by E IV?
    (A) Radiant energy in the form of photons
    (B) Chemical energy being stored in the bonds of glucose
    (C) Chemical energy in the form of ATP
    (D) Chemical energy released by glycolysis

Practice

GO ON TO THE NEXT PAGE.

22. What cellular organelles are represented as A and B, respectively?
    (A) The nucleus and a ribosome
    (B) A mitochondrion and a chloroplast
    (C) A mitochondrion and a ribosome
    (D) A chloroplast and a mitochondrion

23. Which of the following organisms could the cell shown belong to?
    (A) A photosynthetic bacterium
    (B) A photosynthetic protist
    (C) A heterotroph
    (D) A fungus

24. The HSV-1 virus, responsible for cold sores, commonly causes latent infection of cranial nerve cells. When reactivated, the HSV may travel either way along the nerve to infect either the lip or the eye and brain. This reactivation most clearly indicates that
    (A) the virus has entered the lysogenic cycle
    (B) the virus has entered the lytic cycle
    (C) an additional infection with a different microbe has occurred
    (D) the virus has undergone a mutation

25. Multiple crosses involving genes known to occur on the same chromosome produce frequencies of phenotypes that suggest there is a high rate of crossover between these two genes. Which of the following is the most likely explanation for the phenotypic frequencies observed due to crossing over?
    (A) The two genes are far apart from one another.
    (B) The two genes are both recessive.
    (C) The two genes have incomplete dominance.
    (D) The two genes are both located far from the centromere.

26. During maturation of T cells in the thymus, large numbers of T cells that are able to bind to an individual's own MHC (major histocompatibility complex) molecules are killed off. Which of the following is most likely to occur if such cells are allowed to survive?
    (A) The T cells will launch an immune response against the individual's own cells.
    (B) The T cells will be less able to recognize and bind to foreign MHC molecules.
    (C) More T cells than B cells will be produced by the individual's body.
    (D) The individual will be less susceptible to bacterial infections.

27. Which of the following is the likely source of energy for the synthesis of the small organic molecules that presumably predated the first forms of life on Earth?
    (A) Fermentation by bacteria
    (B) Photosynthesis by microscopic algae
    (C) Lightning from frequent storms
    (D) Shifts in ocean currents

GO ON TO THE NEXT PAGE.

28. A scientist places free strands of DNA, which contain a gene that codes for a protein enabling the metabolism of glucose, in a medium containing bacteria that can only survive on the sugar lactose. The scientist heat shocks the bacteria in $CaCl_2$ and lets them recover before plating them in several petri dishes with only glucose as a nutrient source. After several days, there are no signs of bacterial growth in the glucose medium.

    Which of the following is NOT a plausible explanation for the results of the experiment?

    (A) All of the bacteria died from the heat shock treatment.

    (B) The gene for glucose metabolism was not successfully incorporated into any of the bacterial genomes.

    (C) The petri dishes were contaminated with a powerful antibiotic from a previous experiment.

    (D) The DNA strands came from sheep, so the glucose metabolism genes could not be translated in the bacteria.

Questions 29–31

The graph below depicts the distribution of a trait in an animal population.

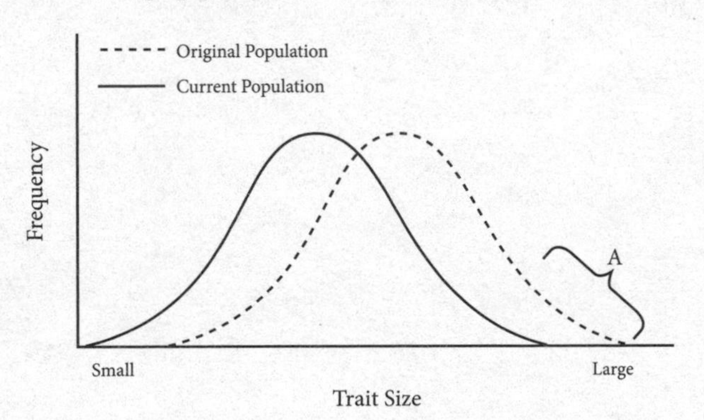

29. The region indicated by the A in the figure depicts

    (A) the variability of sizes in the population

    (B) the mean size of the population

    (C) the phenotypes that were successful in recent environmental conditions

    (D) the phenotypes that fared poorly in recent environmental conditions

30. What type of selection does this figure depict?

    (A) Disruptive

    (B) Directional

    (C) Stabilizing

    (D) Artificial

31. Suppose that in subsequent generations, individuals with the smallest and largest trait sizes had lower reproductive success than individuals with moderate trait sizes. What would the new population curve most probably look like?

    (A) The curve would resemble the original population curve in shape and location.

    (B) The curve would be shifted to the left of the current one.

    (C) The curve would have two peaks, one at smaller sizes and another at larger sizes.

    (D) The curve would be narrower and higher than the current one but mean size would remain the same.

Practice

GO ON TO THE NEXT PAGE.

32. In *Vibrio fischeri*, quorum sensing regulates bioluminescence, the ability to produce light. *V. fischeri* produce and secrete autoinducers, which at high levels and at high cell density causes a regulatory response and the expression of the luxCDABE genes. Which of the following illustrations correctly depicts quorum sensing?

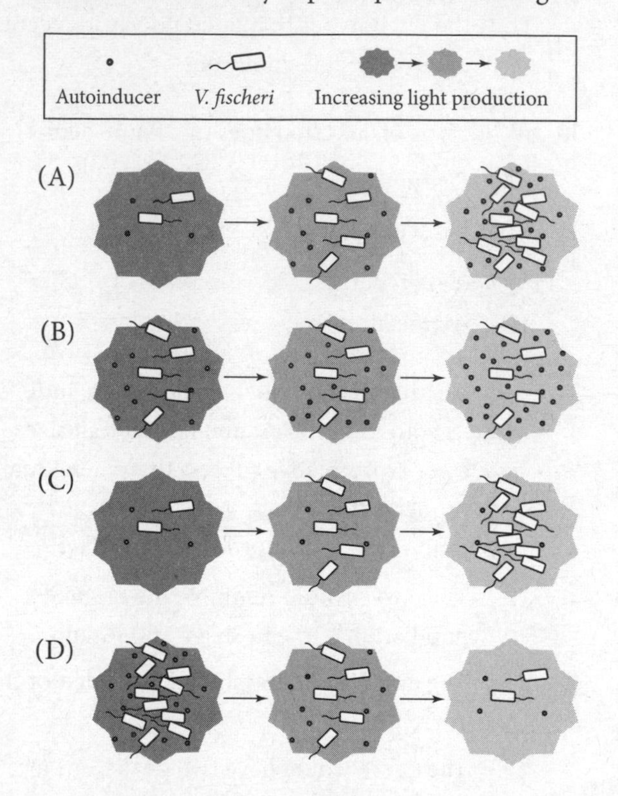

33. Animals in the phylum *Echinodermata*, such as sea stars and sand dollars, are thought to be more closely related to the phylum *Chordata* (which includes humans and other vertebrates) than to other animal phyla. Which of the following observations provides the best justification for this conclusion?

    (A) *Echinodermata* and *Chordata* are the most abundant animal phyla.

    (B) During development, the anus forms prior to the mouth in both phyla.

    (C) All species in both phyla have a common ancestor.

    (D) Neither phyla includes obligate anaerobes.

34. Most pharmacological agents, including common drugs like acetaminophen, can exhibit toxic effects at sufficiently high doses, while at lower doses they may be harmless or even therapeutically useful. Which of the following best characterizes why most drugs exert toxic effects at sufficiently high dosages?

    (A) Artificial substances are inherently toxic, so high enough doses of any such substance will result in negative effects.

    (B) At sufficiently high doses the metabolic processing of most drugs changes, thereby creating toxic metabolites that are not present at lower doses.

    (C) Most drugs interfere with normal biological functioning, and higher doses can cause an excessive degree of interference that manifests as toxicity.

    (D) Even pharmaceutically pure drugs contain trace amounts of impurities, and at high doses these impurities are present in sufficiently great quantities to cause toxic side effects.

GO ON TO THE NEXT PAGE.

35. The protozoan parasite that causes African sleeping sickness, *Trypanosoma brucei*, is capable of changing the proteins it expresses on its surface to evade its host's immune system. Unless aggressive medical treatment is pursued after infection, the parasite may remain in the host permanently. *T. brucei* is an extracellular parasite usually spread by bites from the tsetse fly, which feeds on the blood of humans and other mammals. Which mechanism of immunity does *T. brucei*'s ability to change surface proteins most directly subvert?

(A) Nonspecific immunity, because the parasite can change to avoid inflexible mechanisms of host immunity

(B) Cell-mediated immunity, because killer T cells cannot recognize the cell surface proteins on *T. brucei* that would indicate to the immune system that a foreign organism is present

(C) Phagocytosis, because white blood cells cannot safely ingest the parasite without becoming infected themselves

(D) Humoral immunity, because antibodies in the blood cannot effectively bind to *T. brucei* well enough to trigger a robust immune response to the parasite

36. Both skin cells and nerve cells arise from embryonic cells with the same genetic information, though these two cell types are radically different in both form and function. How is it that two cell types within an organism can arise from the same progenitor cell, yet perform vastly different specialized functions?

(A) Because the cells contain the same DNA, they also contain the same mRNA transcripts. However, these transcripts are expressed by ribosomes in different ways, leading to differences between cells.

(B) The cells contain all the same DNA, but different transcription factors and other regulatory proteins promote the synthesis of different sets of mRNA in each cell type.

(C) During development, genetic information is lost as cells develop into their mature forms, thereby silencing genes that will not be useful to the cell's intended function.

(D) The proteins created within the cell are ultimately the same, but they perform different functions based on the signals that the cell receives from neighboring cells.

Practice

GO ON TO THE NEXT PAGE.

37. The illustration below depicts the regulation of erythropoiesis (formation of red blood cells).

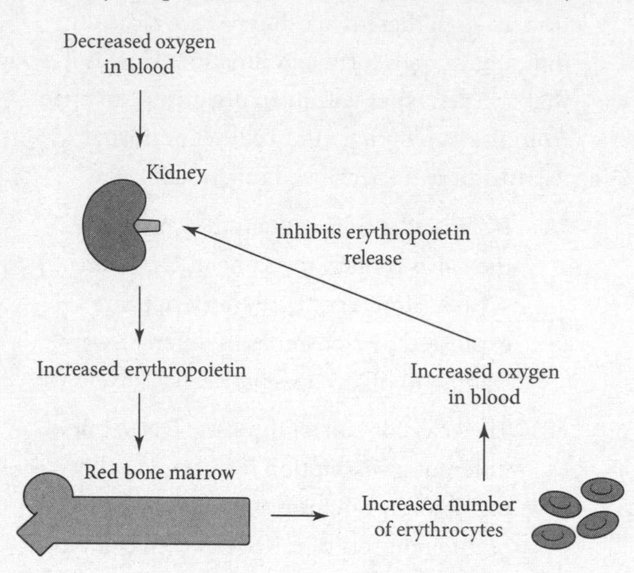

Suppose kidney cells are NOT able to detect low oxygen levels in the blood due to pathological conditions. Which of the following would be the most likely result?

(A) An increase in erythropoietin and an increase in the number of erythrocytes

(B) Tissue hypoxia due to cells not receiving enough oxygen

(C) The return to homeostasis by relieving the initial stimulus

(D) Inhibition of erythropoietin release

Questions 38–41

Indicators are chemicals that change color in the presence of specific substances such as proteins, monosaccharides, and lipids. A more intense color indicates a greater concentration of macromolecule. Benedict's test identifies molecules that have a ketone or aldehyde group (such as found in aldoses and ketoses), the Biuret test determines the presence of peptide bonds, and Sudan III is a dye used for staining triglycerides. The following table shows the negative and positive results of each test.

Indicator	Color	
	Negative	Positive
Benedict's	blue	orange
Biuret	blue	purple
Sudan III	dark red	orange

Researchers used Benedict's solution, Biuret reagent, and Sudan III to identify unknown solutions. The following data were obtained.

	Benedict's	Biuret	Sudan III
Solution 1	orange	purple	orange
Solution 2	blue	blue	dark red
Solution 3	orange	blue	dark red
Solution 4	blue	purple	dark red
Solution 5	blue	blue	orange
Solution 6	yellow-orange	purple	dark red
Solution 7	blue	pink	dark red

38. Which indicator(s) test(s) for the presence of lipids?

(A) Benedict's

(B) Biuret

(C) Sudan III

(D) Benedict's and Sudan III

GO ON TO THE NEXT PAGE.

39. Which of the following conclusions is best supported by the data?
    (A) Solution 3 contains glucose because Benedict's test is positive.
    (B) Solution 3 contains albumin because Benedict's test is positive.
    (C) Solution 4 contains glucose because the Biuret test is positive.
    (D) Solution 5 contains albumin because the Biuret test is positive.

40. Which of the following statements best explains why solution 7 turned pink when the Biuret reagent was added?
    (A) The concentration of peptide bonds is low, indicating the solution contains short-chain peptides.
    (B) The concentration of peptide bonds is high, indicating the solution contains long-chain peptides.
    (C) The concentration of peptide bonds is low, indicating the solution contains monosaccharides.
    (D) The concentration of peptide bonds is high, indicating the solution contains polysaccharides.

41. Whole milk is composed of water, lactose, fat, protein, and minerals. Based on the data, which of the solutions could be whole milk?
    (A) Solution 1
    (B) Solution 2
    (C) Solution 6
    (D) Solution 7

42. Researchers use vectors to introduce new genetic material into target cells. Which of the following best explains the advantage of plasmids over viruses as DNA delivery vectors?
    (A) Plasmids deliver only a limited amount of genetic material but can provide cells with antibiotic resistance.
    (B) Plasmids have the capacity to deliver larger genes and have a lower probability of toxic effects on non-target cells.
    (C) Plasmids are more efficient in transferring DNA to host cells and generally do not elicit an immune response.
    (D) Plasmids are able to target specific types of cells and can be manufactured on a large scale.

43. A cell at equilibrium has no free energy for metabolic reactions that keep it alive. The diagram below shows a metabolic pathway in a cell in which one reaction feeds the next.

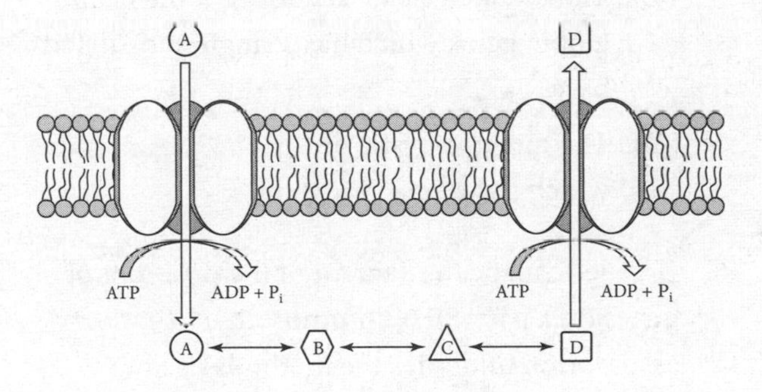

Which of the following best explains how the cell stays out of equilibrium?
    (A) The cell consumes ATP to increase the concentration of molecule A in the cell.
    (B) The cell produces ATP to increase the concentration of molecule B in the cell.
    (C) The cell increases the concentration of molecule C in the cell to generate ATP.
    (D) The cell uses energy to increase the concentration of molecule D in the cell.

GO ON TO THE NEXT PAGE.

44. A certain drug acts by binding to enzymes at positions other than the active site, causing enzymatic activity to decrease. This is an example of:

(A) noncompetitive inhibition

(B) competitive inhibition

(C) allosteric inhibition

(D) non-allosteric inhibition

45. The sequence of nucleotides in DNA codes for production of proteins, and that code is carried to the ribosome as mRNA. Which of the following protein structures represents the longest strand of nucleotide bases?

(A) The gene for hemoglobin, a protein, which consists of 20 amino acids

(B) The normal chloride channel protein gene associated with cystic fibrosis, which consists of 63 nucleotide bases

(C) The mutated sickle cell form of the hemoglobin gene, which has a single substituted base

(D) The mutated cystic fibrosis mRNA strand, which has a base insertion

46. A genetic map is a diagram of the positions of genes on a particular chromosome. What assumption underlies the methods used to construct a genetic map?

(A) Recombination frequencies are directly proportional to the distance between genes on a chromosome.

(B) Recombination frequencies are inversely proportional to the distance between genes on a chromosome.

(C) Linked genes never cross over.

(D) Recessive genes are less common than dominant genes.

Questions 47–48

Free energy (G) is defined as $\Delta G = \Delta H - T\Delta S$ where H is enthalpy, T is temperature in Kelvin, and S is entropy. The ΔG during a reaction indicates whether or not the reaction occurs spontaneously.

Living organisms use free energy to grow, maintain organization, and reproduce. Examples of reactions that occur in cells include breaking down glucose, converting pyruvate to lactate, and/or synthesizing ATP:

glucose + $O_2 \rightarrow CO_2 + H_2O$	$\Delta G = -2880$ kJ/mol of glucose
glucose $\rightarrow$ lactate + H^+	$\Delta G = -195.8$ kJ/mol
pyruvate + NADH + $H^+ \rightarrow$ lactate + NAD^+	$\Delta G = -25.1$ kJ/mol
ADP + $P_i \rightarrow$ ATP + H_2O	$\Delta G = 57$ kJ/mol

47. Which of the following statements best describes how cells use energy from the breakdown of glucose via cellular respiration to increase organization in the system?

(A) Cells can couple the endergonic reaction with a process that increases entropy.

(B) Cells can couple the endergonic reaction with a process that decreases entropy.

(C) Cells can couple the exergonic reaction with a process that increases entropy.

(D) Cells can couple the exergonic reaction with a process that decreases entropy.

GO ON TO THE NEXT PAGE.

48. Living organisms do not violate the second law of thermodynamics. Which of the following pairs of cellular processes can a cell couple to maintain order?

(A) Breakdown of glucose via cellular respiration and formation of lactate from glucose
(B) Formation of lactate from glucose and formation of lactate from pyruvate
(C) Formation of lactate from pyruvate and breakdown of glucose via cellular respiration
(D) Breakdown of glucose via cellular respiration and ATP synthesis

49. In *Drosophila*, the alleles for red eye color (R) and straight wings (W) are dominant, and the alleles for white eye color (r) and curly wings (w) are recessive. A cross between two flies produces progeny comprised of 607 red-eyed flies with straight wings and 202 red-eyed flies with curly wings. Assuming neither of these two genes is sex-linked, which of the following are most likely to be the genotypes of the parents?

(A) RRWW × RRWW
(B) RRWW × RRWw
(C) RrWw × RRWw
(D) RrWw × RrWw

50. According to Darwin's theory of natural selection, which of the following illustrates the evolution of bacteria in the presence of antibiotics?

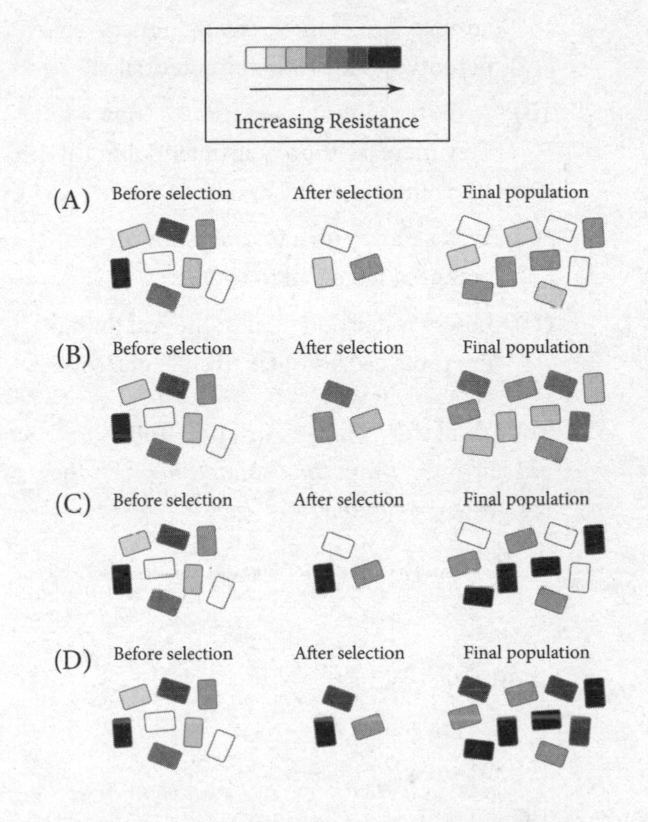

51. Transpiration is a process by which water moves through plants from soil to atmosphere. Water potential defines the flow of water in a system (the tendency of water to move from one area to another). The table lists water potential for a plant system.

Component	Water Potential (MPa)
Soil	−0.3
Roots	−0.3 – −0.6
Stem	−0.6 – −0.8
Leaves	−0.8 – −7.0
Atmosphere	−10.0 – −100.0

Practice

GO ON TO THE NEXT PAGE.

Which of the following statements is most consistent with the information in the table?

(A) When water evaporates from a leaf into the atmosphere, it moves from higher water potential to lower water potential.

(B) As water molecules evaporate from a leaf, they increase the water potential in the stem and cause the uptake of water.

(C) Water moves up a gradient of water potential from soil to roots.

(D) Lower water potential in the soil than in the roots causes water to enter the roots.

52. The graph shows the growth of *Paramecium aurelia* and *Paramecium caudatum* when they are grown separately.

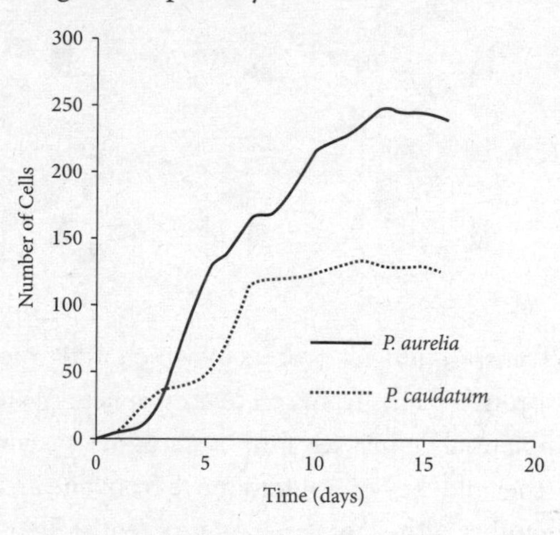

Which of the following best illustrates how competition affects the growth of *P. aurelia* and *P. caudatum* when they are grown together in the same niche?

(A)

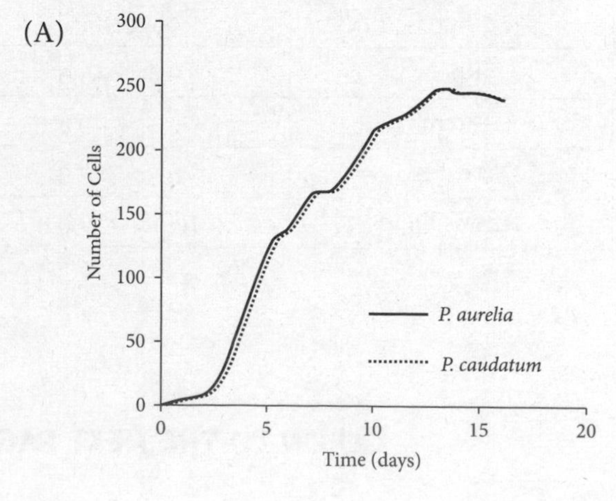

(B)

(C)

(D)

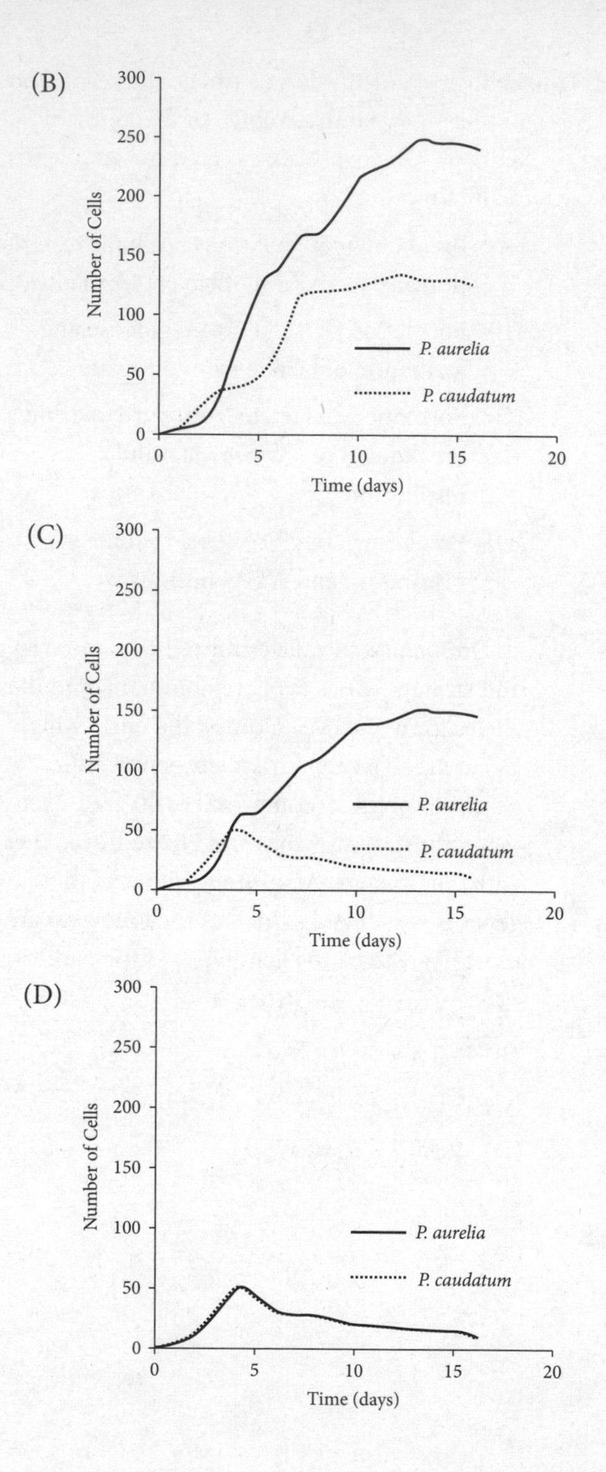

GO ON TO THE NEXT PAGE.

Questions 53–56

In a drug efficacy experiment, samples of *E. coli*, *P. chrysogenum*, *P. bursaria*, and *S. cerevisiae* were exposed to the drug, allowing researchers to observe its effects on bacteria, fungi, protozoa, and yeast, respectively. Samples were plated on their respective growth plates. An additional set of samples that were not treated with the drug was plated as well. The results of the experiment are summarized below. The shaded area represents extensive growth.

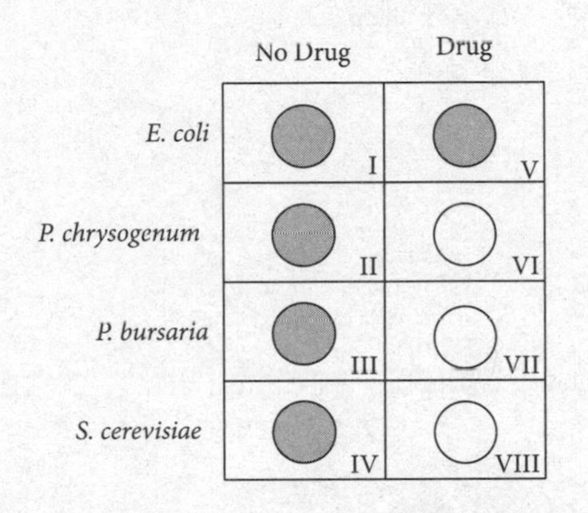

53. Various drugs function by targeting a specific structure within a cell. Because different types of cells contain different components, this allows drugs to act with increased selectivity. Which of the following cellular structures could be targeted specifically to affect eukaryotes while not affecting prokaryotes?

    (A) Cell membrane
    (B) Chromosome
    (C) Endoplasmic reticulum
    (D) Ribosome

54. Based on the results of the experiment, the least likely cellular target of the drug is the organism's

    (A) mitochondria
    (B) Golgi apparatus
    (C) lysosomes
    (D) cell wall

55. The researchers perform a second set of experiments, focusing on multicellular organisms. They collect data on the effects of the drug on different types of cells in each organism. They find that the drug has the greatest effect on cells with a high concentration of rough ER and Golgi bodies, a moderate amount of mitochondria and smooth ER, few lysosomes, a cell wall, and no cilia. The likely primary function of these cells is which of the following?

    (A) Locomotion
    (B) Protein synthesis
    (C) Storage
    (D) Transport

56. Which of the following is an advantage that eukaryotic organisms have over prokaryotic organisms?

    (A) Eukaryotes have specialized organelles that increase the efficiency of their specific processes.
    (B) Eukaryotes possess flagella that allow them to propel themselves toward specific stimuli.
    (C) Eukaryotes do not possess a cell wall and so can better control what enters and exits the cell.
    (D) Eukaryotes do not possess a nuclear envelope around their DNA, decreasing the likelihood of harmful mutations.

Practice

GO ON TO THE NEXT PAGE.

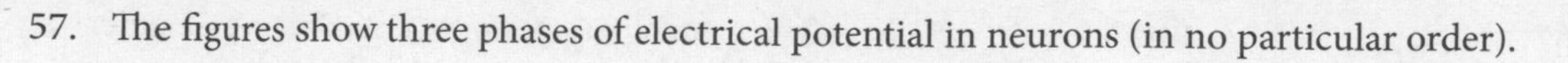

57. The figures show three phases of electrical potential in neurons (in no particular order).

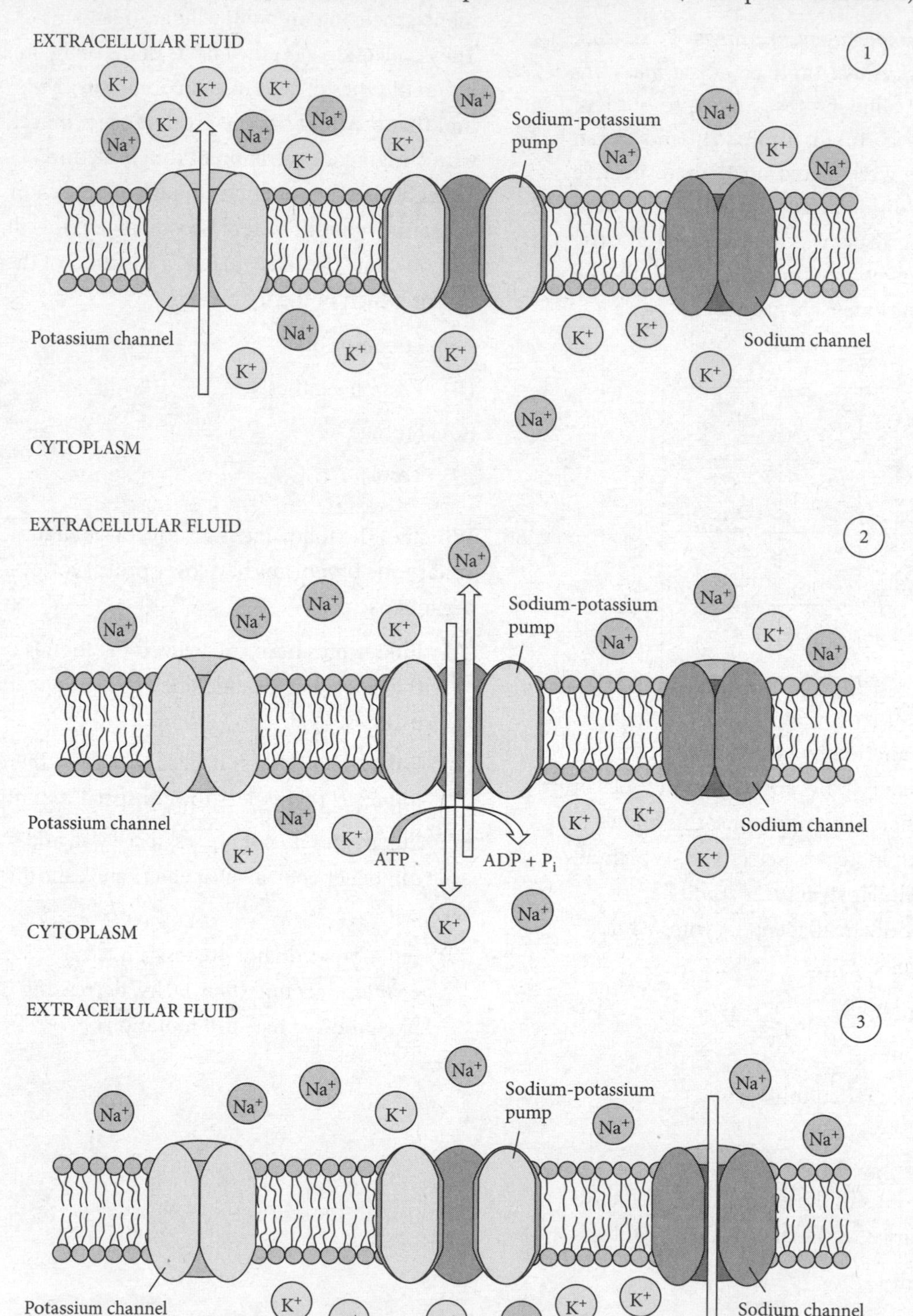

GO ON TO THE NEXT PAGE.

Which of the following correctly places the three phases on the graph of action potential?

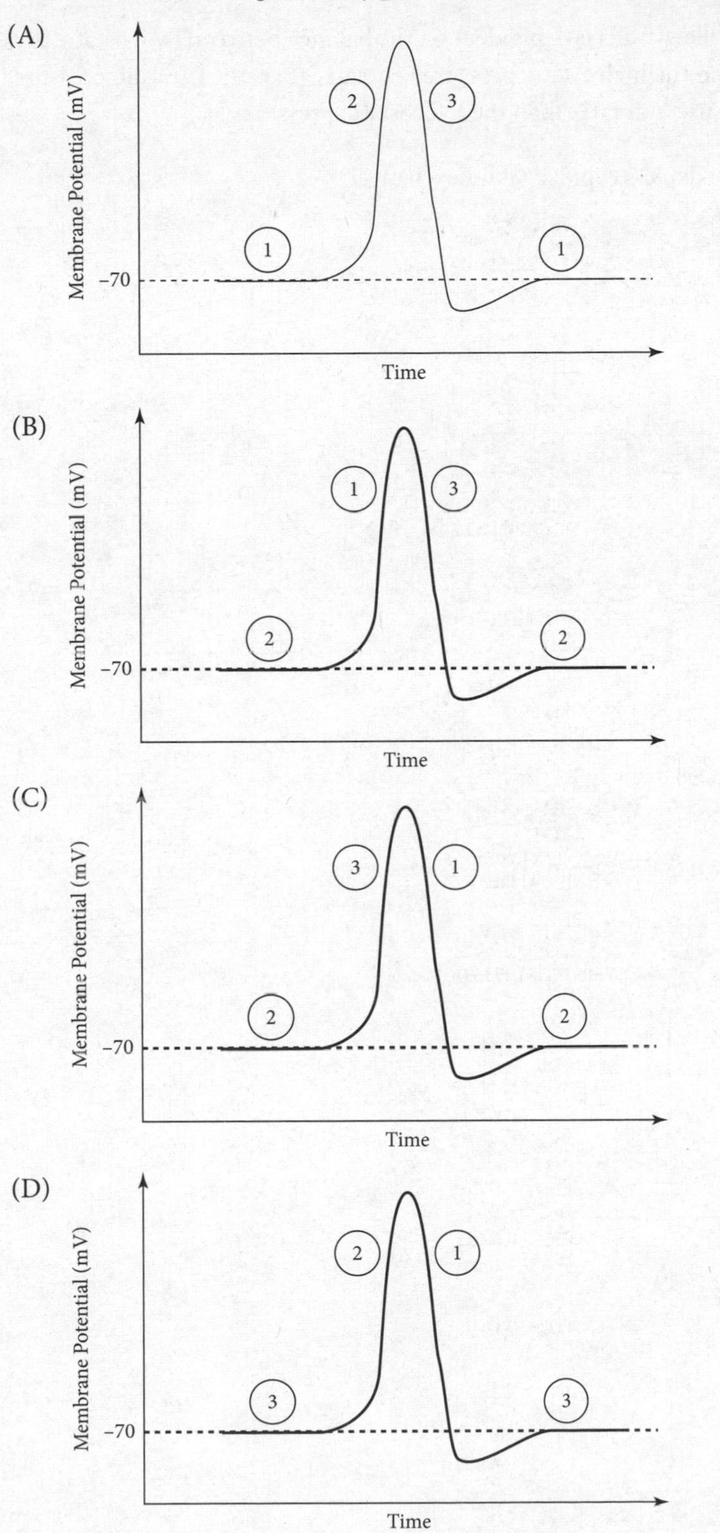

Practice

GO ON TO THE NEXT PAGE.

Questions 58–59

Starling's law states that transport across capillary walls is dependent on the balance between hydrostatic pressure and osmotic pressure. Filtration occurs when the hydrostatic pressure is greater than the osmotic pressure, and reabsorption occurs when the osmotic pressure is greater than the hydrostatic pressure.

58. Which of the following most accurately depicts capillary fluid exchange?

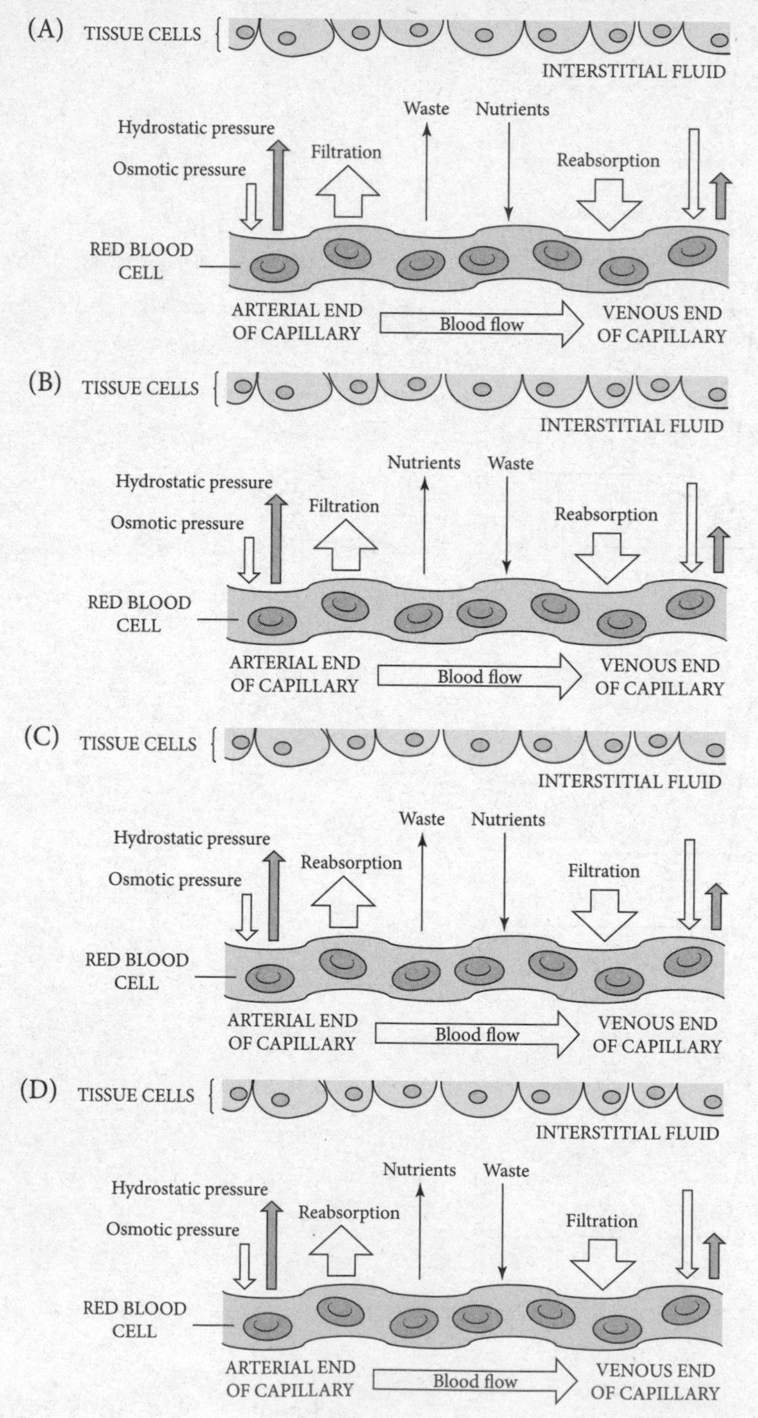

GO ON TO THE NEXT PAGE.

59. Increasing blood pressure is the body's response to tissues not receiving sufficient oxygen and nutrients. Methylsulfonylmethane (MSM) increases the permeability of cell membranes. Which of the following best explains how MSM affects cell pressure and fluid exchange?

(A) MSM increases blood pressure to deliver sufficient oxygen and nutrients to tissues.

(B) MSM decreases blood pressure to balance cell pressure and promotes transport of oxygen and nutrients to tissues.

(C) MSM decreases blood pressure, causing tissues to become overloaded with carbon dioxide and waste.

(D) MSM restores cell pressure balance by hardening cell membranes to restrict transport of oxygen and nutrients.

60. A scientist performs an experiment under laboratory conditions to study the activity of a specific enzyme. Data about substrate and product concentration over time is collected and analyzed. Which of the following statements correctly explains why the indicated experimental variable must be carefully maintained to ensure that accurate data is obtained?

(A) The cofactor concentration must be maintained at low levels because an overabundance will cause them to competitively inhibit the substrate.

(B) The temperature must be held as high as possible because enzymes have their greatest catalytic activity at high temperatures.

(C) The pH must be held constant because changes in pH can cause the enzymes to undergo a structural change, rendering them unusable.

(D) The enzyme concentration must be maintained at high levels because increasing enzyme concentration will linearly increase the rate of reaction.

GO ON TO THE NEXT PAGE.

Questions 61–63

The following diagram presents a simplified version of the nitrogen cycle.

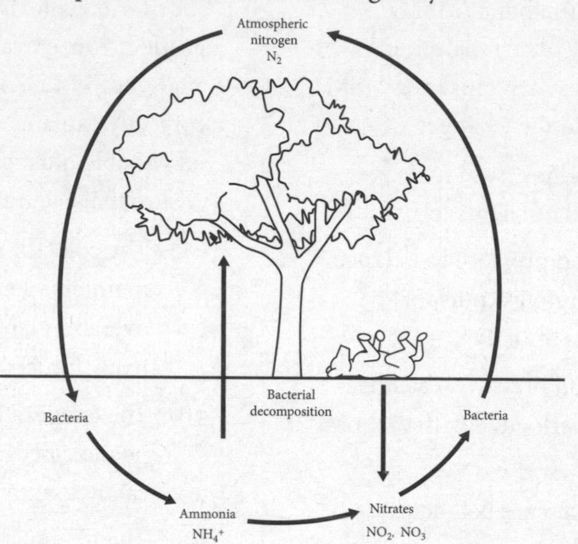

61. Unlike some species of bacteria and plants, animals are more limited in the forms of nitrogen that they can utilize in their cells. Which of the following is a form of nitrogen that animals are capable of utilizing?
    (A) N_2
    (B) NH_4^+
    (C) Amino acids
    (D) NO_3^-

62. Nitrifying bacteria are a group of organisms that are capable of converting ammonia (NH_3) or ammonium (NH_4^+) to nitrate (NO_3^-) through a series of enzyme-catalyzed reactions. This conversion is most analogous to
    (A) the aerobic metabolism of glucose to carbon dioxide
    (B) the reduction of carbon dioxide in photosynthesis
    (C) the synthesis of polypeptides in translation
    (D) the hydrolysis of ATP in muscle cells

63. Nitrogen is the mineral nutrient that contributes the most to plant growth and crop yields. Which of the following best explains why plants can suffer from nitrogen deficiencies, despite the fact that the atmosphere is nearly 80% nitrogen?
    (A) Free nitrogen in the atmosphere cannot be used by plants.
    (B) Plants can only utilize nitrogen in its gaseous state.
    (C) The nitrogen available to plants is dependent upon the breakdown of rock.
    (D) 80% of the atmosphere is only a fraction of the total nitrogen demand of plants.

END OF PART A
DO NOT STOP
PLEASE CONTINUE TO PART B

SECTION I
Part B

Directions: Part B consists of six questions requiring numeric answers. Calculate the correct answer for each question, and enter on the line provided on the answer sheet.

121. A germ cell has 48 chromosomes after DNA replication but before undergoing meiosis. During meiosis I, the homologs for a particular chromosome do not segregate and all the genetic material for that chromosome stays inside one of the two daughter cells. After undergoing meiosis II, what is the minimum possible number of chromosomes that will be present in one of the gametes?

122. In a given population, 7,192 people have attached earlobes and 4,881 people have unattached earlobes. The allele for unattached earlobes is recessive. Assuming that the population is in Hardy-Weinberg equilibrium, what fraction of the population is heterozygous? Give your answer as a decimal and round to the nearest hundredth.

123. On your summer break, you conducted a mark and recapture investigation to determine the size of the population of crickets in your yard. On your first attempt, you capture, mark, and release 107 crickets. On your second attempt, you capture 81 crickets, 17 of which are marked. Based on this data, what should be your estimate of the total population size to the nearest whole number?

124. The ocean food web below shows how carbon flows between organisms and the surrounding environment. The arrows at the far left indicate how much carbon is lost via respiration. Arrows pointing upward indicate how much carbon is consumed by species at higher trophic levels, while the curved arrows pointing down indicate how much carbon is lost to decomposition by bacteria. Note that some of the sources of carbon loss for bacteria and small fish are not depicted.

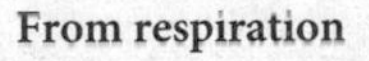

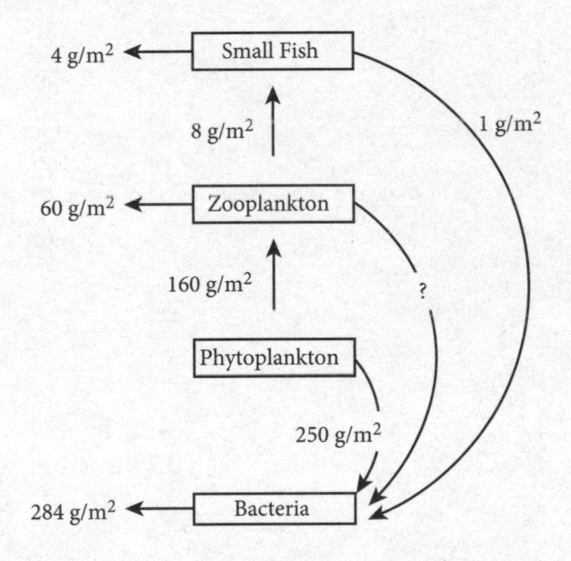

How much carbon (in g/m^2) from zooplankton is used by decomposers? Give your answer to the nearest whole number.

GO ON TO THE NEXT PAGE.

125. The data table below displays the population growth for a population of bacteria that reproduces every 12 hours. Calculate the mean rate of population growth (individuals per hour) between 36 and 48 hours. Give your answer to the nearest whole number.

Hours	Number of Individuals
0	20
12	60
24	180
36	540
48	1,620
60	4,860
72	14,580
84	43,740
96	44,299
108	44,800

126. In a rare breed of African cichlids, the allele for yellow fins (*Y*) is dominant over the allele for orange fins (*y*). Yellow and orange cichlids were bred in captivity to sell into the aquarium trade. The breeders counted 426 yellow babies and 328 orange babies. Calculate the chi-square value for the null hypothesis that the yellow parent was heterozygous for yellow fins. Round your answer to the nearest tenth.

END OF SECTION I

IF YOU FINISH BEFORE TIME IS CALLED, YOU MAY CHECK YOUR WORK ON SECTION I ONLY.

DO NOT GO ON TO SECTION II UNTIL YOU ARE TOLD TO DO SO.

SECTION II
Reading Period—10 minutes
80 Minutes—8 Questions

Directions: Questions 1 and 2 are long free-response questions that should require about 20 minutes each to answer. Questions 3 through 8 are short free-response questions that should require about 6 minutes each to answer. Read each question carefully and write your response on scratch paper. Answers must be written out. Outline form is not acceptable. It is important that you read each question completely before you begin to write.

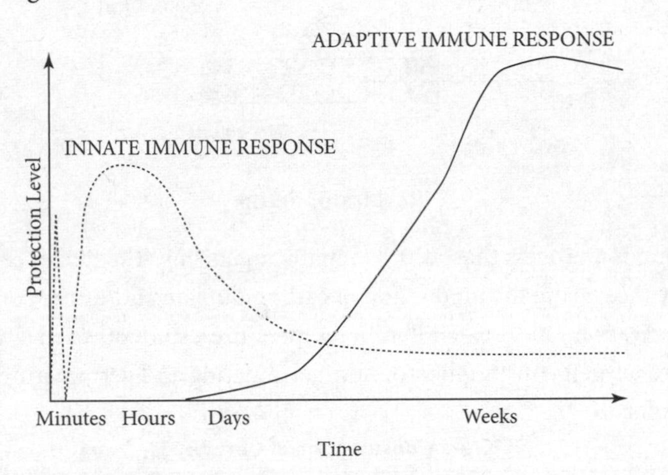

Figure 1. Immune System Response

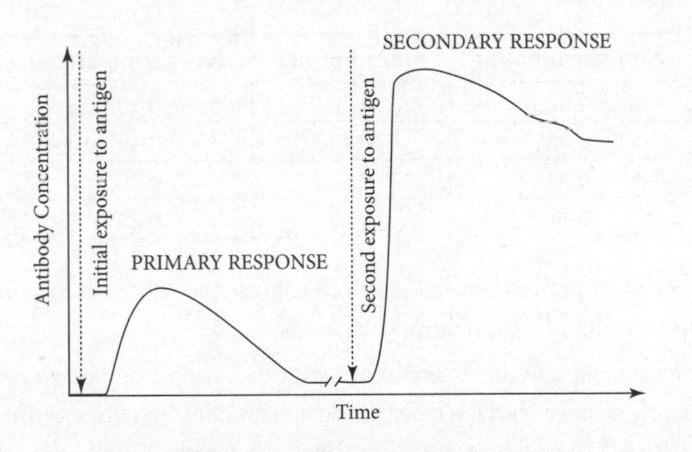

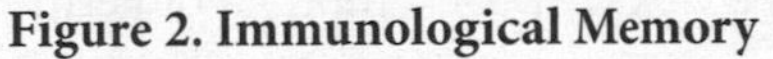
Figure 2. Immunological Memory

1. The mammalian immune system consists of innate immunity and adaptive immunity (Figure 1). Innate immunity involves nonspecific defense mechanisms such as physical and chemical barriers that respond to antigen (bacteria or virus) exposure. Adaptive immunity involves antigen-specific immune response and occurs if the innate immune system is unsuccessful in destroying the antigen. The adaptive immune system has the unique ability to remember an encounter with an

Practice

GO ON TO THE NEXT PAGE.

antigen. Upon a subsequent encounter with the same antigen, immunological memory allows the immune system to mount a response that is faster and greater (Figure 2). Artificially acquired immunity can be induced by vaccination.

(a) **Describe** how the innate and the adaptive immune systems respond to a laceration, such as an injury from falling on a barbed-wire fence.

(b) **Describe** how vaccination is based on immunological memory and **explain** the purpose of a booster dose (the re-exposure to an antigen after initial immunization).

(c) **Provide** two reasons why the flu vaccination has a fairly low success rate.

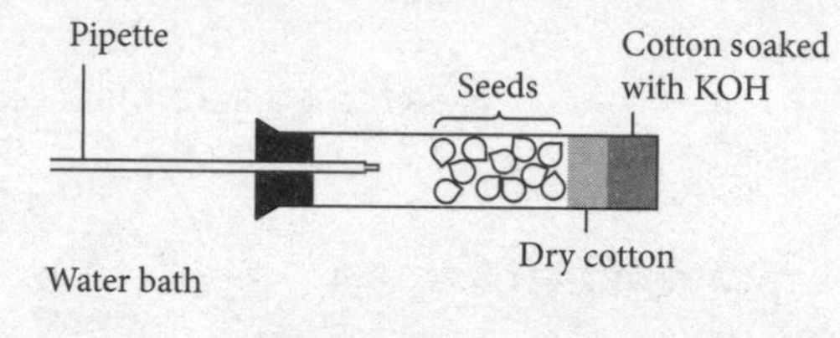

Respirator Setup

2. Cellular respiration is a process that occurs in many organisms. The rate of cellular respiration can be determined by measuring the production of carbon dioxide and/or the consumption of oxygen. To investigate respiration rate at two different temperatures, students used a respirator (see diagram above) to measure oxygen consumption of non-germinating and germinating corn seeds, recording the following results:

Oxygen Consumption of Corn Seeds

Time (min)	O_2 consumed (mL)			
	10°C		25°C	
	Non-germinating	Germinating	Non-germinating	Germinating
5	0.02	0.20	0.10	0.40
10	0.05	0.40	0.15	0.80
15	0.08	0.60	0.20	1.20
20	0.11	0.80	0.25	1.60

(a) **Construct** an appropriately labeled graph to analyze the effect of temperature over time on oxygen consumption of germinating corn seeds.

(b) Based on the data, **explain** the differences in oxygen consumption between non-germinating and germinating corn seeds. **Describe** the most likely effect of temperature on the rate of cellular respiration. **Predict** the likely oxygen consumption for germinating corn at 20°C after 20 min.

(c) **Propose** an appropriate control treatment for the experiment, and **describe** how the control treatment would increase the validity of the results.

(d) The tetrazolium test is an alternative test for seed viability, in which seeds are incubated in tetrazolium, rinsed, and then evaluated for color. In the presence of hydrogen ions, tetrazolium is reduced from a colorless compound to a red compound. **Predict** the color of non-germinating and germinating corn seeds. **Justify** your response.

GO ON TO THE NEXT PAGE.

3. A scientist conducts an experiment with penicillin-sensitive bacteria in which he adds a plasmid containing a gene that confers penicillin resistance. Following a protocol that elicits normal growth and uptake of the plasmid DNA, the scientist then adds bacteria to four new plates, as shown here.

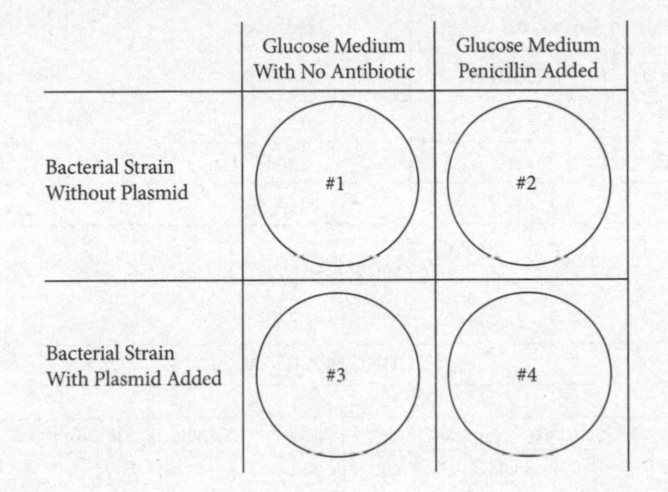

(a) **Describe** the protocol the scientist likely used to encourage the uptake of the plasmid DNA.

(b) **Predict** the growth patterns the scientist should expect to see on the plates and **provide** reasoning for your prediction.

4. Growth curves are models used to track the rise and fall of populations. The graph shown here depicts the population growth curves for three different populations of an organism.

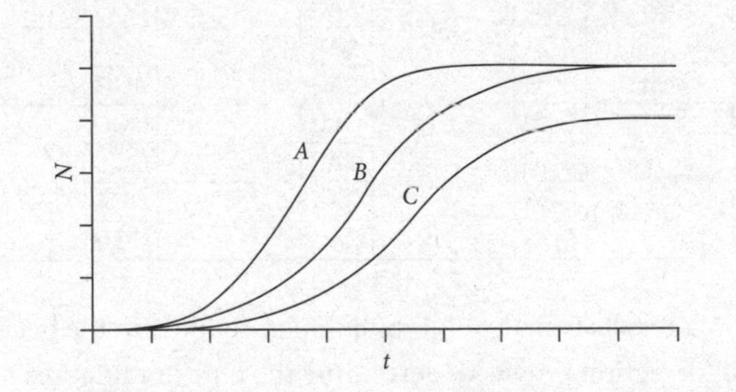

(a) Using the information presented in the graph, **determine** the carrying capacities of the three populations relative to one another and **give support** for your answer.

(b) Based on the graph, **determine** the relative growth rates of the three populations. **Justify** your response.

Practice

GO ON TO THE NEXT PAGE.

5. Point mutations that involve the substitution of a base can result in a silent, missense, or nonsense mutation. Scientists investigated three different point mutations in the genes of hemoglobin. The results are shown in the table below.

Mutation Location		**Before**	**After**
Globin	**Codon**	**Mutation**	**Mutation**
beta	6	GAA	GUA
beta	39	CAG	UAG
alpha	142	UAA	CAA

CODON TABLE

FIRST POSITION	SECOND POSITION								THIRD POSITION
	U		C		A		G		
	code	amino acid	code	amino acid	code	amino acid	code	amino acid	
U	UUU	phe	UCU	ser	UAU	tyr	UGU	cys	U
	UUC		UCC		UAC		UGC		C
	UUA	leu	UCA		UAA	STOP	UGA	STOP	A
	UUG		UCG		UAG	STOP	UGG	trp	G
C	CUU	leu	CCU	pro	CAU	his	CGU	arg	U
	CUC		CCC		CAC		CGC		C
	CUA		CCA		CAA	gln	CGA		A
	CUG		CCG		CAG		CGG		G
A	AUU	ile	ACU	thr	AAU	asn	AGU	ser	U
	AUC		ACC		AAC		AGC		C
	AUA		ACA		AAA	lys	AGA	arg	A
	AUG	met	ACG		AAG		AGG		G
G	GUU	val	GCU	ala	GAU	asp	GGU	gly	U
	GUC		GCC		GAC		GGC		C
	GUA		GCA		GAA	glu	GGA		A
	GUG		GCG		GAG		GGG		G

(a) Sickle cell anemia results from a point mutation at codon 6 in the beta globin of hemoglobin. Based on the information provided, **determine** the type of mutation that causes sickle cell anemia. **Justify** your response.

(b) Beta-thalassemia is an inherited blood disorder in which a shortened globin chain results in a functionally useless protein. Based on the information provided, **determine** which point mutation causes beta-thalassemia. **Provide** reasoning for your answer.

GO ON TO THE NEXT PAGE.

6. A pedigree shows the genetic relationships between members of a family and can be used to analyze the inheritance of a trait. Males are indicated by squares, and females are indicated by circles. Shaded shapes represent individuals who exhibit the trait. Refer to the following pedigree.

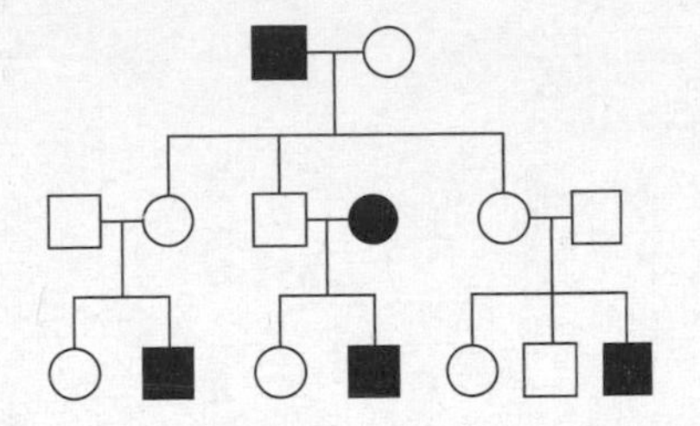

   (a) **Explain** the inheritance pattern of the allele(s) represented in the pedigree. **Provide** evidence to support your answer.

   (b) **Identify** the genotype for the grandmother at the top of the tree.

7. DNA makes up the genetic code for all living organisms. A series of processes and functions must occur within the cell to carry out the genetic instructions encoded in DNA. Transcription is the first step in gene expression and is followed by translation.

   (a) **Describe** the initiation stage in transcription.

   (b) **Describe** the elongation stage in transcription.

   (c) **Describe** the termination stage in transcription.

8. Scientists modeled enzyme kinetics for competitive and noncompetitive inhibition using a Lineweaver-Burk plot (see graph), where *V* represents the enzyme's reaction rate, [S] is the substrate concentration, and [I] is the inhibitor concentration.

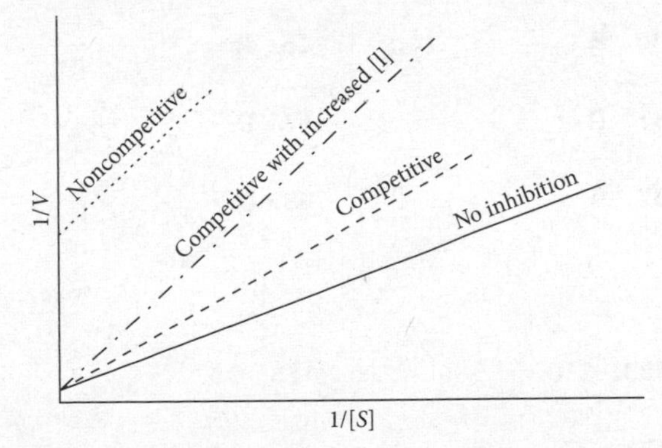

   A competitive inhibitor binds to the same site as the substrate, and a noncompetitive inhibitor binds to a site different from the substrate's binding site.

   (a) Using the graph, **determine** how increasing substrate concentration affects inhibition for each type of inhibitor: competitive and noncompetitive.

   (b) Based on the information, **explain** the effect of inhibitor concentration on the enzyme's reaction rate in competitive inhibition.

STOP—END OF EXAM

ANSWER KEY

Section I, Part A

1. A	17. D	33. B	49. C
2. B	18. D	34. C	50. D
3. D	19. D	35. D	51. A
4. A	20. C	36. B	52. C
5. D	21. C	37. B	53. C
6. A	22. D	38. C	54. D
7. B	23. B	39. A	55. B
8. C	24. B	40. A	56. A
9. C	25. A	41. A	57. C
10. B	26. A	42. B	58. B
11. B	27. C	43. A	59. B
12. C	28. D	44. C	60. C
13. B	29. D	45. D	61. C
14. D	30. B	46. A	62. A
15. C	31. D	47. D	63. A
16. A	32. A	48. D	

Section I, Part B

121. 23	123. 510	125. 90
122. 0.46	124. 92	126. 12.7

Section I, Part A Number Correct: ______________

Section I, Part B Number Correct: ______________

Section II Points Earned: ______________

Enter your results to your Practice Test 1 assignment to see your 1–5 score and view detailed answers and explanations by logging in at kaptest.com.

Haven't registered your book yet? Go to kaptest.com/booksonline to begin.

Practice Test 2

Section I, Part A

1 A B C D
2 A B C D
3 A B C D
4 A B C D
5 A B C D
6 A B C D
7 A B C D
8 A B C D
9 A B C D
10 A B C D
11 A B C D
12 A B C D
13 A B C D
14 A B C D
15 A B C D
16 A B C D
17 A B C D
18 A B C D
19 A B C D
20 A B C D
21 A B C D
22 A B C D
23 A B C D
24 A B C D
25 A B C D
26 A B C D
27 A B C D
28 A B C D
29 A B C D
30 A B C D
31 A B C D
32 A B C D
33 A B C D
34 A B C D
35 A B C D
36 A B C D
37 A B C D
38 A B C D
39 A B C D
40 A B C D
41 A B C D
42 A B C D
43 A B C D
44 A B C D
45 A B C D
46 A B C D
47 A B C D
48 A B C D
49 A B C D
50 A B C D
51 A B C D
52 A B C D
53 A B C D
54 A B C D
55 A B C D
56 A B C D
57 A B C D
58 A B C D
59 A B C D
60 A B C D
61 A B C D
62 A B C D
63 A B C D

Section I, Part B

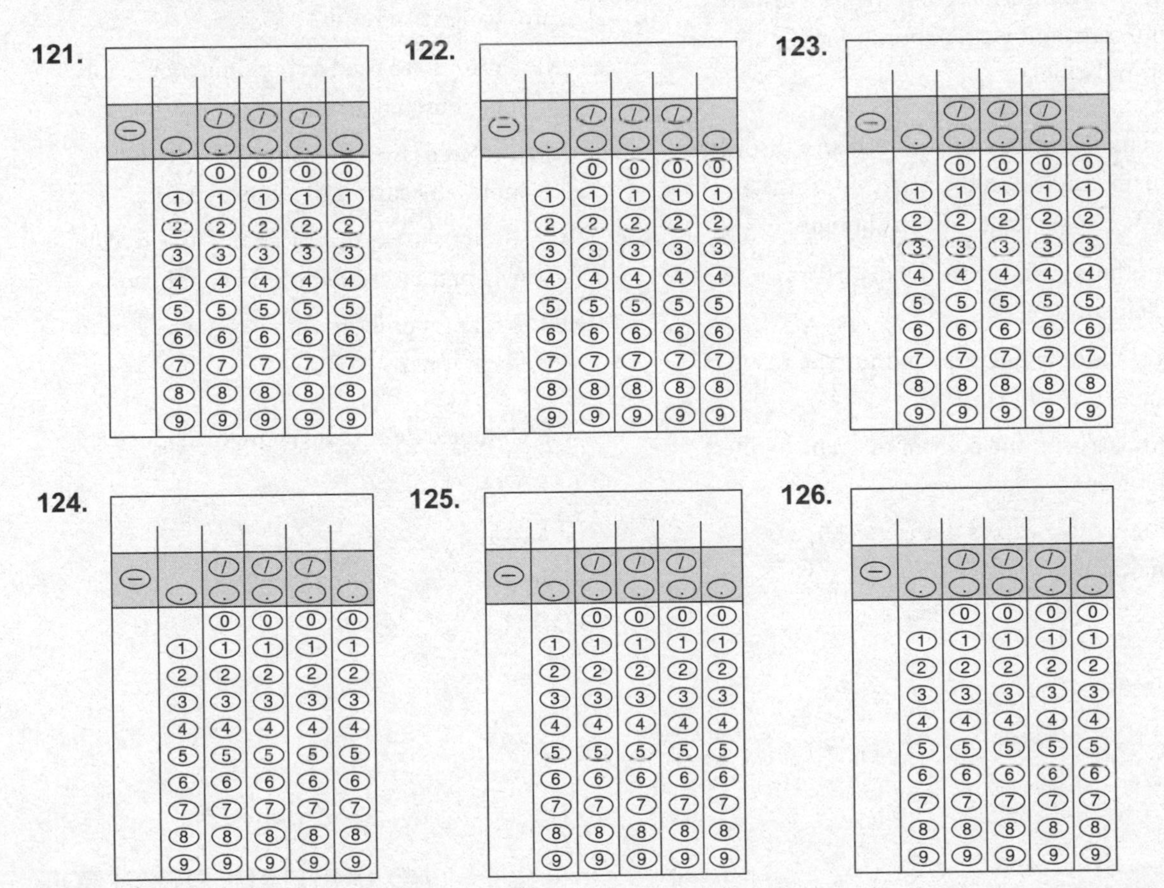

SECTION I
90 Minutes—69 Questions
Part A

Directions: Each of the questions or incomplete statements below is followed by four suggested answers or completions. Select the answer that is best in each case and enter the appropriate letter in the corresponding space on the answer sheet. When you have completed part A, you should continue on to part B.

1. A botanist observes that a mature root cell can dedifferentiate in tissue culture and give rise to a diversity of plant cells. Which of the following best explains this observation?

   (A) Root cells contain all the genes necessary to produce a variety of cells.

   (B) Each type of cell has a unique genetic blueprint.

   (C) The tissue culture transferred proteins necessary for plant differentiation to the root cell.

   (D) mRNA transcripts from the root cell are translated only after appropriate stimulation.

2. In human beings, sex-linked recessive disorders are usually carried on the X chromosome and most often affect males. This is because

   (A) mothers always pass the disorders on to their sons

   (B) it takes only one copy of the gene to affect males

   (C) it takes only one copy of the gene to affect females

   (D) it takes two copies of the gene to affect males

Questions 3–4

The following are the net reactions of photosynthesis and aerobic cellular respiration.

$$6\ CO_2 + 6\ H_2O + \text{energy} \rightarrow C_6H_{12}O_6 + 6\ O_2$$

$$C_6H_{12}O_6 + 6\ O_2 \rightarrow 6\ CO_2 + 6\ H_2O + \text{ATP}$$

3. During aerobic cellular respiration, the C-H bond in glucose is broken and electrons are ultimately transferred to oxygen. Which of the following best explains why the concentration of electron-poor hydrogens (H^+) does not drastically change as a result?

   (A) Protons are used as building blocks for macromolecules.

   (B) Protons associate with ATP, which normally carries a negative charge.

   (C) Protons are transported out of the cell, where they are removed via diffusion.

   (D) Protons combine with oxygen anions to form water.

4. Which molecule is reduced to form glucose?

   (A) CO_2

   (B) ATP

   (C) O_2

   (D) H_2O

GO ON TO THE NEXT PAGE.

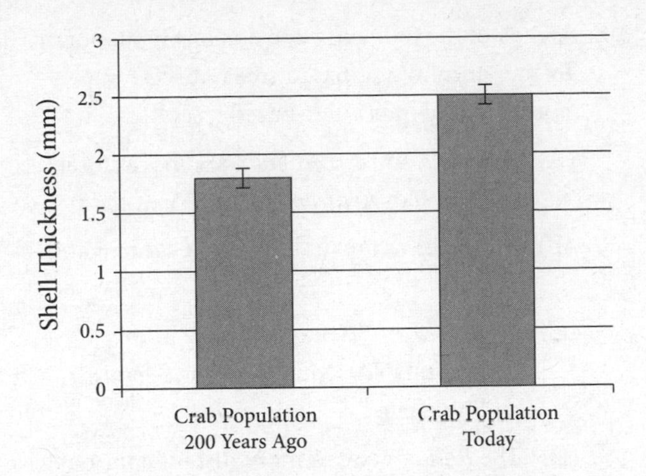

5. Crab shells are composed mainly of calcium carbonate, and their thickness determines how vulnerable crabs are to predators and environmental factors. Over the past 200 years, ocean acidification has caused a 25 percent increase in acidity. One study compared average shell thickness of a crab population 200 years ago to that of a crab population today. The data in the graph best support which of the following claims?

   (A) Average shell thickness decreased because ocean acidification lowered the pH of the seawater and caused shells to dissolve.

   (B) Average shell thickness decreased because predators selected thick-shelled crabs over thin-shelled crabs.

   (C) Average shell thickness increased because thin-shelled crabs were more easily penetrated by predators than were thick-shelled crabs.

   (D) Average shell thickness increased because ocean acidification lowered the levels of available carbonate ions for shell production.

6. *Naegleria fowleri* causes a fatal form of meningitis. The infectious form of this organism inhabits fresh water in warm climates, often in the sediment of lakes. It can infect humans when they swim in infested lakes, allowing entry through the nose. *N. fowleri* has a true membrane-bound nucleus and cellular organelles. It is a unicellular, heterotrophic organism that lacks a cell wall and moves via pseudopodia. What type of organism is it?

   (A) Bacteria

   (B) Virus

   (C) Protozoan

   (D) Fungus

7. What is the probability that a mother who is a carrier for cystic fibrosis, an autosomal recessive disorder, will have an affected child if the father is genotypically normal?

   (A) 0%

   (B) 25%

   (C) 50%

   (D) 100%

8. The sodium potassium pump is an ATPase that pumps 3 Na^+ out of the cell and 2 K^+ into the cell for each ATP hydrolyzed. Cells can use the pump to help maintain cell volume. Which of the following would most likely happen to the rate of ATP consumption immediately after a cell is moved to a hypotonic environment?

   (A) It would remain the same.

   (B) It would decrease.

   (C) It would increase.

   (D) It would increase, then decrease.

GO ON TO THE NEXT PAGE.

Questions 9–12

Two teams of scientists created the following pair of phylogenetic trees based on available data.

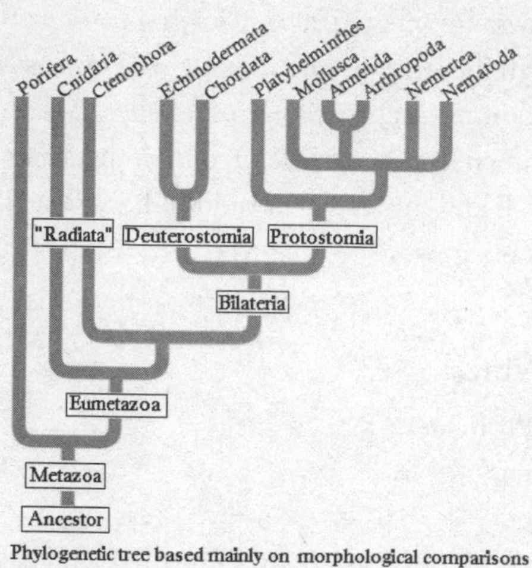

Phylogenetic tree based mainly on morphological comparisons

Phylogenetic tree based mainly on morphological comparisons

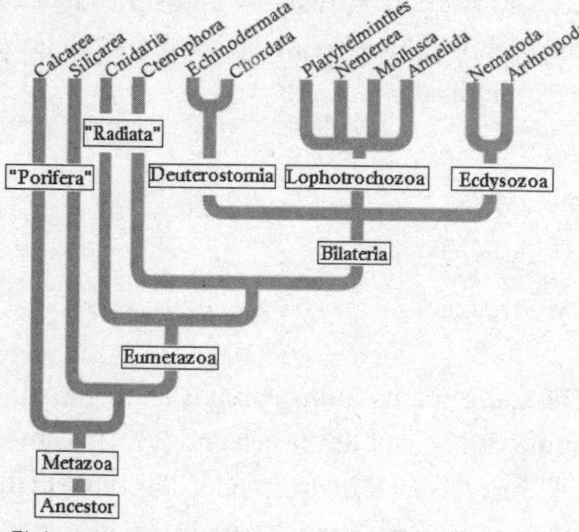

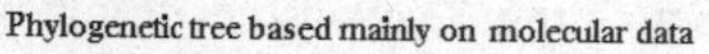

Phylogenetic tree based mainly on molecular data

9. Which of the following indicates the most substantial difference between the two phylogenetic trees?

   (A) Molecular differences among *Porifera* undetected by visual comparison

   (B) *Protostomia*'s replacement by two sister taxa in the molecular tree

   (C) The relative timeline of *Deuterostomia* divergence

   (D) The convergence of the entire animal kingdom on a common ancestor

10. Which of the following conclusions is supported by the morphology-based tree but NOT supported by the molecular-based tree?

    (A) *Annelida* and *Arthropoda* are more closely related than *Arthropoda* and *Nematoda.*

    (B) The *Porifera* are divided into *Calcarea* and *Silicarea.*

    (C) *Nemertea* and *Nematoda* have more common DNA sequences than *Mollusca* and *Nemertea.*

    (D) The *Radiata* have a more distant common ancestor than the *Bilateria.*

11. If the scientists who created the morphology-based tree relied mainly on phenotypic comparisons of adult and developing organisms, while those who created the molecular-based tree compared homologous *hedgehog* genes (a gene important for development), then which of the following statements would both teams of scientists most likely agree upon?

    (A) As the number of shared features increases so does the likelihood of independent evolution.

    (B) The fossil record is the ultimate authority in phylogenetics.

    (C) The molecular-based tree is the result of modern technology and is unlikely to be altered.

    (D) Neither phylogenetic tree completely and accurately describes actual evolutionary history.

GO ON TO THE NEXT PAGE.

12. The phylum *Cnidaria* consists of over 10,000 aquatic species, with cnidocytes (specialized cells used mainly for capturing prey) being the distinguishing characteristic. Which of the following is most likely to be classified as a Cnidarian?
    (A) An organism lacking digestive, circulatory, and nervous systems that feeds by drawing in water through pores
    (B) An organism with clearly defined sides (i.e., top/bottom, left/right)
    (C) A planktonic organism capable of responding to the environment equally from all directions
    (D) An organism that actively stalks its prey with coordinated complex movements

13. A foreign object originally located next to a tree is eventually "consumed" by the growing tree. As the tree grows taller, the height of the object in the tree is unchanged. These observations support which of the following statements?
    (A) Trees get taller by growing at the branch tips.
    (B) Vertical growth and horizontal growth are independent.
    (C) Once a certain maximum is attained, vertical height is relatively constant.
    (D) Height is increased via cell proliferation in the root system.

14. It is hypothesized that negative pressure in the phloem and cohesion among water molecules are responsible for the bulk movement of water against gravity in vascular plants. Which of the following fluid-filled cylinders best illustrates cohesion?

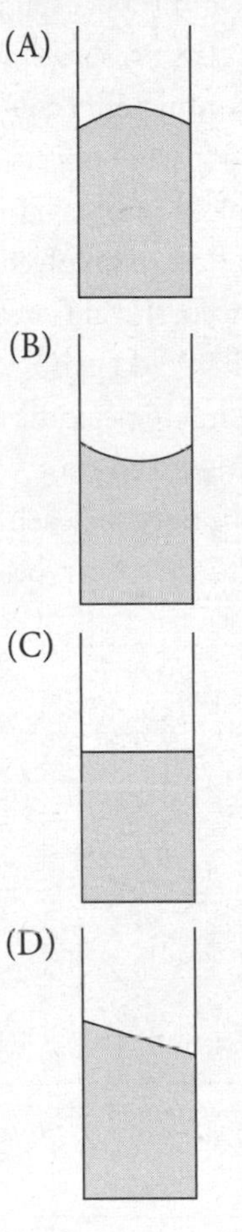

GO ON TO THE NEXT PAGE.

Questions 15–17

A scientist studying the ecology of cities found that, in developed landscapes, plant roots were not colonized by mycorrhizal fungi to the same degree that they were in a nearby nature preserve. In addition, she found that rates of photosynthesis and root respiration were much higher in plants in the preserve than for plants in city landscapes. She conducted a controlled greenhouse experiment to see what effects mycorrhizal colonization had on plant photosynthesis and respiration. Her experimental design involved growing 10 plants in soil rich in mycorrhizal fungal elements and 10 in the same soil that had been sterilized to remove the fungi. She made periodic measurements of plant photosynthesis and root respiration and calculated the mean rates for each experimental treatment. Her results are shown below.

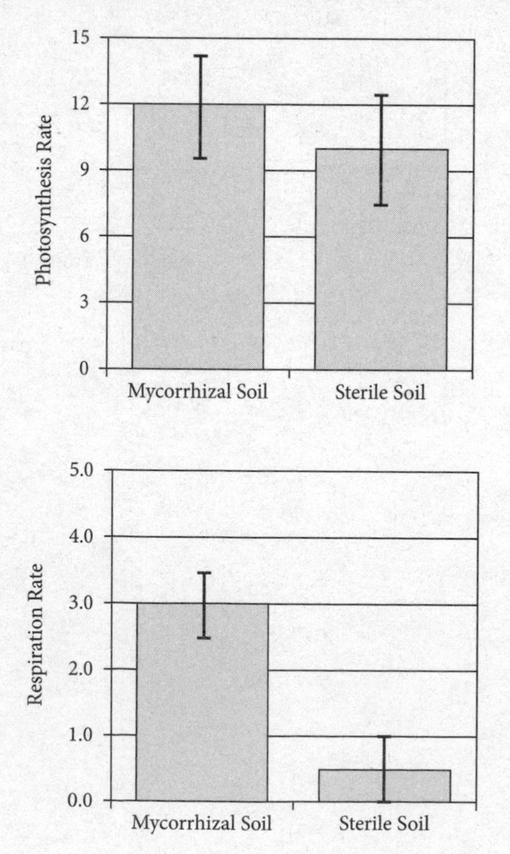

15. What conclusion can most reasonably be drawn from the data?
    (A) The presence of mycorrhizae increased photosynthesis and respiration rates significantly.
    (B) The presence of mycorrhizae increased photosynthesis but not respiration.
    (C) The presence of mycorrhizae increased respiration but not photosynthesis.
    (D) The presence of mycorrhizae had no effect on photosynthesis or respiration.

16. Which of the following provides the most plausible reason for why mycorrhizae have the influence they do on plant metabolic processes?
    (A) Mycorrhizae are important plant pathogens.
    (B) Mycorrhizae are important plant parasites.
    (C) Mycorrhizae are important plant predators.
    (D) Mycorrhizae are important plant symbionts.

17. Assuming that further experimentation showed conclusively that plants in cities had reduced rates of photosynthesis and respiration due to lack of colonization by mycorrhizae, what important biogeochemical cycle of an ecosystem would be most affected?
    (A) The nitrogen cycle
    (B) The water cycle
    (C) The hydrological cycle
    (D) The carbon cycle

Practice

GO ON TO THE NEXT PAGE.

18. In some organisms, features that have no function become vestigial and are ultimately lost. In many cave-dwelling animals, organs such as the eyes have been lost while other sense organs have increased in size. Which of the following hypotheses to explain the loss of nonfunctioning organs would NOT be considered correct according to the contemporary understanding of evolution?
    (A) Mutations causing the reduction in the size of nonfunctional organs become fixed by genetic drift.
    (B) There is natural selection against organs that are not used because the organs interfere with other, more important bodily functions.
    (C) The development of a nonfunctional organ requires energy expenditures that would be better spent on building other tissues or maintaining other traits.
    (D) All organs are maintained or eliminated as a result of how much they are used.

19. The Cdk inhibitor p16 binds to Cdk4/cyclin D complexes, which are normally responsible for allowing cells to pass through the restriction point from G_1, the first growth phase of the cell cycle, into S phase, when chromosome replication occurs. Underexpression of p16 protein could lead to
    (A) uncontrolled cell division
    (B) cessation of mitosis
    (C) increased inhibition of Cdk4/cyclin D complexes
    (D) overexpression of p53 protein

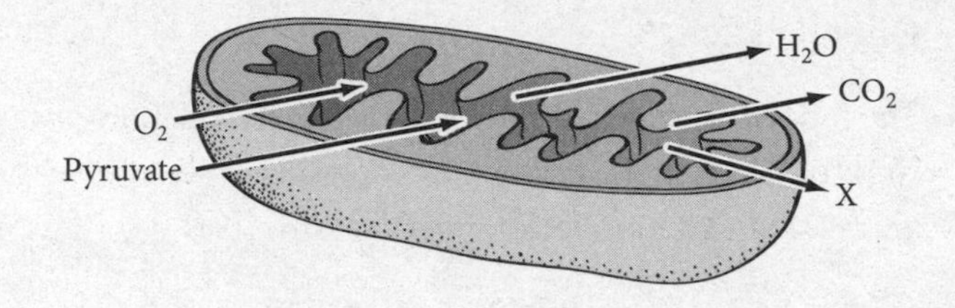

20. In the diagram pictured above, the letter *X* represents
    (A) glucose
    (B) $NADP^+$
    (C) ATP
    (D) ADP

21. Some plants can reproduce through self-pollination, in which a plant is fertilized by its own pollen. Self-pollination in plants is an example of
    (A) asexual reproduction, because a single parent is involved
    (B) asexual reproduction, because offspring are genetically identical to the parent plant
    (C) asexual reproduction, because offspring are genetically unique
    (D) sexual reproduction, because offspring are produced via fusion of gametes

Practice

GO ON TO THE NEXT PAGE.

Questions 22–23

The following diagram shows the feedback relationships between levels of the hormones insulin and glucagon and a number of digestive processes.

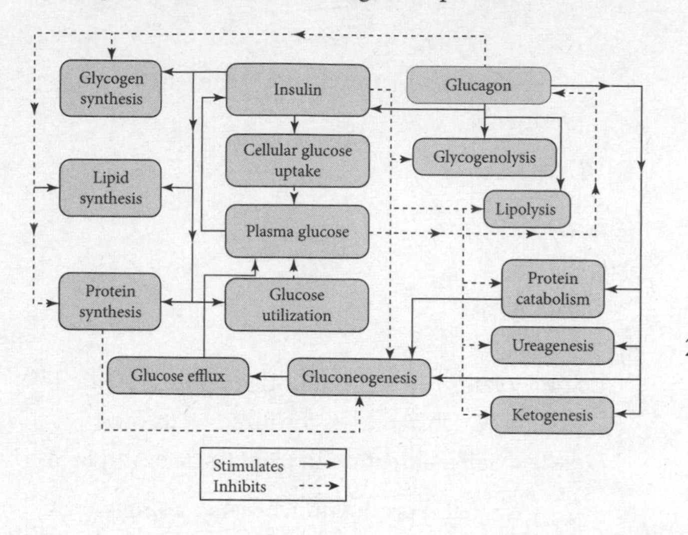

22. Based on the data in the figure above, the most reasonable conclusion is that
    (A) protein is the body's preferred energy source
    (B) insulin and glucagon act antagonistically
    (C) insulin and glucagon are produced by beta and alpha cells, respectively
    (D) brain cells are able to uptake glucose without insulin

23. A person with an inability to synthesize insulin would be expected to show
    (A) low glucagon levels and low glucose levels
    (B) high glucagon levels and low glucose levels
    (C) low glucagon levels and high glucose levels
    (D) high glucagon levels and high glucose levels

24. The respiratory quotient (RQ) is calculated as the ratio of carbon dioxide produced to the oxygen consumed for the complete combustion of a given fuel source. The RQ for carbohydrates is around 1.0, while the respiratory quotient for lipids is around 0.7. In resting individuals, the RQ is most likely to be
    (A) 1.2
    (B) 1.0
    (C) 0.8
    (D) 0.6

25. In a food chain that consists of grass → grasshoppers → spiders → mice → snakes → hawks, the organism(s) that possess the most biomass within the community is (are) the
    (A) grass
    (B) grasshoppers
    (C) mice
    (D) snakes

26. Biotechnology is used for a number of applications in medicine, such as the manufacture of drugs and essential biological compounds. How are bacteria typically used to produce human insulin?
    (A) They are grown on media rich in sugar, which stimulates insulin production in the bacteria.
    (B) The DNA sequence that codes for human insulin production is inserted into the bacterial genome.
    (C) Human pancreas cells are grown in culture with bacteria and transformation occurs.
    (D) Specific bacteriophage viruses are used to produce the correct mutation in the bacterial genome.

GO ON TO THE NEXT PAGE.

27. Chimpanzees demonstrate insight learning when presented with a problem. Stacking boxes and climbing on them to bat down a suspended banana with a stick demonstrates the ability to

(A) combine unrelated experiences to solve novel problems

(B) associate a new stimulus with a particular reward or punishment

(C) retain memories with no immediate consequence and use them later

(D) learn to ignore stimuli that have no positive or negative associations or consequences

28. What is the primary danger of a population relying on an agricultural monoculture for its dietary staple?

(A) People get tired of eating the same thing.

(B) Children develop aversions to foods that appear too often in their diet.

(C) Allergies develop after repeated exposure to the same food.

(D) Genetic similarity makes an entire crop vulnerable to a single pest or pathogen.

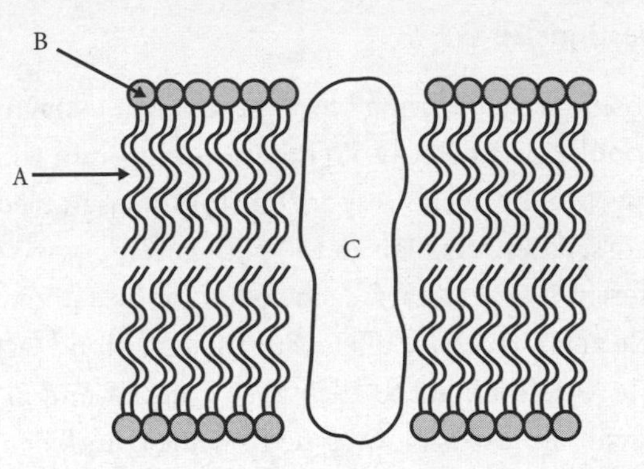

29. The diagram above depicts a cross-section of a cell membrane. Which of the following accurately describes the diagram?

(A) The region of the membrane labeled A is nonpolar.

(B) The region of the membrane labeled B is hydrophobic.

(C) The structure labeled C is a complex carbohydrate.

(D) Charged ions such as Na^+ diffuse directly through the membrane bilayer.

Practice

GO ON TO THE NEXT PAGE.

Questions 30–31

Consider the following blood group data taken from a population in Hardy-Weinberg equilibrium with respect to the alleles responsible for different blood factors. All individuals in the population possess two different blood factors, each coded for by a dominant allele and a recessive allele. For the first blood factor, allele *R* is dominant to allele *r*, so both *RR* and *Rr* individuals test as blood type R, while *rr* individuals test as blood type r. For the second blood factor, allele *F* is dominant to allele *f*, so both *FF* and *Ff* individuals test as blood type F, while *ff* individuals test as blood type f. The frequencies observed for blood type in the population are as follows:

Blood Types	Frequency
R and F	0.60
R and f	0.15
r and F	0.24
r and f	0.01

30. Based on the data provided, the frequencies of the *r* allele and the *F* allele, respectively, are

(A) 0.6 and 0.5

(B) 0.25 and 0.25

(C) 0.5 and 0.6

(D) 0.25 and 0.84

31. Given the information above, all of the following statements concerning this population are true EXCEPT that

(A) there are no new mutations arising among the blood type alleles

(B) there is significant gene flow between this population and others

(C) the population is large in size

(D) there is no positive selection for the *R* allele

32. A certain plant is grown in darkness and observed to produce tall stems with non-expanded leaves. After being transported into daylight, the same plant develops broad, green leaves; short, sturdy stems; and long roots. Which of of following provides the best explanation for these observations?

(A) There is minimal energy expenditure until sunlight is detected.

(B) Sunlight stimulates the elongation of stems and proliferation of leaves.

(C) The plant is exhibiting a specialized response to mechanical stress.

(D) The plant normally sprouts underground.

33. A patient's parents both have a disease that is caused by a sex-linked dominant allele. This disease was passed down to the patient's mother from the patient's grandfather. The chance that the patient will have this disease is

(A) 25%

(B) 50%

(C) 75%

(D) 100%

34. Which of the following relationships is NOT an example of symbiosis?

(A) A tick sucking blood from a dog

(B) A clownfish living in the tentacles of a sea anemone for protection

(C) Ants farming aphids to feed on their honeydew

(D) Cows eating grass and leaving manure behind as fertilizer

GO ON TO THE NEXT PAGE.

35. Human white blood cells can "crawl" to damaged tissues to contact and phagocytize bacteria at the site of the damage. Which of the following processes facilitates this type of movement?
    - (A) Addition of phospholipids to the cell plasma membrane
    - (B) Rapid formation and deformation of actin filaments in the cytoskeleton
    - (C) Beating of cilia against vessel and tissue surfaces
    - (D) Whip-like motion of the cell's flagellum

36. Evolution may be defined as changes over time in the allele frequency of a population or species. All of the following are examples of evolutionary processes EXCEPT
    - (A) artificial selection in domestic dogs
    - (B) populations in Hardy-Weinberg equilibrium
    - (C) mutations that decrease reproductive fitness
    - (D) natural selection between different colors of insects

37. Organisms that reproduce sexually exhibit zygotic, gametic, or sporic meiosis. One way to determine the type of life cycle an organism has is by
    - (A) observing embryonic development
    - (B) comparing the diploid and haploid forms of the organism
    - (C) determining when in the life cycle fertilization occurs
    - (D) determining if gametes are multicellular or unicellular

38. Which of the following shows the correct order of hierarchy from simple to complex?
    - (A) Molecule → tissue → cell → organism
    - (B) Cell → tissue → organ → organism
    - (C) Organism → species → biosphere → ecosystem
    - (D) Molecule → cell → organism → tissue

39. Production of purple kernels in *Zea mays* (corn) is dominant over yellow kernels, and smooth kernels are dominant over wrinkled. The two traits are passed on independently of one another. Two heterozygous corn plants with purple, smooth kernels are crossed. Out of 160 of their offspring plants, how many would be expected to have yellow and smooth kernels?
    - (A) 120
    - (B) 90
    - (C) 60
    - (D) 30

40. Organisms transform energy when they ingest food, breaking it down into nutrients used to build tissues and make repairs. Each energy transfer increases the universe's level of
    - (A) order
    - (B) stability
    - (C) disorder
    - (D) energy

GO ON TO THE NEXT PAGE.

41. According to the ABC hypothesis for the functioning of organ identity genes, three classes of genes (*A*, *B*, and *C*) are responsible for the spatial pattern of floral parts. Sepals develop from the region where only *A* genes are active. Petals develop where both *A* and *B* genes are expressed. Stamens arise where *B* and *C* genes are active and carpels arise where only *C* genes are expressed. Furthermore, it is observed that if *A*-gene or *C*-gene activity is missing, then the activity of the other spreads throughout.

    Which of the following could be the floral morphology for a mutant lacking *C*-gene activity?

    (A) Sepal-petal-stamen-carpel-stamen-petal-sepal

    (B) Carpel-stamen-carpel-stamen-carpel

    (C) Sepal-carpel-sepal

    (D) Sepal-petal-sepal-petal-sepal

42. A model of the simple nervous system of the sea snail, *Aplysia*, is shown here. Note that a nociceptor is a pain receptor.

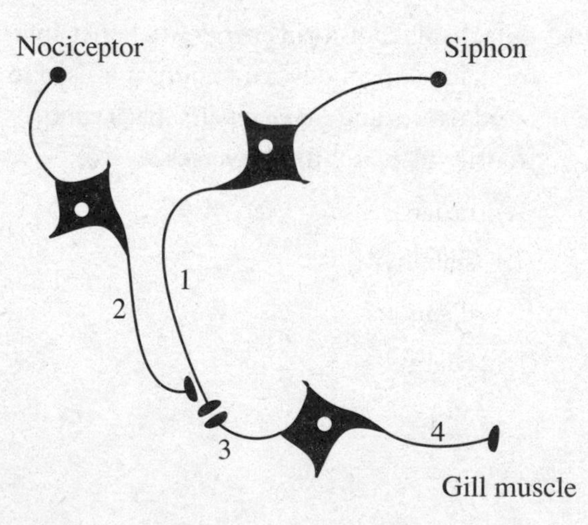

Which of the following points indicate neural axons?

(A) 1 and 2 only

(B) 1 and 3 only

(C) 1, 2, and 4 only

(D) 2, 3, and 4 only

Questions 43–44

The figure illustrates three types of point mutations involving a nucleotide base substitution.

	No mutation	Point mutations		
		Silent	Nonsense	Missense
DNA	ATG TAC	ATA TAT	ATC TAG	GTG CAC
RNA	UAC	UAU	UAG	CAC
AMINO ACID	TYR	TYR	STOP	HIS
PROTEIN	normal	normal	incomplete	faulty

43. Hemoglobin is a tetramer, which consists of four subunits (two α and two β). Sickle cell anemia (abnormally shaped red blood cells) is caused by a mutation in the 20th nucleotide of the hemoglobin β gene, in which glutamic acid is changed for valine. Which of the following describes the result of the substitution?

    (A) The mutation has no effect on the structure of hemoglobin.

    (B) The cell stops the production of hemoglobin.

    (C) The properties of the amino acid are not altered, so hemoglobin does not lose its function.

    (D) The structure and function of hemoglobin are altered.

GO ON TO THE NEXT PAGE.

44. A frameshift mutation is caused by an insertion or deletion of one or more nucleotides in the DNA sequence. Which would most likely be the result of inserting one nucleotide instead of substituting a base in the DNA sequence?

    (A) The frameshift mutation alters the R-groups in RNA.

    (B) The amino acid sequence changes after the insertion.

    (C) The insertion causes a missense mutation, resulting in a nonfunctional protein.

    (D) The frameshift mutation affects the RNA sequence but not the amino acid sequence.

45. Enzymes are regulated in a number of different ways, including through competitive and noncompetitive inhibition. Competitive inhibitors of enzymes can be reversed by

    (A) increasing the pH above the enzyme's optimal range

    (B) increasing the concentration of substrate

    (C) adding noncompetitive inhibitors

    (D) lowering the temperature below the enzyme's optimal range

46. Clumps of overlapping mammalian cells would indicate a problem with cell cycle regulation because

    (A) normal cells use cyclin-dependent kinases for regulation

    (B) normal cell division is stimulated by active anaphase-promoting complex

    (C) normal cells show density-dependent inhibition and anchorage dependence

    (D) normal cell division is stimulated by the presence of growth factor

47. Deep-diving air breathers (e.g., seals, whales, penguins) have numerous adaptations that allow them to dive to great depths and remain underwater for long periods of time. Which of the following would be LEAST likely to be found in a deep diver?

    (A) Decreased O_2 storage in the lungs and increased O_2 storage in the blood

    (B) High concentrations of myoglobin

    (C) A large volume of blood per body mass ratio

    (D) Efficient use of buoyancy to aid locomotion

48. Organisms rely upon the physical and chemical properties of water to facilitate a variety of bodily functions. For instance, human beings sweat when overheated in order to regulate temperature. Sweat cools the human body because

    (A) water has a high heat of vaporization

    (B) water has a low heat of vaporization

    (C) water dissolves salts

    (D) water adheres to heat receptors

49. Bacteriophages are a class of viruses that infect bacteria. Upon entering a bacterium, a phage can enter into one of two reproductive cycles: the lytic and the lysogenic. The main difference between the lytic and lysogenic cycles of phage reproduction is that

    (A) the lysogenic cycle alters the bacteria, while the lytic cycle does not

    (B) the lytic cycle alters the bacteria, while the lysogenic cycle does not

    (C) the lysogenic cycle kills the host cell, while the lytic cycle does not

    (D) the lytic cycle kills the host cell, while the lysogenic cycle does not

GO ON TO THE NEXT PAGE.

50. Eukaryotic cells generate up to 36 ATP per molecule of glucose during aerobic respiration, but only 2 ATP per molecule during fermentation. Aerobic respiration produces much more ATP per glucose molecule than fermentation because

(A) aerobic respiration uses glycolysis to oxidize glucose

(B) oxygen is necessary to release energy stored in pyruvate

(C) fermentation uses NAD^+ as the oxidizing agent in glycolysis

(D) it requires energy to transport molecules into mitochondria

51. A certain species of bird has beaks of variable length, with an average length of 10 centimeters and a standard deviation of 2 centimeters. It feeds primarily on a species of insect that burrows underground, with most of the insects located 8 to 10 centimeters beneath the surface. If the insect starts burrowing to a greater depth, which of the following population distributions for beak length is most likely to result?

(A) A mean of 10 centimeters with a standard deviation of 4 centimeters

(B) A mean of 8 centimeters with a standard deviation of 2 centimeters

(C) A mean of 12 centimeters with a standard deviation of 2 centimeters

(D) A mean of 10 centimeters with a standard deviation of 1 centimeter

52. An object known as the Murchison meteorite struck the Earth in Australia in 1969. Analysis of the rock shows it contains at least 50 amino acids, 19 of which are found in living organisms on Earth, as well as several nucleotides. Assume such amino acid-containing meteorites have existed throughout the history of the solar system. Of the following statements, it is most reasonable to conclude that one or more meteorites

(A) are the sole source of all life on Earth

(B) cannot be the origin of amino acids on Earth

(C) contained complex biological polymers, such as proteins and nucleic acids

(D) contributed some of the precursors of life, but never contained any living organisms

53. The *cecum* is a portion of the large intestine located near the junction of the small intestine and the large intestine. The following table lists the diets of several vertebrates, as well as the average length of the cecum as measured in 20 individuals of that species.

Species	Average Cecum Length	Diet
1	40.1 cm	herbivore
2	5.7 cm	omnivore
3	6.8 cm	carnivore
4	30.2 cm	ruminant

Based on the table, which of the following conclusions is most plausible?

(A) The cecum can become a vestigial structure in meat eaters, since it is shorter in species 2 and 3 than in species 1 and 4.

(B) The cecum evolved to have an important role in the digestion of protein, since it is shorter in species 2 and 3 than in species 1 and 4.

(C) Species 1 is more closely related to species 4 than to species 2, because 1 and 4 are both herbivores.

(D) Species 2 and 3 have a recent common ancestor, since the average cecum lengths in species 2 and 3 are approximately equal.

GO ON TO THE NEXT PAGE.

54. Most biomass pyramids show a rapid decrease in biomass as trophic level increases. In aquatic systems, however, this pattern may be reversed so that one observes a larger standing crop of consumers compared with producers. Which of the following offers the best explanation for this pattern?

    (A) Aquatic producers tend to have larger body sizes than terrestrial producers.
    (B) Water is an easier medium to live in, so aquatic organisms require less food.
    (C) Biomass in aquatic systems cannot be measured accurately.
    (D) Phytoplankton is rapidly consumed, but it has a high turnover rate.

Questions 55–57

Sickle cell anemia is caused by mutant hemoglobin DNA, which is more common in humans with African ancestry than in those with European ancestry. The sickle cell allele creates an altered mRNA codon that produces hemoglobin containing valine rather than glutamic acid. If a person inherits both alleles for the sickle cell trait, that person's hemoglobin will polymerize under low oxygen conditions (i.e., elevated physical activity). This can result in brain damage, paralysis, kidney failure, and other very serious physiological problems.

55. Based on the information provided, the mutation for sickle cell hemoglobin is most likely an example of

    (A) a base-pair substitution
    (B) a frameshift mutation
    (C) a silent mutation
    (D) a mutagen

56. Heterozygotes for the sickle cell trait have an increased resistance to malaria. If malaria were eradicated and effective treatment for sickle cell anemia made universally available, what would be the expected effect on the sickle cell allele?

    (A) The frequency of the allele would remain roughly constant.
    (B) The frequency of the allele would decrease.
    (C) The frequency of the allele would increase.
    (D) The frequency of homozygous individuals would decrease, but the frequency of heterozygous individuals would increase.

57. Genetic mutations, such as the one found in the sickle cell anemia allele, can be found in all organisms, from the simplest prokaryotes to the most complex eukaryotes. However, these mutations have a greater impact on the genetic diversity of bacteria populations than on that of human populations because

    (A) human sexual reproduction recombines existing alleles
    (B) bacteria reproduce more rapidly than humans
    (C) new bacteria are generated through sexual reproduction
    (D) genetic mutations are much rarer in humans than in bacteria

Practice

GO ON TO THE NEXT PAGE.

58. Substances that are formed as intermediates or products during a biochemical reaction are called "metabolites." Certain drugs, called "antimetabolites," possess chemical similarities that allow them to mimic metabolites and participate in normal biochemical reactions, but are different enough that they interfere with overall cellular function. These drugs are commonly used as antibacterial or anticancer agents. Which of the following statements offers the most plausible explanation of how these drugs work?
   (A) The antimetabolite binds to the active site of an enzyme and directly inhibits it, acting as a competitive inhibitor.
   (B) The antimetabolite binds to the active site of an enzyme, allowing the reaction to proceed and produce an unusable end product.
   (C) The antimetabolite binds to the enzyme at a location that is not the active site, changing the structure and function of the enzyme.
   (D) The antimetabolite binds to the metabolite, keeping it from reaching and binding to the active site of an enzyme.

59. The concept of gradualism was initially used to explain the formation of geologic features over vast stretches of time, but aspects of this idea were later incorporated into Darwin's theory of evolution. Which of the following best describes an idea shared both by geologic gradualism and Darwin's theory?
   (A) Change occurs mainly through catastrophic events.
   (B) Slow and continuous processes can lead to drastic changes.
   (C) Certain heritable traits are gradually favored over others.
   (D) Resources are limited and there is a struggle for existence among individuals.

60. Sensitization allows neurons to respond more to a stimulus than they would have prior to sensitization, for a short period of time. What change in the axon terminal could explain this phenomenon?
   (A) The axon terminal is permanently depolarized.
   (B) The axon terminal is permanently hyperpolarized.
   (C) The axon terminal remains depolarized longer following an action potential.
   (D) The axon terminal remains polarized longer following an action potential.

61. HIV (human immunodeficiency virus), the virus that causes AIDS (acquired immune deficiency syndrome), is classified as retrovirus. This means that
   (A) pieces of its RNA genome are spliced into a single strand of RNA before translation
   (B) it reverses its morphology from type C to type D as it enters cells
   (C) it transcribes its RNA genome into a DNA genome
   (D) the promoter regions for RNA polymerase binding are located downstream of genes rather than upstream of them

GO ON TO THE NEXT PAGE.

62. When first exposed to a microbe, antibodies are produced to slow the infection. If a second exposure occurs, the levels of antibody in the bloodstream will be

(A) higher than before because the body is having a secondary response

(B) lower than before because the body has already fought off this infection

(C) the same as before because the body knows exactly how much antibody to produce to fight the infection

(D) nonexistent because the body has already fought off the infection and does not need to do it again

---

63. The table shows properties of water, isopropanol, and benzene.

Liquid	Molecular Formula	Boiling Point (°C)	Melting Point (°C)	Specific Heat Capacity (kJ/kg°C)
Water	H_2O	100.0	0.0	4.18
Isopropanol	C_3H_8O	82.6	−89.0	2.68
Benzene	C_6H_6	80.1	5.5	1.73

Based on the information provided in the table, which of the following best explains why living systems depend on the properties of water, rather than those of other liquids?

(A) Water is composed of hydrogen and oxygen, the two most important essential elements found in living organisms.

(B) Water's high specific heat capacity enables water to buffer temperature changes in living systems.

(C) Water's melting point at 0°C enables water to act as a solute for chemical reactions.

(D) Water's high boiling point enables living organisms to thrive at higher temperatures.

END OF PART A
DO NOT STOP
PLEASE CONTINUE TO PART B

SECTION I
Part B

Directions: Part B consists of six questions requiring numeric answers. Calculate the correct answer for each question, and enter on the line provided on the answer sheet.

121. If a population in Hardy-Weinberg equilibrium has 500 individuals and 127 have the *bb* genotype, assuming simple dominance of the *B* allele, what is the frequency of the *Bb* genotype? Round your answer to the nearest hundredth.

122. You complete an experiment in which a piece of potato is placed in an open container of 0.32 M sucrose solution at a temperature of 27°C. After a few minutes, you measure the mass of the potato again and determine that no change in mass has occurred.

    Calculate the water potential of the solutes within the piece of potato. Round your answer to the nearest hundredth.

123. A scientist extracts DNA from the nucleus of cells and sequences it. The scientist determines that 27% of the nucleotide bases are guanine. What percentage of the bases are thymine?

124. If the pH of a solution has decreased from 6 to 2, by what factor has the H^+ ion concentration changed?

125. In a particular species of guppy, tails can be either long or short and either feathered or straight. A mating between a short feathered-tailed female and a short straight-tailed male produces 30 short straight-tailed guppies, 42 short feathered-tailed guppies, 10 long straight-tailed guppies, and 14 long feathered-tailed guppies. Calculate the chi-square value for the null hypothesis that the short feathered-tailed guppy was heterozygous for the feathered-tail allele. Round your answer to the nearest hundredth.

126. What is the probability that the genotype *rrss* will be produced by a cross in which the genotypes of the parents are both *RrSs*? Give your answer as a fraction.

END OF SECTION I

IF YOU FINISH BEFORE TIME IS CALLED, YOU MAY CHECK YOUR WORK ON SECTION I ONLY.

DO NOT GO ON TO SECTION II UNTIL YOU ARE TOLD TO DO SO.

SECTION II
Reading Period: 10 minutes
80 Minutes—8 Questions

Directions: Questions 1 and 2 are long free-response questions that should require about 20 minutes each to answer. Questions 3 through 8 are short free-response questions that should require about 6 minutes each to answer. Read each question carefully and write your response on scratch paper. Answers must be written out. Outline form is not acceptable. It is important that you read each question completely before you begin to write.

1. The one gene-one enzyme hypothesis states that one gene directly produces one enzyme, which directly affects one step in a metabolic pathway (Figure 1).

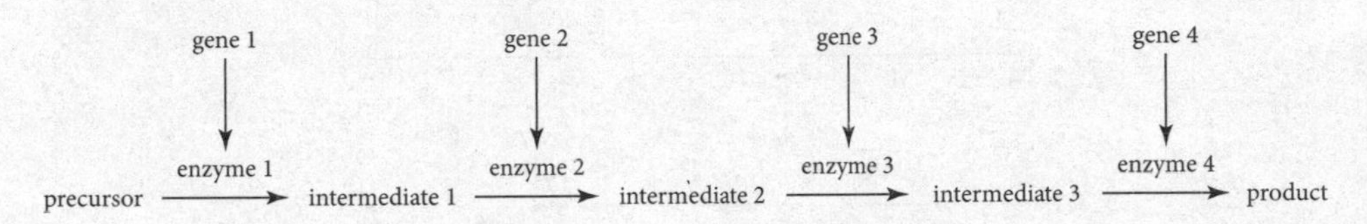

Figure 1. One gene-one enzyme hypothesis

Researchers performed an experiment to investigate the pathway of arginine (an essential amino acid). *Neurospora* mutants, each mutated in a different gene, were grown on minimal medium (sugar, salts, and vitamins) and minimal medium with an additional supplement. The results are shown in Table 1. If a mutant is supplied with the compound it is unable to produce, it will grow.

Mutant	Minimal Medium (MM)	MM + Arginine	MM + Citrulline	MM + Argininosuccinate	MM + Ornithine
1	no	yes	no	yes	no
2	no	yes	yes	yes	yes
3	no	yes	no	no	no
4	no	yes	yes	yes	no

(a) **Propose** an appropriate control treatment for the experiment, and **describe** how the control treatment would increase the validity of the results.

(b) **Explain** how genetic mutations affect enzyme synthesis.

(c) Using the data in the table, **identify** the intermediates of the arginine pathway on the template provided. **Provide** reasoning for the intermediates on the pathway.

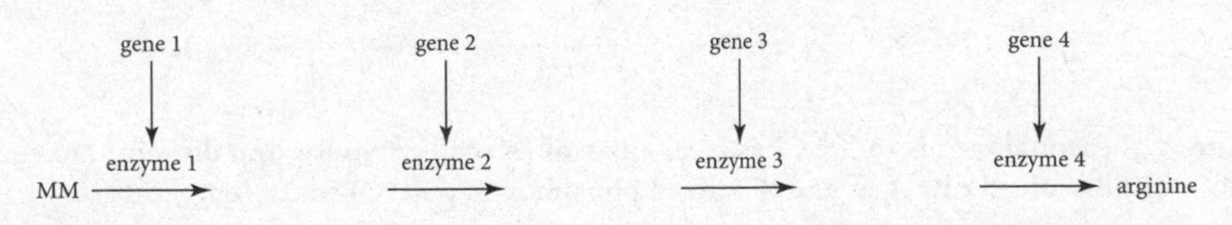

GO ON TO THE NEXT PAGE.

2. Consider the following graphs of seasonal trends in the Chesapeake Bay.

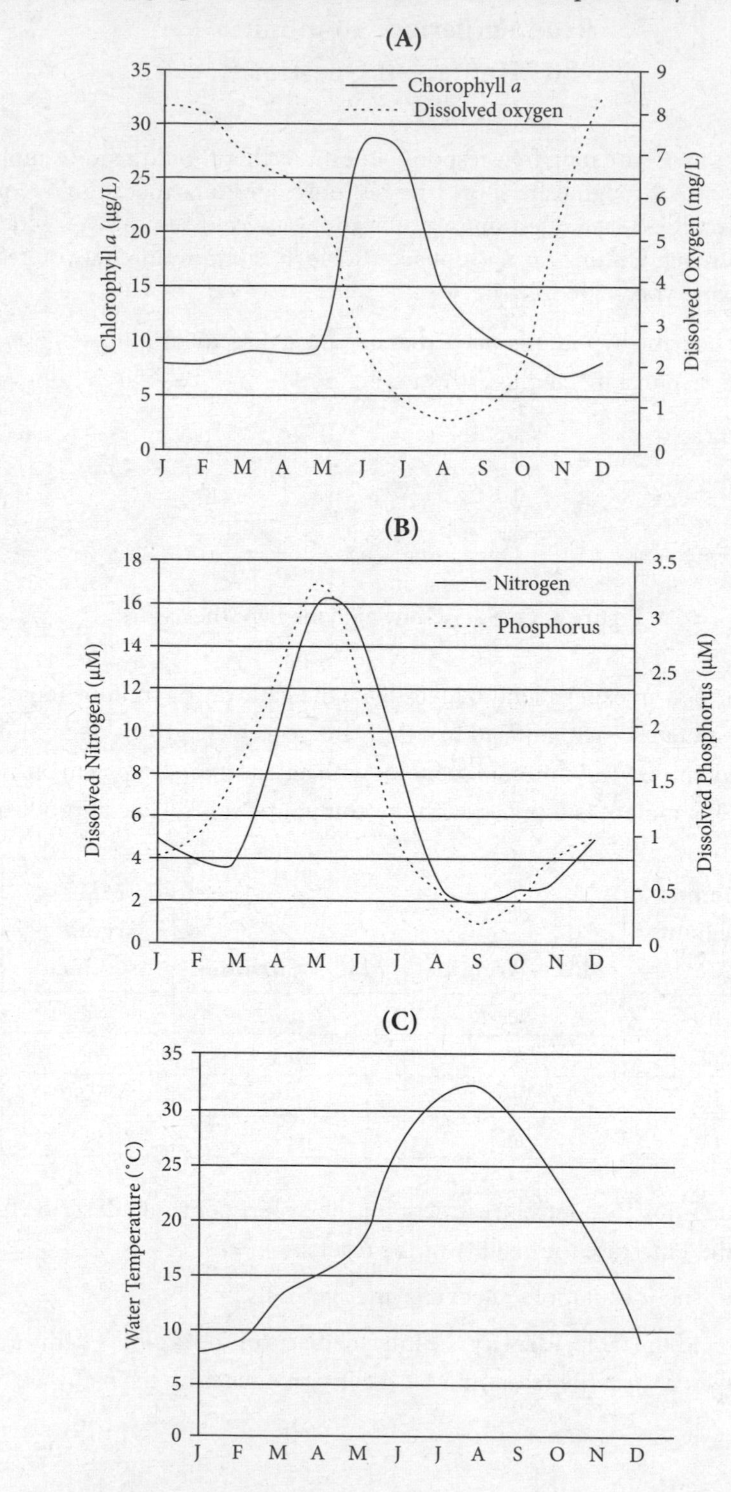

Figure 1. Seasonal trends in the Chesapeake Bay of (A) chlorophyll *a* and dissolved oxygen, (B) dissolved nitrogen and dissolved phosphorus, and (C) water temperature

GO ON TO THE NEXT PAGE.

Harmful algae blooms fueled by excess nutrients (nitrogen and phosphorus) from agricultural fields, sewage treatment plants, industrial facilities, and the atmosphere disrupt marine ecology. Examples of organisms affected include underwater grasses, oysters, and fish. The graphs show the seasonal trends of algal blooms (as measured by chlorophyll *a*), dissolved oxygen, dissolved nitrogen, dissolved phosphorus, and water temperature in the Chesapeake Bay.

(a) In spring, rain and melting snow cause large volumes of water to flow into the bay. Based on an analysis of the data, **explain** the seasonal trends of algal blooms, of dissolved nitrogen, and of dissolved phosphorus.

(b) When algal blooms die, they sink to the bottom of the bay and are decomposed by bacteria. **Explain** what most likely causes the level of dissolved oxygen to dip in the summer months.

(c) Pick TWO of the listed organisms affected by algal blooms and **describe** how algal blooms impact their ecology.

(d) Some algae blooms can produce toxic chemicals. **Describe** how toxins may affect fish-eating birds.

(e) **Propose** one treatment method and **explain** how it would reduce algal blooms.

3. All living organisms contain genetic information that provides several functions inherent to the individual organism and to the perpetuation of its species. Causes of biodiversity are genetic drift, genetic mutation, and genetic shuffling.

   (a) Two examples of genetic drift are the bottleneck effect and the founder effect. **Describe** the cause and effect of each.

   (b) **Describe** ONE genetic shuffling step that occurs in meiosis.

4. In a long-term project studying the interactions of several species of animals on an isolated island, scientists counted the number of individuals of each species visiting a site on the island over the course of several days, every summer for 100 years. The results from that study are shown in the following graph.

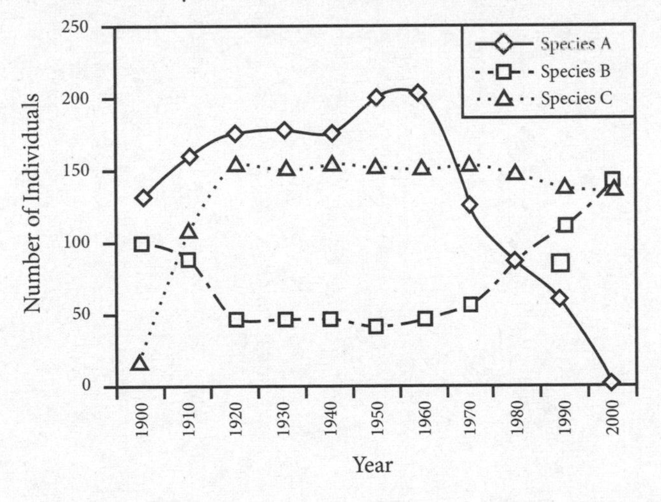

   (a) Based on the data, **propose** TWO hypotheses about the ecological relationships among Species A, B, and C.

   (b) **Provide** reasoning to support the hypotheses you proposed in part (a).

GO ON TO THE NEXT PAGE.

Table 1. Tonicity and relative osmolarity

Tonicity of Solution	Relative Osmolarity	
	Extracellular fluid	**Intracellular fluid**
Hypertonic	higher	lower
Hypotonic	lower	higher
Isotonic	equal	equal

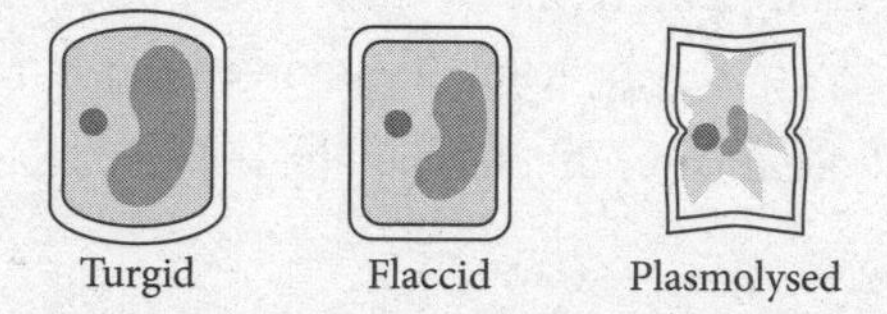

Figure 1. Plant cell states in a variety of water potentials

5. Tonicity describes the effect of a solution on a cell and is related to osmolarity, the total concentration of all solutes in the solution. The table describes the osmolarity of extracellular and intracellular fluids of cells in hypertonic, hypotonic, and isotonic solutions. Depending on the tonicity, a plant cell may be turgid, flaccid, or plasmolysed (Figure 1).
   (a) A plant cell is placed in a solution with higher water potential. Based on the information, **identify** the state of the plant cell and **provide** reasoning for your conclusion.
   (b) **Explain** what would happen to an animal cell placed in a hypotonic solution.
   (c) **Propose** a situation in which plant cells may experience plasmolysis.

Practice

GO ON TO THE NEXT PAGE.

Table 1. Laboratory Test Results

	Patient Value	Normal Value
Hematology		
Hemoglobin	15	13–18 gm/dl
White blood cell count	36,000	5,000–10,000 μl/mm3
% Neutrophils	97	48–73%
% Lymphocytes	3	18–48%
Platelet count	175,000	150,000–350,000/ml
Blood Chemistry		
Glucose	444	70–110 mg/dl
Blood urea	87	7–18 mg/dl
Ethanol	0.1	0–0.1 mg/dl
Blood pH	7.0	7.35–7.45

Table 2. Causes of Abnormal Test Results

Test	Higher than Normal	Lower than Normal
Hemoglobin count	bone marrow dysfunction	anemia cancer
White blood cell count	infection diabetes	cancer
% Neutrophils	infection stress	infection vitamin deficiencies
% Lymphocytes	autoimmune disease blood cancer infection	infection
Platelet count	anemia cancer infection	autoimmune disease
Blood glucose	diabetes	insulinoma
Blood urea	kidney failure diabetes	liver damage pregnancy
Blood pH	severe dehydration endocrine disorder	liver failure low blood sugar

6. A patient who has an elevated temperature and slightly decreased blood pressure has a normal electrocardiogram (EKG) and no obvious signs of trauma on the body. Table 1 shows the patient's laboratory test results, and Table 2 shows causes of levels higher or lower than normal in the body.
   (a) Based on the information provided, **identify** the likely diagnosis for this patient. **Justify** your answer.
   (b) **Explain** why a doctor might request bacterial blood and urine tests for the patient.

Practice

GO ON TO THE NEXT PAGE.

7. The following pedigree shows the occurrence of a genetic disorder through several generations and several lineages of a family. The squares represent males and the circles represent females; those that are half-shaded represent individuals heterozygous for, but unaffected by, the defective gene. The symbols G1, G2, and G3 represent the first, second, and third generations subsequent to the initial mating pair. The symbols L1, L2, and L3 represent the three lineages of descent from the initial generation.

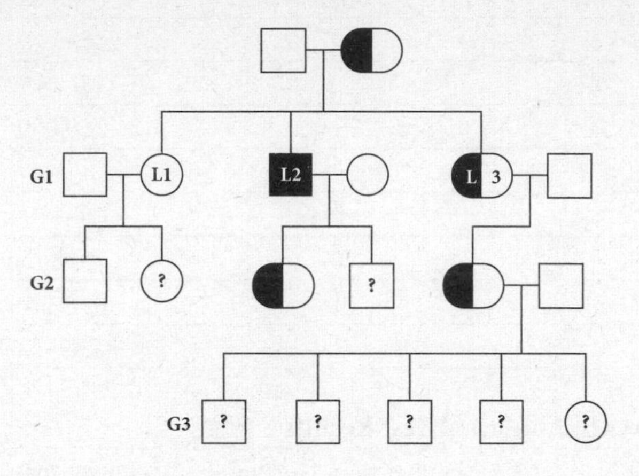

   (a) **Identify** the most probable inheritance pattern of the disorder shown by the pedigree. **Justify** your response.

   (b) **Calculate** the likelihood that a male who descended from the initial mating pair expresses the disorder.

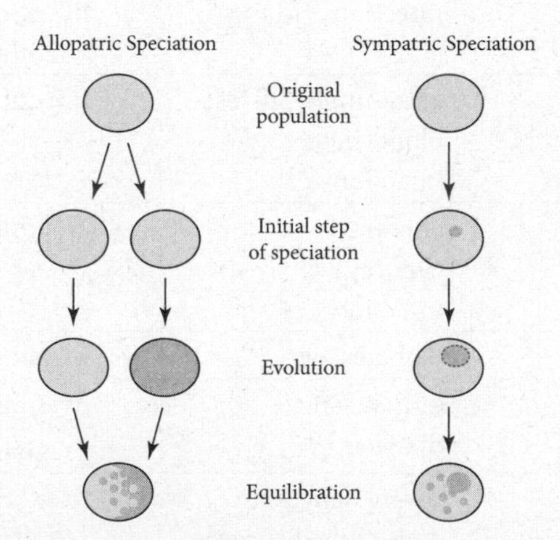

Figure 1. Allopatric and Sympatric Speciation

8. Speciation involves the evolutionary formation of new and genetically distinct species. Figure 1 shows two types of speciation: allopatric and sympatric.

   (a) **Describe** the initial step in allopatric speciation.

   (b) **Describe** the initial step in sympatric speciation.

   (c) **Explain** whether allopatric or sympatric specification is the more common mode of speciation.

STOP—END OF EXAM

ANSWER KEY

Section I, Part A

1. A
2. B
3. D
4. A
5. C
6. C
7. A
8. C
9. B
10. A
11. D
12. C
13. A
14. A
15. C
16. D
17. D
18. D
19. A
20. C
21. D
22. B
23. C
24. C
25. A
26. B
27. A
28. D
29. A
30. C
31. B
32. D
33. C
34. D
35. B
36. B
37. B
38. B
39. D
40. C
41. D
42. C
43. D
44. B
45. B
46. C
47. D
48. A
49. D
50. B
51. C
52. D
53. A
54. D
55. A
56. A
57. B
58. B
59. B
60. C
61. C
62. A
63. B

Section I, Part B

121. 0.50
122. −7.98
123. 23
124. 10000
125. 2.67
126. 1/16

Section I, Part A Number Correct: ______________

Section I, Part B Number Correct: ______________

Section II Points Earned: ______________

Enter your results to your Practice Test 2 assignment to see your 1–5 score and view detailed answers and explanations by logging in at kaptest.com.

Haven't registered your book yet? Go to kaptest.com/booksonline to begin.

ANSWERS AND EXPLANATIONS

CHAPTER 3

Test What You Already Know

1. D Learning Objective: 3.1

The process labeled "A" shows carbon moving from animals to carbon dioxide in the atmosphere, which must be cellular respiration. Animals use cellular respiration to synthesize ATP from molecules in food, giving off CO_2 as a byproduct. Choice (D) is thus correct. Choice A is incorrect because it describes decomposition, in which dead plants and animals are broken down by bacteria. Choice B is incorrect because it describes the process of photosynthesis, which actually removes carbon dioxide from the atmosphere and which is not performed by animals. Choice C is incorrect because it describes the formation of fossil fuels.

2. D Learning Objective: 3.2

Living systems require a continuous input of energy to maintain their ordered state. During the process of maintaining order (dynamic homeostasis), some energy is transferred to the surroundings, thereby increasing the total entropy of the universe. The tendency of the universe to increase in entropy over time is described in the second law of thermodynamics, making choice (D) correct. Choices A and B are incorrect because energy can never be created or destroyed, but only converted into other forms (according to the first law of thermodynamics). Choice C is incorrect because the first law does not pertain to entropy.

3. C Learning Objective: 3.3

Water will reach a higher point if adhesion is greater in the vessel. Adhesion would be greater in the negatively-charged lining of the flowering plant's vessels, while adhesion would be minimal to a hydrophobic surface, such as the lining of *Myrothamnus*. Choice (C) is thus correct. Choices A and B are incorrect because the dye will reach a higher level in the flowering plant. Choice D is incorrect because the hydrophobic lining only affects adhesion and not cohesion, which describes the attractive forces between the water molecules themselves.

4. B Learning Objective: 3.4

Since the molecule is single-stranded, it is likely RNA. DNA is usually double-stranded. Although certain viruses have single-stranded DNA, the scientist has already determined that the nucleic acid does not have viral origins. Because RNA contains ribose (a pentose), a phosphate group, and the bases adenine, uracil, cytosine, and guanine, choice (B) is correct. Choice A is incorrect because only DNA contains thymine. Choice C is incorrect because ribose is a pentose (5-carbon sugar), not a tetrose (4-carbon sugar). Choice D is incorrect because nucleic acids do not contain fatty acids.

5. D Learning Objective: 3.5

The hydrophilic portion of the protein would be found toward the outside of the plasma membrane extending into the cytoplasm of the cell or outward into the extracellular matrix. In the intact cell, the hydrophobic portion would be contained within the hydrophobic lipid region of the membrane. Choice (D) is thus correct. Choice A is incorrect because the interior of the cell contains cytoplasm, an aqueous solution, which would repel the hydrophobic portion of the protein. Choice B is incorrect because the interior of the phospholipid bilayer is hydrophobic and would repel the hydrophilic portion of the protein. Choice C is incorrect because the hydrophobic portion of the protein would most likely be embedded in the membrane, not on its surface, and because the hydrophilic portion would not necessarily extend to the outside of the cell but could also extend inside.

Test What You Learned

1. C Learning Objective: 3.2

According to the graph, the energy of the products is higher than the energy of the reactants, which means that energy is absorbed during the reaction. Such a reaction is described as endergonic, making choice (C) correct. Choice A is incorrect because energy is absorbed, not released, in an endergonic reaction. Choices B and D are incorrect because the reaction is not exergonic.

2. B Learning Objective: 3.3

Panting works in an analogous fashion to sweating, with both processes using the evaporation of water as a cooling mechanism. Because water has a high heat of vaporization, a lot of energy is absorbed (and removed from the reindeer) in the process of converting liquid water to vapor. Choice (B) is therefore correct. Choice A is incorrect because the surfaces of the nose and mouth are too small to allow much heat to radiate away. Choices C and D are incorrect because absorbing heat from outside the reindeer would actually increase body temperature, not decrease it.

3. A Learning Objective: 3.5

Starch and glycogen are more similar in structure than either is to cellulose. Humans do not have the necessary enzyme to break the bonds connecting glucose molecules in cellulose, even though some organisms do have these enzymes. Choice (A) is thus correct. Choice B is incorrect because the glucose monomers are identical in all three polymers. Choice C is incorrect because some herbivorous mammals (e.g., cows, horses, and goats) are capable of digesting cellulose, due to the symbiotic bacteria that inhabit their digestive tracts. Choice D is incorrect because cellulose does not have more associated water molecules than other glucose polymers and because it is not necessarily larger than the others.

4. B Learning Objective: 3.4

Triacylglycerols are hydrophobic and compact, containing more energy per gram than glycogen. Glycogen is hydrophilic and takes up more space. Thus, determining whether the unknown molecules are hydrophobic or hydrophilic and ascertaining their energy density would be sufficient to distinguish between the two types of compounds. Choice (B) is correct. Choice A is incorrect because both types of compounds contain oxygen. Choice C is incorrect because both types of compounds can be found in skeletal muscle. Even though knowing the molar mass of an unknown substance can help to identify it, choice D is incorrect because determining the mass of the material would be insufficient without also knowing the number of moles present.

5. A Learning Objective: 3.1

Reduction is the gain of electrons whereas oxidation is the loss of electrons. (You can also think of oxidation as an increase in the number of bonds to oxygen and reduction as a decrease in the number of bonds to oxygen.) During photosynthesis, electrons are given off from water and accepted by derivatives of carbon dioxide. This results in the production of glucose, which supplies the plant with food, and oxygen gas is a byproduct. Choice (A) is thus correct. Choice B is incorrect because carbon dioxide is reduced and water oxidized. Choice C is incorrect because oxygen is not a food source. Choice D is incorrect because it combines the errors from choices B and C.

CHAPTER 4

Test What You Already Know

1. B Learning Objective: 4.1

The RNA World Hypothesis proposes that self-replicating RNA molecules arose early in Earth's history and became increasingly organized. These self-replicating structures were called protobionts. Choice (B) is thus correct. Choice A is incorrect because DNA is actually more stable than RNA, which may explain why most organisms now use DNA genomes rather than RNA genomes. Choice C is incorrect because a number of polymers are believed to have preceded the existence of cells and their protobiont precursors, but which one happened to be first does not necessarily have any impact on whether protobionts were RNA-based or DNA-based. Choice D is incorrect because the RNA World Hypothesis maintains that the first self-replicating molecules were purely RNA-based, not requiring DNA at any point in their replication.

2. C Learning Objective: 4.2

According to Darwin's theory of natural selection, fit individuals are those that are able to survive and reproduce successfully, passing their genes on to the next generation. Because the question stem says to assume that rates of reproduction are effectively equal, the only relevant criterion for fitness will be the rate of survival. Because mismatched females had the highest survival rate, they have the greatest fitness, making choice (C) correct. Choices A, B, and D are incorrect because all of these groups demonstrated lower rates of survival.

3. B Learning Objective: 4.3

Because larger birds are more likely to survive, there is selection for larger size. This means that birds may become larger in future generations unless other selective pressures against larger size arise. Choice (B) is thus correct. Choice A is incorrect because no information is given about the availability of food. Choice C is incorrect because no information is given concerning the reproductive cycles of the birds. Choice D is incorrect because non-human animals do not simply decide to change their behavior as a species in response to environmental circumstances; if behavior in a species of non-human animals changes, it is almost always the result of selective pressures that favor one kind of behavior over another, with those organisms that behave in the way that is selected against eventually being removed from the gene pool.

Test What You Learned

1. D Learning Objective: 4.3

Given the great variation in size among the species, it is plausible to believe that multiple genes control the size of the birds. Consequently, it is quite probable that the species would respond relatively quickly to selective pressure for a particular size, making choice (D) correct. Choices A and C are incorrect because the question stem states that differences in color and beak depend on single genes with only two distinctive alleles. Thus, it is improbable that intermediate forms would develop easily. Choice B is incorrect because none of the birds currently possess fur and it would be highly unlikely that this completely novel trait would arise in the species simply as a result of random mutations. It is far more probable that the species would develop thicker feathers in a colder environment rather than an entirely different type of protective covering.

2. B Learning Objective: 4.1

The atmosphere of the Earth was very low in oxygen content when life arose, so early life had to be able to thrive in an anaerobic environment. Choice (B) is thus correct. Choice A is incorrect because life requires self-replicating molecules (such as DNA or RNA), but monomers like nucleotides and amino acids are too simple to replicate themselves. Choice C is incorrect because the early Earth is actually believed to have had higher temperatures than are experienced today, so the first cells had to function in a hot environment. Choice D is incorrect because biologists widely agree that life originated on Earth only once, so that all living organisms share a common ancestor.

3. A Learning Objective: 4.2

In order for adaptation to a particular environmental circumstance to occur, genetic variation in a relevant trait must exist or arise through mutation. Because the increased carbon dioxide levels make their environment more acidic, the cod should possess variability in their ability to survive in acidic conditions, so that genes promoting greater survival in an acidic environment can be selected for. Choice (A) is thus correct. Choice B is incorrect because animals do not metabolize carbon dioxide; rather, CO_2 is a waste product of the metabolism of glucose and other food sources. Choice C is incorrect because learning plays a relatively small role in the ability of simple animals like fish to survive. Choice D is incorrect because no data is provided about temperature, so there is no basis for concluding that a cooler environment would help the cod to survive.

CHAPTER 5

Test What You Already Know

1. D Learning Objective: 5.1

All prokaryotes lack a true nucleus, while all eukaryotes contain one, so the cell isolated by the biologists must be a prokaryote. Choice (D) is correct. Choice A is incorrect because eukaryotes possess a true nucleus, but the isolated cell did not. In addition, some prokaryotes are quite large, so size does not immediately reveal the identity of a cell. Choice B is incorrect because the lack of a nucleus is enough to conclude that the cell is prokaryotic. Choice C is incorrect because viruses are not, properly speaking, organisms and because viruses lack their own ribosomes, instead using those of the host cells that they occupy.

2. A Learning Objective: 5.2

The contractions of the heart are powered by ATP, so choice (A) is correct. Choice B is incorrect because muscle cells need large amounts of mitochondria to contract, not to synthesize proteins. While muscles do contain large amounts of proteins, turnover is slow and proteins are continuously made. Choice C is incorrect because liver cells do contain some mitochondria, just fewer than other tissue types, as can be seen in the graph. Choice D is incorrect because the amount of mitochondria varies in accordance with the function of the tissue.

3. D Learning Objective: 5.3

Based on the information provided, cell membranes require a significant proportion of unsaturated fatty acids in order to maintain fluidity at low temperatures. Thus, *B. psychrophilus* needs to contain a larger proportion of unsaturated fatty acids, because it tends to be found at lower temperatures. Choice (D) is correct. Choice A is incorrect because there is a correlation found in the table: temperature is inversely correlated with unsaturated fatty acid content. Choice B is incorrect because bacteria in colder environments require additional fluidity, not additional rigidity. Choice C is incorrect because the high temperatures that *G. stearothermophilus* is exposed to require greater rigidity, not greater fluidity, which is why the species contains such a small proportion of unsaturated fatty acids.

4. B Learning Objective: 5.4

Molecular oxygen is a small, hydrophobic molecule that diffuses through the layer of surfactants and the thin epithelium in an alveolus. The higher the concentration of oxygen, the greater will be the rate of diffusion, as seen in Graph A. Choice (B) is thus correct. Choice A is incorrect because the membrane is permeable to molecules like oxygen and because oxygen diffuses down its concentration gradient. Choices C and D are incorrect because they indicate the wrong graph and because oxygen neither binds to a carrier nor flows by osmosis (osmosis specifically describes the diffusion of water).

5. D Learning Objective: 5.5

Albumin does not permeate either type of membrane due to its size. The dialysis tube selects only according to size, so NaCl is small enough to cross it. A red blood cell membrane, in contrast, selects according to both hydrophobicity and size; because NaCl is polar, it is not permeable across the plasma membrane, though it can be transported with carrier proteins. Choice (D) is thus correct. Choice A is incorrect because carrier proteins can allow NaCl to exit a red blood cell. Choice B is incorrect because albumin is not permeable across a cell membrane (it would require an alternative mechanism of transport). Choice C is incorrect because a red blood cell membrane is not permeable to NaCl.

Test What You Learned

1. D Learning Objective: 5.4

The results in the table for chlorophenyl hydrazone suggest that the protein requires energy in the form of ATP to transport sucrose. Thus, the most justifiable conclusion is that the protein is involved in the active transport of sucrose, choice (D). Choices A and C are incorrect because the protein requires ATP to be effective. Choice B is incorrect because the protein is inhibited by a similar sugar (maltose) but not by a dissimilar sugar (raffinose), which suggests that the protein is specific to sucrose.

2. C Learning Objective: 5.5

The pharmacist's dilution of albumin was correct: when a solution of 25% albumin is diluted to contain five times as much water, it will have one-fifth of the original concentration, or 5% albumin. However, the pharmacist should have used sterile saline solution instead of the distilled water, to ensure that there was an adequate concentration of NaCl. Because the resulting solution was hypotonic to the red blood cells, water from the solution osmosed into the cells, causing them to burst, as in choice (C). Choices A and B are incorrect because the pharmacist did not err in his dilution of the albumin. Choice D is incorrect because the solution is hypotonic to the red blood cells, not hypertonic.

3. C Learning Objective: 5.3

Because the center of the cell membrane consists of hydrophobic fatty acid tails, amino acids with nonpolar (hydrophobic) side chains would be far more likely to be found there than amino acids with charged (hydrophilic) side chains. A hydrophobic compound would be more stable (have a lower free energy) while a hydrophilic compound would be less stable (have a higher free energy) within a membrane. Thus, Graph A must correspond to charged amino acid side chains and Graph B to nonpolar amino acid side chains. Choice (C) is correct. Choice A is incorrect because it reverses the results. Choices B and D are incorrect because the length of the side chains is irrelevant; only the polarity of the side chain matters.

4. B Learning Objective: 5.1

Animal cells do not contain cell walls but many prokaryotes do contain such structures. Thus, a drug targeting cell walls specifically (such as penicillin) would be relatively safe to use in humans. Choice (B) is correct. Choices A and C are incorrect because only eukaryotic cells contain organelles such as the nucleus and Golgi apparatus, so these drugs would be harmless against bacteria while being toxic to humans. Choice D is is incorrect because both eukaryotes and prokaryotes contain phosopholipids in their plasma membranes, so this drug would be harmful to both bacteria and human cells.

5. D Learning Objective: 5.2

The graph indicates that most of the labelled carbon ends up in the chloroplasts, suggesting that carbon fixation primarily takes place there. Choice (D) is thus correct. Choice A is incorrect because the proportion of labeled carbon varies greatly between compartments, with very little found in the cytosol and mitochondria. Choice B is incorrect because the fraction of labeled carbon found in the cytosol is quite low. Choice C is incorrect because PGA is not produced in the mitochondria and because very little labeled carbon is found there.

CHAPTER 6

Test What You Already Know

1. D Learning Objective: 6.1

Structure 2 is the lipid envelope and structure 3 is the capsid. The lipid envelope provides a protective coating for the capsid, which contains the RNA, the genetic material of the influenza virus. Choice (D) is thus correct. Choice A is incorrect because the nucleoprotein is actually structure 1, the genetic material of the virus. Choice B is incorrect because it reverses the identities of structures 2 and 3. Choice C is incorrect because the neuraminidase is structure 4, a surface enzyme that facilitates viral replication.

2. D Learning Objective: 6.2

In step 1, the influenza virus attaches to the host cell and prepares for infection. In step 2, the virus is engulfed by the epithelial cell and, in step 3, the virus is released and enters the nucleus, where it will begin replicating by using the resources of the host cell. Only in step 4 does the viral mRNA begin synthesizing new viral proteins. Choice (D) is thus correct.

3. A Learning Objective: 6.3

Minor mutations that alter the antigens of the virus are quite common and, if enough of these mutations accumulate, can make an immune system exposed to an earlier form of the virus unable to recognize the new form. Choice (A) is thus correct. Choice B is incorrect because major mutations resulting from two strains infecting a host at the same time are quite uncommon. Choice C is incorrect because the most successful viruses from a previous year would still be recognizable by an immune system that was inoculated against them. Choice D is incorrect because viruses do not use meiosis to reproduce, so crossing over cannot occur in viruses.

4. C Learning Objective: 6.4

Plasmids may contain genes for antibiotic resistance or other characteristics that can help bacteria survive in adverse circumstances, but they are independent of the bacterial chromosome that contains genes for essential functions. Choice (C) is thus correct. Choice A is incorrect because genes for growth control are generally contained in the bacterial chromosome. Choice B is incorrect because some bacteria contain no plasmids but are still able to reproduce. Choice D is incorrect because plasmids are not the main DNA molecules found in bacteria; the circular bacterial chromosome contains genes for essential enzymes.

5. B Learning Objective: 6.5

Genetically modified vegetables can be bred to be more resistant to disease and to last much longer than organic and other non-GMO vegetables. Choice (B) is thus correct. Choice A is incorrect because transportation is generally the same for both GMO and non-GMO vegetables. Choice C is incorrect because people disagree about taste, so there is no objective basis for saying that genetically modified vegetables taste better. Choice D is incorrect because genetically modified seeds are typically much more expensive than other seeds.

6. C Learning Objective: 6.6

Penicillin is an antibiotic, so it would reduce the growth of *E. coli* unless some of the bacteria contained, for instance, plasmids with antibiotic resistance. In any case, it is safe to conclude that there would be less bacterial growth on the plate with penicillin. Choice (C) is correct. Choices A and B are incorrect because they suggest a different outcome. Choice D is incorrect because there is no reason to suspect that the plate without penicillin would show less growth than normal.

Test What You Learned

1. A Learning Objective: 6.6

E. coli bacteria are single-celled organisms that reproduce easily with very short generation times, allowing for many generations of offspring to emerge in a relatively short time. This makes them prime candidates for the study of genetic transformations. Choice (A) is thus correct. Choices B, C, and D are all complex eukaryotic organisms that reproduce sexually with longer generation times than bacteria, making it more difficult to study the impact of a transformed gene across generations.

2. B Learning Objective: 6.2

If a cell loses its ability to function and cannot be repaired, then it could lyse (burst open) or undergo apoptosis (programmed cellular death). Whether it does either depends on the severity of the damage to the cell. Choice (B) is thus correct. Choice A is incorrect because the cell will remain intact if it is capable of repairing the damage. Choices C and D are incorrect because they only present one of the two viable options for a damaged cell.

3. C Learning Objective: 6.4

Plasmids are small, circular DNA molecules that carry genes that can enhance bacterial survival. The exchange of plasmids between bacteria in processes of horizontal gene transfer allows for the spread of traits like antibiotic resistance. Plasmids can also be transferred vertically when a bacterium divides and replicates not only its chromosome but also any plasmids it contains. Choice (C) is thus correct. Choice A is incorrect because flagella are structures used for movement, not gene transmission. Pili are hair-like structures that allow for adhesion to other cells. Though they can be used in the process of conjugation, such horizontal gene transfer cannot occur without the actual genes, often contained in plasmids, which make plasmids the more important structure for spreading antibiotic resistance. Thus, choice B is incorrect. Choice D is incorrect because chlorosomes are complexes used in photosynthesis.

4. A Learning Objective: 6.1

A sick student working in the school cafeteria could potentially infect many other students by contaminating food with the virus. The best advice for such a student would be to remain at home until all symptoms subside, at which point the student would likely no longer be infectious. Choice (A) is thus correct. Choices B and D are incorrect because antibiotics are only effective against bacterial infections. Choice C is incorrect because over-the-counter flu medications will at best alleviate symptoms but will not stop someone from being infectious.

5. D Learning Objective: 6.5

Pathway 1 is transformation (genetic transfer from loose genes in the extracellular environment), pathway 2 is transduction (genetic transfer using a virus), and pathway 3 is conjugation (genetic transfer using a bacterial pilus). Choice (D) is thus correct. The other answer choices mislabel one or more of the pathways.

6. B Learning Objective: 6.3

RNA viruses lack the proofreading ability of DNA polymerases. As such, they are more likely to mutate than DNA viruses. Choice (B) thus presents the best explanation for Ebola's higher mutation rate. Choice A is incorrect because it reverses the two viruses. Choice C is incorrect because retroviruses tend to mutate more quickly than DNA viruses. Choice D is incorrect because retroviruses tend to mutate about as quickly as other RNA viruses.

CHAPTER 7

Test What You Already Know

1. A Learning Objective: 7.1

Mycorrhizae are important in helping plants obtain limiting nutrients, such as phosphate. The graph shows lower percent root colonization at higher phosphate levels, suggesting that mycorrhizae are less important to petunias under those conditions. Choice (A) is thus correct. Choice B is incorrect because the graph shows higher colonization at low phosphate levels. Choice C is incorrect because there is a clear pattern: mycorrhizae colonization increases as environmental phosphate levels decrease. Choice D is incorrect because it provides a less plausible explanation of the results. There is no reason to suppose that higher levels of phosphate inhibit mycorrhizal growth, but there is reason to expect that petunias would be less accommodating to the mycorrhizae when they have an adequate supply of phosphate.

2. B Learning Objective: 7.2

Xylem is dead tissue that transports water in plants. Because it already contains pores, it can be used as a filter if water is pushed through it. Choice (B) is thus correct. Choice A is incorrect because periderm, also known as bark, would have to be hollowed out to form a straw-like structure, and even then it would not contain pores useful for filtering. Choice C is incorrect because phloem is living vascular tissue, not dead. Choice D is incorrect because the cortex is not used to transport water, so it could not be used as a pump.

3. B Learning Objective: 7.3

Plants have receptors on guard cell membranes that respond to light. These allow the guard cells to become turgid as needed, opening the stomata and allowing more carbon dioxide to diffuse in and oxygen to diffuse out. This increased gas exchange allows for a higher rate of photosynthesis, which would be reflected in higher sugar concentrations measured in the leaves. Choice (B) thus provides the most plausible hypothesis, making it correct. Choice A is incorrect because closing the stomata would actually decrease the rate of photosynthesis since there would be less input of carbon dioxide and less output of oxygen. Choices C and D are incorrect because cortical cells are not involved in the regulation of stomata.

4. C Learning Objective: 7.4

According to the graphs, significantly larger fruit is produced from natural flowering induction in the Smooth Cayenne cultivar, for both the AMI and NMI maturation approaches. Larger fruit growth means that more energy has been invested into growing the fruit. Choice (C) is thus correct. The other choices are incorrect because they fail to identify the methods that produce the largest fruit.

5. D Learning Objective: 7.5

Some plants are capable of having a greater than normal number of chromosomes, a condition known as polyploidy. When these plants self-fertilize, they produce new plants with the same number of chromosomes. Thus, if two diploid gametes came together, they would form a tetraploid plant, from which other tetraploid plants could be produced. Choice (D) is thus correct. Choice A is incorrect because a hexaploid plant would produce triploid gametes. Choice B is incorrect because the loss of entire sets of chromosomes is an extremely rare event; two diploid gametes would normally fuse to form a tetraploid organism. Choice C is incorrect because many polyploid plants are capable of reproduction.

6. C Learning Objective: 7.6

Plants that live in dry environments (xerophytes) have adaptations to reduce their water loss through transpiration because dry conditions cause greater evaporation. Thus, a xerophyte could be expected to see less dramatic changes in transpiration rates when placed in environments with differing humidity levels. Choice (C) is thus correct. Choice A is incorrect because it reverses the difference between the two plants. Choice B is incorrect because a smaller plant would be expected to have a lower absolute rate of transpiration than a larger plant, but not a lower *difference* in transpiration rates at two distinct humidity levels. Choice D is incorrect because humidity matters more than temperature for transpiration; for instance, plants would be expected to have lower transpiration rates in hot, moist environments like tropical rainforests than in hot, dry environments like deserts.

Test What You Learned

1. A Learning Objective: 7.6

In transpiration, plants lose water through their stomata. The only way for the plant to replace the lost water is by pulling it up through its roots and stem by capillary action. A potometer is able to measure the water taken up by a plant's cut stem indirectly by measuring the water lost from a tube connected to the stem. Thus, choice (A) is correct. Choice B is incorrect because a potometer cannot directly measure water taken up by the roots, but instead indirectly measures how much is taken up by a stem. Choice C is incorrect because a potometer does not attach to a plant's leaves. Choice D is incorrect because a potometer measures the amount of water lost, not pressure, and because the water is not actively pushed through the plant's vascular system but drawn up using capillary action.

2. B Learning Objective: 7.3

To be an effective experiment, the sample size must be sufficiently large and the variable under investigation must be the only condition that differs between experimental groups. Because it uses a sizable sample of 200 and divides the group based on lighting direction, the experiment in (B) presents the most effective design for testing the impact of phototropism. Choice A is incorrect because the sample size is too small and because including an actual wall can introduce complicating variables that make it harder to gauge the impact of phototropism alone. Choice C is incorrect because the sample size is way too small and because the variation in lighting direction does not match the conditions of plants growing against a wall. Choice D is incorrect because variation in the wavelengths of light used adds a complicating variable and does not mimic the conditions of plants growing against a wall.

3. D Learning Objective: 7.1

In order to support the hypothesis that *T. melanosporum* obtains its carbon from its host, rather than from the surrounding soil, experimental results should indicate that the radioactive carbon (which is taken up by the tree in photosynthesis) is found both in the tree itself and in the fungus. In addition, there should be very little of the ^{13}C in the surrounding soil, because otherwise the fungus might be obtaining the radioactive carbon from that. Thus, choice (D) is correct. Choice A is incorrect because it suggests that the ^{13}C is somehow getting into the surrounding soil before being taken up by the fungus. Choice B is incorrect because it does not indicate whether the fungus took up the radioactive carbon from the tree. Choice C is incorrect because it says nothing about the carbon in the fungus but, even if the fungus did contain the ^{13}C, the results would not indicate whether this carbon came from the tree or the soil.

4. C Learning Objective: 7.5

The results indicate significant levels of gene flow within 5 meters, as well as smaller amounts of gene flow at greater distances. Though the percentages are not huge, they are large enough to have an impact on the frequency of the herbicide resistance trait, especially if the application of herbicide provides a selective pressure. Choice (C) is correct. Choice A is incorrect because even a small percentage of gene flow can have a significant effect, especially when the selective pressure of a herbicide is present. Choice B is incorrect because the researchers planted seeds with the herbicide resistance genes, so it is already clear that such genes can be found in both pollen and seeds. Choice D is incorrect because the trait can spread over large distances given enough time and the appropriate selective pressure.

5. A Learning Objective: 7.4

The most likely outcome is that fruits are larger when the plants have access to a greater supply of nutrients, such as when the soil is fertilized. Choice (A) is thus correct. Choice B is incorrect because it suggests the opposite. Choices C and D are incorrect because there is not enough information to determine whether natural or artificially induced flowering would enhance the availability of energy and nutrients for fruit growth.

6. B Learning Objective: 7.2

As water moves through pores from one cell to another, the size of the pores limits which dissolved materials can pass through. In this way, the tissue acts as a filter. Because they contain more cells, gymnosperm tissues provide more opportunities for filtration than angiosperm tissues with the same stem length. Choice (B) is thus correct. Choice A is incorrect because there is less passage through pores due to the smaller number of cells in angiosperm vascular tissue, so there would be less filtration. Choice C is incorrect because the path followed by the water (and the number of pores it passes through) matters more than the distance traveled. Choice D is incorrect because monocots and dicots are both angiosperms with only slight differences in vascular tissue.

CHAPTER 8

Test What You Already Know

1. D Learning Objective: 8.1

In order to determine if bacteriocin contributed to an organism's nonspecific response, there would need to be some evidence that native bacteria that produced the bacteriocin could inhibit the growth of bacteria not native to the organism. Choice (D) proposes exactly such an experiment, making it correct. Choice A is incorrect because antibodies are an aspect of specific immune defense, not nonspecific. Choice B is incorrect because this experiment would only provide information about the bacterial response to antibiotics, but would demonstrate nothing about the effects of bacteriocin on nonspecific response. Choice C is incorrect because it would only provide information about the structure of bacteriocin and not about its effects on nonspecific response.

2. A Learning Objective: 8.2

According to the figure, C3b binds to red blood cells far less frequently when complexed with Ecb than when isolated. This suggests that Ecb serves to lessen immune response by interfering with the ability of C3b to bind to cells. Choice (A) is thus correct. Choice B is incorrect because Ecb decreases the activity of C3b. Choice C is incorrect because Ecb decreases the likelihood of opsonization. Choice D is incorrect because the interference that Ecb causes actually lessens immune response and increases the likelihood of *S. aureus* infection.

3. C Learning Objective: 8.3

At day 20, according to the graph, the humoral immune response was greater, with more antibody production in the control group than in the antibiotic-treated group. This suggests that immune response was enhanced by exposure to the pathogen. Choice (C) is thus correct. Choices A and B are incorrect because the production of antibodies is part of humoral response, not cell-mediated response. Choice D is incorrect because the response was more substantial in the control group, not the treatment group.

4. B Learning Objective: 8.4

The increased stimulation of B cells will lead to greater production of B cells, which in turn will cause an increased ability to produce antibodies. Choice (B) is thus correct. Choice A is incorrect because B cells are not phagocytic. Choice C is incorrect because B cells do not directly kill cancer cells. Choice D is incorrect because B cells mature in the bone marrow, not the thymus, before migrating to other parts of the body. (T cells mature in the thymus.)

5. A Learning Objective: 8.5

Antibodies target specific antigens. When the influenza virus mutates from year to year, its antigens become sufficiently different that existing antibodies no longer recognize the virus. This means that a new vaccine is required for adequate immune defense. Choice (A) is thus correct. Choice B is incorrect because it does not provide a reason for requiring a new vaccine; even if influenza is unusually virulent, a vaccine against it could provide defense. Choice C is incorrect because influenza can be treated by medicines and because a response to medicines is not required for vaccine development. (For instance, rabies does not respond to medicines, but vaccines have been developed for it.) Choice D is incorrect because the influenza virus can be recognized by antibodies and can be prevented with standard vaccination.

6. C Learning Objective: 8.6

Memory B cells persist after exposure to a pathogen and there is a greater probability that older individuals will have had at least one exposure to malaria. This explains why the frequency of IgG-secreting memory B cells is higher in older individuals on the graph. Choice (C) is thus correct. Choice A is incorrect because younger individuals actually show less immunity than older individuals in the graph. Choice B is incorrect because there is no reason provided to explain why older individuals are more likely to develop malaria than younger individuals. Choice D is incorrect because memory B cells only form after exposure to pathogens, so they will not simply increase in number if there is no exposure.

Test What You Learned

1. D Learning Objective: 8.5

The horizontal axis on the bar graph represents the control group (untreated) and the different strains of *B. subtilis*. The far left bar shows that the M2e protein alone is not sufficient to cause substantial antibody production, but the three bars on the right show that antibodies can be produced when the M2e protein is expressed on some *B. subtilis* endospores. Choice (D) is thus correct. Choice A is incorrect because there was a significant amount of variability between the different strains, with the strains on the right half of the graph producing far greater responses. Choice B is incorrect because the body can still respond to a vaccination even if it uses genetically modified bacteria. Choice C is incorrect because the antibody production only increased significantly when M2e was combined with some of the strains of spores.

2. C Learning Objective: 8.4

According to the table, patients with acute HBV infection often had activated T cells specific for other viruses as well. Choice (C) is thus correct. Choice A is incorrect because T cells are involved in cell-mediated immunity, not humoral immunity. Choice B is incorrect because some patients lacked activated T cells specific for HBV. Choice D is incorrect because no information was provided on vaccination status.

3. B Learning Objective: 8.3

The humoral immune response involves the production of antibodies that respond to antigens. Thus, an increase in antibody diversity would enable the humoral defense to respond to a wider variety of invading microbes. Choice (B) is correct. Choice A is incorrect because it is cytotoxic T cells, not antibodies, that directly kill virally infected cells. Choice C is incorrect because antibodies are secreted by B cells, not helper T cells. While antibodies participate in activating helper T cells, choice D is incorrect because antibodies have a much larger role in the humoral immune system.

4. A Learning Objective: 8.1

The clear areas on the plate represent areas without bacterial growth. The SakA(+) plate has more clearing than the SakA(–) plate, suggesting that bacteria that produce the bacteriocin SakA can inhibit the growth of other bacteria. This likely contributes to the nonspecific immune defense of the mice. Choice (A) is thus correct. Choice B is incorrect because the clear zones are more present in the SakA(+) plate. Choices C and D are incorrect because the results suggest that SakA does contribute to nonspecific response.

5. D Learning Objective: 8.6

Vaccination causes an immune response that includes the production of memory cells that recognize the viral strain from the vaccine. Exposure to the same strain years later will produce a rapid immune response because of these memory cells. Choice (D) is thus correct. Choice A is incorrect because innate immunity is not affected by vaccination. Choice B is incorrect because immune response is faster, not slower, to a repeat exposure. Choice C is incorrect because immune response is generally more effective during a repeat exposure.

6. B Learning Objective: 8.2

The harsh environment in the stomach quickly kills a wide range of bacteria and other pathogens, including pathogens to which the organism has not been previously exposed. Choice (B) is thus correct. Choice A is incorrect because pathogens in the stomach have already passed through the mouth and esophagus, so they are not being killed immediately upon entry. Also, not all pathogens are killed by the stomach, so there remains some possibility of infection. Choices C and D are incorrect because stomach acid is an aspect of nonspecific response; it does not have the capacity to target specific pathogen strains.

CHAPTER 9

Test What You Already Know

1. D Learning Objective: 9.1

According to the diagrams, the same transducing cascade is employed in both skeletal and smooth muscles, up until the point when protein kinase A is activated. In skeletal muscles, it leads to the breakdown of glycogen, while in smooth muscles, it causes less interaction between myosin and actin. Thus, protein kinase A has different targets in the two muscle types, making choice (D) correct. Choice A is incorrect because epinephrine attaches to the β_2 adrenergic receptor in both skeletal and smooth muscles. Choice B is incorrect because cAMP is the second messenger in both types of muscles. Choice C is incorrect because glycogen metabolism is not affected in smooth muscle.

2. B Learning Objective: 9.2

Cells in the late stage of growth secrete a signal molecule in the medium. If the hypothesis is correct, a supernatant conditioned by these cells would induce quorum sensing in cells that have not yet reached high density. Thus, choice (B) presents the most effective way to test the hypothesis. Choice A is incorrect because early cells would not secrete much of the signal molecule into the supernatant. Choices C and D are incorrect because a fresh medium would not contain the signaling molecule.

3. C Learning Objective: 9.3

According to the lower right image in the diagram, only dimers bound to ligands carry enough phosphate groups to be activated fully. Choice (C) is thus correct. Choice A is incorrect because full phosphorylation occurs only when the ligand is bound to the dimer. Choice B is incorrect because the dimer without the ligand can be partially phosphorylated, as is clear from the image on the upper right. Choice D is incorrect because the monomers cannot be phosphorylated and because the signaling cascade will not be initiated without a signal.

4. D Learning Objective: 9.4

The question stem states that the cholera toxin locks a G-protein into being bound to GTP. According to the top image in the diagram, the G-protein subunits cannot reform until GTP has been hydrolyzed back to GDP. Thus, the cholera toxin interferes with this stage, making choice (D) correct. Choice A is incorrect because this is not a typical stage in the signaling cycle and because cholera does not bind to the receptor, but uses a different mechanism to enter the cell. Choice B is incorrect because the G-protein complex is still able to dissociate into subunits in the presence of the toxin. Choice C is incorrect because adenylyl cyclase is continuously activated in the presence of the cholera toxin, as stated in the question stem.

5. B Learning Objective: 9.5

According to the results, only cells in close proximity to the infected cell express an antiviral response. This suggests a type of paracrine signaling. Choice (B) is thus correct. Choice A is incorrect because endocrine signaling can act over large distances by traveling through the bloodstream, but cells outside of the immediate vicinity of cell A do not express the antiviral response. Choice C is incorrect because electrical signals act quickly, but it takes hours for the antiviral response to be expressed by nearby cells. Choice D is incorrect because the virus is nonreplicating, as noted in the question stem, so cell A cannot shed copies of the virus.

6. A Learning Objective: 9.6

The abnormally high levels of TSH in the patient suggest that her thyroid gland is not producing enough thyroxine, which (through the negative feedback loop) would inhibit levels of TSH. This is confirmed by the fact that administering a synthetic thyroid hormone relieves her symptoms. Choice (A) is thus correct. Choice B is incorrect because, if it were true, she would not have responded to the administration of levothyroxine. Choice C is incorrect because her pituitary gland is definitely responsive, as is clear from her high levels of TSH. Choice D is incorrect because the high levels of TSH diminished after treatment, which shows the feedback loop is capable of functioning normally given enough thyroid hormone.

Test What You Learned

1. A Learning Objective: 9.3

As seen in the figure, if present at a high enough dose acetylcholine can reverse the effect of atropine, suggesting that the two compounds compete for the same site on the receptor. Choice (A) is thus correct. Choice B is incorrect because atropine actually has an inhibitory effect: a higher dose of acetylcholine is required to achieve the same response. Choice C is incorrect because increased acetylcholine reversed the effects of atropine, indicating that it is not an irreversible inhibitor. Choice D is incorrect because there was a substantial effect: the response curve moved to the right.

2. C Learning Objective: 9.1

As explained in the question stem, epinephrine binds to a G-protein-coupled receptor, which initiates a signaling cascade that activates protein kinase A, decreasing interaction between actin and myosin and thereby indirectly decreasing muscle contraction. Choice (C) is thus correct. Choice A is incorrect because acetylcholine binds to an ion-gated channel while epinephrine binds to a G-protein-coupled receptor, so the two could not compete for the same receptor. Choice B is incorrect because protein kinase A does not inhibit glucose synthesis, according to the question stem. Choice D is incorrect because the binding of epinephrine to the receptor activates the adenylyl cyclase, as noted in the question stem.

3. C Learning Objective: 9.2

The tyrosine kinase domain is what makes RTKs distinctive. The presence of sequences encoding a putative tyrosine kinase domain in the protists would be a strong indication that the protein is in fact a receptor tyrosine kinase. Choice (C) is thus correct. Choice A is incorrect because ligand regions are determined by the ligands they bind, so they would not be conserved unless they bind the same ligand. Choice B is incorrect because transmembrane domains could be similar across a wide range of proteins, since spanning the membrane always requires hydrophobic residues. Choice D is incorrect because only the tyrosine kinase domain is essential for determining the evolutionary relationship.

4. A Learning Objective: 9.5

A patient with type II diabetes is resistant to insulin, but the β cells of the pancreas still appropriately respond to high glucose levels. This suggests that both glucose and insulin levels would remain high in such a patient after eating. Only Patient A presents such a response, so he is the most likely to suffer from type II diabetes. Choice (A) is correct. Choice B is incorrect because Patient B shows a normal response, in which glucose and insulin levels spike after eating but then decline with time. Choice C is incorrect because Patient C seems to have type I diabetes, since the consumption of glucose does not trigger an increase in his insulin levels. Choice D is incorrect because only Patient A shows a response consistent with type II diabetes.

5. B Learning Objective: 9.6

LH spikes immediately before ovulation, but otherwise remains at consistently low levels. Thus, it is the most effective target for determining ovulation. Choice (B) is correct. Choice A is incorrect because FSH peaks twice, once early in the cycle and once prior to ovulation, so it would be a less reliable marker of ovulation than LH. Choice C is incorrect because estrogen peaks twice, once before and once after ovulation, making it an unreliable indicator. Choice D is incorrect because the progesterone peak only appears well after ovulation.

6. B Learning Objective: 9.4

The question stem indicates that a higher concentration of acetylcholine (resulting from the administration of a cholinesterase inhibitor) is able to reverse the effects of the poison. This suggests that tubocurarine and acetylcholine compete for the same site on the ion-gated channel. Choice (B) is thus correct. Choice A is incorrect because acetylcholine must be released for a cholinesterase inhibitor (a compound that inhibits cholinesterase, an enzyme that breaks down existing acetylcholine) to be capable of increasing acetylcholine concentration. Choice C is incorrect because increased acetylcholine could reverse the effects of tubocurarine, which would not happen if the tubocurarine simply blocked the Na^+ influx. Choice D is incorrect because the tubocurarine interferes with the effects of acetylcholine, not with the actual contraction mechanism of the muscle.

CHAPTER 10

Test What You Already Know

1. A Learning Objective: 10.1

According to the diagram, structure 1 is a dendrite and structure 4 is part of an axon, specifically the axon terminal. In nervous system signal transmission, an action potential travels down the length of the axon and, upon reaching the axon terminal, causes the release of neurotransmitters, which cross the synapse and stimulate the dendrites of an adjacent neuron or neurons to transmit the signal further. Choice (A) is thus correct. Choice B is incorrect because signals are conducted in one direction, from the axon of one neuron to the dendrites of the next. Choices C and D are incorrect because they reverse the identity of the structures.

2. C Learning Objective: 10.2

The most common fates of neurotransmitters are to be degraded by enzymes present in the synapse or to be taken back up by the terminal of the presynaptic neuron. Choice (C) is thus correct. Choice A is incorrect because neurotransmitters are not absorbed by other neurons. Choice B is incorrect because it is the presynaptic neuron, not the postsynaptic neuron, that can take up excess neurotransmitters. Choice D is incorrect because neurotransmitters will generally not degrade without the help of an enzyme.

3. D Learning Objective: 10.3

During the absolute refractory period of an action potential, sodium channels are deactivated for a short period. As a result, another impulse cannot be transmitted, which limits the number of impulses that can be transmitted over a certain period of time. Choice (D) is thus correct. Choices A and B are incorrect because they wrongly state that another impulse can occur. Choice C is incorrect because the sodium channels, not the potassium channels, are deactivated during the refractory period.

4. B Learning Objective: 10.4

Feeling warm is a result of stimulation of specific tactile neurons (those associated with the sense of touch), so it would be categorized as a species of sensory function. Choice (B) is thus correct. Choice A is incorrect because causing muscles to move would actually be a motor function. Choice C is incorrect because skin cells do not need to move in order to feel warm. Choice D is incorrect because motor functions lead to movement of parts of the body; involving communication between cells is insufficient to make something a motor function.

5. C Learning Objective: 10.5

The MRI scan results indicate that the lesion is found in the parietal lobe, which plays a role in spatial and visual perception (as noted in the question stem). Consequently, it is probable that the lesion led to the fall, because damage to the parietal lobe can negatively impact perception and make a person more susceptible to falling. Choice (C) is thus correct. Choice A is incorrect because the lesion is not in the frontal lobe, nor does the frontal lobe control those functions. Choice B is incorrect because the lesion is not in the occipital lobe, which is located at the very rear of the brain. Choice D is incorrect because the lesion appears to be responsible for the loss of spatial perception, and the question stem suggests that the lesion is likely a result of cancer, not of a traumatic head injury.

6. B Learning Objective: 10.6

Pain arises from external sensory receptors, many of which are found in the skin and organs. Therefore, pain likely evolved to help human beings gather information about the external environment, such as which external objects are capable of causing bodily harm. Choice (B) is thus correct. Choice A is incorrect because interoceptors monitor aspects of the internal environment, such as blood pressure and pH, but they are not associated with pain perception. Choice C is incorrect because it suffers from the same error as choice A, and also because it reverses the proximal and ultimate causes. Choice D is incorrect because it reverses the proximal and ultimate causes.

Test What You Learned

1. D Learning Objective: 10.4

Fear can occur when sensory stimuli are received and processed by the nervous system. It usually involves higher cognitive functions such as emotions, memories, or learning to generate a response. Because of these cognitive components, fear is a higher function involving several distinct parts of the brain, as in choice (D). Choice A is incorrect because perception is but one component of fear. Choice B is incorrect because fear does not necessarily require tactile sensation. Choice C is incorrect because motor functions are only a part of the behavioral response to fear.

2. C Learning Objective: 10.6

As can be seen in the diagram, the lens focuses the light onto the retina, which then converts the light into nerve impulses that travel along the optic nerve to the brain. Choice (C) is thus correct. Choice A is incorrect because light is not focused onto the iris, but rather the retina. Choice B is incorrect because it reverses the roles of the lens and the retina and mischaracterizes the role of the iris. Choice D is incorrect because the lens focuses the light onto the retina, which converts the light energy to nerve impulses.

3. A Learning Objective: 10.5

The hippocampus is a region of the forebrain that regulates the formation and retention of memory. Extensive damage to the hippocampus could cause a patient to lose the ability to form new memories. Choice (A) is thus correct. Choice B is incorrect because the forebrain is also involved in memory formation, so memory would likely be affected. Choice C is incorrect because the hypothalamus does not control memory formation. Choice D is incorrect because auditory and visual processing actually occur primarily in the midbrain, not the forebrain.

4. A Learning Objective: 10.2

According to the diagram, the dopamine reuptake proteins in the presynaptic cells are inhibited by cocaine. This causes an increase in the concentration of dopamine in the synapses, which leads to increased stimulation of the postsynaptic dopamine receptors and the psychoactive effects of the drug. Choice (A) is thus correct. Choice B is incorrect because the dopamine receptors are on the postsynaptic neurons, not the presynaptic neurons, and because cocaine does not inhibit the production of dopamine, but rather inhibits its presynaptic reuptake. Choice C is incorrect because cocaine does not react with dopamine directly, as can be seen in the figure. Choice D is incorrect because there is no indication in the diagram of plaque formation.

5. C Learning Objective: 10.3

Sodium channels are responsible for the depolarization of action potentials in neurons. Therefore, inhibiting them would block the transmission of pain signals, preventing the feeling of pain. Choice (C) is thus correct. Choice A is incorrect because preventing the reuptake of neurotransmitters would increase stimulation of postsynaptic neurotransmitter receptors, causing an increase in action potential production, leading to more pain. Choice B is incorrect because this would allow neurons to repolarize more quickly, reducing the length of the refractory period and enabling neurons to send pain sensations at a faster rate. Choice D is incorrect because it suggests the opposite of what would be necessary to inhibit pain.

6. B Learning Objective: 10.1

The myelin sheath increases the rate at which signals propagate down an axon. The loss of the myelin sheath in neurodegenerative diseases such as multiple sclerosis and encephalomyelitis leads to weakened signal transmission. Choice (B) is thus correct. Choice A is incorrect because there would be an adverse impact from losing the myelin sheath. Choice C is incorrect because it states the opposite of what would occur. Choice D is incorrect because signal transmission is still possible without myelin, such as when signals travel through the unmyelinated dendrites of a neuron.

CHAPTER 11

Test What You Already Know

1. C Learning Objective: 11.1

Muscle tissue during exercise produces more CO_2 because such tissue is rapidly metabolizing glucose to support muscle contractions. This increase in carbon dioxide causes the bicarbonate equation to shift to the right, producing more H^+ and lowering the pH. Because hemoglobin has less affinity for oxygen at a lower pH, as can be seen in the graph, it will give up its oxygen to active muscle tissue more readily. Choice (C) is thus correct. Choice A is incorrect because oxygen binds less tightly to hemoglobin when CO_2 is released, lowering the pH. Choice B is incorrect because pH actually decreases in active muscle tissue. Choice D is incorrect because CO_2 is released by active muscle tissue, not taken up.

2. D Learning Objective: 11.2

In order to transport lipids (which are hydrophobic) in blood (an aqueous solution) some kind of transport molecule is required that can interact simultaneously with nonpolar lipids and polar water molecules. Lipoproteins, which contain both a hydrophobic core and a hydrophilic surface, are ideally suited for this function. Choice (D) is thus correct. Choice A is incorrect because lipids are not water soluble, so any small droplets would coalesce to form large drops. Choice B is incorrect because lipids would not spontaneously form bilayers in the blood. Choice C is incorrect because hydrophilic carbohydrates could not easily bind to hydrophobic lipids.

3. A Learning Objective: 11.3

A tissue with an increased rate of ATP consumption (high metabolic rate) will utilize more oxygen to increase ATP production. Thus, the pressure of oxygen in the tissue will decrease, causing an increase in oxygen released (decrease in binding affinity). According to the graph, at a given pressure of oxygen, myoglobin's fractional saturation of oxygen is higher than hemoglobin's. This indicates that myoglobin has a higher affinity than hemoglobin and is less inclined to release oxygen. Thus, choice (A) is correct. Choice B is incorrect because myoglobin is less inclined to release oxygen and has a higher oxygen-binding affinity than hemoglobin. Choices C and D are incorrect because, in tissue with a high metabolic rate, oxygen will be released and not bound.

4. C Learning Objective: 11.4

As noted in the question stem, many digestive enzymes require a basic pH, but chyme enters the duodenum with an acidic pH. Thus, bicarbonate ions help to neutralize the chyme and raise the pH to a level that allows the digestive enzymes to function. Choice (C) is thus correct. Choice A is incorrect because cofactors are usually metals or small organic compounds; bicarbonate is not a cofactor for any digestive enzymes. Choice B is incorrect because emulsification requires molecules that have both polar and nonpolar regions, unlike bicarbonate. Choice D is incorrect because bicarbonate ions do not stimulate muscle cells.

5. B Learning Objective: 11.5

High sodium levels can cause an increase in blood pressure, so patients with high blood pressure would benefit from a treatment that lowers the levels of sodium in their blood. Because aldosterone increases sodium reabsorption, some mechanism that prevented its release could potentially help patients with hypertension. ACE inhibitors would do exactly this by blocking the conversion of angiotensin I to angiotensin II, thereby preventing the release of aldosterone. Choice (B) is correct. Choice A is incorrect because the Na^+/K^+ pump is used not simply in nephrons but in cells throughout the entire body, so inhibiting the pump would have many negative side effects. Choice C is incorrect because it would actually lead to the release of more aldosterone, causing a greater increase in blood pressure. Choice D is incorrect because going on a high salt diet to increase blood osmolarity would only decrease renin release due to more fluid already being retained in the blood vessels, thereby increasing blood pressure.

Test What You Learned

1. A Learning Objective: 11.4

Bile salts are amphipathic molecules because they contain both polar and nonpolar regions. This allows them to serve as emulsifiers, which allows for the digestion of lipids by lipases (lipid-digesting enzymes). Choice (A) is thus correct. Choice B is incorrect because proteins are denatured by HCl in the stomach, not by bile salts. Choice C is incorrect because fat soluble vitamins would be absorbed by the body, not eliminated from the body. Choice D is incorrect because carbohydrates are hydrophilic and would not readily interact with bile salts.

2. C Learning Objective: 11.5

Albumin is a large protein, which ordinarily would not pass through the filter provided by the glomerulus. The presence of albumin in the urine typically indicates some kind of glomerular damage. Choice (C) is thus correct. Choices A and B are incorrect because uric acid and sodium and potassium ions are typical components of urine. Choice D is incorrect because urine is ordinarily sterile and free of cells. White blood cells would only be present in urine if there were some kind of infection.

3. B Learning Objective: 11.3

Since hemoglobin, a tetramer, exhibits the Bohr effect, a decrease in pH (increase in acidity and H^+) would decrease hemoglobin's oxygen-binding affinity (release of oxygen to be delivered). Since myoglobin, a monomer, does not exhibit the Bohr effect, a decrease in pH would not affect myoglobin's oxygen-binding affinity. Thus, the correct answer is (B). Choice A is incorrect because myoglobin's oxygen-binding affinity would not decrease as a result of a decrease in pH. Choices C and D are incorrect because hemoglobin's oxygen-binding affinity would not increase as a result of a decrease in pH.

4. B Learning Objective: 11.1

When the partial pressure of oxygen is lower, the body responds by producing more red blood cells, so that more hemoglobin is available to bind to inhaled oxygen. Choice (B) is thus correct. Choice A is incorrect because hemoglobin structure is specified by the genes that encode for it, and it would be impossible for all the hemoglobin genes in the body to spontaneously change their structures in the same way. Choice C is incorrect because a few weeks is not enough time to develop lung tissue and diaphragm muscle, as would be necessary to increase respiratory volume. Choice D is incorrect because increasing plasma pH would carry widespread negative consequences.

5. D Learning Objective: 11.2

The diagram shows the blood delivering oxygen, a nutrient, to tissue cells and removing carbon dioxide, a waste product. This type of process occurs in active muscle tissues as ATP is consumed in muscle contractions, requiring the input of oxygen and the removal of carbon dioxide to continue functioning. Choice (D) is thus correct. Choice A is incorrect because the transport of water and nitrogenous wastes is not featured in the diagram. Choice B is incorrect because neither oxygen nor carbon dioxide is a chemical messenger. Choice C is incorrect because the diagram shows the delivery, not retrieval, of nutrients and the retrieval, not delivery, of wastes.

CHAPTER 12

Test What You Already Know

1. D Learning Objective: 12.1

Enzymes can only catalyze one reaction at a time and it takes a certain amount of time for the reaction to occur. Unless the enzyme concentration is increased, the enzymes become saturated at high concentrations of substrate, and the rate stops increasing with increasing substrate concentration. Choice (D) is thus correct. Choice A is incorrect because the graph provides no information about temperature. Choice B is incorrect because the varying concentrations of AR are represented by the different curves in the graph, not by the variation along the *x*-axis, as for H_2O_2. Choice C is incorrect because there is no data provided in the graph that suggests that high H_2O_2 concentrations prevent the active site from changing shape as needed—the reaction rate does not decrease at higher concentrations.

2. B Learning Objective: 12.2

The lower the K_M value, the lower the concentration of substrate necessary to reach maximum velocity. Similarly, the lower the k_{cat} value, the more quickly the enzyme acts. Thus, to be the more efficient cofactor, both values should be lower, as is the case for $NADP^+$. Choice (B) is correct. Choice A is incorrect because it wrongly states that the turnover rate is higher for $NADP^+$. Choices C and D are incorrect because they wrongly name NAD+ as the more efficient cofactor.

3. B Learning Objective: 12.3

Because the *y*-intercept represents the maximum velocity and there are different values for each line, the maximum velocity changes depending on how much of the inhibitor is present. Because increasing the substrate concentration can restore a competitively inhibited reaction to its original maximum velocity, this must instead be a noncompetitive inhibitor. Choice (B) is thus correct. Choice A is incorrect because the *x*-intercept represents a substrate concentration, not the maximum velocity. Choices C and D are incorrect because they mischaracterize the graph (the curves have different *y*-intercepts but the same *x*-intercept).

4. C Learning Objective: 12.4

When the reactants are added, the researchers know the exact concentrations of each. This makes it easy to examine initial reaction rates as a function of concentration. However, as the reaction proceeds, it will be more difficult to determine the exact concentration at any given point. Choice (C) is thus correct. Choice A is incorrect because the reaction rate is affected by multiple variables, so it does not follow the simple pattern suggested. Choice B is incorrect because there is no information provided about whether the reaction generates heat. Choice D is incorrect because the researchers were not acting to save time but to work with information that had already been ascertained.

5. B Learning Objective: 12.5

According to the table, each increase in concentration of 11.1 μM leads to an increase in V_{max} of about 1.5 μM/minute. Thus, V_{max} for the concentration given in the question stem should be around 11.2 μM/minute. Choice (B) is thus correct. Choice A is too high and choices C and D are too low.

Test What You Learned

1. C Learning Objective: 12.4

Because the reaction rate is already close to its maximum at 6.0 mM, the rate would not change much if the substrate concentration was increased further. At most, the rate would increase slightly. Choice (C) is thus correct. Choice A is incorrect because the reaction rate would not increase significantly, but only slightly. Choice B is incorrect because the reaction rate would not decrease. Choice D is incorrect because the change is predictable.

2. B Learning Objective: 12.5

The reaction rate is predicted to increase due to the increase in substrate concentration. This is true unless it is already at the maximum for that pH (which is not the same as the maximum at any pH because each enzyme has an optimal pH at which it functions best). Choice (B) is thus correct. Choice A is incorrect because the rate should increase with more substrate (unless already at maximum). Choice C is incorrect because the reaction rate would increase unless it is already at the maximum value for that pH. The fact that the rate seems unusually favorable at 7.75 is irrelevant and could just be random variation. Choice D is incorrect because there is no guarantee that the rate will increase to the value obtained for a pH of 7.25.

3. D Learning Objective: 12.2

The lowest K_M value in the table occurs for the wild type enzyme with the $NADP^+$ cofactor, which also has the highest value for k_{cat}/K_M. This indicates that the wild type enzyme with $NADP^+$ is far more efficient than any of the alternatives. Choice (D) is thus correct. Choice A is incorrect because the wild type enzyme is more efficient with NAD^+ and reaches maximum velocity at a lower concentration than the mutant. Choice B is incorrect because the wild type enzyme with NAD^+ reaches maximum velocity at a lower concentration and has higher substrate specificity than the mutant. Choice C is incorrect because the wild type enzyme is more efficient with NADP+ and has higher substrate specificity than the mutant.

4. A Learning Objective: 12.3

Because the substrate and the inhibitor compete for the active site in competitive inhibition, it is possible to increase the substrate concentration enough to reach the same maximum velocity with the inhibitor as at a lower substrate concentration without the inhibitor. Choice (A) is thus correct. Choice B is incorrect because a competitive inhibitor binds to the active site. Choice C is incorrect because this is more indicative of a noncompetitive inhibitor. Choice D is incorrect because temperature effects give no indication of whether the inhibitor is competitive.

5. A Learning Objective: 12.1

A certain amount of energy (the activation energy) is needed for the reaction to take place. At higher temperatures, there is more kinetic energy and the increased rate of random collisions between molecules increases the probability that they will collide in the right way to react. Choice (A) is thus correct. Choice B is incorrect because there is more kinetic energy at higher temperatures, not at lower temperatures, and because increased kinetic energy actually raises the likelihood that the substrate will bind to the enzyme. Choice C is incorrect because 30°C is actually a higher temperature than 37°C, so it makes no sense to say that the protein "denatured due to excess heat." Choice D is incorrect because substrate concentration remains constant, according to the question stem.

CHAPTER 13

Test What You Already Know

1. C Learning Objective: 13.1

In chloroplasts, as electrons move through the electron transport chain, protons are pumped from the stroma into the thylakoid lumen, creating a proton gradient and producing NADPH. Protons flow down this gradient back out into the stroma through an ATP synthase to produce ATP. The correct answer is (C). Choice A is incorrect because the flow of protons is reversed. Choice B is incorrect because the locations of the stroma and thylakoid lumen are switched. Choice D is incorrect because ATP and NADPH are being expended instead of produced.

2. A Learning Objective: 13.2

The concentration of NADPH increases in the chloroplast when the plant cell is not producing sufficient ATP to meet cell energy demands. In cyclic photophosphorylation, electrons move from the reaction center in photosystem I (P700), through an electron transport chain, and then back to the same reaction center to produce ATP. This process does not produce oxygen or NADPH. Thus, shifting from noncyclic photophosphorylation to cyclic photophosphorylation will only increase ATP levels. ATP is then used in the Calvin cycle to make sugars to meet cell energy demands. The correct answer is (A). Choice B is incorrect because the opposite is true: ATP levels will increase and oxygen is not produced. Choice C is incorrect because carbohydrates are synthesized in the Calvin cycle. Choice D is incorrect because oxygen and NADPH are produced in noncyclic photophosphorylation.

3. D Learning Objective: 13.3

Lacking mitochondria, red blood cells predominantly produce energy by anaerobic respiration. Thus, red blood cells produce ATP via glycolysis followed by lactic acid fermentation. The correct answer is (D). Choice A is incorrect because aerobic cellular respiration occurs in mitochondria, which red blood cells lack. Choice B is incorrect because the products of anaerobic fermentation are ATP and NAD^+ (glycolysis produces ATP and NADH, and fermentation regenerates NAD^+). Choice C is incorrect because human red blood cells undergo lactic acid fermentation, not alcohol fermentation.

4. A Learning Objective: 13.4

By increasing the permeability of the inner mitochondrial membrane, thermogenin uncouples the electron transport chain from oxidative phosphorylation. Electrons are transferred to O_2 to generate H_2O, but protons that have been pumped out of the mitochondrial matrix into the intermembrane space return to the mitochondrial matrix due to the "leaky" membrane. Thus the proton gradient, which is needed to drive ATP production, is decreased, and energy is released as heat instead of used for ATP production. The correct answer is (A). This mechanism is used by hibernating animals to stay warm. Choice B is incorrect because decreasing the proton gradient would decrease ATP production. Choices C and D are incorrect because thermogenin decreases the proton gradient.

5. D Learning Objective: 13.5

During cellular respiration, cells break down glucose ($C_6H_{12}O_6$) to make ATP. The products (CO_2 and H_2O) have less energy than the reactants, so the process is exergonic. During photosynthesis, plants use sunlight to make glucose from carbon dioxide and water. The products have more energy than the reactants, so the process is endergonic. The correct answer is (D). Choices A and B are incorrect because photosynthesis is not exergonic and cellular respiration is not endergonic. Choice C is incorrect because cellular respiration consumes glucose to make ATP.

6. C Learning Objective: 13.6

DPIP acts as a substitute for $NADP^+$ in photosynthesis. During photosynthesis, DPIP is reduced and becomes colorless, which results in an increase in light transmittance. According to the data, an increase in light transmittance over time is found for the tube exposed to light only. The correct answer is (C). Choice A is incorrect because DPIP changes from blue to clear when reduced during photosynthesis. Choice B is incorrect because the tube exposed to light only had the highest rate of photosynthesis (highest percent transmittance). Choice D is incorrect because the solution in all four tubes was blue at time 0.

7. B Learning Objective: 13.7

Aerobic phases of cellular respiration in eukaryotes occur in mitochondria, which synthesize ATP from ADP. Higher than normal levels of ADP suggest that the transgenic mice are unable to convert ADP to ATP as efficiently because the mutation has impaired their mitochondria. The correct answer is (B). Choice A is incorrect because a higher rate of oxygen consumption in mitochondria would decrease ADP levels (electrons are transferred to oxygen when ADP is phosphorylated to make ATP). Choice C is incorrect because ADP levels would be greater when ATP is utilized (ATP $\rightarrow$ ADP + P_i). Choice D is incorrect because carbon dioxide is released, not consumed, in cellular respiration.

Test What You Learned

1. D Learning Objective: 13.2

In the light-dependent reactions of photosynthesis, light energy splits water, releasing oxygen, and NADPH and ATP are produced. In the light-independent reactions of photosynthesis, carbon dioxide is reduced by NADPH and ATP to make glucose. Thus, the correct answer is (D). Choices A and B are incorrect because photosynthesis consumes water and carbon dioxide and produces oxygen and glucose ($6CO_2 + 6H_2O \rightarrow C_6H_{12}O_6 + 6O_2$). Choice C is incorrect because the light-dependent and light-independent reactions are switched: light-dependent reactions produce NADPH and ATP for light-independent reactions.

2. A Learning Objective: 13.5

Autotrophs convert free energy in sunlight to ATP and NADPH to produce carbohydrates from carbon dioxide in the Calvin cycle. Heterotrophs metabolize carbohydrates by hydrolysis, releasing carbon dioxide in the Krebs cycle while synthesizing ATP, NADH, and $FADH_2$. Thus, the correct answer is (A). Choice B is incorrect because the Calvin cycle and Krebs cycle are switched: autotrophs use CO_2 to make glucose in the Krebs cycle, and heterotrophs produce CO_2 as a waste product in the Calvin cycle. Choices C and D are incorrect because heterotrophs do not use CO_2 to make glucose.

3. C Learning Objective: 13.1

Photosynthesis converts light energy to chemical energy. First, chlorophyll molecules arranged in photosystems embedded within thylakoid membranes of chloroplasts trap energy. The absorbed light energy of the photon is then transferred to an electron, and the excited electron is transferred to a primary electron acceptor. The chlorophyll is oxidized (loses an electron), and the electron is transferred to the electron transport chain. Thus, the correct answer is (C). Choices A and B are incorrect because the light reactions occur in the thylakoid membrane, not the chloroplast membrane. Choice D is incorrect because the direction of the energy transfer arrows is reversed and because the molecule should be a primary electron acceptor, not a donor.

4. B Learning Objective: 13.6

Since the rate of photosynthesis as a function of carbon dioxide concentration behaves similarly to the rate of photosynthesis as a function of light intensity, the graph can be used to determine the relationship between carbon dioxide concentration and the rate of photosynthesis. According to the figure, at low light intensity, the rate of photosynthesis increases as light intensity increases, whereas at high light intensity, the rate of photosynthesis plateaus (an increase in light intensity does not increase the rate of photosynthesis). Thus, light intensity is a limiting factor of photosynthesis at low light intensity. Carbon dioxide has the same effect, so at low carbon dioxide concentrations, carbon dioxide is a limiting factor of photosynthesis. The correct answer is (B). Choice A is incorrect because, at low concentrations, carbon dioxide and the rate of photosynthesis are directly related. Choices C and D are incorrect because, at high concentrations, carbon dioxide does not affect the rate of photosynthesis.

5. B Learning Objective: 13.3

Metabolism of glucose to lactate generates only 2 ATP per molecule of glucose, whereas oxidative phosphorylation generates up to 36 ATP per molecule of glucose. Thus, a high rate of glucose uptake is required to meet energy needs. The correct answer is (B). Choice A is incorrect because the cancer cells convert glucose to lactic acid (normally an anaerobic process) via aerobic glycolysis (glycolysis in the presence of oxygen), which only generates 2 ATP per molecule of glucose. Choice C is incorrect because neither lactic acid fermentation nor mitochondrial oxidative phosphorylation consumes ATP. Choice D is incorrect because although oxygen is present, it is not consumed in aerobic glycolysis.

6. B Learning Objective: 13.7

During anaerobic respiration in yeast (alcoholic fermentation), glycolysis breaks down glucose into pyruvate, which is then converted to carbon dioxide and ethanol. Thus, increasing the amount of pyruvate would, in turn, increase the amount of carbon dioxide and ethanol produced. The correct answer is (B). Choice A is incorrect because pyruvate is oxidized under aerobic conditions and lactic acid is not a product of yeast fermentation. Although choice C is factually true, it does not explain why the solution with sodium fluoride and pyruvate increased respiration, and therefore is incorrect. Choice D is incorrect because, according to the data, sodium fluoride does not promote aerobic respiration and also because aerobic respiration would increase CO_2 production.

7. D Learning Objective: 13.4

The electron transport chain produces a proton gradient that drives ATP production. During aerobic respiration, oxygen is the final electron acceptor in the ETC and is reduced to form water, marking the completion of the ETC. Since bacterial ETCs are usually shorter than mitochondrial ETCs, the proton gradient (used to drive ATP synthesis) that is generated by bacterial ETC will be less than that generated by mitochondrial ETC. Thus, bacterial ETC will produce less ATP per oxygen molecule. The correct answer is (D). Choice A is incorrect because oxygen is not an electron donor but the final electron acceptor. Choice B is incorrect because alcoholic fermentation does not occur under aerobic conditions. Choice C is incorrect because protons, not electrons, are pumped from one side of the membrane to the other to form a proton gradient.

CHAPTER 14

Test What You Already Know

1. B Learning Objective: 14.1

There are many potential reasons that could account for the relative lengths of these stages, so this question can be effectively answered by eliminating flawed answer choices that contain incorrect descriptions of events during the cell cycle. Choice A is incorrect because it suggests that protein synthesis occurs only during the S phase, without addressing that the S (Synthesis) phase is so named because DNA is synthesized during that phase. In fact, protein synthesis occurs throughout interphase, especially during the G_1 and G_2 phases (the first and second Gap phases) while the cell is growing and synthesizing new proteins in preparation for the S phase and the M phase (Mitosis). Choice C is incorrect because the newly synthesized chromosomes made during the S phase must maintain their stability throughout G_2, which is already much longer than either the S or M phases. Furthermore, if the chromosomes did become unstable, then both daughter cells would be affected because the new chromosomes contain both old and newly synthesized strands of DNA. Choice D assumes that because the two phases are drawn to the same proportion that they are exactly the same, though the question stem states that the lengths in the figure are only "roughly proportional" to the relative durations of these stages. Choice D also ignores the various mechanisms that cells possess to regulate the cell cycle, including checkpoints, cyclin-dependent kinases, and other cell signaling factors. Choice (B) correctly describes the purpose of G_1 and G_2 (first and second Gap phases) with an appropriate description of the activities taking place within a cell during that stage.

2. D Learning Objective: 14.2

DNA synthesis is a necessary step in the cell cycle of a mitotically dividing cell, and epithelial cells are a common example of mitotically dividing cells (only a few types of cells do not routinely divide, such as neurons and cardiac muscle cells). Checkpoints in the cell cycle prevent mitosis in the event that certain events do not happen as planned. In order to move from the synthesis phase into the second growth phase before mitosis, DNA synthesis must be completed. However, if DNA synthesis were blocked by a drug, chromosomes could not be copied and the cell would remain unable to complete the synthesis phase and progress toward mitosis, thereby preventing the cell from dividing. Thus, choice (D) is correct. Choices A and B are incorrect because the cells would be affected. Choice C is incorrect because the cells would not even be able to divide once, as noted above.

3. A Learning Objective: 14.3

The cell depicted contains two pairs of chromosomes, so each of its gametes would ordinarily contain only two chromosomes total. Gamete 2 actually contains three chromosomes (one extra), including two copies of the longer chromosome. When gamete 2 combines with a standard gamete during fertilization, it will add one more copy of the longer chromosome, for a total of three copies. Choice (A) is thus correct. Choice B is incorrect because fertilization will lead to three copies of the longer chromosome, not two. Choices C and D are incorrect because gamete 2 has an extra chromosome, not one fewer.

4. D Learning Objective: 14.4

Like other features of meiotic division and fertilization, the primary purpose of sexual reproduction is to increase the genetic fitness of a population by increasing the extent of genetic diversity within the population as a whole. Choice (D) most directly references this purpose, making it correct. Though there are error-checking mechanisms in place to ensure the integrity of chromosomes, crossing over is not one of those mechanisms, so choices A and B are both incorrect. Although C provides an explanation in terms of the genetic fitness of progeny, the "best" genes are selected by natural selection at an organismal level and not at a molecular level due to crossing over.

5. C Learning Objective: 14.5

The first generation offspring should all have the genotype *Ry* (because they receive an *R* allele from the homozygous red parent and a *y* allele from the homozygous yellow parent). Thus, when a *Ry* (red) apple tree is crossed with a *yy* (yellow) apple tree, half of the offspring should be *Ry* (red) and the other half should be *yy* (yellow). Out of 10 offspring, 5 would be red and 5 yellow. This eliminates choices A and B. To determine whether this is statistically significant, calculate chi-square as follows:

phenotype	*o*	*e*	*o–e*	$(o–e)^2$	$(o–e)^2/e$
red	6	5	1	1	0.2
yellow	4	5	−1	1	0.2
					$\chi^2 - 0.4$

Because 0.4 is less than 3.84, the threshold for $p = 0.05$ with 1 degree of freedom, the result is not statistically significant, making (C) correct. Choice D is incorrect because it misrepresents the chi-square value.

Test What You Learned

1. C Learning Objective: 14.5

The results without insecticide suggest that three-quarters of the cells are ordinarily expected to be in interphase and one-quarter in mitosis. With a sample of 653 total cells, that means that about 490 should be in interphase and 163 in mitosis. This allows the following chi-square calculation:

phase	*o*	*e*	*o–e*	$(o–e)^2$	$(o–e)^2/e$
interphase	575	490	85	7225	14.745
mitosis	78	163	−85	7225	44.325
					$\chi^2 = 59.07$

Because 59.07 is greater than 3.84 (the significance threshold for $p = 0.05$ with 1 degree of freedom), the effect of the insecticide is statistically significant. Choice (C) is thus correct. Choice A is incorrect because it suggests the effect is not significant. Choices B and D are incorrect because they wrongly suggest that the chi-square value is less than 3.84.

2. D Learning Objective: 14.3

Nondisjunction is an error in the separation of chromosomes, which can occur during meiosis (in either meiosis I or meiosis II) and which leads to genetic abnormalities. It produces gametes with either one missing or one extra chromosome. Choice (D) is thus correct. Choice A is incorrect because chromosomes normally separate to opposite poles of the cell during meiosis. Recombination involves the exchange of material between chromosomes but does not itself alter the number of chromosomes in gametes, so B is also incorrect. Choice C is incorrect because chromatids are supposed to separate during anaphase II; only the failure to separate properly could cause a discrepancy in chromosome number.

3. A Learning Objective: 14.2

Except in a very small number of cell types (e.g., erythrocytes and platelets), mitosis results in complete copies of genetic information being passed along to both daughter cells equally, so (A) is correct. Most differences in the structure and function of daughter cells are attributable primarily to the different mRNA and protein contents between the two cells, and not to the presence, absence, or direct modification of genetic material. Thus, choices B, C, and D are incorrect.

4. A Learning Objective: 14.1

During the 3 stages of interphase (G_1, S, and G_2), there is pronounced growth, DNA replication, and crucial preparation for cellular division. DNA replication errors during this process, if not detected and fixed, would result in mutations being passing on to daughter cells. Choice (A) is thus correct. Choice B is incorrect because unrepaired mutations would be passed on. Choices C and D are incorrect because cells do in fact grow during interphase.

5. B Learning Objective: 14.4

The most likely explanation for an offspring having a genetically influenced trait that neither of its parents possesses is that the parents were carriers (heterozygous) for the trait. Choice (B) is thus correct. Choice A is incorrect because it misidentifies crossing over as "genetic drift" and because crossing over only results in the exchange of genetic information between chromosomes, so it could not bring into being a new allele for red hair. Choice C is incorrect because the Y chromosome predominantly contains genes that influence primary and secondary sexual characteristics in males, so it is unlikely to influence hair color. Choice D is incorrect because gene flow can only alter the allelic frequencies of a population as a whole but cannot change the genotypes of existing members of a population, such as the parents described in the question stem.

CHAPTER 15

Test What You Already Know

1. A Learning Objective: 15.1

Energy enters the ecosystem from the Sun. Moss captures some of this energy through photosynthesis. The primary consumer, the reindeer, feeds on moss. The wolf, the secondary consumer, obtains energy from the reindeer. The wolf will eventually die and be consumed by decomposers. Choice (A) is thus correct. The other answer choices are incorrect because they list the components in the wrong order.

2. C Learning Objective: 15.2

Photosynthesis is an endergonic process, in which the products have more free energy than the reactants. Choice (C) is thus correct. Choice A is incorrect because not all of the solar energy is transformed; some of the energy is lost as heat. Choice B is incorrect because the solar energy is also converted into mechanical energy in the movements of paramecia and amoebae. Choice D is incorrect because photosynthesis is not exergonic.

3. C Learning Objective: 15.3

Activating enzymes that break down starch into simple sugars would enhance fermentation by the yeast. Choice (C) is thus correct. Choices A and B are incorrect because the yeast are only capable of fermenting simple sugars, not starch, so neither experimental treatment would lead to much fermentation. Choice D is incorrect because germinating seedlings would only deplete starch reserves, leaving even fewer carbohydrates available for fermentation.

4. D Learning Objective: 15.4

The conditions of the experiment create an artificial H^+ gradient across the thylakoid membranes. The flow of H^+ is what allows ATP synthase in the membranes to produce ATP. Thus, the experiment supports the hypothesis in (D). Choice A is incorrect because no NADPH was added to the flasks. Choice B is incorrect because the experiment was conducted in darkness. Choice C is incorrect because the membranes were not damaged in the experiment.

5. B Learning Objective: 15.5

The platelet behavior described in the question stem is a classic example of positive feedback, in which one response (the activation of the platelets) leads to more of the same effect. Choice (B) is thus correct. Choice A is incorrect because this is positive feedback, not negative feedback. Choice C is incorrect because there is a separate negative feedback loop (described in the question stem) that stops platelet activation. Choice D is incorrect because platelet activation promotes activation of inactive platelets, but does not cause more platelets to be produced.

6. B Learning Objective: 15.6

Chemicals released during the infection reset the hypothalamus to a higher temperature setpoint. Physiological responses such as shivering raise the core temperature to match the new setpoint. Thus, even when the child has a temperature higher than normal, she will experience symptoms like chills and shivering until her body reaches the new setpoint of 40°C. Choice (B) is thus correct. Choice A is incorrect because sweating only occurs when physiological temperature is above the setpoint. Choice C is incorrect because the child will feel cold, not hot, until her body reaches the adjusted setpoint. Choice D is incorrect because these are responses to dehydration, such as might happen when someone sweats profusely, which is not expected in this set of circumstances.

Test What You Learned

1. D Learning Objective: 15.4

Endergonic reactions happen all the time in cells because they are coupled with exergonic reactions. Reaction X, which is endergonic, could proceed in a cell if coupled with reaction Y, which is exergonic. Choice (D) is thus correct. Choice A is incorrect because endergonic reactions can take place in living cells. Choice B is incorrect because reaction Y still needs to reach its activation energy; some energy input would be required. Choice C is incorrect because enzymes can only lower activation energy; they cannot change the free energy of products or reactants.

2. B Learning Objective: 15.2

The cell and its surroundings form a closed system. The total entropy of the system increases even as the entropy of the cell decreases. Choice (B) is thus correct. Choice A is incorrect because the release of water molecules does not increase entropy. Choice C is incorrect because entropy actually decreases when complex molecules are synthesized. Choice D is incorrect because the laws of thermodynamics apply to all physical systems, including organisms.

3. C Learning Objective: 15.3

With few exceptions, all energy enters the biosphere as light from the Sun through the process of photosynthesis. After the asteroid impact, debris blocking the Sun prevented photosynthesis, taking out the foundation of the food web. Choice (C) is thus correct. Choice A is incorrect because energy flow does not stop during periods of cold temperatures, but only diminishes. Choice B is incorrect because these events do not have an impact on the energy levels in the biosphere. Choice D is incorrect because energy flow would not stop if most organisms were killed, but would just be reduced.

4. C Learning Objective: 15.1

Energy flows from the Sun to a producer, then to a consumer, then to the breakdown of energy-rich carbohydrates by catabolism, and finally to ATP supporting sliding filaments in the muscles. Energy is lost at each step as heat dissipated to the surrounding environment. The traditional way of representing this loss of energy is a pyramid, as in (C). Choice A is incorrect because some energy is dissipated into the environment as heat with each step. Choice B is incorrect because energy does not circulate in this scenario. Choice D is incorrect because energy does not accumulate at each level, but diminishes since some is lost as heat with each step.

5. B Learning Objective: 15.6

Body size was larger when temperatures were colder and smaller when temperatures were warmer. Thus, body size decreases when temperature increases, as in (B). Choice A is incorrect because body size is inversely correlated with ambient temperature. Choice C is incorrect because it states the opposite of what occurs. Choice D is incorrect because body size both increased and decreased over the course of woodrat evolution.

6. A Learning Objective: 15.5

Thyroid hormone production is regulated by a negative feedback loop. If insufficient iodine is available to produce thyroxine, the anterior pituitary will not be inhibited and will continue producing TSH. This is what leads to the formation of the goiter. Choice (A) is thus correct. Choice B is incorrect because the lack of iodine does not stimulate cell division. Choice C is incorrect because the thyroxine precursors do not form new compounds in the gland. Choice D is incorrect because the gland does not become enlarged in order to increase iodine filtration.

CHAPTER 16

Test What You Already Know

1. D Learning Objective: 16.1

RNA contains phosphate, the sugar ribose, and the nitrogenous bases adenine, cytosine, guanine, and uracil. Only choice (D) includes the appropriate sugar and an appropriate RNA base that is distinct from the bases of DNA. Choices A and B are incorrect because they contain the sugar deoxyribose, a component of DNA. Choice C is incorrect because RNA does not contain thymine, a base found only in DNA.

2. B Learning Objective: 16.2

Because the *lac*$^-$ cells contain everything necessary to construct enzymes that can ferment lactose except for the DNA with the *lac*$^+$ genes, the only cells that would be adversely affected would be those incubated with extract B, which loses those genes due to the action of the DNases. Because those cells would be unable to ferment lactose, they would appear as colorless. Choice (B) is thus correct. Choice A is incorrect because the *lac*$^-$ cells contain all the proteins and amino acids necessary to construct lactose-fermenting enzymes, provided they have the proper genes (which they can get from extract A), so they would appear blue. Choice C is incorrect because the *lac*$^-$ cells contain all the RNA molecules and nucleotides necessary to transcribe and translate the lactose-fermenting enzymes, provided they have the proper genes (which they can get from extract C), so they would also appear blue. Choice D is incorrect because the cells incubated with extract B will appear colorless, as explained above.

3. D Learning Objective: 16.3

Short strands of DNA would ordinarily not be found in the cellular environment, so something must be preventing these short strands from being joined together into longer strands. Joining nucleotides together requires the creation of phosphodiester bonds, so if enzymes that created these bonds (ligases) were deficient in the cell, it could explain the existence of these short strands. Choice (D) is thus correct. Choice A is incorrect because DNA must be unwound in order to be replicated, and it is clear that some replication is occurring since the newly labeled nucleotides are finding their way into DNA strands. Choice B is incorrect because there were no single-stranded DNA molecules (as indicated by the unchanged results after adding mung bean nuclease), so DNA polymerase I must have filled in the gaps on the new strands. Choice C is incorrect because new strands were in fact synthesized.

4. C Learning Objective: 16.4

In order for the same gene to produce two distinct mRNA transcripts, different exons from the gene must be spliced together for each protein. Thus, alternative splicing in different tissues explains the existence of these proteins that come from the same gene, making (C) correct. Choice A is incorrect because mutations occur largely at random, so it would be highly improbable to have two distinct proteins consistently produced from the same gene as a result of separate mRNA mutations. Choice B is incorrect because the mRNA sequences are identical at some points, which would not be possible if both the sense and antisense strands were being used. Choice D is incorrect because the proteins have differing amino acid sequences, which would not result simply from post-translational modifications. Moreover, D would not explain why the mRNA sequences are themselves different.

5. A Learning Objective: 16.5

Messenger RNA is read in the 5′ to 3′ direction, so to convert the codons to amino acids, simply move from left to right and match each one to the table. This yields a sequence of ser-ala-his-leu-tyr-val, as in choice (A). Choice B is incorrect because it mistakenly interprets the third and fifth codons in the sequence. Choice C is incorrect because it provides the sequence if an antisense mRNA were translated (UCG becomes AGC, GCA becomes CGU, etc.), but mRNA is read directly by the ribosome without the creation of an antisense strand. Choice D is incorrect because it provides the sequence if the mRNA strand were read from 3′ to 5′, but this is the opposite of the direction of reading in translation.

6. D Learning Objective: 16.6

Because phosphorylation directly modifies existing proteins, it must be regulated after translation occurs, making it a post-translational modification. Choice (D) is correct. Choices A, B, and C are incorrect because these are all processes that occur before a protein has been constructed.

7. C Learning Objective: 16.7

The enzyme would break the plasmid at the points labeled "HaeIII" in the diagram. This would create two fragments, one with a length of 900 base pairs and the other with a length of $400 + 700 = 1{,}100$ base pairs. Choice (C) is thus correct. Choices A and B are incorrect because they fail to account for the total length of one of the strands. Choice D is incorrect because it presents the results if both HaeIII and EcoRI were used.

Test What You Learned

1. C Learning Objective: 16.3

In order for DNA polymerases to begin their work, small RNA primers must first attach to the single strands of DNA after they have been separated by helicase. Choice (C) is thus correct. Choice A is incorrect because mRNA does not bind to DNA at random to form sequences. Choice B is incorrect because tRNA does not bind to DNA, but rather to mRNA at ribosomes. Choice D is incorrect because there is nothing in the description of the isolation procedure that suggests the researcher is adding ribonucleotides.

2. A Learning Objective: 16.4

The insertion of an estrogen response element upstream of the initiation site would make the expression of tubulin become estrogen responsive. Choice (A) is thus correct. Choice B is incorrect because the baseline expression of tubulin would remain unchanged; the only effect would be to make expression estrogen responsive.

Choice C is incorrect because the mutation does not occur in the region that encodes for tubulin, but rather upstream of that region. Choice D is incorrect because termination takes place downstream of the initiation site, not upstream.

3. C Learning Objective: 16.2

Given that the extract contains all the cellular machinery necessary for translation of insulin except for mRNA, the insulin will only be produced when pancreatic mRNA (some of which will code for insulin) is added to the mixture. Choice (C) is thus correct. Choice A is incorrect because the gene for insulin cannot be transcribed and translated without mRNA. Choice B is incorrect because the mixture already contains ribosomes and because mRNA is needed. Choice D is incorrect because the DNA is not necessary; the mRNA transcripts are sufficient, given that all of the machinery for translation is already present in the mixture.

4. B Learning Objective: 16.5

The conversion of the fourth codon from UAU to UAA, changes it from a tyr to a stop signal. This is a nonsense mutation that will cause the sequence to be prematurely terminated, generating a shorter peptide sequence. Choice (B) is thus correct. Choice A is incorrect because the amino acids in the sequence remain unchanged; the only difference is that the fourth codon becomes a stop signal. Choice C is incorrect because the protein is made shorter, not longer. Choice D is incorrect because a change in fact occurred: the peptide sequence was made shorter.

5. A Learning Objective: 16.6

The most likely reason for an mRNA sequence to be far longer than an active protein is some type of post-translational modification that excises a segment of the protein. Choice (A) thus provides the best explanation. Choice B is incorrect because insulin is secreted into the bloodstream, as noted in the question stem, not embedded in a membrane. Choice C is incorrect because the question stem suggests that the excised region is an inactive peptide; a dimer would be split into two active proteins. Choice D is incorrect because it would be redundant and a waste of energy to split up the protein into amino acids and then reassemble it.

6. D Learning Objective: 16.1

To find which conclusion is most probable, use process of elimination to rule out impossible or unlikely conclusions. The percentages given in the question stem add up to 72%, so the remaining base (whether uracil or thymine) would have to constitute 28% of the nucleotides. This eliminates choices A and B. If the molecule were double-stranded, it would be expected that there would be equal amounts of cytosine and guanine, as well as equal amounts of adenine and either thymine or uracil. Because the percentages are unequal, choice C can also be ruled out. Thus, the most probable conclusion is that this a single-stranded RNA molecule, choice (D).

7. B Learning Objective: 16.7

The EcoRI enzyme would only cleave the plasmid at one location, which would not separate the plasmid into multiple fragments, but would leave a single linear strand of DNA 2,000 base pairs long. Choice (B) is thus correct. Choice A is incorrect because it is too short. Choices C and D are incorrect because the plasmid is cleaved in only one location, not two, as would be required to have two fragments.

CHAPTER 17

Test What You Already Know

1. A Learning Objective: 17.1

According to the question stem, "shell development is heavily influenced by genetics," so it would be unlikely for handedness of the shell to vary entirely based on environmental factors. This is precisely what is suggested in choice (A), in which the environment seems to be the sole determinant of handedness. Because (A) offers the least likely outcome, it is correct. Choice B is incorrect because this result is consistent with sinistral being the dominant trait. Choice C is incorrect because this result is consistent with dextral being the dominant trait. Choice D is incorrect because this result is consistent with a more complex non-Mendelian inheritance pattern, with environmental factors having limited influence.

2. B Learning Objective: 17.2

According to the question stem, the gene is off by default but can be expressed when an activator binds and turns it on. This is positive control, as indicated in choice (B). Choice A is incorrect because the gene is not expressed throughout development, but is off by default (constitutive expression means that a gene is continually expressed). Choice C is incorrect because the default for the gene is off and it is turned on by an activator; in negative control, a gene is on by default and is turned off with a repressor. Choice D is incorrect because the gene is controlled by an activator, not a repressor; in addition, operons are not always controlled by repressors.

3. C Learning Objective: 17.3

According to the table, workers express more genes relating to responding to stimuli than do queens. This suggests that they may need greater use of their sense organs, perhaps for activities such as foraging. Choice (C) is thus correct. Choice A is incorrect because queens express more genes relating to biological regulation. Choice B is incorrect because queens express considerably more genes relating to developmental processes. Choice D is incorrect because workers express more genes relating to metabolic processes.

4. B Learning Objective: 17.4

In order for phenotypic changes to be similar, the genes affected in both fruit flies and honeybees should also be similar. Choice (B) is thus correct. Choice A is incorrect because the regulation of different genes is unlikely to produce similar phenotypic changes in the two species. Choices C and D are incorrect because royalactin is described in the question stem as increasing juvenile hormone levels, so it is unlikely to act as a hormone itself.

Test What You Learned

1. B Learning Objective: 17.3

The question stem specifies that "bee morphology is driven by an external chemical signal." One possible source of this external chemical signal could be the food that is fed to larvae; larvae fed different foods could develop different body types. Choice (B) is thus correct. Choice A is incorrect because the question stem specifies that the differences in morphology are driven by an external chemical signal, not a temperature differential. Choice C is incorrect because it is highly unlikely that random mutations could produce the consistent differences in morphology between queens and workers. Choice D is incorrect because the question stem specifies an external chemical signal is responsible, not a difference in the use of body parts.

2. D Learning Objective: 17.1

In both of the examples provided in the question stem, the direction of coiling in the mother determines the direction of coiling in all offspring. This suggests that something from the mother, perhaps mRNA transcribed from her genes, is definitive for determining shell handedness. Choice (D) is thus correct. Choice A is incorrect because males and females are actually crossed in the experiment, so the species could not be parthenogenetic. Choice B is incorrect because the coiling direction of the mother seemed to have a definitive influence on coiling direction in offspring, so environmental factors could not be the only determinant. Choice C is incorrect because the shell handedness of the males had no influence on the offspring.

3. A Learning Objective: 17.2

According to the question stem, the Hoxc9 regulatory complex represses forelimb development in the posterior areas of the body, which correspond to the caudal region. Thus, the complex is exerting negative control in the caudal region, which will lead to forelimb development in the rostral region. Choice (A) is thus correct. Choice B is incorrect because the negative control is actually exerted in the caudal region, not the rostral region. (If negative control were exerted in the rostral region, then forelimb development would be inhibited in the rostral and allowed to progress in the caudal region instead.) Choices C and D are incorrect because repression is a type of negative control, not positive control.

4. C Learning Objective: 17.4

Because abd-A increases much more in workers than in queens, it is likely to be important for the differences in worker hindlimbs. Choice (C) is thus correct. Choice A is incorrect because queens actually express atx-2 the most at the L4 stage. Choice B is incorrect because dac is expressed at a relatively constant level in queens, but it appears to decrease in workers. Choice D is incorrect because the difference in dll expression between queens and workers is relatively minimal.

CHAPTER 18

Test What You Already Know

1. B Learning Objective: 18.1

The only DNA that showed double strand breaks was the decondensed DNA, and these breaks became more substantial as the radiation intensity increased. This suggests that condensation provides protection from radiation damage. Choice (B) is thus correct. Choice A is incorrect because it reverses the results for condensed and decondensed DNA. Choice C is incorrect because only the decondensed DNA experienced damage. Choice D is incorrect because decondensed DNA was the only type to show DSBs.

2. D Learning Objective: 18.2

A mutation in the operator region would most likely prevent the repressor from recognizing the operator, thereby stopping it from binding. With no repressor bound, the genes would be continuously expressed. Choice (D) is thus correct. Choice A is incorrect because only the operator region would be affected and because this would cause the genes to be expressed. Choice B is incorrect because the genes for lactose metabolism proteins would be unaffected by a mutation to the operator region. Choice C is incorrect because the repressor would not itself be changed by a mutation to the operator region and because the mutation is more likely to prevent the binding of the repressor, rather than cause it to bind too tightly.

3. C Learning Objective: 18.3

The most likely explanation for a trait to appear in the offspring of parents who both lack that trait is that the gene for that trait is recessive and that both parents are heterozygous (carriers). Choice (C) is thus correct. Choice A is incorrect because it provides a less likely explanation of the result; non-deleterious mutations are quite rare, so it is far more probable that the parents were simply carriers of a recessive gloving gene. Choice B is incorrect because a dominant allele is always expressed, so it would not be possible in this circumstance for both parents to lack gloving. Choice D is incorrect because a recessive allele will not be expressed unless it is received from both parents.

4. A Learning Objective: 18.4

If men require only one copy of the mutant allele to have malaria resistance, but women must be homozygous (i.e., require two copies), then the malaria resistance allele must be X-linked recessive. Choice (A) is thus correct. Choice B is incorrect because the results provide no information about the geographic distribution of the allele. Choice C is incorrect because males with the mutant allele have less G6PD activity, not more. Choice D is incorrect because the results indicate that even homozygous females and hemizygous males display some amount of G6PD activity, so it is not true that they produce no functional G6PD.

Test What You Learned

1. C Learning Objective: 18.2

According to Table 1, even a 1% EMS dose led to a substantial number of mutations. Table 2 indicates that missense mutations are the most common. Indeed, it is more likely that missense mutations (which change the identities of one or more amino acids without otherwise altering protein structure) would be more likely to produce favorable results than mutations which altered the structure more dramatically. Choice (C) is thus correct. Choice A is incorrect because splicing mutations are relatively rare, as indicated in Table 2, and because a high dose is not needed to induce mutations (and too high of a dose could produce too many mutations, which is more likely to be deleterious to the plant). Choice B is incorrect because silent mutations have no impact on amino acid sequences in a protein, so there is no chance that they could lead to improvements. Choice D is incorrect because nonsense mutations add a premature stop codon, which truncates proteins in a way that usually renders them inoperable; thus, these mutations are less likely to result in benefits.

2. D Learning Objective: 18.4

In a codominant inheritance pattern, an organism will display both of the traits if its genome includes both of the codominant alleles. Thus, individuals with type AB blood will produce red blood cells with both A and B antigens. Choice (D) is thus correct. Choice A is incorrect because each allele codes for an antigen separately. There is no way to combine the characteristics of both antigens into a single hybrid. Choice B is incorrect because AB individuals will display both A and B antigens on all red blood cells. Choice C is incorrect because it is safe to predict that all of the individual's red blood cells will contain both antigens.

3. A Learning Objective: 18.1

RNA polymerase must be able to bind to a promoter (a specific DNA sequence) in order for transcription to occur. This is only possible when the DNA is relatively loosely packed. Thus, euchromatin will be transcribed more readily than heterochromatin, making choice (A) correct. Choice B is incorrect because polymerase will access euchromatin more easily than heterochromatin. Choices C and D are incorrect because DNA polymerase is operative in DNA replication, not in transcription.

4. A Learning Objective: 18.3

Because the agouti and full-color traits are dominant, according to the question stem, then it is quite likely that the male is homozygous dominant for both traits if all of the offspring were agouti and full-color. Choice (A) is thus correct. Choice B is incorrect because a male that was heterozygous for both traits would likely lead to a greater mix of offspring phenotypes when mated with the specified female. Choices C and D are incorrect because a parent that was homozygous dominant for both traits would only produce offspring that are agouti and full-color.

CHAPTER 19

Test What You Already Know

1. D Learning Objective: 19.1

The circumstances described in the question stem provide an example of disruptive selection, in which there are selective pressures in favor of extreme phenotypes and against moderate phenotypes. This is likely to produce a body size distribution in which there are more small and large rabbits and fewer medium-sized rabbits. Choice (D) is thus correct. Choice A is incorrect because it suggests that only small body size is selected for, but large body size is also selected for. Choice B is incorrect because it suggests that only large body size is selected for, but small body size is also selected for. Choice C is incorrect because it presents an example of stabilizing selection, in which the mean is favored over the extremes, the opposite of the scenario described in the question stem.

2. B Learning Objective: 19.2

In order to be present in all three domains (Bacteria, Achaea, and Eukarya), glycolysis must be a primitive process. The presence of glycolysis in the common ancestor of the three domains would explain its prevalence among all organisms. Choice (B) is thus correct. Choice A is incorrect because the efficiency of glycolysis is not at issue in the question stem, but rather its prevalence in many organisms. Choice C is incorrect because it is more likely that glycolysis appeared once during the early development of life than it is that it appeared independently many times during evolution. Choice D is incorrect because glycolysis does not require the presence of mitochondria and because Bacteria and Archaea are domains of prokaryotes, which lack mitochondria.

3. B Learning Objective: 19.3

When the founders of an isolated population happen to have a particular trait in large abundance, subsequent generations of that population will likely continue to have that trait at a higher frequency due to genetic drift. Because more of the founding population had large beaks and because there were no selective pressures on beak size, it is reasonable to expect that more birds on the island will have large beaks. Choice (B) is thus correct. Choice A is incorrect because beak size does not experience selection in any particular direction since the question stem stated that food was available for many beak types. Choice C is incorrect because the founding population had a different distribution of traits than the mainland population. Choice D is incorrect because there is no selective advantage for large beak size as a result of the availability of food for any beak type; more large-beaked birds are expected due to a founder effect.

4. C Learning Objective: 19.4

According to the biological species concept, two populations are distinct species if they are unable to interbreed to produce fertile offspring. Because the resistant buffalo grass cannot interbreed with the nonresistant, the resistant grass has undergone speciation. Choice (C) is thus correct. Choice A is incorrect because speciation does not require geographic isolation; distinct species can occupy the same geographical area. Choice B is incorrect because distinct populations can look very similar while still being separate species. Choice D is incorrect because the way that human beings use a population of organisms is irrelevant to whether that population is a distinct species.

5. C Learning Objective: 19.5

Although there is a lot of data in the table, the only relevant information concerns the comparison of yeast, a fungus, to other species. The table shows that yeast has fewer base substitutions in comparison to animal species (fruit flies, sea urchins, rats, and humans) than in comparison to plant species (soybeans and maize). Thus, the classification of Fungi as distinct from Plantae is appropriate and choice (C) is correct. Choice A is incorrect because yeast are more closely related to humans than to maize, based on the table. Choice B is incorrect because yeast are more closely related to animal species than to soybeans. Choice D is incorrect because the revision was appropriate.

6. A Learning Objective: 19.6

The best way to support the hypothesis that snakes and lizards share a common ancestor would be to find more intermediate forms, such as snakes with hind limbs, in the fossil record. Choice (A) is thus correct. Choice B is incorrect because it provides an example of Lamarckian evolution, in which the use (or disuse) of a trait causes it to develop (or atrophy); this conception of evolution has been discredited. Choice C is incorrect because it provides less solid evidence of common ancestry than does (A); it is not much different than the evidence already provided in the question stem concerning embryonic hind limb buds in snakes. Choice D is incorrect because this morphological similarity does not necessarily indicate a close relationship between the species; all reptiles are covered in scales.

7. D Learning Objective: 19.7

In the situation described in the question stem, parrotbills will experience a selective pressure in favor of being able to recognize the cuckoo eggs, so they can remove them from their nests and avoid the costs of parasitism. To counteract this, there will be a strong selective pressure on the cuckoos to have eggs as similar as possible to parrotbill eggs, to make it more difficult for the parrotbills to recognize the parasitic eggs. Choice (D) is thus correct. Choice A is incorrect because unique eggs would make it easier for parrotbills to recognize the intruders. Choice B is incorrect because the size of the eggs is not at issue in the question stem; if anything, eggs that are the same size as the parrotbill eggs (thereby making them harder to recognize as parasitic) would be selected for. Choice C is incorrect because a uniform, parrotbill-like appearance would be favored over greater variation in egg appearance.

8. A Learning Objective: 19.8

When researchers choose rats to breed based upon one characteristic, such as running ability, there is also selection on related traits and closely linked genes. Choice (A) is thus correct. Choice B is incorrect because the question stem states that the rats were specifically selected for running capacity and not multiple other traits. Choice C is incorrect because some of the characteristics, such as body weight and susceptibility to heart failure, are distinct phenotypes. Choice D is incorrect because the researchers were selecting for running ability, not reduced fitness.

9. B Learning Objective: 19.9

A consistent change in a specific direction across multiple generations suggests the existence of a selective pressure, which is making members of the species with the dominant trait fitter than those who lack the trait. Choice (B) is thus correct. Choice A is incorrect because random variation is equally likely to cause changes in any direction; three generations of changes in the same direction suggests something other than randomness. Choice C is incorrect because it would be unlikely to see large numbers of mutations that happened to result in a higher frequency of the dominant trait. Choice D is incorrect because only small populations are heavily affected by genetic drift, but the population in question is large.

Test What You Learned

1. A Learning Objective: 19.5

The most likely explanation for nearly identical pathways and identical compounds appearing in all living organisms is that these emerged early in the course of evolution in a common ancestor. Choice (A) is thus correct. Choice B is incorrect because convergent evolution would be unlikely to lead to the same result in all organisms. Choice C is incorrect because the question stem stipulates that identical compounds (ATP, NADH, $FADH_2$) are used by all organisms. Choice D is incorrect because the compounds listed are not the products of larger molecules being broken down.

2. C Learning Objective: 19.3

The question stem notes that the island features finches with large beaks, while the mainland includes finches with a wide array of beak sizes. It also notes that boats regularly carry travelers from the mainland. It is reasonable to conclude, therefore, that these boats could also bring finches with them, including small-beaked finches that can take advantage of food sources underutilized by the large-beaked population. This would be an example of gene flow. Choice (C) is thus correct. Choice A is incorrect because visitors feeding the finches would not necessarily cause the population to diversify. Choice B is incorrect because a novel mutation is not necessary to introduce small beaks into the population; small beaks already exist on the mainland. Choice D is incorrect because there is no connection suggested in the question between the destruction of the finch habitat and a selective pressure for smaller beaks.

3. C Learning Objective: 19.9

With such similar answer choices, this question can best be answered using the process of elimination. As noted in the question stem, TT and TC represent specific alleles, so the values in the table are allelic frequencies. This eliminates choices B and D, which wrongly suggest that these are genotypic frequencies. In Hardy-Weinberg calculations, allelic frequencies are represented by the variables p and q, making choice A incorrect and choice (C) the correct answer.

4. C Learning Objective: 19.2

The function of photolyases, as described in the question stem, suggests that they would emerge relatively early in evolution, while cryptochromes would evolve later, most likely in a common ancestor of plants and animals. The similarities between the two suggest that cryptochromes may have come about due to mutations in the genes for photolyases. Choice (C) is thus correct. Choice A is incorrect because the shared sequences between the two proteins suggest they are homologous, sharing a common lineage. Choice B is incorrect because there is not extensive convergence between the functions of photolyases and cryptochromes. Choice D is incorrect because not all light receptor proteins have identical structures.

5. B Learning Objective: 19.8

Artificial selection is typically used to select for polygenic traits (ones with multiple genetic influences) that experience a wide range of variation. The most straightforward way to do this is to breed only those animals that possess the desired traits (or that possess them to the greatest extent). Selectively breeding cows with above-average milk production would ensure that milk production increases in subsequent generations. Choice (B) is thus correct. Choice A is incorrect because it is unlikely that a characteristic as complex as milk production can be tied to a single gene. Choice C is incorrect because milk production is unlikely to be governed only by a single gene with a simple Mendelian inheritance pattern. Choice D is incorrect because increasing genetic diversity will not necessarily lead to higher milk production.

6. B Learning Objective: 19.4

Mass extinctions, such as the one described in the question stem, open a large number of niches for exploitation by the surviving species, often leading to adaptive radiation. Choice (B) is thus correct. Choice A is incorrect because organisms are not spontaneously generated; they evolve from ancestral organisms that survive and reproduce. Choice C is incorrect because animal life eventually recovered; otherwise animals would not exist today. Choice D is incorrect because some species went extinct, as indicated by the figure.

7. D Learning Objective: 19.6

If the same genes controlled the expression of scales, feathers, and hair, it would suggest that all three shared a common origin, the hypothesis presented in the question stem. Choice (D) is thus correct. Choice A is incorrect because distinct structures with analogous functions can be the result of convergent evolution and are not necessarily the product of divergence from a common ancestor. Choice B is incorrect because it is a negative claim that provides no support for the hypothesis. Choice C is incorrect because analogous structures often result from convergent evolution, so this does not necessarily suggest common ancestry.

8. A Learning Objective: 19.1

The question stem describes two selective pressures on the trees: high winds will select against any tree significantly taller than the others, which will be more susceptible to being blown down, while dense foliage selects against any tree significantly shorter than the others, which will be unable to get enough sunlight for photosynthesis. Because average heights are favored, the distribution should form a bell curve, as suggested in choice (A). Choice B is incorrect because tall heights make the trees susceptible to high winds that can uproot them. Choice C is incorrect because short heights make it harder for the trees to get enough sunlight. Choice D is incorrect because both tall and short trees have disadvantages, meaning that average values are favored over the extremes.

9. D Learning Objective: 19.7

The results of the study show that eggs with greater contrast are far more likely to be rejected than those with less contrast. This suggests that there is coevolution between the parrotbill and cuckoo populations, with a constant selective pressure for cuckoo eggs to resemble parrotbill eggs. Choice (D) is thus correct. Choice A is incorrect because there is an obvious selective pressure on the cuckoos to have eggs that resemble those of the parrotbills, so the effects are predictable. Choice B is incorrect because there is selection on coloration: cuckoo eggs that resemble parrotbill eggs are favored over those that do not. Choice C is incorrect because the taller bars indicate that contrasting eggs are more likely to be rejected; contrast should actually decrease with subsequent generations.

CHAPTER 20

Test What You Already Know

1. B Learning Objective: 20.1

In the phylogenetic tree, the two most closely related groups will have the closest shared branch, representing the most recent common ancestor. Among the options listed in the answer choices, the groups with the closest common ancestor are the *Bucconidae* and the *Galbulidae*, choice (B). All the other choices feature groups with less recent common ancestors.

2. D Learning Objective: 20.2

In the cladogram, the branch that diverges to form the *Sarcopterygii* and the *Actinopterygii* is labeled *Osteichthyes*, so it is true that an *Osteichthyean* was the common ancestor of those two clades. Choice (D) is thus correct. Choice A is incorrect because the *Sarcopterygii* branch extends to the end of the cladogram, suggesting that members of this clade still exist. Choice B is incorrect because the *Percomorpharia* and the *Ovalentariae* are relatively close on the cladogram, with a recent common ancestor. Choice C is incorrect because the *Actinopterygii* has many more branches in the cladogram (and thus greater diversity) than the *Sarcopterygii*.

3. A Learning Objective: 20.3

Unusual traits are more likely to develop once in a common ancestor than to develop independently on multiple occasions. Thus, it is more likely that a common ancestor led to a shared trait in groups *f*, *g*, and *h* than that the same unusual trait would evolve independently for group *e* and for the common ancestor of *a* and *c*. Choice (A) is thus correct. Choice B is incorrect because a cladogram shows phylogenetic relationships but does not necessarily reflect morphological similarities. Choice C is incorrect because cladograms do not represent "progress"; the organisms in group *x* were likely just as well adapted to their environment as any of the later organisms. Choice D is incorrect because not all traits are preserved in the descendants of particular groups.

Test What You Learned

1. A Learning Objective: 20.2

According to the cladogram, *Serpentes* and *Iguania* share a relatively recent common ancestor (the point labeled *Toxicofera*), while the common ancestor between *Gekkota* and *Serpentes* is only found at the first branching of the tree. Choice (A) is thus correct. Choice B is incorrect because *Scincidae* is monophyletic, deriving not from multiple ancestors but from a single common ancestor. Choice C is incorrect because some lizard clades (such as the *Iguania*) are more closely related to snakes and mosasaurs than they are to other lizards, such as the *Gekkota*. Choice D is incorrect because this cladogram is not organized according to timescale, but rather according to common ancestry; extinct species are not localized to a specific part of the diagram.

2. B Learning Objective: 20.3

The groups that branch earliest lack tube building, suggesting that tube building evolved in a common ancestor of the *Ampithoe*, *Caprella*, and *Aroidae* but was secondarily lost in the *Caprella*. Choice (B) is thus correct. Choice A is incorrect because tube building is only found in groups that emerge from a later common ancestor. Choice C is incorrect because more than half of the groups in the diagram lack temperature tolerance, so it seems improbable that it was possessed by the progenitor of all of these groups. Choice D is incorrect because the *Caprella* lack temperature tolerance.

3. C Learning Objective: 20.1

Phylogenetic trees always branch out from a common ancestor, so the leftmost branch represents the common ancestor of all birds. Because it is the common ancestor, it represents the point at which birds first evolved from other vertebrate ancestors (most likely reptiles, based on fossil and molecular evidence to date). Choice (C) is thus correct. Choice A is incorrect because the branch does not indicate relationships; the structure of the tree itself is what indicates the relationships between taxa. Choice B is incorrect because unresolved and unknown species are not featured on the tree. Choice D is incorrect because this tree provides no information about whether particular species are extinct.

CHAPTER 21

Test What You Already Know

1. C Learning Objective: 21.1

Venus flytraps have adapted to detecting small animal movements, with the innate behavior of snapping shut when an animal is detected, so prey might be consumed. The plant will thus only respond when the researcher simulates animal movements by rapidly stimulating the plant in several locations. Choice (C) is thus correct. Choice A is incorrect because snapping shut must be a rapid response in order to capture prey effectively; the accumulation of compounds would take too much time, giving prey an opportunity to escape. Choice B is incorrect because increasing the pressure had no effect. Choice D is incorrect because the Venus flytrap is exhibiting an innate behavior; lacking a proper nervous system, it is too simple of an organism to learn and become sensitized to specific responses.

2. C Learning Objective: 21.2

As noted in the question stem, mycorrhizae are a class of fungi that are beneficial to the vines. Using a fungicide will harm the damaging mold but also the helpful mycorrhizae. The fertilizers are necessary to counteract the loss of the mycorrhizae, which help plants to acquire nutrients. Choice (C) is thus correct. Choice A is incorrect because the nutrients in the soil (nitrogen, phosphorus, and minerals) are simple compounds and ions that would not be affected by a chemical treatment. Choice B is incorrect because fungicides only damage fungus, not plants. Choice D is incorrect because mold does not assist in nutrient uptake; on the contrary, it is a parasite that consumes nutrients in the vines' grapes.

3. B Learning Objective: 21.3

By making the land impenetrable to wildlife and preventing native plants from flourishing, the blackberries function as an invasive species. Their most probable effect would be to decrease biodiversity, as invasive species tend to do. Choice (B) is thus correct. Choice A is incorrect because the negative characteristics of the plants, such as the difficulty that thorny brambles pose for harvesting, make them less valuable as a food source. Choice C is incorrect because the blackberries are likely to have the opposite effect, as described in the question stem. Choice D is incorrect because the plants function as an invasive species.

4. A Learning Objective: 21.4

As noted in the question stem, females are far more likely to rear offspring than males, and the data in the figure shows that females are responsible for most of the alarm calls. Because making an alarm call not only alerts other squirrels to the presence of a predator, but also reveals the location of the squirrel making the call (thereby making the caller a potential target for the predator), it is an example of an altruistic behavior. Since females are making most of these calls, the function is most likely to protect the young that females tend to raise. Choice (A) is thus correct. Choice B is incorrect because the frequency of calling behavior is affected by age and sex, so it is unlikely to be a behavior intended to benefit the entire group; more probably, it is a behavior that females use to alert the young they are rearing. Choice C is incorrect because the alarm calls would only divert the predator's attention to the squirrel making the call and not to nearby squirrels (who would likely run away and hide in response to the call). Choice D is incorrect because the calls are adapted to alert other squirrels, not to warn the predator. Squirrels are small animals; rather than defending themselves, they can better avoid being eaten by running and hiding.

5. D Learning Objective: 21.5

According to the graph, 100% of the fruit flies eventually gathered close to the cotton tip with banana (while far fewer gathered close to the tip without it). This suggests that the banana smell serves as a positive chemotactic stimulus. Choice (D) is thus correct. Choices A and C are incorrect because the movement is clearly not random, irrespective of whether a chi-square analysis is performed; all of the flies gathered near the banana-coated tip, but only half near the non-coated tip. Choice B is incorrect because the results show the opposite, that the fruit flies were attracted to the smell.

Test What You Learned

1. B Learning Objective: 21.2

If a parasite is resistant to a host's nonspecific immune system, then stimulating an immune response can aid the parasite by eliminating competing parasites. Choice (B) is thus correct. Choice A is incorrect because stimulating the immune system will not make the host significantly more vulnerable to infection, even though it does require energy. Choice C is incorrect because this stimulation does, in fact, present an indirect advantage to the parasite by allowing it to outcompete parasites that are vulnerable to the host's nonspecific immune system. Choice D is incorrect because stimulation of the nonspecific immune system usually leads to an inflammatory response, which will not increase the amount of nutrients at the site of inflammation.

2. D Learning Objective: 21.5

According to the data in the graph, fruit flies stay farther away from an ammonia-soaked cotton tip than from a non-soaked tip. This suggests that ammonia is a volatile compound that repels the flies, choice (D). Choice A is incorrect because the data shows that the flies do respond to ammonia, moving away from it. Choice B is incorrect because the data shows that the flies are still capable of moving; the percentage close to the ammonia-soaked tip fluctuates with time. Choice C is incorrect because more flies prefer to approach the non-soaked tip, indicating that the ammonia actually repels them.

3. A Learning Objective: 21.4

A volatile compound is one that evaporates easily, allowing it to become airborne and spread out over significant distances. Thus, jasmonate is the most likely candidate for a communication molecule, making choice (A) correct. Choice B is incorrect because large sugars are unlikely to become airborne, so there is no way for them to allow for communication between plants. Choices C and D are incorrect because these water soluble compounds are unlikely to become airborne, so they would not make for effective communication between plants.

4. B Learning Objective: 21.3

Wolves are predators, so if they disappear from an ecosystem, it should lead to an increase in the populations of animals that they preyed upon. The cause of the erosion most likely has to do with effects caused by one of these prey species, as they increased in number. Choice (B) is correct because it presents a plausible scenario, in which an increase in elk populations leads to a decrease in vegetation, which in turn would contribute to riverbank erosion. Choice A is incorrect because it gives an explanation of why the presence of wolves could cause erosion, but the question stem specifies that it was the disappearance of wolves that caused this. Choice C is incorrect because predators hunt other animals; they do not generally feed on vegetation. Choice D is incorrect because erosion occurred after the wolves disappeared; if the dams help combat erosion, then erosion should not be a problem after a beaver predator is removed and more dams can be built.

5. C Learning Objective: 21.1

In order to detect whether the octopus recognizes faces, the volunteers could conduct an experiment in which none of their faces are visible (by using masks to cover them) and observe whether this has any impact on drenching frequency. If the results showed that Michael is drenched less often when wearing a mask, it suggests that the octopus can in fact recognize faces. Choice (C) is thus correct. Choice A is incorrect because it excludes the volunteer whose face is most likely recognizable by the octopus. Choice B is incorrect because it only serves to provide feedback to the octopus, but does not reveal whether it might recognize faces. Choice D is incorrect because it tests the hypothesis that the octopus splashes at specific times, not the hypothesis that it recognizes faces.

CHAPTER 22

Test What You Already Know

1. D Learning Objective: 22.1

The population may grow very rapidly at first as it accesses new resources but, eventually, a larger population will make resources more limited. This will form a logistic growth curve, as suggested in correct choice (D). Choice A is incorrect because exponential growth cannot continue forever; given a large enough population, resources will start to become limited. Choices B and C are incorrect because species are capable of thriving in new environments if conditions are suitable (and the circumstances in the question stem suggest that they are).

2. A Learning Objective: 22.2

Because many more birds were captured in the time immediately following Hurricane Iris, it is not reasonable to conclude that the birds experienced a high rate of mortality. Choice (A) is thus correct. Choice B is incorrect because this could explain why more birds were captured immediately after the hurricane. Choice C is incorrect because a change in foraging habits could lead to more birds being captured. Choice D is incorrect because a tendency to spread out further could also lead to more of the birds being captured.

3. C Learning Objective: 22.3

The question stem explains that rinderpest makes wildebeest sick, so a reduction in rinderpest should lead to an increase in the wildebeest population. Because the wildebeest are grazers, an increase in their population should lead to a decrease in the amount of grass, which in turn will make it harder for fires to spread. Choice (C) is thus correct. Choice A is incorrect because more wildebeests will result in less grass, not more grass. Choices B and D are incorrect because the wildebeest population should increase when rinderpest prevalence decreases.

4. B Learning Objective: 22.4

Plants obtain energy from the Sun, so the role of fertilizer is not to provide energy but to provide other nutrients that the plant needs, such as nitrogen, which is used to build proteins and other organic molecules. Choice (B) is thus correct. Choice A is incorrect because plants do not derive energy directly from the fertilizer and because the nitrogen is not consumed, but is recycled through the nitrogen cycle. Choice C is incorrect because plants do not take up energy directly from the fertilizer. Choice D is incorrect because nitrogen does not slow down plant metabolism.

5. A Learning Objective: 22.5

Small populations tend to experience a lot of inbreeding, which increases their susceptibility to deleterious recessive alleles. By introducing the Texas panthers, the researchers were able to increase the genetic diversity in the population, increasing the chances that the population will survive. Choice (A) is thus correct. Choice B is incorrect because the genetic differences from the Texas subspecies actually helped to promote the survival of the population, so the genetic differences did matter. Choice C is incorrect because there is no reason to suspect that the Texas panthers would be better adapted to the environment than the Florida panthers that had already evolved there. Choice D is incorrect because the population was not beyond saving.

6. D Learning Objective: 22.6

According to the data provided, vertebrates are able to reach levels above the reference values, while macroinvertebrates and plants always remain at or below the reference values. This suggests that vertebrates can colonize wetlands easily, but that it is not so easy for plants and macroinvertebrates. Choice (D) is thus correct. Choice A is incorrect because the macroinvertebrates actually have lower initial mean response ratios than plants. Choice B is incorrect because the animal populations represented in Figure A do eventually reach stable levels. Choice C is incorrect because macroinvertebrate density did reach the reference levels, even though it did drop off later.

7. B Learning Objective: 22.7

The plots showed relatively stable or decreasing net biomass growth, but three plots is probably not sufficient to make generalizations. Choice (B) is thus correct. Choices A and C are incorrect because none of the locations showed an increase in biomass growth. Choice D is incorrect because there was some change in the biomass growth rate in the three locations.

Test What You Learned

1. C Learning Objective: 22.4

Ammonia can be taken up by plants and transformed into other forms of nitrogen, which can then be taken up by other organisms. Because ammonia plays a role in the nitrogen cycle, choice (C) is correct. Choice A is incorrect because ammonia is not broken down in cellular respiration. Choice B is incorrect because organisms do not break down ammonia to release heat. Choice D is incorrect because ammonia is not a major source of energy for producers.

2. A Learning Objective: 22.1

According to the graphs, the inner shelf and mid shelf coral populations were relatively stable with slight increases during the period without disturbances. However, the outer shelf population had a period of rapid growth followed by a drop after disturbances. Choice (A) is thus correct. Choice B is incorrect because the outer shelf does not exhibit a logistic (S-shaped) growth curve since the population drops off rather than plateauing. Choice B also says that the inner and mid shelf lacked disturbance, which is false based on the data. Choices C and D are incorrect because none of the figures showed classic exponential or logistic growth curves.

3. D Learning Objective: 22.7

Increasing the light intensity will cause the plants to engage in additional photosynthesis, increasing their biomass. This, in turn, will cause increases in the biomass of the caterpillars and birds. Choice (D) is thus correct. Choice A is incorrect because trophic pyramids are only very rarely inverted. Choice B is incorrect because the other two levels would also increase in biomass. Choice C is incorrect because all the levels would increase in biomass.

4. C Learning Objective: 22.3

Increased rinderpest prevalence would lower wildebeest populations, increasing the amount of grass (because wildebeests feed on the grass) and providing more material to start and spread fires, thereby decreasing the number of trees. Choice (C) is thus correct. Choice A is incorrect because more rinderpest means fewer wildebeests, which in turn leads to more grass. Choice B is incorrect because fire reduces the tree population. Choice D is incorrect because the figure suggests that humans have only an adverse effect on the elephant population.

5. B Learning Objective: 22.2

Light levels are an abiotic factor, while prey availability is a biotic factor. Choice (B) thus presents an experimental design that investigates exactly one abiotic and one biotic factor. Choice A is incorrect because it only investigates biotic factors. Choice C is incorrect because it only investigates abiotic factors. Choice D is incorrect because it manipulates the biotic and abiotic factors simultaneously, making it impossible to determine the effects of each factor individually.

6. D Learning Objective: 22.6

The spread of antibiotic resistance is typically a direct result of excessive use of antibiotics in medicine, rather than a result of human travel and trade. Antibiotics kill nonresistant bacteria, leaving behind only those bacteria able to resist them. The surviving bacteria then reproduce, passing on antibiotic resistance genes to all offspring. Choice (D) is thus correct. The remaining answer choices all present unintended consequences of human transportation.

7. B Learning Objective: 22.5

In order for selection to favor some individuals in the prey species over others, there must already be genetic variation with respect to the trait being selected for. Choice (B) is thus correct. Choice A is incorrect because selection can occur even when a trait is relatively rare; it need not be ubiquitous. Choice C is incorrect because non-Mendelian traits are also subject to selection. Choice D is incorrect because the plot lines depicted in Figure A are sufficient to produce the phenotypic changes in Figure B.

APPENDIX

AP Biology Equations and Formulas

STATISTICAL ANALYSIS AND PROBABILITY

Mean

$$\bar{x} = \frac{1}{n}\sum_{i=1}^{n} x_i$$

Standard Deviation

$$s = \sqrt{\frac{\sum (x_i - \bar{x})^2}{n-1}}$$

Standard Error of the Mean

$$SE_{\bar{x}} = \frac{s}{\sqrt{n}}$$

Chi-Square

$$\chi^2 = \sum \frac{(o-e)^2}{e}$$

s = sample standard deviation (i.e., the sample-based estimate of the standard deviation of the population)

$\bar{x}$ = mean

n = size of the sample

o = observed results

e = expected results

Degrees of freedom equal the number of distinct possible outcomes minus one.

CHI-SQUARE TABLE

p value	Degrees of Freedom							
	1	2	3	4	5	6	7	8
0.05	3.84	5.99	7.82	9.49	11.07	12.59	14.07	15.51
0.01	6.64	9.21	11.34	13.28	15.09	16.81	18.48	20.09

LAWS OF PROBABILITY

If A and B are mutually exclusive, then $P(\text{A or B}) = P(\text{A}) + P(\text{B})$

If A and B are independent, then $P(\text{A and B}) = P(\text{A}) \times P(\text{B})$

HARDY-WEINBERG EQUATIONS

$p^2 + 2pq + q^2 = 1$

$p + q = 1$

p = frequency of the dominant allele in a population

q = frequency of the recessive allele in a population

METRIC PREFIXES

Factor	Prefix	Symbol
10^9	giga	G
10^6	mega	M
10^3	kilo	k
10^{-2}	centi	c
10^{-3}	milli	m
10^{-6}	micro	μ
10^{-9}	nano	n
10^{-12}	pico	p

Mode = value that occurs most frequently in a data set

Median = middle value that separates the greater and lesser halves of a data set

Mean = sum of all data points divided by number of data points

Range = value obtained by subtracting the smallest observation (sample minimum) from the greatest (sample maximum)

RATE AND GROWTH

Rate

$$\frac{dY}{dt}$$

Population Growth

$$\frac{dN}{dt} = B - D$$

Exponential Growth

$$\frac{dN}{dt} = r_{max} N$$

Logistic Growth

$$\frac{dN}{dt} = r_{max} N\left(\frac{K-N}{K}\right)$$

dY = amount of change

dt = change in time

B = birth rate

D = death rate

N = population size

K = carrying capacity

r_{max} = maximum per capita growth rate of population

Temperature Coefficient Q_{10}

$$Q_{10} = \left(\frac{k_2}{k_2}\right)^{\frac{10}{T_2 - T_1}}$$

Primary Productivity Calculation

mg O_2/L × 0.698 = mL O_2/L

mL O_2/L × 0.536 = mg carbon fixed/L

(at standard temperature and pressure)

T_2 = higher temperature

T_1 = lower temperature

k_2 = reaction rate at T_2

k_1 = reaction rate at T_1

Q_{10} = the factor by which the reaction rate increases when the temperature is raised by ten degrees

Water Potential (Ψ)

$$\Psi = \Psi_P + \Psi_S$$

Ψ_P = pressure potential

Ψ_S = solute potential

The water potential will be equal to the solute potential of a solution in an open container because the pressure potential of the solution in an open container is zero.

The Solute Potential of the Solution

$$\Psi_S = -iCRT$$

i = ionization constant (this is 1.0 for sucrose because sucrose does not ionize in water.)

C = molar concentration

R = pressure constant (R = 0.0831 liter bars/mole K)

T = temperature in Kelvin (°C + 273)

SURFACE AREA AND VOLUME

Volume of a Sphere

$$V = \frac{4}{3}\pi r^3$$

Volume of a Rectangular Solid

$$V = lwh$$

Volume of a Right Cylinder

$$V = \pi r^2 h$$

Surface Area of a Sphere

$$A = 4\pi r^2$$

Surface Area of a Cube

$$A = 6s^2$$

Surface Area of a Rectangular Solid

$$A = \Sigma \text{ (surface area of each side)}$$

r = radius

l = length

h = height

w = width

s = length of one side of a cube

A = surface area

V = volume

Σ = sum of all

Dilution (used to create a dilute solution from a concentrated stock solution)

$$C_i V_i = C_f V_f$$

i = initial (starting)

C = concentration of solute

f = final (desired)

V = volume of solution

Gibbs Free Energy

$$\Delta G = \Delta H - T\Delta S$$

ΔG = change in Gibbs free energy

ΔS = change in entropy

ΔH = change in enthalpy

T = absolute temperature (in Kelvin)

pH $= -\log_{10}[H^+]$

For guidance on selecting the right study plan for you, see Ch. 2.

COMPREHENSIVE REVIEW IS BEST IF YOU HAVE AT LEAST 2 MONTHS TO PREP.

COMPREHENSIVE REVIEW

For each chapter, complete:

- Pre-Quiz
- All Topics
- Post-Quiz
- Online Quiz

WEEK 1 (8–10 hours)

Chapters 1–2

Practice Test 1

Chapters 3–4

WEEK 2 (8–10 hours)

Chapters 5–8

WEEK 3 (8–10 hours)

Chapters 9–11

WEEK 4 (6–8 hours)

Chapters 12–13

WEEK 5 (6–8 hours)

Chapters 14–15

WEEK 6 (6–8 hours)

Chapters 16–17

WEEK 7 (6–8 hours)

Chapters 18–19

WEEK 8 (10–12 hours)

Chapters 20–22

Chapter 23

Practice Test 2

(cont.)

For guidance on selecting the right study plan for you, see Ch. 2.

TARGETED REVIEW IS BEST IF YOU HAVE AT LEAST A MONTH TO PREP.

TARGETED REVIEW

WEEK 1 (8–10 hours)

Chapters 1–2

Practice Test 1

Chapters 3–6:

Pre-Quizzes

Target Topics:

____

____

____

____

____

____

____

____

Post-Quizzes

WEEK 2 (8–10 hours)

Chapters 7–13:

Pre-Quizzes

Target Topics:

____

____

____

____

____

____

____

____

____

____

____

____

Post-Quizzes

(cont.)

KAPLAN

For guidance on selecting the right study plan for you, see Ch. 2.

TIME CRUNCH REVIEW IS BEST IF YOU HAVE LESS THAN A MONTH TO PREP.

TIME CRUNCH REVIEW

DAY 1 (1–2 hours)

Ch. 3–4 High Yield Topics:

Free Energy in Living Systems

Darwin and Natural Selection

Ch. 3–4 Post-Quizzes

DAY 2 (2–3 hours)

Ch. 5–6 High Yield Topics:

Prokaryotes versus Eukaryotes

Organelles

Membrane Transport Mechanisms

Plasmids

Transduction

AP Biology Lab 8

Ch. 5–6 Post-Quizzes

DAY 3 (1–2 hours)

Ch. 7–8 High Yield Topics:

Nonspecific Defenses

Antibodies

Ch. 7–8 Post-Quizzes

DAY 4 (1–2 hours)

Ch. 9–10 High Yield Topics:

General Principles of Signaling

Hormones

Action Potentials

Ch. 9–10 Post-Quizzes

DAY 5 (1–2 hours)

Ch. 11–12 High Yield Topics:

Binding

Enzyme Regulation

AP Biology Lab 13

Ch. 11–12 Post-Quizzes

DAY 6 (3 hours)

Practice Test 1

(cont.)

KAPLAN

TIME CRUNCH REVIEW

DAY 7 (2–3 hours)

Ch. 13–14 High Yield Topics:

Photosynthesis

Anaerobic Respiration

Aerobic Respiration

AP Biology Labs 5, 6, and 7

Mitosis

Meiosis

Ch. 13–14 Post-Quizzes

DAY 8 (2–3 hours)

Ch. 15–16 High Yield Topics:

Energy and Thermodynamics

Feedback Mechanisms

Nucleic Acids

DNA Replication

Transcription

Translation

Ch. 15–16 Post-Quizzes

DAY 9 (1–2 hours)

Ch. 17–18 High Yield Topics:

Gene Regulation and Cell Spec.

Chromosomes

Mendelian Inheritance

Ch. 17–18 Post-Quizzes

DAY 10 (2–3 hours)

Ch. 19–20 High Yield Topics:

Genetic Drift and Gene Flow

Speciation and Extinction

AP Biology Lab 2

Phylogenetic Trees

Cladograms

Ch. 19–20 Post-Quizzes

DAY 11 (2–3 hours)

Ch. 21–22 High Yield Topics:

Competition and Cooperation

AP Biology Lab 12

Population Dynamics

Environment Dynamics

AP Biology Lab 10

Ch. 21–22 Post-Quizzes

DAY 12 (3 hours)

Practice Test 2

TARGETED REVIEW

WEEK 3 (8–10 hours)

Chapters 14–19:

Pre-Quizzes

Target Topics:

____

____

____

____

____

____

____

____

____

____

____

____

Post-Quizzes

WEEK 4 (8–10 hours)

Chapters 20–22:

Pre-Quizzes

Target Topics:

____

____

____

____

____

____

____

____

Post-Quizzes

Chapter 23:

Selected Sample FRQs

Practice Test 2

COMPREHENSIVE REVIEW

Reviewing Your Practice Tests

To get the most out of the Comprehensive Review plan, be sure to complete the following steps after taking each Practice Test:

- Use the Answer Key to tally up your total number of correct multiple-choice and grid-in questions.
- Use the scoring rubrics and sample essays (found online with the Practice Test Answers and Explanations) to self-score your free-response questions.
- Enter your raw scores into the online scoring tool to determine what your scaled score (1–5) would likely be with a similar performance on Test Day.
- Review the online Answers and Explanations for multiple-choice and grid-in questions, both for those that you missed (to see how to get to the correct answers) and for those you got correct (to ensure you solved them efficiently).
- Spend extra time reviewing topics on which you frequently missed questions.